SATELLITE COMMUNICATIONS

SATELLITE COMMUNICATIONS

SECOND EDITION

Robert M. Gagliardi

Department of Electrical Engineering
University of Southern California

VNR VAN NOSTRAND REINHOLD
New York

Library of Congress Catalog Number 90-40748
ISBN 0-442-22745-0

Printed in the United States of America

Van Nostrand Reinhold
115 Fifth Avenue
New York, New York 10003

Chapman and Hall
2-6 Boundary Row
London, SE1 8HN, England

Thomas Nelson Australia
102 Dodds Street
South Melbourne 3205
Victoria, Australia

Nelson Canada
1120 Birchmount Road
Scarborough, Ontario M1K 5G4, Canada

16 15 14 13 12 11 10 9 8 7 6 5 4 3 2

Library of Congress Cataloging-in-Publication Data

Gagliardi, Robert M., 1934–
Satellite communications / Robert M. Gagliardi.—2nd ed.
p. ca.
Includes bibliographical references and index.
ISBN 0-442-22745-0
1. Artificial satellites in telecommunication. I. Title.
TK5104.G33 1991
621.382'54—dc20 90-40748
CIP

In dedication to the memory of my beloved mother and father,
Louise Musco Gagliardi and Michael Gagliardi,
Hamden, Connecticut

Contents

Preface

This second edition of *Satellite Communications* is a revised, updated, and improved version of the first edition (Van Nostrand, 1984) and has been extended to include many newer topics that are rapidly becoming important in modern and next-generation satellite systems. The first half of the book again covers the basics of satellite links, but has been updated to include additional areas such as Global Positioning and deep space satellites, dual polarization, multiple beaming, advanced satellite electronics, frequency synthesizers, and digital frequency generators. The second half of the book is all new, covering frequency and beam hopping, on-board processing, EHF and optical cross-links, and mobile satellites and VSAT systems. All of these latter topics figure to be important aspects of satellite systems and space platforms of the twenty-first century.

As in the first edition, the objective of the new edition is to present a unified approach to satellite communications, helping the reader to become familiar with the terminology, models, analysis procedures, and evolving design directions for modern and future satellites. The presentation stresses overall system analysis and block diagram design, as opposed to complicated mathematical or physics descriptions. (Backup mathematics is relegated to the appendices where a reader can digest the detail at his own pace.) The discussion begins with the simplest satellite systems and builds to the more complex payloads presently being used.

The book is intended for students, engineers, and scientists working in the area of satellite communications. It can be used in the classroom, for continuing education or in-house courses, and as an at-home or on-the-job self-reading aid. A newcomer to the satellite field should benefit from this unified treatment, beginning with the simplest concepts and extending to the more complex. The practicing engineer will find the text useful for review, for aiding in analytical studies and performance evaluation, and for addressing issues posed by newer satellite technologies.

The material is presented at a senior or first-year graduate level and assumes the reader has some basic electrical engineering background. Previous knowledge of communication systems is advantageous, but not necessary. Key background needed would be Fourier transforms, some electromagnetic theory and electronics, and elementary probability and noise theory.

Chapter 1 introduces and summarizes the various types of satellite links and their key parameters and constraints. Chapter 2 serves as a basic review of modulation, decoding, and coding and emphasizes those most common in satellite links. (Appendices A and B act as a backup for this review.) Chapter 3 reviews and applies basic link power analysis to satellites, pointing out the key design equations, system tradeoffs, and inherent problem areas. Chapter 4 examines the satellite payload and electronics, discussing the modern technologies and signal processing limitations. Chapter 5 develops the FDMA format, Chapter 6 the TDMA format, and Chapter 7 the CDMA and spread spectrum format. Advantages and disadvantages of each are derived and highlighted.

The remainder of the text introduces completely new chapters. Chapter 8 covers frequency-hopping systems. Chapter 9 discusses the advantages and technologies of on-board processing, a topic that is becoming increasingly important as satellite payloads get bigger and more sophisticated. Chapter 10 examines the latest technology in satellite crosslinks, including antenna pointing and autotracking. Chapter 11 considers the use of satellites for direct mobile and home broadcasting, and the use of small earth stations (VSATs). An appendix on ranging and position location is included to aid text discussion.

Each chapter contains a homework problem set, with problems ranging from straightforward to moderate difficulty, aiding the student in reviewing and understanding the text presentations. The book includes over 50 tables and 250 figures to simplify discussion and catalog reference material for on-the-job task solutions.

I wish to personally thank Ms. Georgia Lum of the Electrical Engineering Department at USC for typing the manuscript. I also acknowledge the help of the staff of the Communication Science Institute at USC, Ms. Milly Montenegro, Ms. Cathy Cassells, and Ms. Neela Sastry in putting together the new edition. Lastly, I would like to thank the practicing engineers, classroom students, and university instructors who used the first edition for their criticisms, comments, and suggestions that aided me in upgrading this edition.

SATELLITE COMMUNICATIONS

CHAPTER 1
Introduction

The use of orbiting satellites is an integral part of today's worldwide communication systems. As the technology and hardware of such systems continue to advance significantly, it is expected that satellites will continue to play an ever-increasing role in the future of long-range communications. Each new generation of satellites has been more technologically sophisticated than its predecessors, and each has had a significant impact on the development and capabilities of military, domestic, and international communication systems. This progress is expected to continue into the next century, and the capability to transfer information via satellites may well surpass our present-day expectations.

To the communication engineer, satellite communications has presented a special type of communication link, complete with its own design formats, analysis procedures, and performance characterizations. In one sense, a satellite system is simply an amalgamation of basic communication systems, with slightly more complicated subsystem interfacing. On the other hand, the severe constraints imposed on system design by the presence of a spaceborne vehicle makes the satellite communication channel somewhat special in its overall fabrication. In this book we address some of the important features of modern satellite communications and the corresponding design approaches that have evolved.

1.1 HISTORICAL DEVELOPMENTS OF SATELLITES

Long-range communications via modulated microwave electromagnetic fields were first introduced in the 1920s. With the rapid development of microwave technology, these systems quickly became an important part of our terrestrial (ground-to-ground) and near-earth (aircraft) communication systems. However, these systems were, for the most part, restricted to line-of-sight links. This meant that two stations on Earth, located over the horizon from each other, could not communicate directly, unless by ground transmission relay methods. The use of tropospheric and ionospheric scatter to generate reflected skywaves for the horizon links tends to be far too unreliable for establishing a continuous system.

In the 1950s a concept was proposed for using orbiting space vehicles for relaying carrier waveforms to maintain long-range over-the-horizon communications. The first version of this idea appeared in 1956 as the Echo satellite—a metallic reflecting balloon placed in orbit to act as a passive reflector of ground transmissions to complete long-range links. Communications across the United States and across the Atlantic Ocean were successfully demonstrated in this way. In the late 1950s new proposals were presented for using active satellites (satellites with power amplification) to aid in relaying long-range transmissions. Early satellites such as Score, Telstar, and Relay verified these concepts. The successful implementation of the early Syncom vehicles proved further that these relays could be placed in fixed (geostationary) orbit locations. These initial vehicle launchings were then followed by a succession of new generation vehicles, each bigger and more improved than its predecessors. (See Table 1.1 and Figure 1.1.)

Today satellites of all sizes and capabilities have been launched to serve almost all the countries of the world. Satellites now exist for performing many operations, and present development is toward further increase in their role, as will be discussed in subsequent chapters.

1.2 COMMUNICATION SATELLITE SYSTEMS

A satellite communication system can take on several different forms; Figure 1.2 summarizes the basic types. System I shows an uplink from a ground-based earth station to satellite, and a downlink from satellite back to ground. Modulated carriers in the form of electromagnetic fields are propagated up to the satellite. The satellite collects the impinging electromagnetic field and retransmits the modulated carrier as a downlink to specified earth stations. A satellite that merely relays the uplink carrier as a downlink is referred to as a *relay satellite* or *repeater satellite*. More

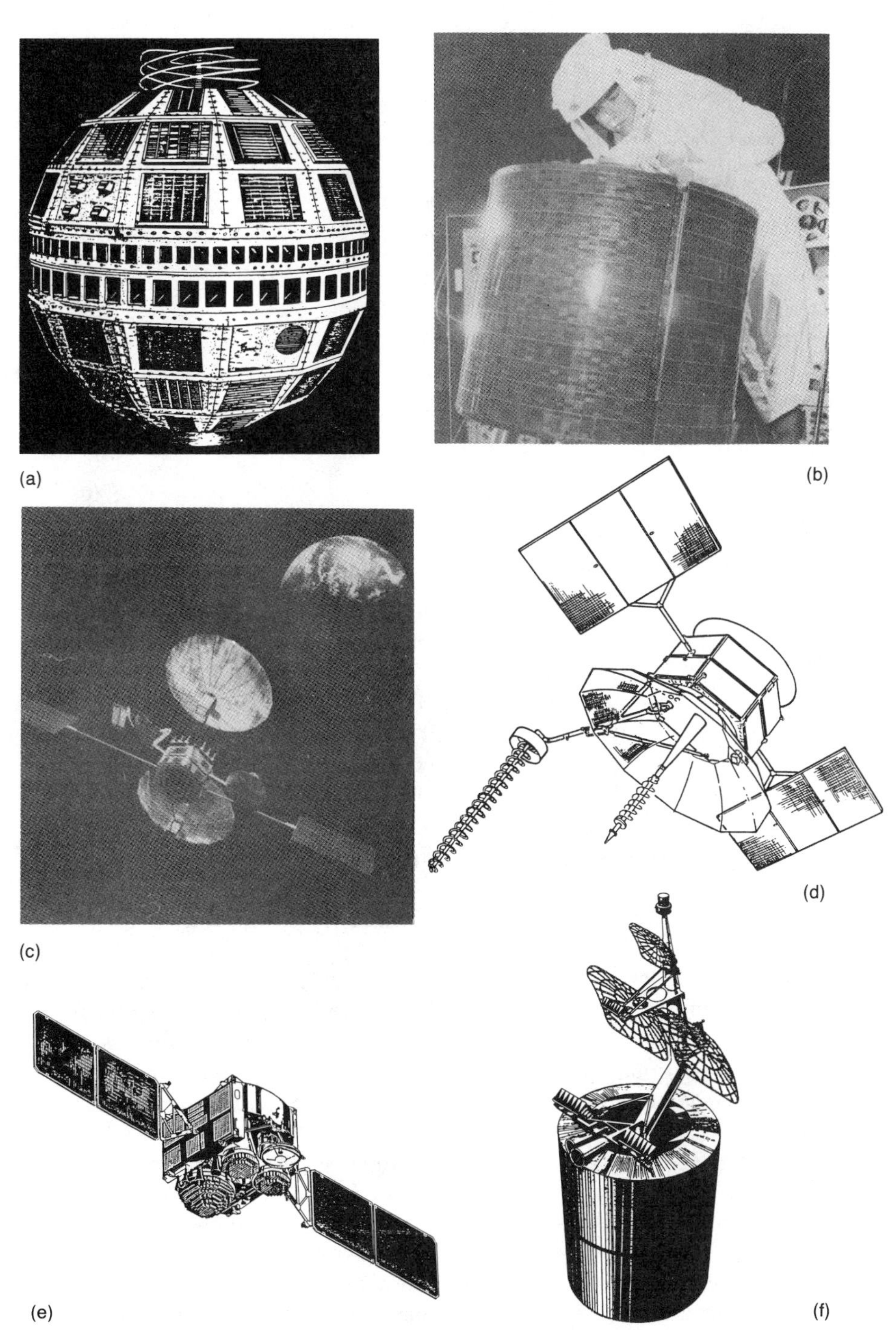

FIGURE 1.1 Communication satellites. (a) Telstar. (b) Early Bird. (c) Fleetsatcom. (d) TDRSS. (e) DSCS-III, (f) Intelsat VI.

TABLE 1.1 Early communication satellites

Satellite	*Launch Date*	*Launch Weight (lb)*	*Number of Transmitters*	*Total RF Bandwidth (MHz)*
Echo	1956	100	0	—
Score	1958	90	1	4
Courier	1960	500	2	13.2
Telstar	1962	170	2	50
Syncom	1963	86	2	10
Intelsat I	1965	76	2	50
II	1966	190	2	130
III	1968	270	4	500
IV	1971	1400	12	450
ATS, A, B, C, D	1966–1969	700	2	50
Telesat	1972–1975	1200	12	500
Westar	1974	1200	12	500
Globecom	1975	1400	24	1000
DSCS, I, II, III	1980–1981	2300	6	500
TDRSS	1980	1600	30	1200
Intelsat V	1981–1983	2000	35	2300
VI	1989	3600	77	3366

commonly, since the satellite *transmits* the downlink by *responding* to the uplink, it is also called a *transponder*. A satellite that electronically operates on the received uplink to reformat it in some way prior to retransmission is called a *processing satellite*.

System II shows a satellite crosslink between two satellites prior to downlink transmission. Such systems allow communication between earth stations not visible to the same satellite. By spacing multiple satellites in proper orbits around the Earth, worldwide communications between remote earth stations in different hemispheres can be performed via such crosslinks.

System III shows a satellite relay system involving earth stations, near-earth users (aircraft, ships, etc.), and satellites. An earth station communicates to another earth station or to a user by transmitting to a relay satellite, which relays the modulated carrier to the user. Since an orbiting satellite will have larger near-earth visibility than the transmitting earth station, a relay satellite allows communications to a wider range of users. The user responds by retransmitting through the satellite to the earth station. The link from earth station to relay to user is called the

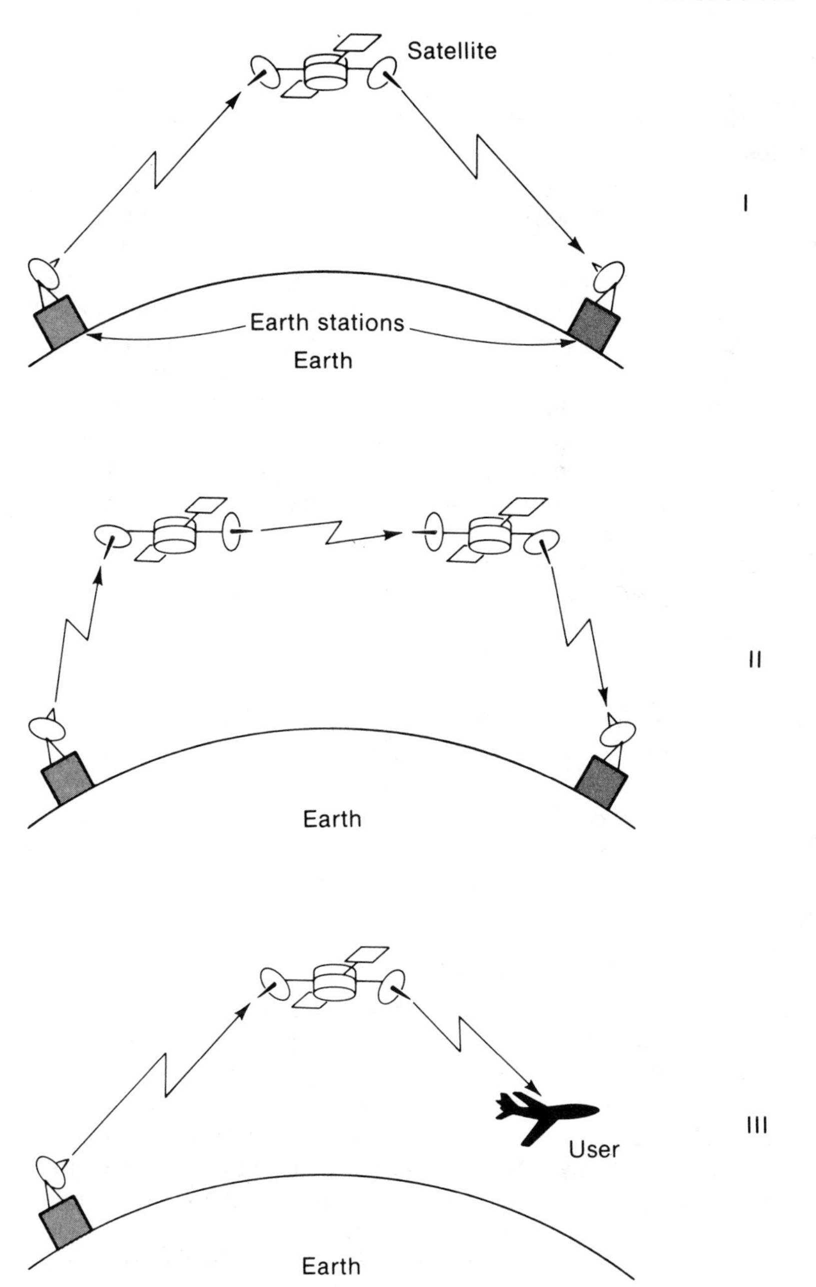

FIGURE 1.2 Satellite systems. (I) Ground–ground; (II) ground–cross-link–ground; (III) ground–user relay.

forward link, while the link from user to satellite to Earth is called the *return link*.

The satellite systems of Figure 1.2 can perform a wide variety of functions, besides the basic operation of completing a long-range communication link. Today's satellites are also used for navigation and position location, terrain observations, weather monitoring, and deep-space exploration, and are an integral part of wide area distribution networks. Figure 1.3a shows a satellite navigation system, in which signals from multiple satellites can be received simultaneously by a moving or stationary receiver and processed instantaneously to determine its location and velocity. This forms the basis of the Global Positioning Satellite (GPS) system in which a network of orbiting satellites are continually available to provide the ranging signals for authorized users anywhere in the world. Figure 1.3b shows a satellite serving as a terrestrial observation vehicle in which weather, terrain, or agricultural information can be collected by cameras and monitors and transmitted to earth-based locations. Figure 1.3c shows a satellite as a primary interconnection between a vast network of moving vehicles and fixed-point earth stations, with voice, data, or command information being exchanged. This is the basis of the forthcoming mobile satellite systems, to be discussed in Chapter 11.

The use of space vehicles to probe the outer universe by returning television and scientific data (Figure 1.3d) has been carried out successfully for several decades. Although simpler in structure and limited in communication capability, these vehicles represent, again, a form of communication satellite whose design principles are similar to those of Figure 1.2. After deriving the key satellite link equations in Chapter 3, the deep-space channel can be viewed as a special case to which the basic analysis can be applied.

Earth stations form a vital part of the overall satellite system, and their cost and implementation restrictions must be integrated into system design. Basically, an earth station is simply a transmitting or receiving or both power station operating in conjunction with an antenna subsystem. Earth stations are usually categorized into large and small stations by the size of their radiated power and antennas. Larger stations may use antenna dishes as large as 10–60 m in diameter, while smaller stations may use antennas of only 3–10 m in diameter, which can be roof-mounted. The current trend is toward *very small aperture terminals* (VSATs), using 1–3 ft (0.3–0.9 m) antennas that can be attached to land vehicles or even manpacks. Large stations may often require antenna tracking and pointing subsystems continually to point at the satellite during its orbit, thereby ensuring maximum power transmission and reception. A given earth station may be designed to operate as a transmitting station only, as a

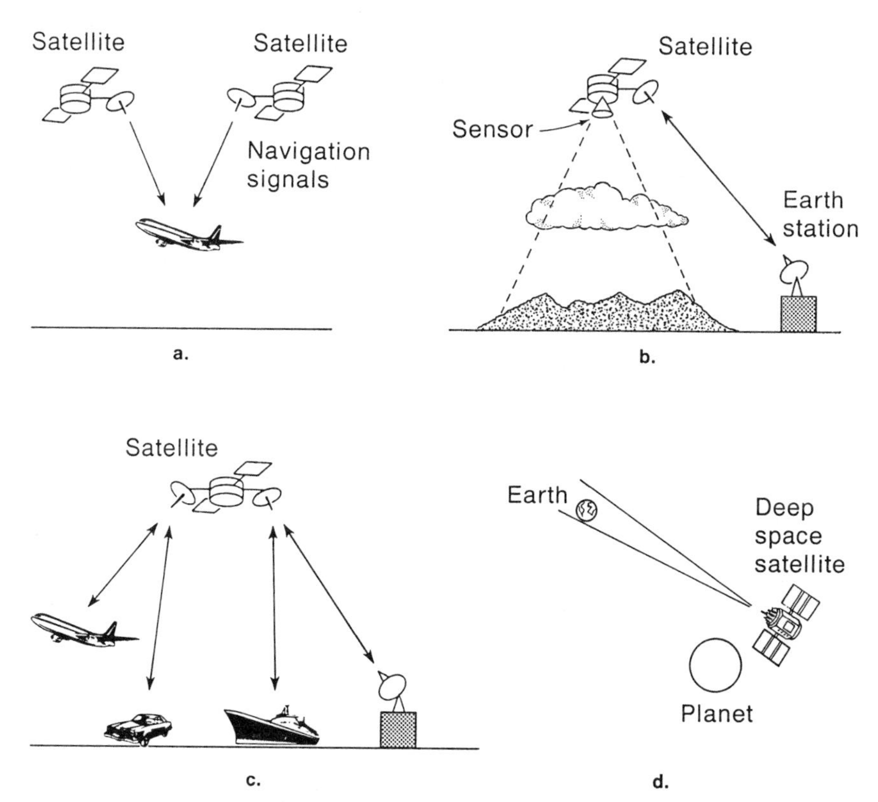

FIGURE 1.3 Satellite uses. (a) Navigation and position location; (b) terrain and weather observation; (c) data networking; (d) deep-space exploration.

receiving station only, or as both. An earth station may transmit or receive single or multiple television signals, voice, or data (teletype, commands, telemetry, etc.) information, as well as ranging (navigation) waveforms, or perhaps a combination of all these items.

The internal electronics of an earth station is generally conceptually quite simple. In a transmitting station (Figure 1.4a), the baseband information signals (telephone, television, telegraph, etc.) are brought in on cable or microwave link from the various sources. The baseband information is then multiplexed (combined) and modulated onto intermediate-frequency (IF) carriers to form the station transmissions, either as a single carrier or perhaps a multiple of contiguous carriers. If the information from a single source is placed on a carrier, the format is called *single channel per carrier (SCPC)*. More typically, a carrier will contain the

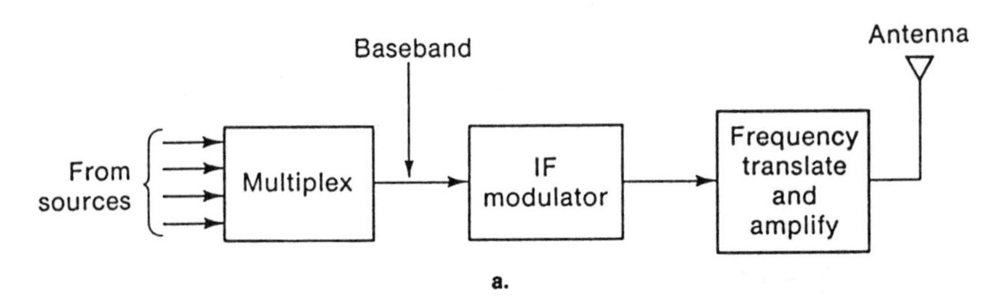

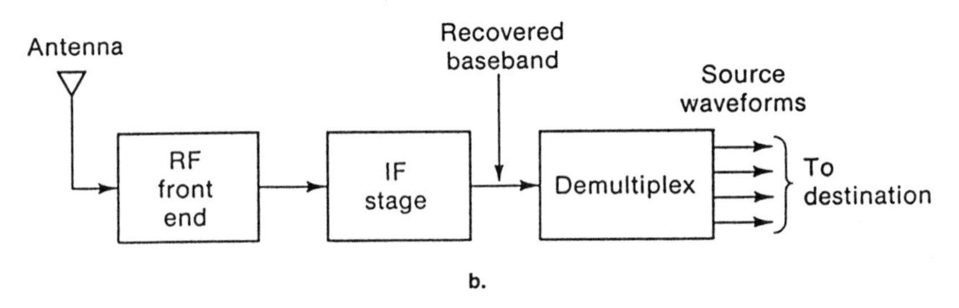

FIGURE 1.4 Earth–station block diagram. (a) Transmitting; (b) receiving.

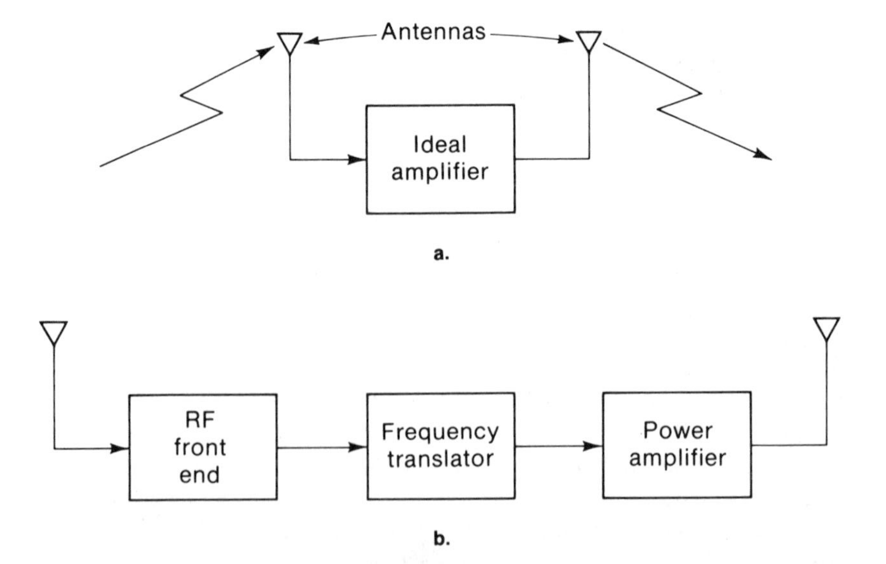

FIGURE 1.5 Satellite block diagram. (a) Ideal; (b) repeater.

multiplexed information from many sources, as in telephone systems. The entire set of station carriers is then translated to radio frequencies (RF) for power amplification and transmission. A receiving earth station corresponds to a low-noise wideband RF front end followed by a translator to IF (Figure 1.4b). At the IF, the specific uplink carriers wishing to be received are first separated, then demodulated to baseband. The baseband is then demultiplexed (if necessary) and transferred to the destination. In some applications an earth station may itself operate in a transponding mode, in which received satellite signals are used to initiate a retransmission from the station to the satellite.

1.3 COMMUNICATION SATELLITES

A *communication satellite* is basically an electronic communication package placed in orbit. The prime objective of the satellite is to initiate or aid communication transmission from one point to another. In modern systems this information most often corresponds to voice (telephone), video (television), and digital data (teletype).

A satellite transponder must relay an uplink or forward link electromagnetic field to a downlink, or a return link. If this relay is accomplished by an orbiting passive reflector, as, for example, in the case of the Echo satellite, the power levels of the downlink will be extremely low owing to the total uplink–downlink propagation loss (plus the additional loss of a nonperfect reflector). An active satellite repeater aids the relay operation by being able to add power amplification at the satellite prior to the downlink transmission. Hence, an ideal active repeater would be simply an electronic amplifier in orbit, as sketched in Figure 1.5a. Ideally, it would receive the uplink carrier, amplify to the desired power level, and retransmit in the downlink. Practically, however, trying to receive and retransmit an amplified version of the same uplink waveform at the same satellite will cause unwanted feedback, or *ringaround*, from the downlink antenna into the receiver. For this reason satellite repeaters must involve some form of frequency translation prior to the power amplification. The translation shifts the uplink frequencies to a different set of downlink frequencies so that some separation exists between the frequency bands. This separation allows frequency filtering at the satellite uplink antenna to prevent ringaround from the transmitting (downlink) frequency band. In more sophisticated processing satellites the uplink carrier waveforms are actually reformatted or restructured, rather than merely frequency-translated, to form downlink. Frequency-band separation also allows the same antenna to be used for both receiving and transmitting, simplifying the satellite hardware.

The frequency translation requirements in satellites means that the ideal amplifier in Figure 1.5a should instead be reconstructed as in the diagram in Figure 1.5b. The satellite contains a receiving front end that first collects and filters the uplink. The collected uplink is then processed so as to translate or reformat to the downlink frequencies. The downlink carrier is then power-amplified to provide the retransmitted carrier. The details of the hardware circuitry of a transponder of this type will be discussed in Chapter 4. As more sophisticated satellites have evolved, the basic transponder model of Figure 1.5b has been modified and extended to more complicated forms. These modifications will also be examined in subsequent chapters.

In addition to the uplink repeating operation, communication satellites may involve other important communication subsystems as well (Figure 1.6). Since satellites may have to be monitored for position location, a *turnaround ranging subsystem* is often required on board. This allows the satellite to return instantaneously an uplink ranging waveform for tracking from an earth station (see Appendix C). In addition, communication satellites must have the capability of receiving and decoding command words from ground-control stations. These commands are used for processing adjustments or satellite orientation and orbit control. Most satellites utilize a separate satellite downlink to specific ground-control points for transmitting command verification, telemetry, and engineering "housekeeping" data. These uplink and downlink subsystems used for tracking, telemetry, and command (TT&C) are usually combined with the uplink processing channels in some manner. This means that, although they are not part of the mainline communication link, their design and performance does impact on the overall communication capability of the entire system.

The details of the fabrication of a satellite space vehicle are beyond the scope of this text. However, some aspects of satellite construction indirectly affect communication performance and therefore should be accounted for in initial system design. Primary power supply for all the communication electronics is generally provided by solar panels and storage batteries. The amount of primary power determines the usable satellite power levels for processing and transmission through the conversion efficiency of the electronic devices. The higher the primary power, the more power is available for the downlink retransmissions. However, increased solar panel and battery size adds additional weight to the space vehicle. Thus, there is an inherent limit to the power capability of the communication system.

Another important requirement in any orbiting satellite is attitude stabilization. A satellite must be fabricated so it can be stabilized (oriented)

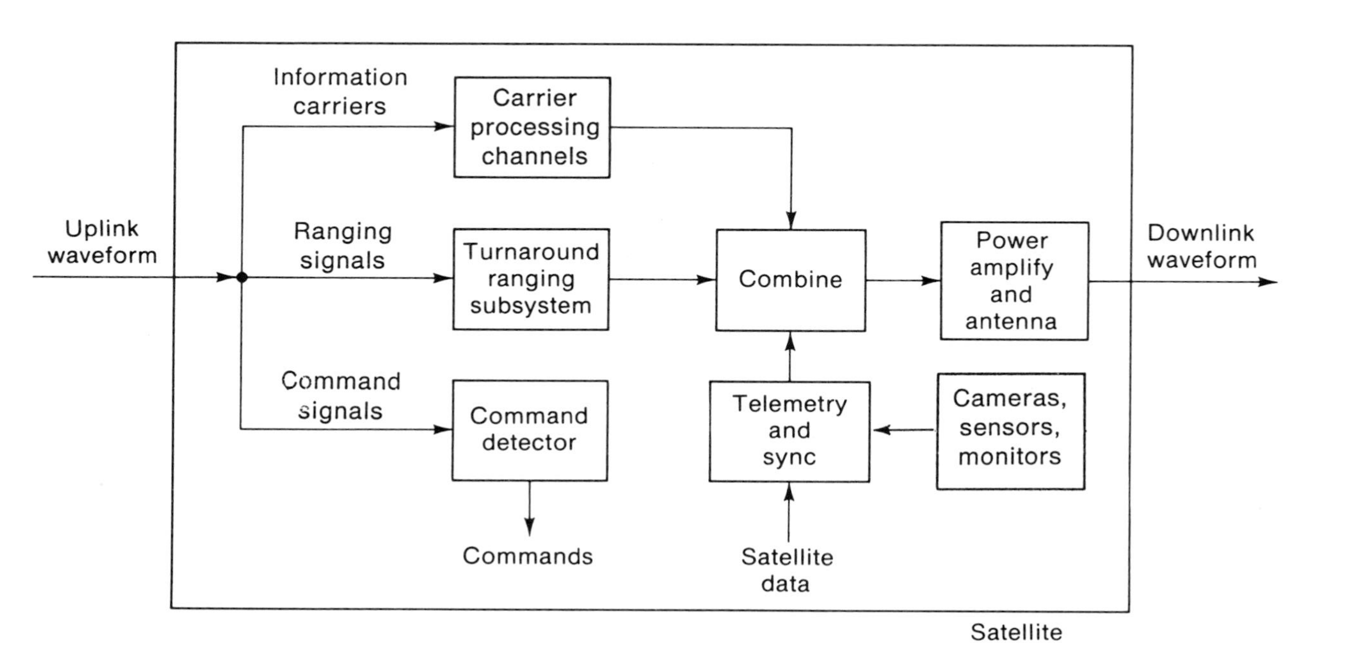

FIGURE 1.6 Satellite subsystems.

in space with its antennas pointed in the proper uplink and downlink directions. Satellite stabilization is achieved in one of two basic ways. Early satellites were stabilized by physically spinning the entire satellite (*spin-stabilized*) in order to maintain a fixed attitude axis. This means all points in space will be at a fixed direction relative to that axis. However, if the entire satellite is spinning, the antennas and solar panel must be *despun* so they continually point in the desired direction. This despinning can be accomplished either by placing the antennas and panels on platforms that are spun in the opposite direction to counteract the spacecraft spin or by using multiple elements that are phased so that only the element in the proper direction is activated at any time. Again we see an inherent limitation to both antenna and solar panel (power) size.

The second stabilization method is carried out via internal gyros, through which changes in orientation with respect to three different axes can be sensed and corrected by jet thrusters (three-axis stabilization). As requirements on satellite antennas and solar panels increased in size, it became correspondingly more difficult to despin, and three-axis stabilization became the preferred method. Spin stabilization has the advantage of being simpler and providing better attitude stiffness. However, spinning is vulnerable to bearing failures, cannot be made redundant, and favors wide diameter vehicles, which may be precluded by launch vehicle size. Also, when despinning multiple-element antennas and solar panels, only a fraction of each can be used at any one time, thus reducing power efficiency. Three-axis stabilization tends to favor vehicles with larger antennas and panels, and favors operation where stabilization redundancy is important.

Attitude stabilization also determines the degree of orientation control, and therefore the amount of error in the ability of the satellite to point in a given direction. Satellite downlink pointing errors are therefore determined by the stabilization method used. Both methods previously described can be made to produce about the same pointing accuracy, generally about a fraction of a degree. We shall see subsequently that pointing errors directly affect antenna design and system performance, especially in the more sophisticated satellite models being developed.

Satellite power amplifiers provide the primary amplification for the retransmitted carrier, and are obviously one of the key elements in a communication satellite. Power amplifiers, besides having to generate sufficient power levels and amplification gain, have additional requirements for reliability, long life, stability, high efficiency, and suitability for the space (orbiting) environment. These requirements have sufficiently been met by the use of *traveling-wave-tube amplifiers* (*TWTAs*), either of the cavity-coupled or helix type (Howes and Morgan, 1976; Strauss et al.,

1977). TWTAs have been developed extensively, their theory of operation is well understood, and they have been implemented successfully in all types of space missions. For this reason TWTAs have emerged as the universal form of both earth-station and satellite power amplifiers. Even as increased demands on power amplifiers will push them to higher power levels and higher frequencies, the TWTAs will undoubtedly continue to be the dominant amplification device. Their continual development has already produced sufficient power levels well into the 30-GHz frequency range.

It is expected that there will be continued effort to develop smaller lighter-weight solid-state amplifiers, such as *gallium arsenide field-effect transistors* (GA FET) for future satellite operations (Di Lorenzo, 1978; Hoefer, 1981; Liechti, 1976). These devices, however, have not been established in higher-power operating modes with reasonably sized bandwidths. They most likely will have future applications with appropriate power-combining, or lower-power, multiple-source operations. FET operation is generally confined to upper frequencies of about 30 GHz. For projected amplification above 30 GHz, *impact avalanche transit time* (IMPATT) diode amplifiers are rapidly developing as a capable medium-power amplifier. Such diodes have been developed at frequencies up to about 60 GHz, and it appears they will become extendable to 100-GHz operation in the near future.

1.4 ORBITING SATELLITES

Typical paths of an orbiting satellite are sketched in Figure 1.7. The satellite encircles the Earth with an orbit that can be *equatorial*, *polar*, or *inclined*. The *subsatellite* point is the instantaneous verticle projection on the Earth's surface of the satellite point in orbit. The subsatellite point therefore maps the satellite orbit along the Earth. Equatorial orbits circle the Earth with subsatellite points along the equator. Polar orbits have satellite points that pass through the poles.

An orbit may be elliptical or circular. For a circular orbit, a satellite must achieve a velocity of

$$v_s = \sqrt{\frac{g_0}{r_E + h}} \tag{1.4.1}$$

where r_E is the Earth's radius ($r_E = 3954$ m $= 6378$ km), h is the orbit altitude, and g_0 is the gravitational coefficient,

$$g_0 = 1.4 \times 10^{16} \text{ ft}^3/\text{s}^2 \tag{1.4.2}$$

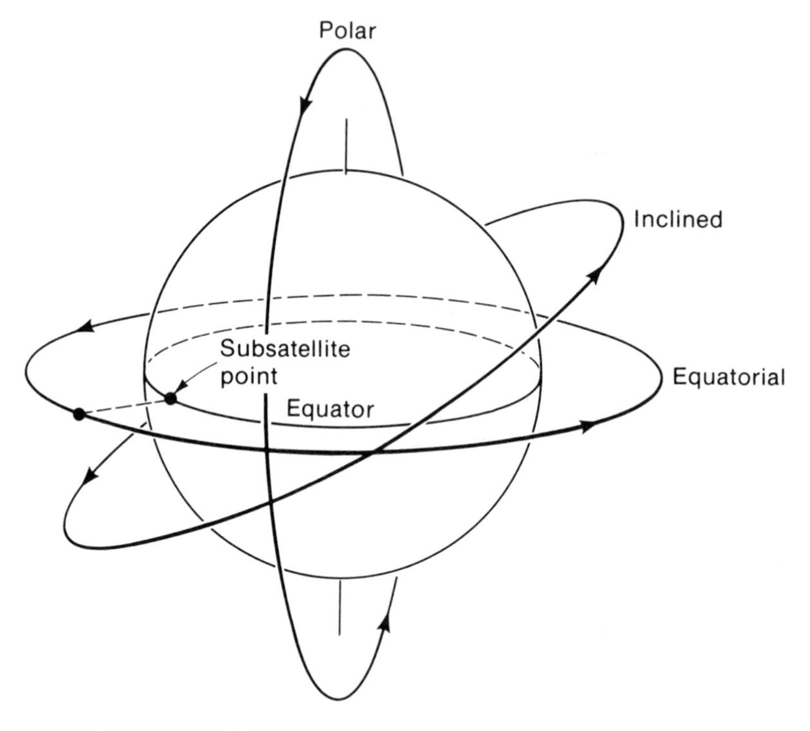

FIGURE 1.7 Satellite orbits.

The time t_s for an orbiting satellite to complete one revolution (orbit period) satisfies $\omega_s t_s = 2\pi$, where ω_s is the angular velocity in rad/s. Since $v_s = (r_E + h)\omega_s$, we have

$$t_s = \frac{2\pi}{v_s}(r_E + h)$$

$$= \frac{2\pi}{\sqrt{g_0}}(r_E + h)^{3/2} \tag{1.4.3}$$

Figure 1.8 plots Eqs. (1.4.1) and (1.4.3) as a function of the orbit altitude h. As the orbit altitude increases, the required satellite velocity decreases, while the orbit period increases. Satellites with altitudes in the approximate range of 100–1000 miles are referred to as *low Earth orbiters* (LEO), and circle the Earth every few hours.

If the satellite orbits in an equatorial plane at exactly the same angular velocity as the Earth rotates, it will appear to be fixed at a specific point

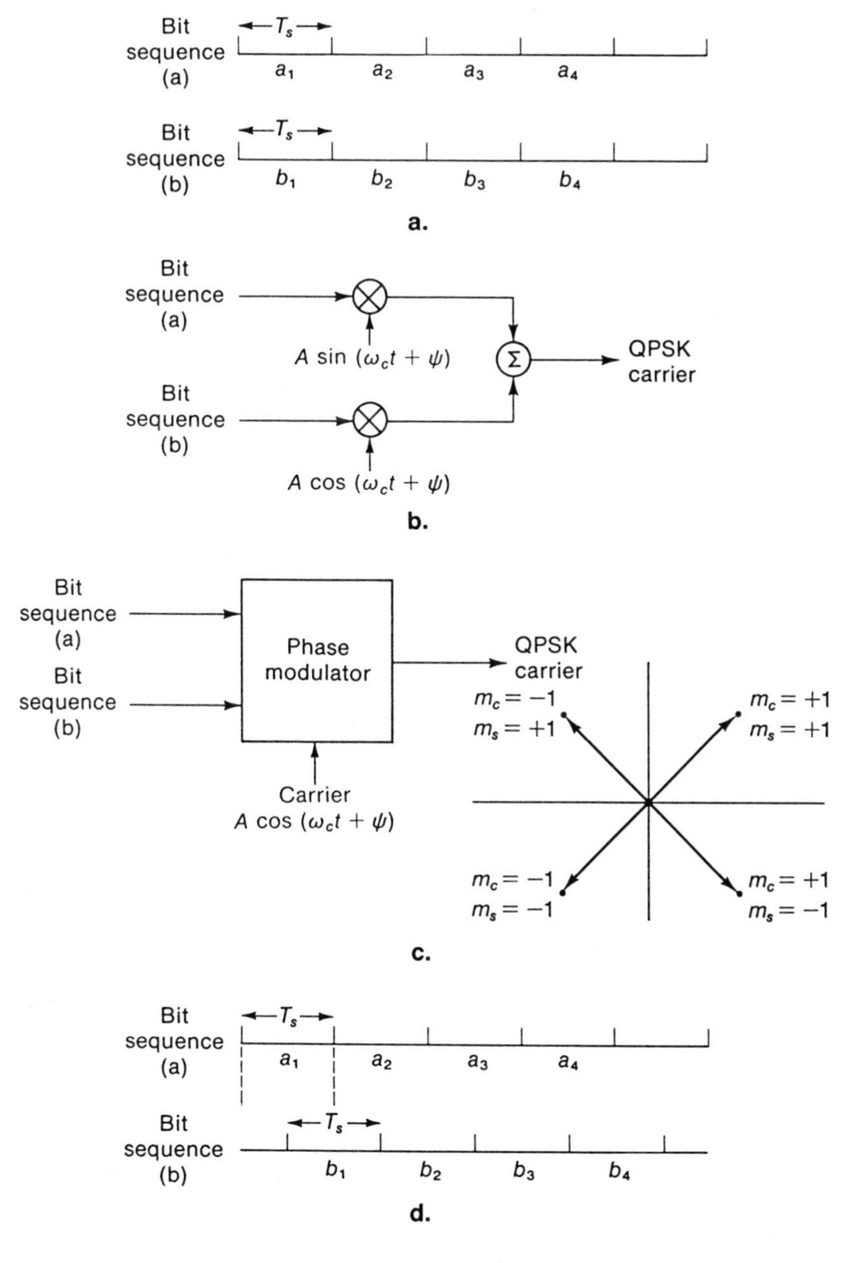

FIGURE 2.8 QPSK carriers. (a) Bit sequence alignment; (b) carrier generator using quadrature amplitude modulation; (c) carrier generator using quadrature phase shifting; (d) bit sequence alignment for OQPSK.

Since both $m_c(t)$ and $m_s(t)$ are either ± 1 at every t for either NRZ or Manchester waveforms, $\alpha(t) = \sqrt{2}$ at every t, and the QPSK carrier also has a constant envelope. Equations (2.3.9) and (2.3.10) also show that QPSK carriers can be formed either by combining separate BPSK quadrature carriers (see Figure 2.8b) or by phase shifting a carrier according to the ratio $m_s(t)/m_c(t)$. It can be easily verified that the phase angle θ in Eq. (2.3.11b) can take on only one of the four phase angles (45°, 135°, −135°, −45°) during each symbol time, depending on the sign of $m_c(t)$ and $m_s(t)$. Thus, a phase-modulation generator of QPSK, as shown in Figure 2.8c, corresponds to hopping the carrier phase between these angles, dependent on the bits of the data sequences of $m_c(t)$ and $m_s(t)$.

When the QPSK carrier is written as in Eq. (2.3.9), we see from its equivalent form in Eq. (2.3.10) that it has carrier power $P_c = (A\sqrt{2})^2/2 = A^2$, while each quadrature component has power $A^2/2$. Conversely, if the carrier amplitude in Eq. (2.3.10) was A, the carrier power would be $A^2/2$ and the components would have power $(A/\sqrt{2})^2/2 = A^2/4$. In other words, each quadrature component of a QPSK carrier always has half the carrier power.

The spectrum of the QPSK carrier is given by the extended version of Eq. (2.3.8) as

$$S_c(\omega) = \frac{A^2}{4}\left[S_{m_c}(\omega \pm \omega_c) + S_{m_s}(\omega + \omega_c)\right] \tag{2.3.12}$$

which sums the individual spectra of $m_c(t)$ and $m_s(t)$ and shifts to the carrier frequency. If identical bit waveforms $p(t)$ are used in each quadrature component, the QPSK carrier has the same spectrum as BPSK. This means the additional quadrature modulation is obtained "free of charge" as far as spectral extent is concerned. In the same spectral band as BPSK, a QPSK carrier can carry an additional data sequence. Since satellite links are invariably bandwidth-constrained, this doubling of the bit rate is extremely advantageous. Note that QPSK has a main hump bandwidth of $2/T_s$ Hz, or equivalently, $2/2T_b = 1/T_b$ Hz, depending on whether it is expressed in symbol times or carrier bit times.

A modified form of QPSK uses bit waveforms on the quadrature channels that are shifted relative to each other; this is referred to as *offset QPSK* (OQPSK). When the QPSK is generated via direct phase modulation, as in Figure 2.8c, offsetting limits the total phase shift that must be provided at each bit change. In standard OQPSK, NRZ encoding is used, and the offset is taken as one-half of a bit time, producing the carrier waveform

$$\begin{aligned} c(t) = {} & A \sum a_k p(t - kT_s) \cos(\omega_c t + \psi) \\ & + A \sum b_k p(t - \tfrac{1}{2}T_s - kT_s) \sin(\omega_c t + \psi) \end{aligned} \tag{2.3.13}$$

where $\{a_k\}$ and $\{b_k\}$ are the bit sequences being transmitted. Figure 2.8d shows the time relation of the two bit sequences in OQPSK. With offsetting, the bit changes of one bit occur at the midpoints of the quadrature bit. Nevertheless, as long as these sequences are independent, the spectrum of OQPSK is given by Eq. (2.3.12). Therefore, OQPSK has the same spectral shape as QPSK and BPSK. An advantage in this is the limited phase shift that must be imparted during modulation. Also, as we shall see subsequently, offsetting also provides spectral and interference advantages during nonideal decoding and nonlinear processing.

Unbalanced QPSK (UQPSK) is a modified version of QPSK in which the power levels and bit rates of each quadrature component are not identical. The UQPSK carrier has the form

$$c(t) = Am_c(t) \cos(\omega_c t + \psi) + Bm_s(t) \sin(\omega_c t + \psi) \tag{2.3.14}$$

where $A \neq B$, and $m_c(t)$ and $m_s(t)$ are NRZ binary waveforms with bit times T_A and T_B, respectively. We define the component powers $P_A = A^2/2$, $P_B = B^2/2$, and the unbalanced power ratio,

$$r = \frac{A^2}{B^2} = \frac{P_A}{P_B} \tag{2.3.15}$$

The total carrier power is again

$$P_c = P_A + P_B \tag{2.3.16}$$

and each component power can be written as a fraction of the total power

$$\begin{aligned} P_A &= \left(\frac{r}{r+1}\right) P_c \\ P_B &= \left(\frac{1}{r+1}\right) P_c \end{aligned} \tag{2.3.17}$$

Unbalanced QPSK can be considered a form of unequal, parallel BPSK transmission, in which the orthogonality of the quadrature carriers maintains the separability. The total carrier power P_c uses the power division in Eq. (2.3.15) to transmit the simultaneous bit rates. The spectrum is again given by Eq. (2.3.12), and therefore will be determined by the BPSK component with the highest bit rate.

Frequency Shift Keying Constant envelope binary signaling can also be generated from frequency modulation. The NRZ bit waveform is used to shift the carrier frequency between two modulation frequencies according to the bit waveform. The encoding is referred to as *frequency shift keying* (FSK). The modulated carrier has the form

$$c(t) = A \cos\left(\omega_c t + 2\pi\Delta_f \int m(t)dt + \psi\right) \qquad (2.3.18)$$

where $m(t)$ is the bit waveform sequence. This carrier corresponds to the frequency modulation of a carrier at ω_c with the baseband waveform $\Delta_f m(t)$.

Equation (2.3.18) differs from Eq. (2.1.1b) for analog modulation because the digital waveform $m(t)$ can be precisely adjusted in amplitude and waveshape. The carrier spectrum is somewhat difficult to compute, but can be estimated by noting that for NRZ bit waveforms $c(t)$ corresponds to sequences of bursts of a carrier at each of the modulation frequencies $f_c \pm \Delta_f$. Since each burst lasts for a bit time of T_b s, each such burst produces a spectrum centered at the modulation frequency, with a spectral width corresponding to that of a gating pulse of width T_b s. Hence, the spectrum of an FSK carrier will appear approximately as a combination of two such spectra, each centered at each of the modulation frequencies, producing the result shown in Figure 2.9. Note that the shape of the spectrum depends on the deviation Δ_f. As the deviation is increased, the two frequencies are further separated, and the ability to distinguish the two frequencies (decode the bit) is improved. However, the resulting carrier spectrum will have a wider main lobe, as shown in Figure 2.9.

2.4 SPECTRAL SHAPING

In modern satellite communications there is usually concern about the spectral tails (spectrum outside the main hump) interfering with other carriers. Hence, there is often an interest in reducing the spectral tails generated in digital carrier modulation. These tails, as depicted in Figures 2.7 and 2.9, are caused by the pulsed, rectangular waveshape of the NRZ or Manchester modulation. One way to reduce tails is to modify the bit waveshapes by rounding off the pulse edges. Therefore, there is an interest

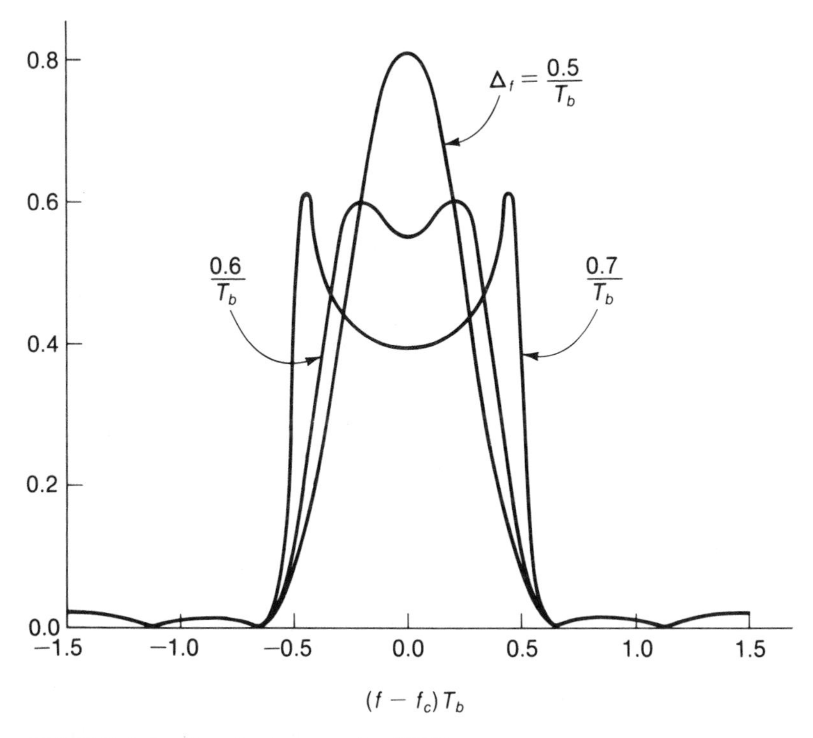

FIGURE 2.9 Carrier spectra for FSK.

in finding bit waveshapes that produce carrier spectral tails that decay faster than NRZ modulation, while still maintaining the constant envelope carrier condition.

In binary signaling this objective can be met by defining frequency-modulated carriers of the form of Eq. (2.3.18), where $m(t)$ is again given by Eq. (2.3.1), except the bit waveforms $p(t)$ are chosen to be smoother in time than the NRZ waveforms used in standard FSK. This avoids the rapid phase changes that tend to expand bandwidth. The smoother frequency modulation, along with proper choice of the frequency deviation, can be used to reduce the overall spectral spreading. For example, a convenient waveform is the *raised-cosine* (RC) pulse in Figure 2.10:

$$p(t) = \frac{1}{2}\left[1 - \cos\left(\frac{2\pi t}{T_b}\right)\right], \qquad 0 \le t \le T_b \tag{2.4.1}$$

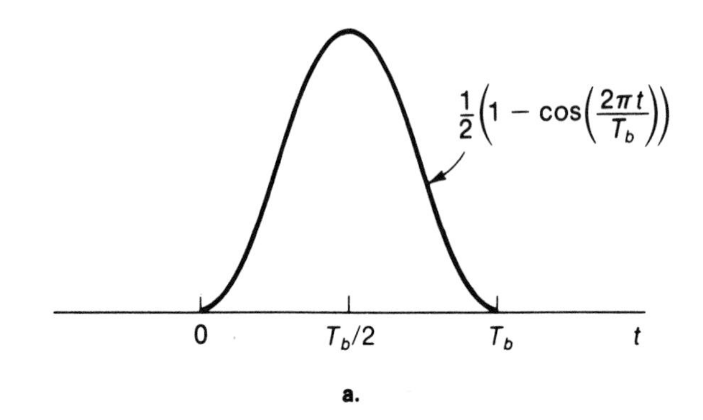

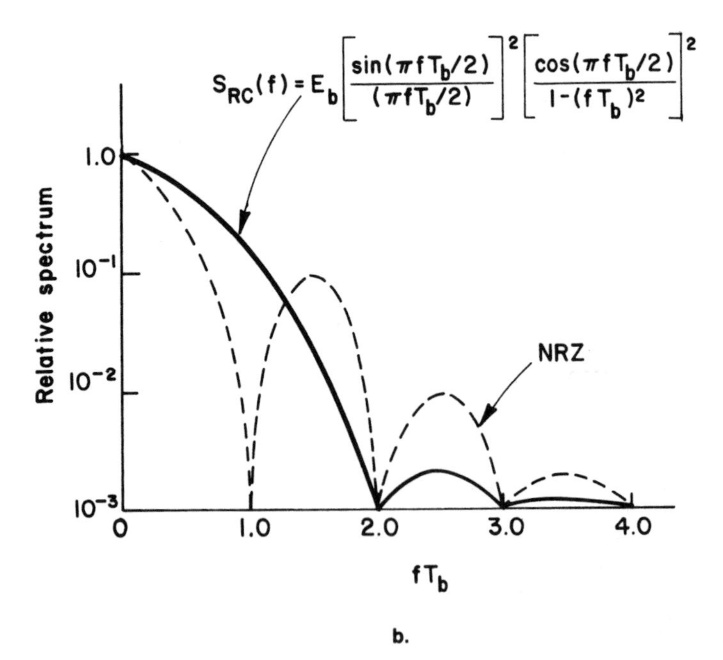

FIGURE 2.10 Raised cosine pulse. (a) Time waveform; (b) frequency function.

The ± 1 data bits multiply each such $p(t)$, forming $m(t)$ as a sequence of binary-modulated RC pulses. Such pulses have a frequency spectrum that decays much faster than NRZ pulses outside $1/T_b$ Hz, as shown in Figure 2.10b. Figure 2.11 shows how the out-of-band spectrum of an FM carrier in Eq. (2.3.18) is reduced from the standard BPSK and FSK spectra, when frequency-modulated with binary RC pulses having various deviations Δ_f.

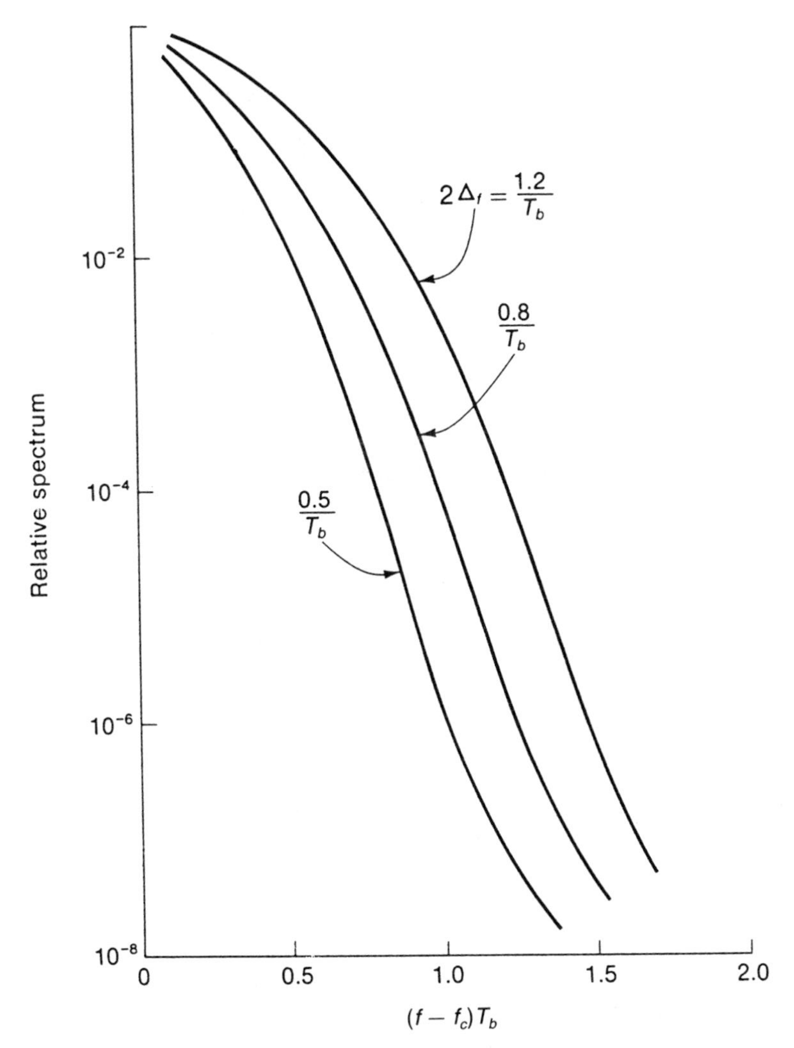

FIGURE 2.11 Carrier spectrum for FSK with binary raised cosine bit pulses (Δ_f = frequency deviation from carrier).

Frequency-shifted carriers have the property that the phase of the carrier traces out a continuous phase trajectory in time, according to the modulation sequence, as shown in Figure 2.12. (Recall carrier phase is the integral of the carrier frequency modulation.) This allows binary frequency-modulated carriers to be decoded by phase demodulation, using processing that attempts to track these phase functions. This phase demodulation has decoding advantages over standard frequency detec-

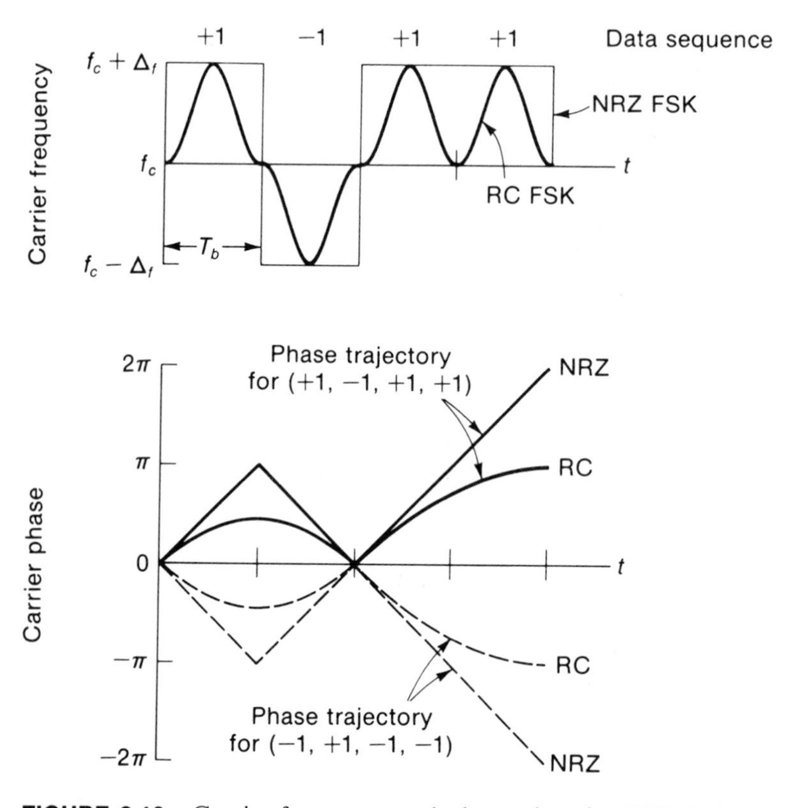

FIGURE 2.12 Carrier frequency and phase plots for FSK (NRZ = rectangular FM pulses; RC = raised cosine FM pulses).

tion. For this reason FM digital carriers of this type are often referred to as *continuous phase* FSK (CPFSK). The shape of the phase trajectories depends on the FM pulse waveform, while the separation between two different trajectories depends on the deviation Δ_f. If a smoother frequency modulation is used, as with RC pulses, the phase trajectories tend to become compressed and harder to distinguish (Figure 2.12). Likewise, decreasing the deviation Δ_f to reduce the spectral lobe, as shown in Figure 2.11, also impairs the ability to decode, either by tracking-phase functions or by direct frequency detection.

Carrier bandwidths with FSK using RC pulses can be further reduced by using pulses extended over longer time lengths while retaining the same bit rate. Figure 2.13a shows several RC pulses with their time length extended over several bits. Data are transmitted by using one of these

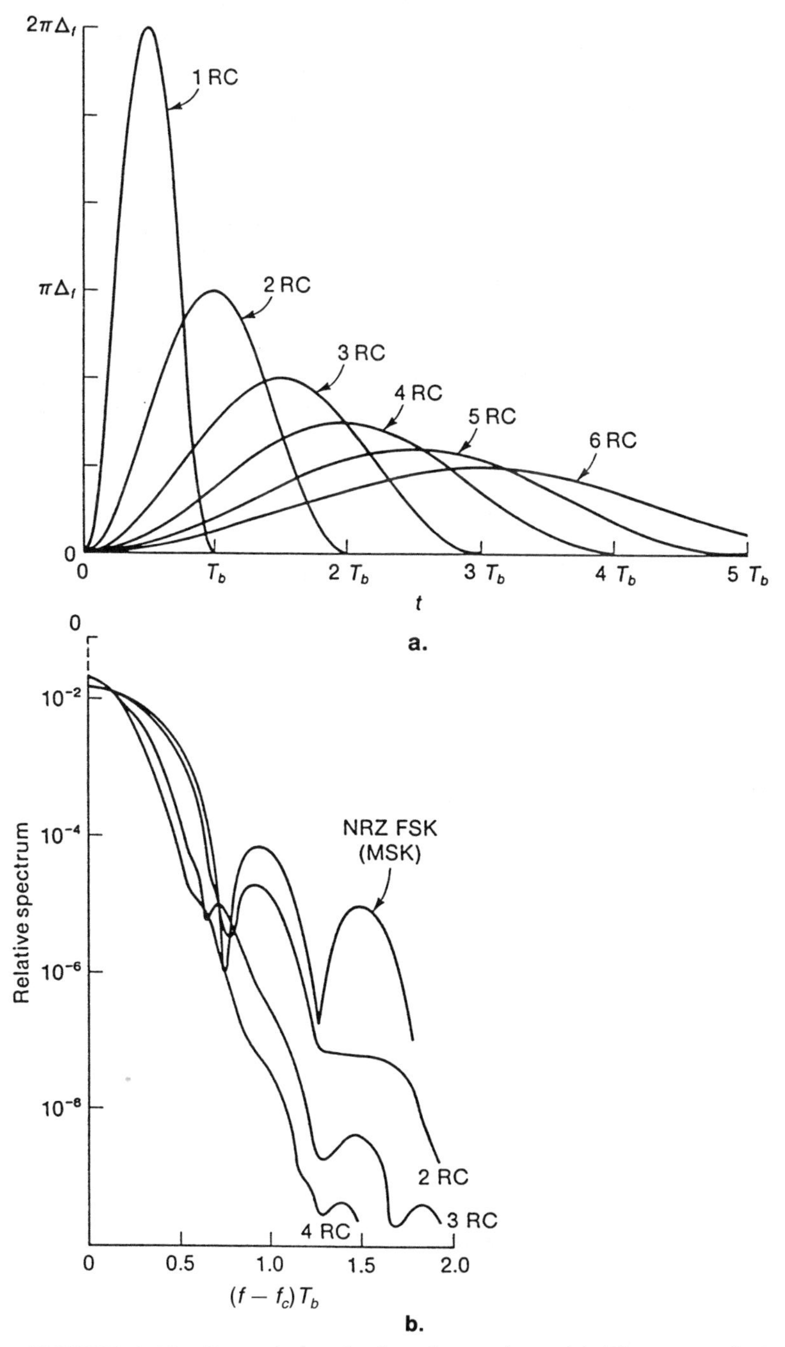

FIGURE 2.13 Extended raised-cosine pulses. (a) Time waveform; (b) carrier spectrum for FSK with length L, denoted LRC (from Aulin et al., 1981) $2\Delta_f T_b = 0.5$.

waveforms as the bit waveform $p(t)$ frequency-modulating the carrier. The longer RC bit waveform will produce carrier spectra with even less spectral spreading, as depicted in Figure 2.13b. The longer the RC pulse, the less the spectral extent. However, when the RC bit waveform is extended over multiple bit times, intersymbol interference occurs in which the phase of one bit period overlaps into the adjacent bit periods. This complicates the corresponding CPFSK phase trajectories and, as we shall see in Section 2.5, results in reduced phase decoding performance.

In QPSK waveforms generated from quadrature multipliers, the use of nonrectangular pulse shapes will generally cause nonconstant envelope carriers to occur. However, offsetting allows use of particular filtered pulse shapes that still retain constant envelopes. One example of filtered OQPSK is the class of *minimum shift-keyed* (MSK) carriers (Gronemeyer and McBride, 1976). Here, an OQPSK format is used, with NRZ bits replaced by half-period sine waves (Figure 2.14a). The MSK carrier is generated from pulse-shaped quadrature multipliers (Figure 2.14b) and produces the bit waveform

$$p(t) = \sin\left(\frac{\pi}{T_s} t\right), \qquad 0 \leq t \leq T_s \tag{2.4.2a}$$

and

$$\begin{aligned} p(t - \tfrac{1}{2}T_s) &= \sin\left(\frac{\pi}{T_s}(t - \tfrac{1}{2}T_s)\right) \\ &= \cos\left(\frac{\pi}{T_s} t\right), \qquad 0 \leq t \leq T_s \end{aligned} \tag{2.4.2b}$$

The MSK carrier is then formed as

$$\begin{aligned} c(t) = {} & A \sum_k a_k \cos\left[\frac{\pi}{T_s}(t - kT_s)\right] \cos(\omega_c t + \psi) \\ & + A \sum_k b_k \sin\left(\frac{\pi}{T_s}(t - kT_s)\right) \sin(\omega_c t + \psi) \end{aligned} \tag{2.4.3}$$

Note that the bit waveforms constituting $m_c(t)$ and $m_s(t)$ still do not overlap. Also at any t, $m_c^2(t) + m_s^2(t) = 1$, so that MSK carriers retain a constant envelope, while producing spectra in Eq. (2.3.12) that depend on the modified $p(t)$ of Eq. (2.4.2). The resulting MSK spectrum is shown in Figure 2.14c superimposed with OQPSK. Although the main spectral hump has been widened slightly for MSK, a significant reduction has

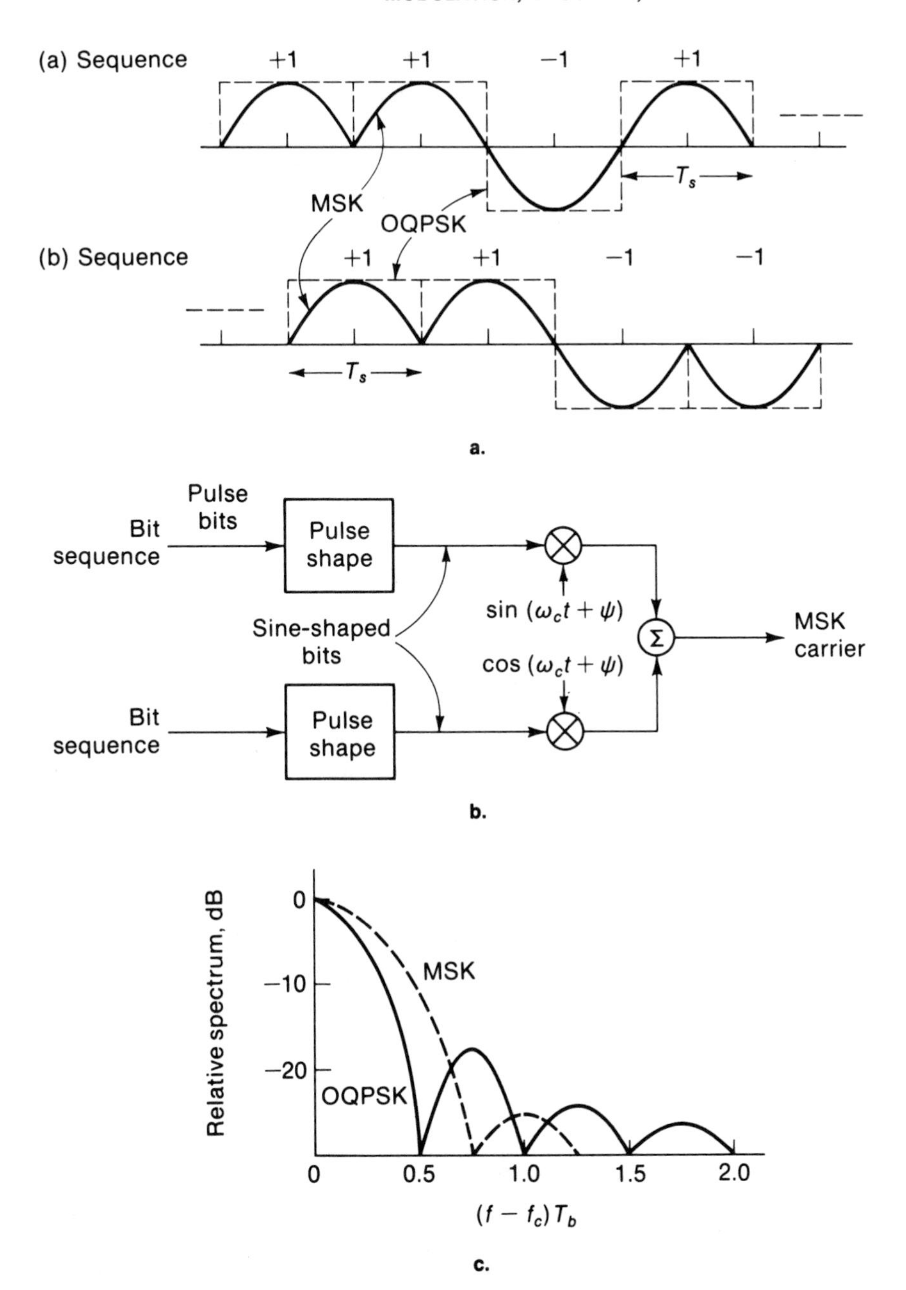

FIGURE 2.14 Minimum shift-keyed (MSK) signaling. (a) Bit waveforms; (b) encoder; (c) spectral comparison with QPSK. *Note*: T_s = symbol time, $T_b = T_s/2$ = QPSK bit time.

occurred in the out-of-band spectral tails. This makes MSK carriers somewhat more favorable in QPSK systems, where out-of-band spectra control and constant envelope carriers are jointly required.

By trigonometrically expanding, the MSK carrier can also be written as

$$c(t) = A\cos[\omega_c t + d_k(\pi/T_s)(t - kT_s) + x_k], \qquad kT_s \le t \le (k+1)T_s \tag{2.4.4}$$

where

$$\begin{aligned} d_k &= a_k, & kT_s/2 \le t \le (k+1)T_s/2 \\ &= b_k, & (k+1)T_s/2 \le t \le (k+2)T_s/2 \end{aligned} \tag{2.4.5a}$$

$$x_k = x_{k-1} + \frac{\pi k}{2}(d_{k-1} - d_k) + \psi \tag{2.4.5b}$$

This now has the form of a carrier that is frequency-shifted between $(f_c + \frac{1}{2}T_s)$ and $(f_c - \frac{1}{2}T_s)$ at each half bit (offset) time, according to the data bits. The carrier corresponds therefore to an FSK carrier with NRZ frequency-modulating bit waveforms and a specific deviation of $\Delta_f T_s = \frac{1}{2}$. In terms of carrier bit time, $T_b = T_s/2$, this is equivalent to $\Delta_f T_b = \frac{1}{4}$. The separation between the two signaling frequencies in MSK is therefore $2\Delta_f = \frac{1}{2}T_b = R_b/2$, or one half of the bit rate. Thus, MSK is, in fact, a form of a CPFSK carrier, having this particular modulation and deviation. This means, too, that the FSK curves with $\Delta_f = \frac{1}{4}T_b$ in Figure 2.13b can be similarly labeled as MSK. Note that we can view MSK as either the simultaneous quadrature modulation of two offset bits, or as a CPFSK modulation of the same two bits in sequence.

2.5 DIGITAL DECODING

The decoding operation is the process required to reconstruct the data bit sequence encoded onto the carrier. The lowest probability of decoding a carrier bit in error (bit-error probability) occurs if phase-coherent decoding is used. Phase-coherent decoding requires the decoder to use a referenced carrier at the same frequency and phase as the received modulated carrier during each bit time. That is, in order to decode optimally each bit, the decoder must provide in each bit interval a carrier at the exact frequency ω_c with the exact phase ψ used in the description of each encoded carrier. In addition, the exact timing of the beginning and ending of each bit must be known. This frequency, phase, and timing coherency must be obtained via a separate synchronization subsystem operating in conjunction with the receiver decoder, as shown in Figure 2.15. The synchronization subsystem must extract the required reference and timing from the received modulated carrier itself. These sync subsystems are discussed in Appendix B.

Phase-coherent decoding can be used with the phase-modulated carriers previously considered. Decoders for BPSK and QPSK carriers are shown in Figure 2.16. In BPSK, the phase-coherent carrier reference is first used to multiply (demodulate) the received carrier waveform. The demodulated waveform is then multiplied by the bit waveform $p(t)$ and integrated over each bit time. This effectively corresponds to the correlation of the received RF waveform with the reference waveform $p(t)\cos(\omega_c t + \psi)$ over each bit. The sign of the integrated bit output is then used to decode the bit. In QPSK the received carrier is simultaneously correlated with each phase-referenced quadrature carrier, each with its own bit waveform. Separate symbol integrations in each quadrature arm are then used to decode each quadrature bit simultaneously. Since the bit waveforms are arbitrary, this decoder will have the same form for OQPSK, UQPSK, and MSK carriers as well, with the waveforms $p_c(t)$ and $p_s(t)$ matching the transmitted bit waveform shape.

The probability of making a bit error (PE) with these decoders is computed in Appendix A. Table 2.4 lists the resulting bit-error probability expressions for decoding in the presence of additive white Gaussian RF noise, along with those expressions of some alternative encoding formats that are to be discussed. The plots of these equations are shown in Figure 2.17. Note that the bit-error probability depends only on the parameter

$$\frac{E_b}{N_0} = \frac{\text{Bit energy of the modulated carrier at decoder input}}{\text{Spectral level of additive noise at decoder input}} \tag{2.5.1}$$

This parameter can be written directly in terms of the decoder carrier power P_c and the carrier bit time T_b as

$$\frac{E_b}{N_0} = \frac{P_c T_b}{N_0} \tag{2.5.2}$$

The fact that the entire class of phase-coherent digital systems, when operating in an additive Gaussian white noise environment, has error probabilities dependent on only the single parameter in Eq. (2.5.2) is an extremely significant result in decoding theory. It means that digital performance can be immediately determined from knowledge of only

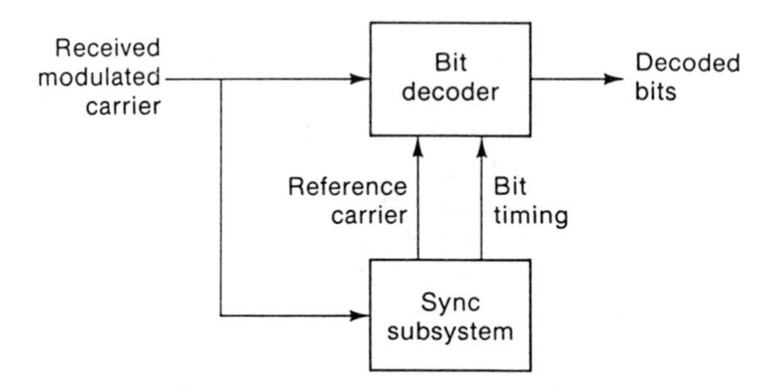

FIGURE 2.15 Bit decoding subsystem.

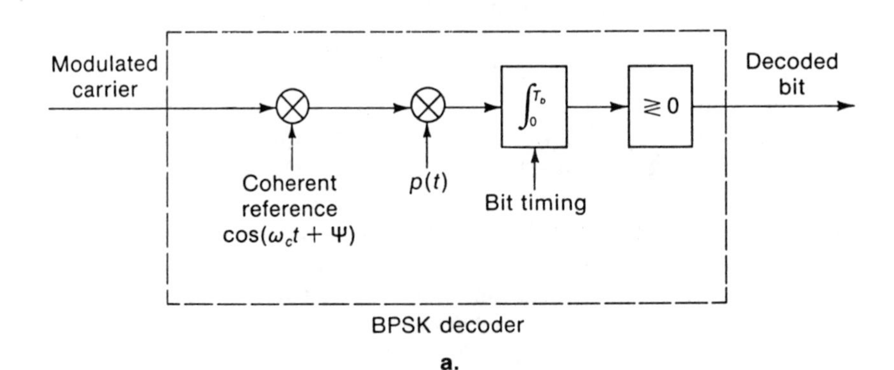

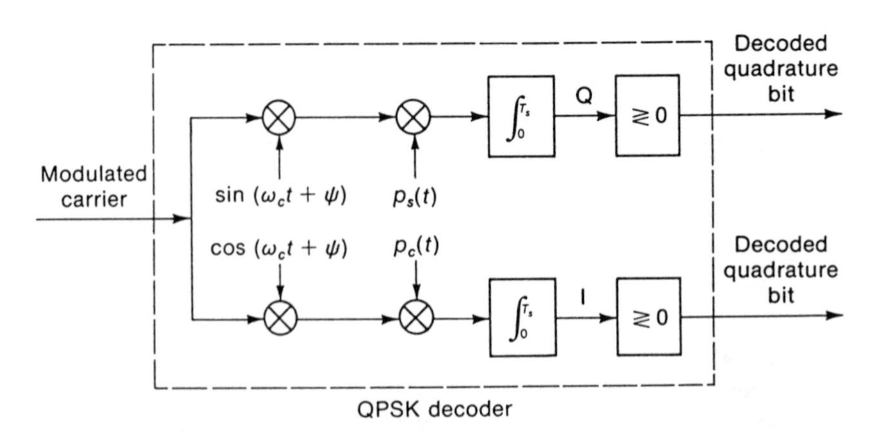

FIGURE 2.16 BPSK and QPSK decoders.

TABLE 2.4 Summary of decoding bit-error probabilities

Binary Format	*Bit-Error Probability* (*PE*)[a]
BPSK, QPSK, MSK (phase-coherent decoding)	$Q\left(\sqrt{\frac{2E_b}{N_0}}\right)$
FSK (phase-coherent frequency detection)	$Q\left(\sqrt{\frac{E_b}{N_0}}\right)$
DPSK	$\frac{1}{2}e^{-E_b/N_0}$
Noncoherent FSK	$\frac{1}{2}e^{-E_b/2N_0}$

[a] E_b = Bit energy

$$Q(x) = \frac{1}{2\pi}\int_0^{\infty} e^{-t^2/2}\,dt$$

decoder carrier power and noise levels. This fact greatly simplifies satellite system analyses. Note that decoding of all binary waveforms previously described, operating with perfect phase coherence, will theoretically produce the same PE as long as they operate with the same E_b/N_0. Thus, one waveform is as good as any other in terms of error probability, and those with advantages of spectral shapes or hardware simplification become increasingly important in satellite systems.

For the BPSK carrier in Eq. (2.3.7), $P_c = A^2/2$, and $E_b = A^2T_b/2$. For a QPSK carrier P_c is again $A^2/2$, and the carrier symbol energy is $A^2T_s/2$. This energy is divided between the two quadrature components, and each operates with decoding energy $A^2T_s/4 = A^2T_b/2$, where T_b is the QPSK carrier bit time ($T_b = T_s/2$). Thus, bit-error probability for either QPSK quadrature decoder can be obtained by reading off the value of PE on a BPSK curve at the value P_cT_b/N_0. Even though a QPSK (or OQPSK) system sends two bits during a symbol time, the PE of each quadrature bit can be obtained from the BPSK curve in Figure 2.17 at the corresponding E_b/N_0.

The distinction among the bit rate actually transmitted on the carrier, the bandwidth required, and the symbol time often leads to confusion in comparing BPSK and QPSK. Table 2.5 summarizes these basic relationships. Note that QPSK can transmit twice the bit rate as BPSK in a given carrier bandwidth, but will require twice the carrier power for the same decoding PE. However, for the same bit rate, both BPSK and QPSK require the same carrier power, but BPSK uses twice the carrier bandwidth.

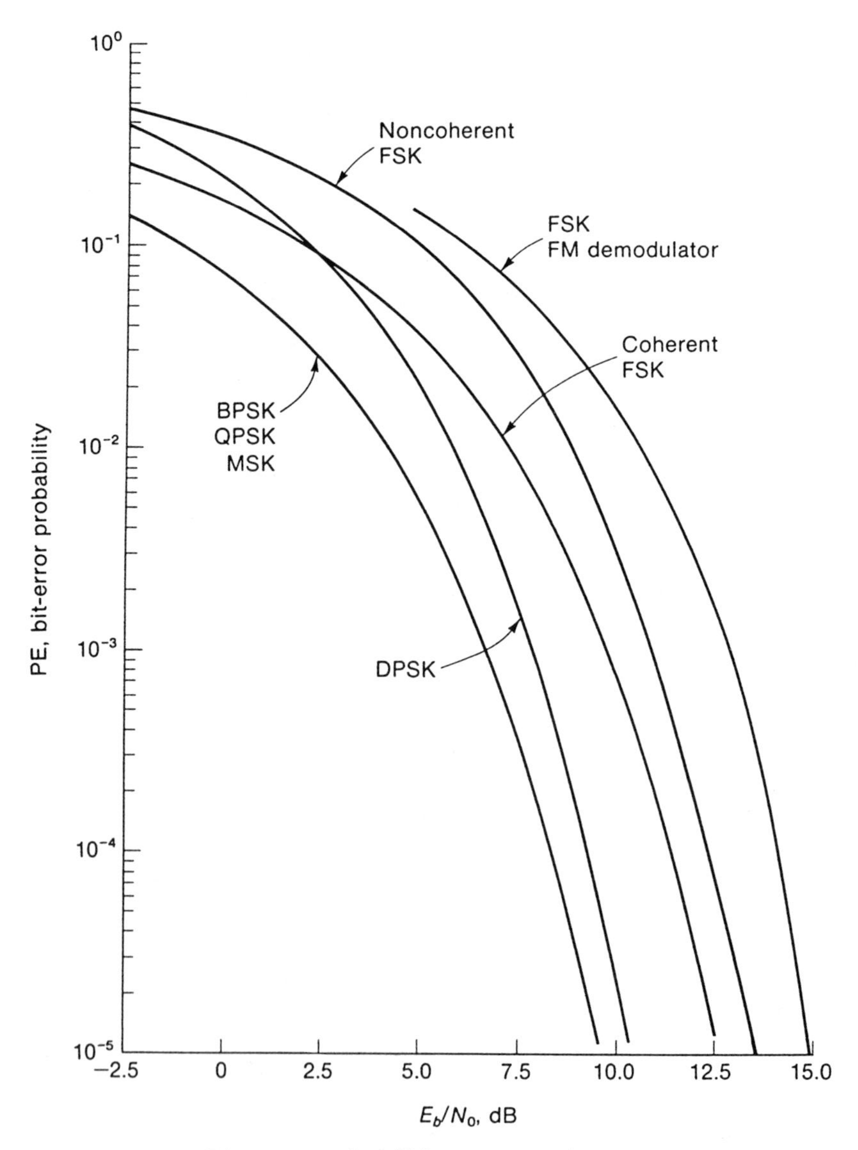

FIGURE 2.17 Bit error probabilities versus E_b/N_0.

MSK carriers, being a form of OQPSK, are also decoded by quadrature correlators using pulse matching to the sinusoidal bit waveforms in Eq. (2.4.3). The bit-error probability depends on the energy of each quadrature

TABLE 2.5 Summary of BPSK and QPSK identities

BPSK Identities
Power in carrier $= P_c$
RF bandwidth $B_c = 2/T_b = 2R_b$
Carrier energy per bit $= P_c T_b = P_c/R_b$
$= 2P_c/B_c$

QPSK Identities
Power in carrier $= P_c$
Power in each component $= P_c/2$
Symbol time per component $= T_s$
Symbol energy per component $= P_c T_s/2$
Carrier bit time $T_b = T_s/2$
RF bandwidth $B_c = 2/T_s = 1/T_b = R_b$
Carrier energy per carrier bit $= P_c T_b$
$= P_c/R_b$
$= P_c/B_c$

pulse component, which is also given by $P_c T_b$. Hence, a perfectly matched MSK decoder produces the same PE as a QPSK decoder with the same carrier symbol energy. This MSK decoding can also be viewed as a form of phase detection in which the carrier phase trajectory over a symbol time is used to decode each offset quadrature bit. This interpretation follows since MSK is also a form of CPFSK with NRZ frequency pulses, and the phase trajectory extends continuously over a symbol time. If a smoother frequency modulation is used (e.g., raised-cosine pulses for spectral control), the resulting carrier phase trajectories will be harder to decode, as we showed in Figure 2.12. If the RC pulse is extended over several bits, the phase intersymbol interference appears as a form of correlative encoding (Deshpande and Wittle, 1981) (i.e., insertion of memory). This can be decoded by introducing elaborate phase algorithms that attempt to unravel the interbit overlap (Cahn, 1974; McLane, 1983; Scharf et al., 1980; Ungerboeck, 1974). Estimates of bit-error performance for those algorithms have been made via performance bounds. Figure 2.18 shows these bounds for various CPFSK raised-cosine signaling, along with the previous result for MSK. We note that little degradation

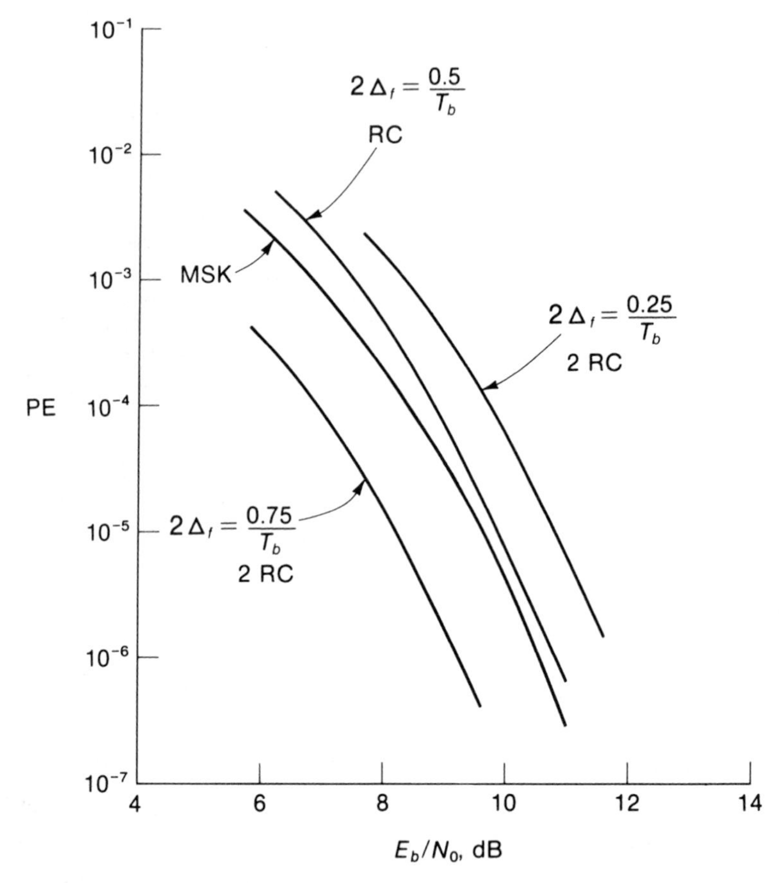

FIGURE 2.18 CPFSK bit error probability bounds for binary cosine FSK (from Aulin et al., 1981).

relative to MSK is predicted by this encoding, while the encoding simultaneously obtains the spectral reduction. Hence, CPFSK carriers with time-extended raised-cosine pulse can be made to perform comparable to MSK, but at the expense of implementing the decoding algorithms. An assessment of the required decoder hardware is discussed in McLane (1983).

Since MSK can also be viewed as an FSK carrier frequency-shifted during each bit time (not symbol time), it can also be decoded by phase-coherent frequency detection (see Appendix A) operating over each subsequent carrier bit time. Since only carrier energy in a bit time (instead of symbol time) is used for distinguishing the two frequencies, the decoding will be 3 dB poorer, as evidenced by the curve labeled "Coherent FSK" in Figure 2.17. This degradation can be reduced to about 1 dB for this type of decoding by using an optimal frequency separation of $0.71/T_b$ (Gagliardi, 1988, Chapter 7) instead of the $0.5/T_b$ separation used in MSK.

The QPSK decoder in Figure 2.16 also decodes unbalanced QPSK (UQPSK) if the quadrature arms are matched to each of the transmitted bit rates. The quadrature correlator separates the orthogonal carriers, and the bit decoders are then synchronized to each individual bit rate. Since the bit times are unequal, the component bit energies will be

$$E_A = P_A T_A, \qquad E_B = P_B T_B \tag{2.5.3}$$

where P_A and P_B are given in Eq. (2.3.17). For equal bit energy in each channel, $E_A = E_B$, we require $P_A T_A = P_B T_B$, or,

$$r = \frac{P_A}{P_B} = \frac{T_B}{T_A} = \frac{R_A}{R_B} \tag{2.5.4}$$

Thus, for equal bit energy per channel, the UQPSK power ratio should equal the data ratio. That is, the higher-rate channel should have the most power. We can therefore transmit unequal bit rates with the same bit energies by using UQPSK with the proper power ratio.

All of the previous decoders assumed perfect phase referencing. If the phase referencing is not perfect (i.e., the decoder carrier used for the correlations in Figure 2.16 does not have the same exact phase as the input modulated carrier), the decoding performance is degraded. This can be observed from Figure 2.19a, which shows how the signal portion of a BPSK decoder is altered when a phase reference error ψ_e is present. The resulting correlation and bit integration produces a signal term whose amplitude is degraded by $\cos \psi_e$. When ψ_e is small, the degradation is minimal, but as $|\psi_e| \rightarrow 90°$, the signal component can vanish, or even change sign if $|\psi_e| > 90°$. For a given phase error ψ_e, the decoded bit energy is $E_b \cos^2 \psi_e$, and the BPSK bit-error probability in Table 2.4 becomes

$$\mathrm{PE}(\psi_e) = Q[(2E_b/N_0)^{1/2} \cos \psi_e] \tag{2.5.5}$$

Hence, a phase-reference error causes the bit-error probability to increase from its minimal value. Typically, the phase reference is generated from noisy phase referencing circuitry, so that ψ_e evolves as a random variable during each bit time. This means that PE in Eq. (2.5.5) is really a conditional PE, conditioned on the random variable ψ_e, and the resulting

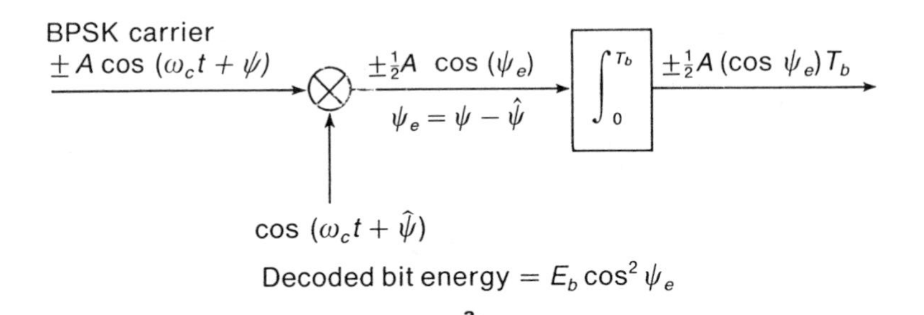

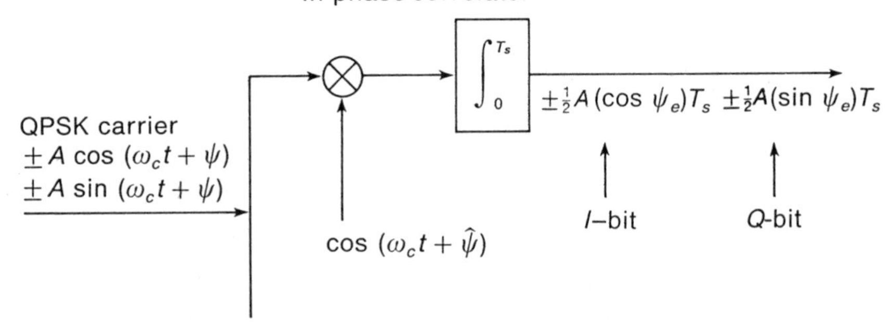

FIGURE 2.19 Effect of phase reference errors on signal decoding term. (a) BPSK decoder; (b) in-phase arm of a QPSK decoder.

average bit-error probability must be determined by averaging over the randomness in the phase error ψ_e. Hence, the bit error probability is computed from

$$\text{PE} = \int_0^{2\pi} \text{PE}(\psi_e) p(\psi_e) d\psi_e \tag{2.5.6}$$

The probability density of the phase error, $p(\psi_e)$, is usually taken as a zero mean Gaussian variable with rms phase σ_e. When Eq. (2.5.6) is evaluated for this Gaussian case, Figure 2.20 shows how the BPSK PE varies from the ideal BPSK case as the rms reference phase error increases. Note that phase-reference errors can produce bit-error probabilities several orders of magnitude higher than that predicted by the ideal case. Phase errors cause a "bottoming" or "flooring" of the PE performance that cannot be reduced even if the bit energy E_b is significantly increased.

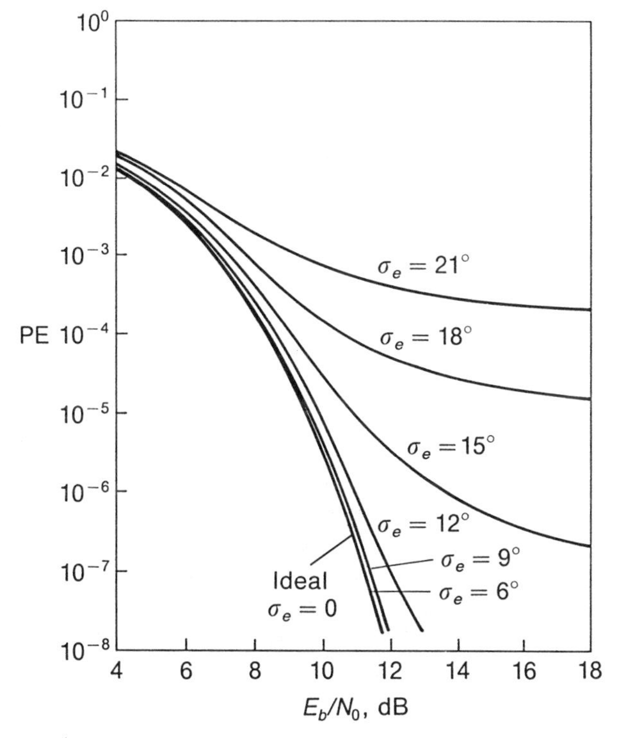

FIGURE 2.20 BPSK PE versus E_b/N_0 with phase referencing errors (σ_e^2 = variance of phase error).

As a rule of thumb, if PE performance is to be maintained at about 10^{-5}, rms phase-reference errors must be limited to about 12° or less.

In QPSK decoding, a phase-reference error is more critical. As shown in Figure 2.19b, reference error not only degrades the decoding signal of each desired bit, but will also cause a cross-coupling of the orthogonal bit. Figure 2.21 shows the effect of a Gaussian phase error ψ_e, with rms value σ_e, on the resulting QPSK decoding. The plot shows a higher degradation than with BPSK for the same phase-error variance. Hence, QPSK systems are degraded more by phase-reference errors than BPSK systems, and generally must be designed with better referencing accuracy.

In offset QPSK, bit transitions in the other channel may occur at the middle of the bit interval of each channel. If no transition occurs, performance during that bit is identical to QPSK. When a transition occurs, the cross-coupling term (involving $\sin \psi_e$) changes polarity at midpoint, and the integrated cross-coupling during the first half bit is

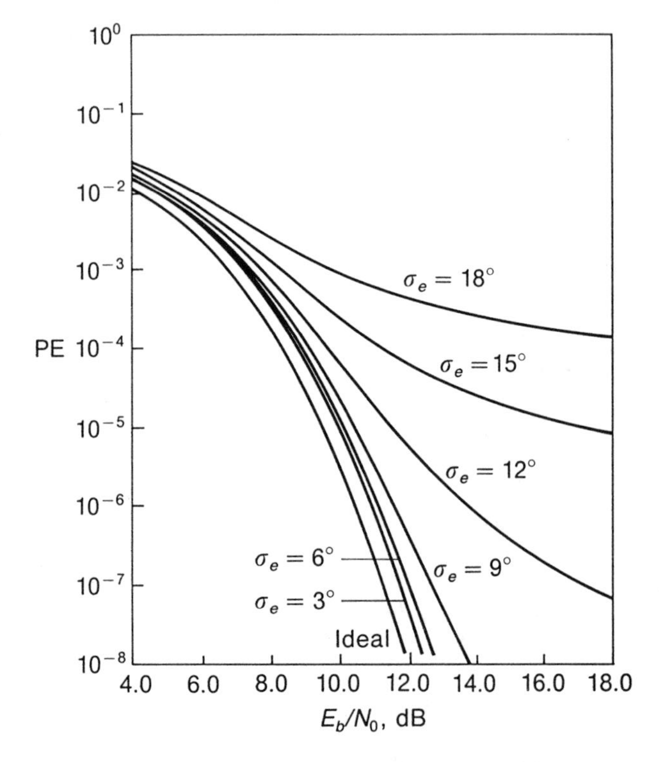

FIGURE 2.21 QPSK PE versus E_b/N_0 with phase-referencing errors (σ_e^2 = variance of phase error).

cancelled by that during the second half of the interval. Performance is identical to that with no interference at all, and therefore is the same as for BPSK with phase-reference error ψ_e. Hence, the resulting OQPSK PE with phase-referencing errors is the average of the PE with BPSK and with QPSK (Figures 2.20 and 2.21) at the same rms phase error σ_e. This shows that offsetting the bits will produce slightly less PE degradation than standard QPSK with the same phase referencing variance.

Decoding performance of unbalanced QPSK with phase-referencing errors is now slightly more complicated, because the cross-coupling involves different bit times and power levels. Consider Figure 2.22, showing $R_A > R_B$ so that there are $r = T_B/T_A$ bits in the A channel for every bit in the B channel. Assuming an integrated-and-dump decoder in each channel, and a phase-referencing error occurs, each A bit will have either a $\pm B$ bit superimposed, with the power level of the weaker channel. We see that the higher power channel in UQPSK has a reduced crosstalk

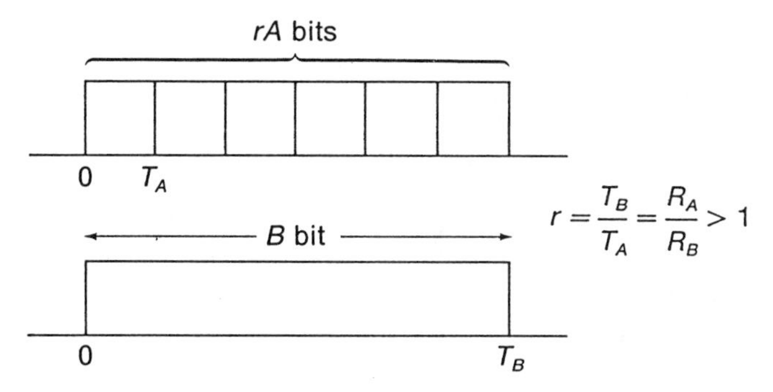

FIGURE 2.22 Unbalanced QPSK bit alignment.

due to the power imbalance. The phase error couples in a lower-power bit onto a higher-power bit, and its effect is therefore reduced.

In the B channel, the integration over a bit time is over r separate A bits coupled in by the phase error. The effect will depend on the number of A bits that match or mismatch the B bit. Since these occur equally likely, PE must be computed by averaging over all possible A bit sequences. Thus, the reference error cross-couples the higher-power A bits onto the lower-power B channel but has the advantage that the cross-coupled bits effectively are averaged over all the possible bit patterns, which partially compensates for the power levels. That is, the worst case occurs less often. It is therefore not immediately evident whether the A and B channel is more degraded by phase-referencing errors in UQPSK, and both must be evaluated for comparison.

In some satellite systems it may be advantageous to avoid completely the receiver task of having to generate a phase-coherent reference for decoding. Two ways to operate a binary encoder without a phase-coherent reference are to do so either by *differential* encoding or by *noncoherent* frequency-shift keying. In *differential PSK* (DPSK), a BPSK carrier is used, but the carrier phase of each bit is referenced to the previous bit. A binary $+1$ or -1 is encoded during a bit time T_b by using either the same phase or a π shift of the previous bit phase, depending on whether the present bit is identical or opposite to the previous bit. This is equivalent to forming the BPSK carrier in Eq. (2.3.7) using the NRZ sequence in Eq. (2.3.1) with

$$d_k = d_{k-1} a_k \tag{2.5.7}$$

where a_k is the present bit. Decoding can be achieved by comparing the present bit integrations with the previous bit, and determining if they

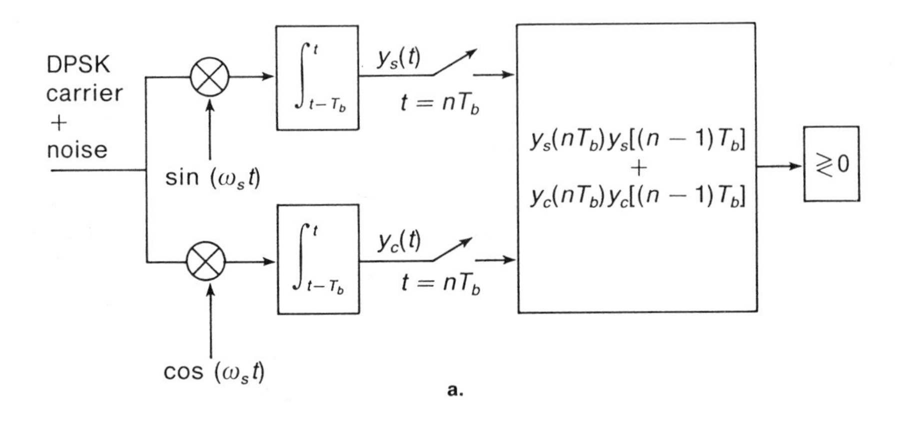

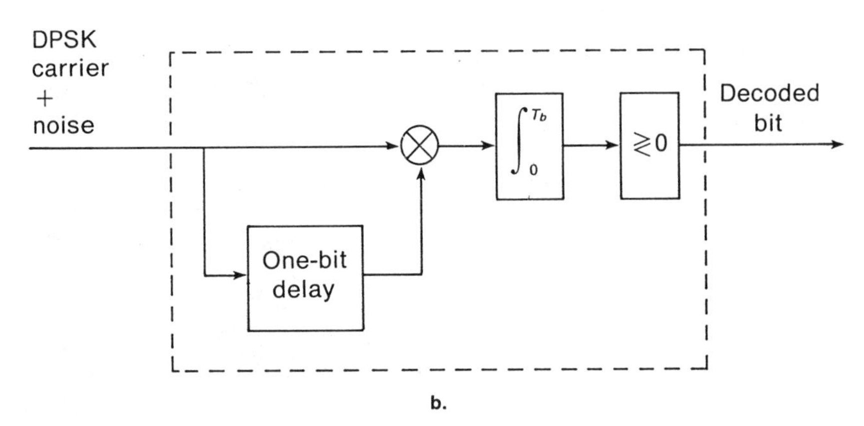

FIGURE 2.23 DPSK decoders. (a) Optimal single-bit decoder; (b) delay-and-correlate decoder.

correlate positively or negatively. Figure 2.23a shows the optimal decoder of this type, which performs the noncoherent present-to-past bit comparison. A quadrature correlator is used, but no phase referencing is required. The correlator generates an in-phase and quadrature bit sample for each bit, and the bit decoding is based on the sample processing as shown. Table 2.4 and Figure 2.17 show the PE performance of this DPSK decoder. Performance is poorer than that for coherent BPSK, but approaches twice the PE of the latter at high E_b/N_0 values. The fact that DPSK can achieve approximate BPSK performance without the need of any phase-referencing subsystem is a prime reason for its popularity.

Some recent results (Divsalar and Simon, 1990) have shown that further DPSK decoding improvement can be achieved by processing with longer memory, i.e., by retaining and using more of the past samples of the received DPSK waveform in the processing block in Figure 2.23a. For example, by using the past two samples (instead of only the previous one) the required E_b/N_0 to produce PE $= 10^{-5}$ can be reduced by another one-half dB, relative to the DPSK curve shown in Figure 2.17. Further increase in the memory during processing can produce DPSK decoding that asymptotically approaches that of an ideal coherent DPSK system (twice the PE of BPSK), provided the unknown carrier phase remains constant throughout the memory interval.

An even simpler DPSK decoder can be used, as shown in Figure 2.23b, which performs a direct delay-and-bit comparison with the previous bit during each bit time. The past bit is delayed and correlated with the present bit to determine if it correlates positively or negatively. The decoding PE for this system is about 2 dB worse than for the decoder in Figure 2.23a (see Appendix A.5).

A second way to transmit binary data without requiring phase-coherent decoding is by the use of the FSK carriers in Eq. (2.3.14), except that pure frequency detection is used. Since the bit modulation is inserted into the frequency of the carrier, the bit can be decoded by making frequency measurements on the received carrier. These frequency measurements can be made without knowledge of the phase. In binary FSK, decoding is accomplished by tuned-matched bandpass filters centered at each of the modulating frequencies, followed by energy detection during each bit time to determine which filter contains the most energy (Figure 2.24). If the frequencies are sufficiently separated, only the correct filter will contain signal carrier energy, while the other filter contains only noise. The resulting bit-error probability is included in Table 2.4 and Figure 2.17. We note that this form of noncoherent frequency decoding again

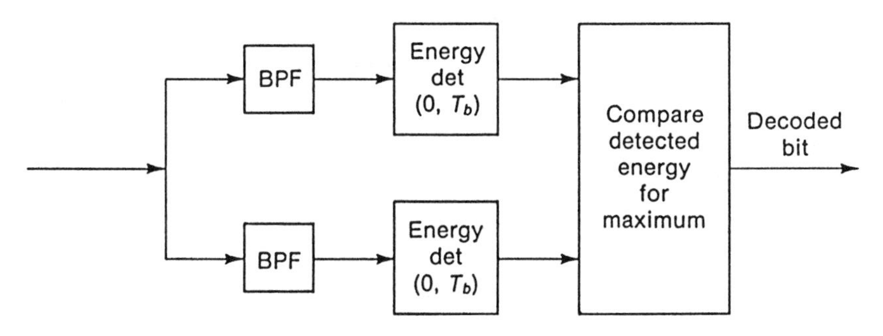

FIGURE 2.24 Noncoherent FSK decoder.

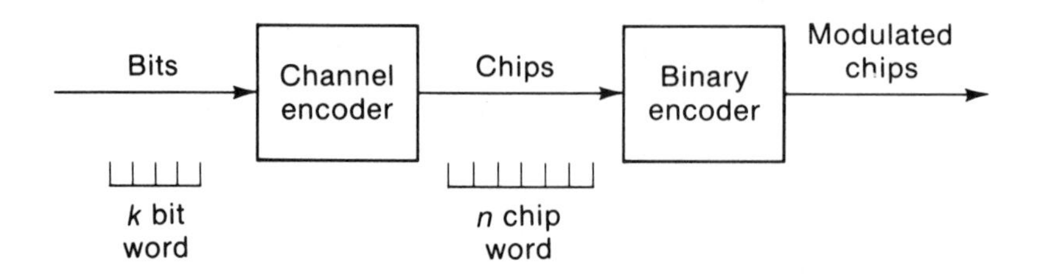

FIGURE 2.25 Error-correction encoding model.

produces poorer (higher) PE at the same carrier bit energy than phase-coherent BPSK. FSK is about 4 dB worse at PE = 10^{-3}, and about 3 dB worse at PE = 10^{-5}.

FSK can also be detected by first frequency demodulating the carrier to recover the baseband $m(t)$. The latter is then followed by bit integrations and sign detections over each bit time to determine bits of $m(t)$. Such systems do about 1 dB worse than bandpass filter-energy detection decoders (Gagliardi, 1988, Chapter 8), as shown in Figure 2.17. These FSK degradations in performance again reflect the price of not implementing phase-referencing circuitry. In low-cost receiver design this trade-off of error performance for hardware simplicity may be both desirable and cost efficient.

2.6 ERROR-CORRECTION ENCODING

Since ideal phase-coherent decoding produces the minimal PE for the carrier waveforms considered, based on processing of the received waveform during each bit time, no other carrier encoding scheme can produce a smaller PE at a given value of E_b/N_0. The phase-coherent curve in Figure 2.17 therefore shows the limiting binary performance. However, we can produce smaller values of PE by operating in such a way that bit processing extends over multiple bit times, leading to further improvement. We can accomplish this by imposing error-correction encoding. This encoding is obtained by inserting a higher level of digital encoding prior to the binary waveform encoding previously described. This additional coding is depicted in Figure 2.25.

Block Encoding

In error-correction block encoding, each block of k data bits is first encoded (mapped) into distinct blocks of n channel bits (called *channel symbols*, or *bauds*, or *chips*), where $n > k$. That is, the chip block size is longer than the data bit block size from which it was generated. There are therefore n/k chips per data bit, and the *code rate* is defined as

$$r = \frac{k}{n} \tag{2.6.1}$$

Each consecutive block of k data bits is converted into a block of n chips, according to the prescribed coding rule. The chips are then sent over the binary channel as if they were data bits. If the n chips are all decoded correctly, the k data bits to which they correspond are immediately identified, and the data bits have effectively been transmitted correctly. If some of the n chips are decoded incorrectly, however, the k data bits may still possibly be correctly identified, since "extra" chips are used to represent the k bits. Some of the chips in a block may be decoded incorrectly without decoding the data bit block incorrectly. If this occurs, the resulting bit-error probability of the data will be less than the chip-error probability of the link, for the same chip carrier power. The number of chips that can be in error before a data bit error occurs is related to the coding rule that maps the k bits into the n chips.

The development of the theory of this mapping is beyond our scope here, but for properly selected code (mapping) rules, the number of allowable chip errors will be related to the difference $n - k$. The larger this difference, the more chip errors can be made before data bit errors occur. Unfortunately, the larger the values of n and k, the more complex is the implementation of the required coding and decoding hardware.

The probability that a transmitted chip will be decoded in error, using the phase-coherent encoding methods of Table 2.4, is given by

$$\epsilon = Q\left(\sqrt{\frac{2E_c}{N_0}}\right) \tag{2.6.2}$$

where E_c is the energy per chip. If the bits are to be sent at a fixed rate, the chips must be sent faster than this rate (shorter chip time and larger carrier bandwidth) since more chips than bits are transmitted. This means the chip time is less than a bit time and

$$E_c = rE_b \tag{2.6.3}$$

where r is again the code rate in Eq. (2.6.1). It can be shown (Viterbi and Omura, 1979) that the resulting bit-error probability is accurately bounded by

$$\text{PE} \leq 2^{-n(r_0 - r)} \tag{2.6.4}$$

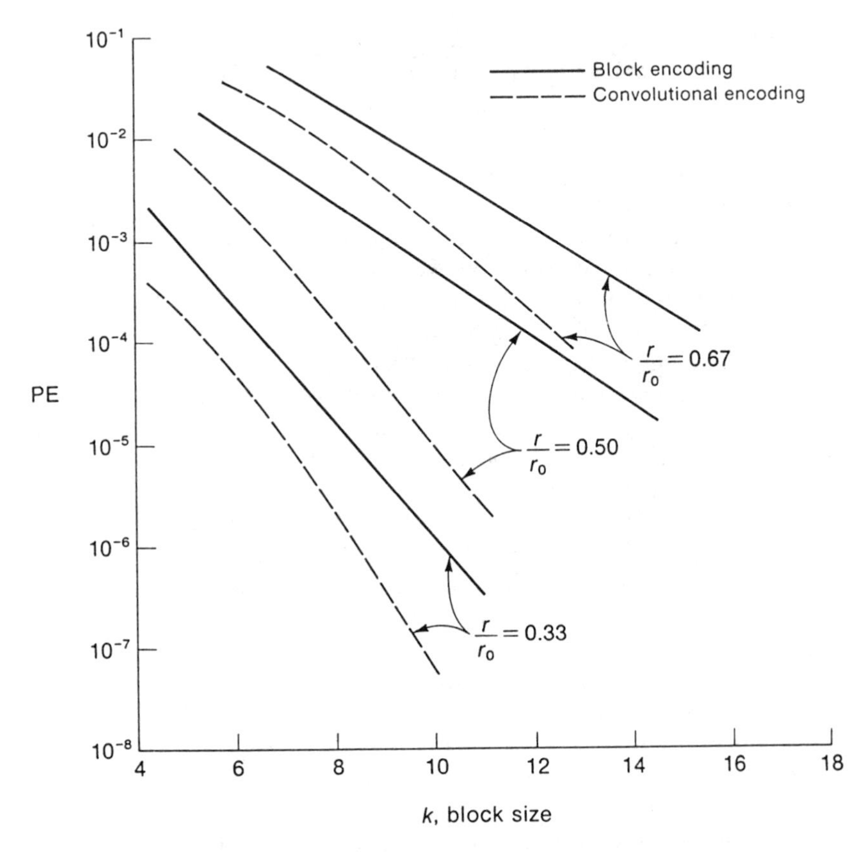

FIGURE 2.26 PE bound versus code block length k and r/r_0 (r = code rate = k/n; r_0 = channel cut-off rate).

where r_0 is the chip *cut-off rate*, defined by

$$r_0 \triangleq 1 - \log_2[1 + \sqrt{4\epsilon(1 - \epsilon)}] \tag{2.6.5}$$

The bit-error probability can therefore be estimated directly from the code parameters. Note that as long as the code rate r is less than the cut-off rate r_0 [the latter dependent on the chip energy through Eqs. (2.6.5) and (2.6.2)] the bit-error probability can be reduced by increasing the chip block length n (and therefore also increasing data block length k, since $k = rn$). This simply reiterates the fact that for a fixed code rate k/n, the larger the values of k and n, the larger is the difference $n - k$ and the more chip errors can be tolerated. The number of computations needed to convert the decoded chips into the decoded bits, however, increases as 2^k. Thus, decoding complexity and required processing speed increases exponentially with error-correction block lengths.

Figure 2.26 plots the bound on Eq. (2.6.4) as a function of block code length k for several values of r/r_0. Typically, $\epsilon \leq 10^{-3}$, for which r_0 in Eq. (2.6.5) is about 1, and the ratio r/r_0 is approximately equal to the code rate r in Eq. (2.6.1). Note the continual improvement in performance as higher levels of coding are introduced at a fixed rate. However, as stated earlier, the number of computations needed also increases exponentially with k. Thus, the price to be paid for PE improvement via error correction is in the decoding complexity that must be implemented. As digital processing capabilities and computation speeds increase through technological advances, higher levels of block lengths can physically be implemented, leading directly to digital performance improvement. It is expected that this push toward faster decoding processing and reduced PE will continue throughout the future of satellite communications.

Convolutional Encoding

An alternative to the block error-correcting coding is the use of convolutional coding (Viterbi, 1967; Viterbi and Omura, 1979). Convolutional coding is another technique for encoding data bits into transmission chips, such that subsequent chip decoding leads to improved bit decoding. Rather than coding fixed-data blocks into fixed-chip blocks, each chip is generated instead through a continuous convolution of previous data bits. This convolution leads again to improved performance when operating at the same code rate. Decoding is achieved by detecting one bit at a time in sequence, based on the processing of sliding sequences of received chips. The optimal decoding is obtained by means of a *Viterbi decoding algorithm* (Forney, 1973; Viterbi, 1967), which "unconvolves" the bits from the decoded chips.

The convolving property of convolutional encoding produces codes that can be decoded with better bit-error performance than previous block codes with the same rate and block length. Since the convolutional encoders are no more complicated than block encoders, convolutional encoding is the preferred method for forward-error correction. Performance analysis is more difficult, however, owing to the interlaced nature of the encoded bits. It has been shown that for a code rate r and *constraint length* k (the number of past bits convolved into each chip), there exists a convolutional code with bit-error probability bounded by

$$\mathrm{PE} \leq \frac{2^{-k(r_0/r)}}{[1 - 2^{-(r_0/r - 1)}]^2} \tag{2.6.6}$$

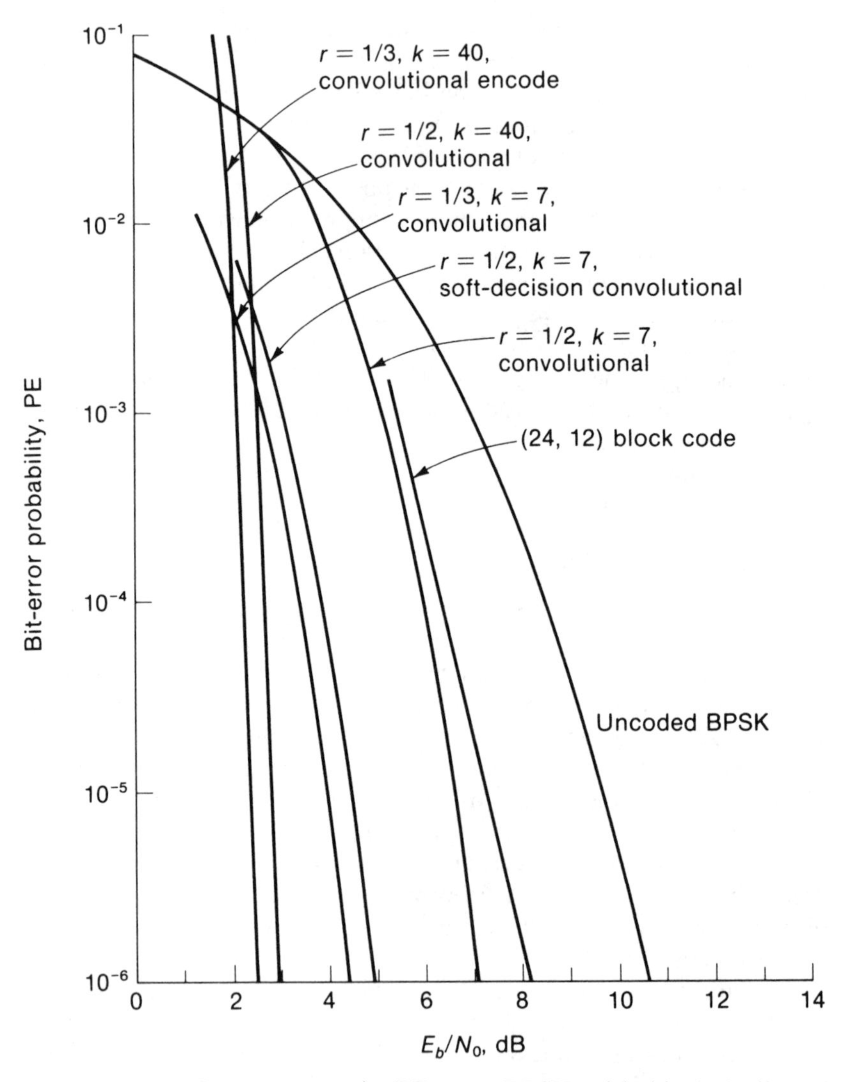

FIGURE 2.27 Improvement in PE over BPSK with block (n, k) and convolutional coding.

where r_0 is again given in Eq. (2.6.5). Figure 2.26 shows Eq. (2.6.6) for the convolutional codes, showing the potential advantage over fixed-length block codes. Convolutional decoding can also be improved by "soft-decisioning" decoding algorithms, which allow the decoder to achieve a finer granularity during the chip processing. This results in an improved chip cut-off rate of approximately

$$r_0 = 1 - \log_2[1 + e^{-rE_b/N_0}] \tag{2.6.7}$$

as compared to Eq. (2.6.5). This soft-decision rate yields the same values of r_0 in (2.6.5) with about 2 dB less E_b. Hence, when entering the convolutional coding curves in Figure 2.26, and reading PE at a specific constraint length k and ratio r/r_0, the resulting performance can be achieved with 2 dB less bit energy if soft-decisioning decoding is used. Soft-decision processing, however, further increases the required decoder complexity.

Forward-error correction, with either block codes or convolutional codes, directly improves the bit-error probability. Figure 2.27 shows the expected improvement over standard BPSK obtained by forward-error correction with various block sizes or constraint lengths k, as a function of channel E_b/N_0. This improvement can be interpreted as a lower PE at the same E_b/N_0, or the same PE at a lower E_b/N_0, as higher levels of error correction are used. The reduction in required E_b/N_0 produced by the coding is referred to as *coding gain*. Again, the receiver processor complexity required in the block decoding or Viterbi decoder algorithms also increases with the length k, placing practical limits on achievable performance.

2.7 BLOCK WAVEFORM ENCODING

The forward-error-correction coding just discussed in Section 2.6 is based on binary transmission, where sequences of binary chips are sent over the link and each is separately decoded. Performance improvement is obtained by inserting extra chips, or redundancy, into the transmitted chip waveform. Another way to achieve improvement in bit-error probability is to use higher levels of data transmission by encoding more than binary information on the transmitted carrier at any one time. This is referred to as *block waveform encoding*. It allows blocks of bits (binary *words*) to be transmitted at one time as one particular waveform. This encoding format is shown in Figure 2.28.

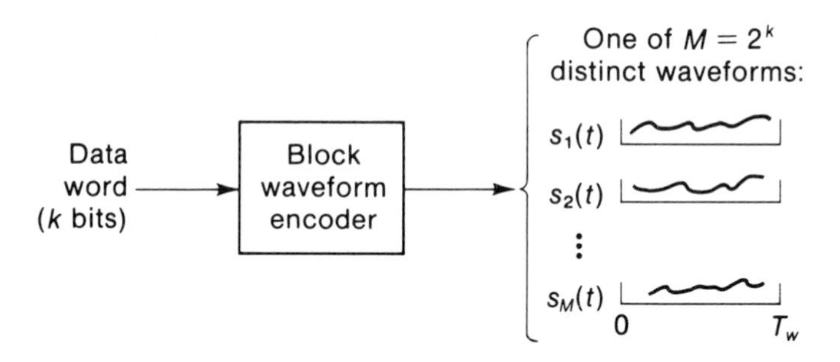

FIGURE 2.28 Block waveform encoding model.

Orthogonal Block Encoding

A common block waveform encoding procedure uses phase coherent orthogonal signals to represent the word. The k bit words are encoded into $M = 2^k$ orthogonal sequences of BPSK chips, which are then transmitted as a single waveform to represent the word (Figure 2.29a). At the decoder, the entire sequence of chips (rather than one chip at a time, as in forward-error correction) is decoded. A single decision concerning the entire received chip sequence is made at the end of the sequence (word) time. The decoder now must have the capability of detecting any of the 2^k sequences that could have been sent for that word. To do this, the decoder must contain a bank of 2^k phase-coherent correlations, each looking for one of the possible chip sequences (Figure 2.29b). The correlator producing the largest output at the end of the word correlation time decodes the transmitted words. In practice this set of parallel correlators is constructed as a single-chip correlator, operating on each received chip, followed by a bank of sum-and-store memory circuits, as shown in Figure 2.29c. Since the correlation of a received waveform with a sequence of BPSK chips is identical to the sum of the individual correlations of each chip, this sum-and-store procedure will generate the same set of 2^k parallel correlation values with simpler hardware.

Orthogonal waveform encoding has the advantage that the orthogonality of the waveforms reduces the possibility of the correct waveform producing a large correlation value in an incorrect correlator. More important, however, it benefits from the fact that a single word decision is made with the combined energy of the entire chip sequence, rather than having separate decisions made on each chip using only the energy of a

$$r = \frac{A^2}{B^2} = \frac{P_A}{P_B} \quad (2.3.15)$$

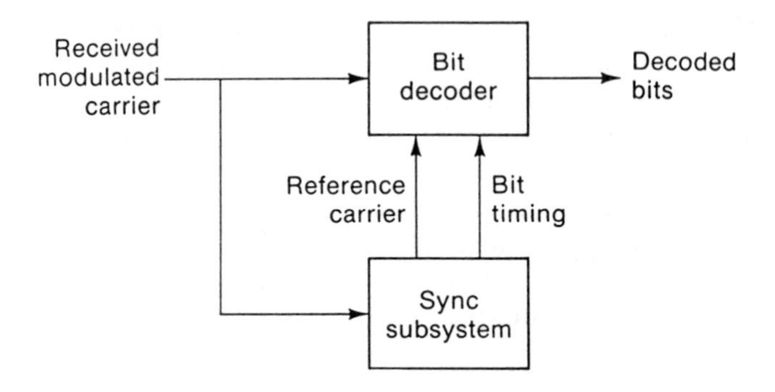

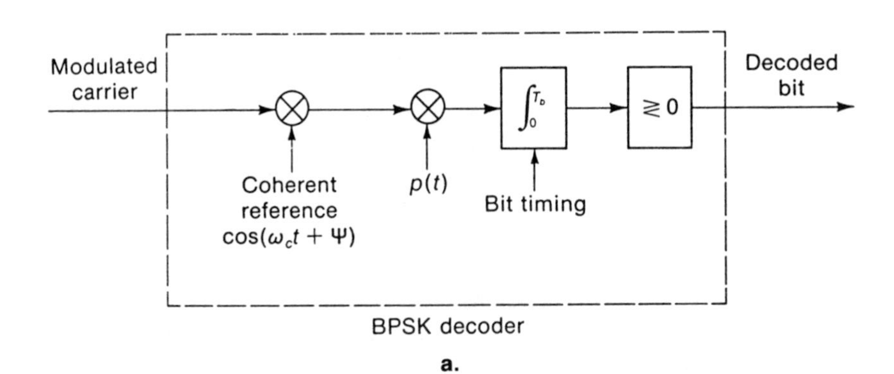

Decoded quadrature bit

Decoded quadrature bit

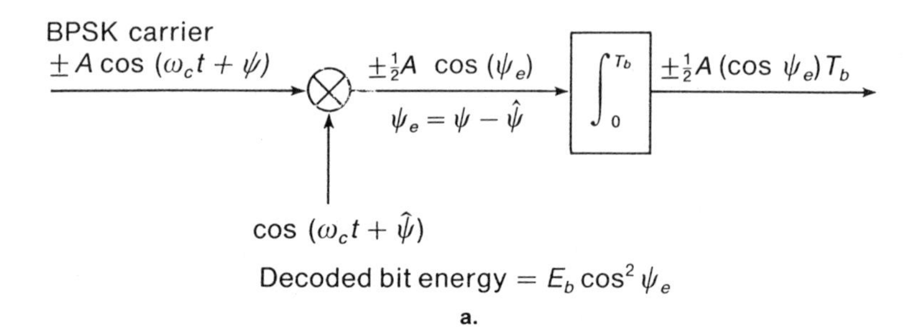

Decoded bit energy $= E_b \cos^2 \psi_e$

QPSK carrier
$\pm A \cos(\omega_c t + \psi)$
$\pm A \sin(\omega_c t + \psi)$

$\pm\frac{1}{2}A(\cos\psi_e)T_s$ $\pm\frac{1}{2}A(\sin\psi_e)T_s$

Decoded bit energy $= E_b[\cos\psi_e \pm \sin\psi_e]^2$

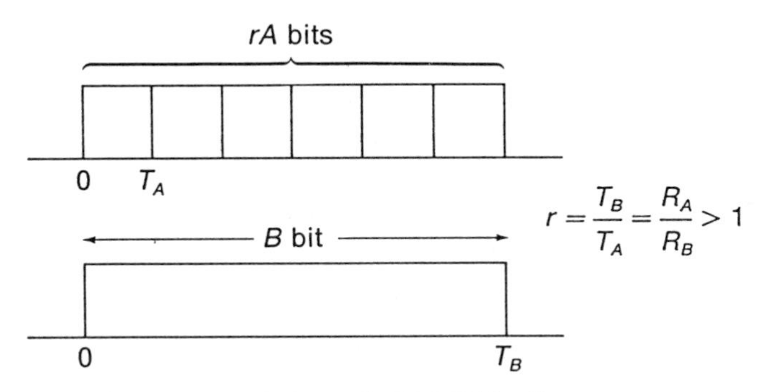

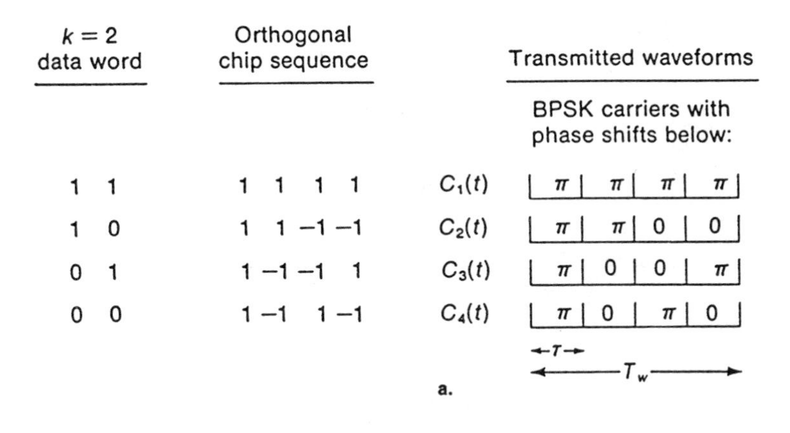

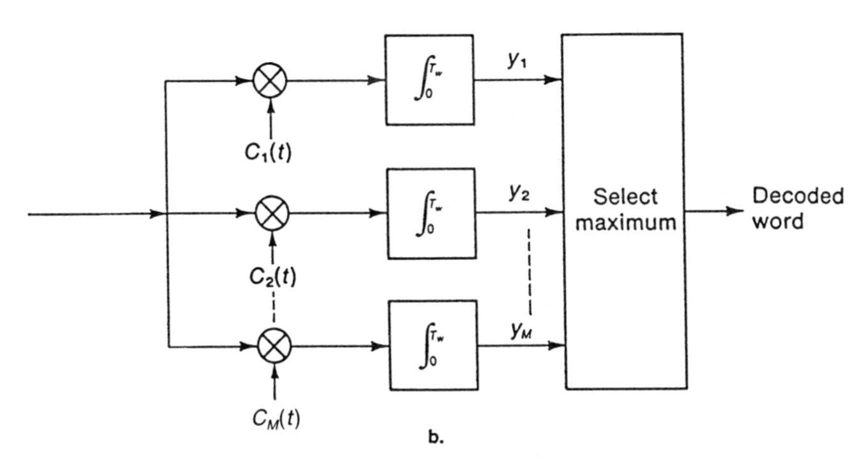

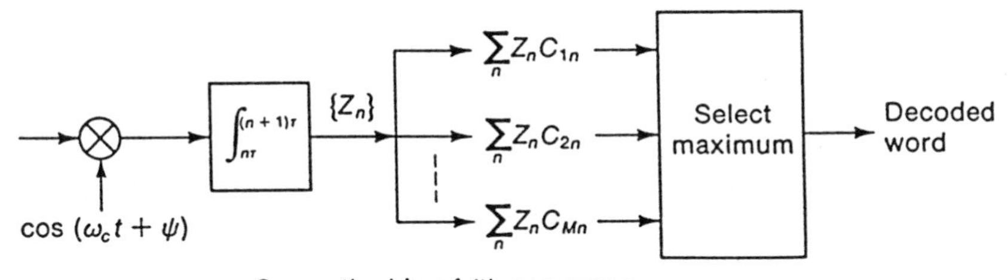

FIGURE 2.29 Waveform encoding with orthogonal BPSK sequences. (a) Encoding example for $k = 2$; (b) required block waveform decoder; (c) alternate decoder model.

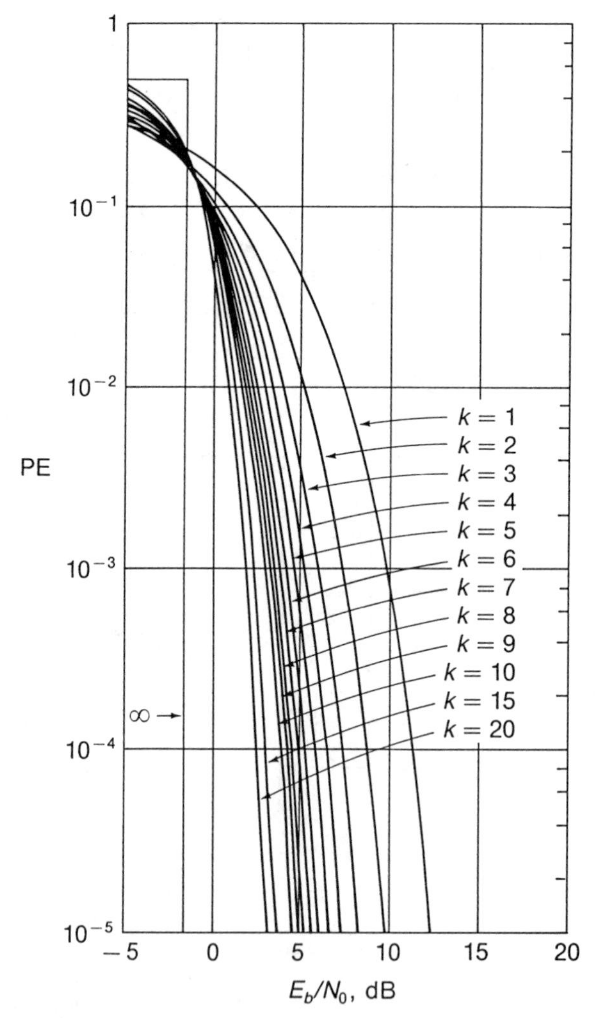

FIGURE 2.30 Phase-coherent orthogonal bit error probability versus E_b/N_0 for values of block length k.

single chip. Hence, in terms of equivalent bit energy, the word decoding energy is

$$E_w = kE_b \tag{2.7.1}$$

where k is the word size. The resulting bit-error probability for orthogonal word encoding is listed in Table 2.6 and is shown in Figure 2.30 for several

TABLE 2.6 Tabulation of word-error probabilities for M-ary block waveform encoding ($M = 2^k$, $k =$ block size)

Block Encoding Format	*Word-Error Probability (PWE)*	*Bit-Error Probability (PE)*
M orthogonal phase-coherent waveforms	$1 - \int_{-\infty}^{\infty} \left[1 - Q\left(y + \sqrt{\frac{2E_w}{N_0}}\right)\right]^{M-1} \frac{e^{-y^2/2}}{\sqrt{2\pi}} dy$	$\frac{M/2}{M-1}$ PWE
MPSK	$\approx 2Q\left[\left(\frac{2E_w}{N_0}\right)^{1/2} \sin\left(\frac{\pi}{M}\right)\right]$	$\approx \frac{\text{PWE}}{\log_2 M}$
MFSK	$1 - e^{-\rho^2/2} \sum_{q=0}^{M-1} (-1)^q \binom{M-1}{q} e^{(\rho^2/2)/1+q}$	$\frac{M/2}{M-1}$ PWE
MASK	$2\frac{M-1}{M} Q\left[\left(\frac{6E_w}{(M^2-1)N_0}\right)\right]^{1/2}$	$\approx \frac{\text{PWE}}{\log_2 M}$
M-CPFSK	$\leq (M-1)Q\left[\left(\frac{d_{\min} E_w}{N_0}\right)^{1/2}\right]$	$\leq \frac{M}{2} Q\left[\left(\frac{d_{\min}^2 E_w}{N_0}\right)^{1/2}\right]$

$E_x =$ word energy $= (\log_2 M)E_b/N_0$
$\rho^2 = 2E_b/N_0$
$Q(x) = (2\pi)^{1/2} \int_x^{\infty} e^{-t^2/2}\, dt$
$d_{\min} =$ minimum distance (see Appendix A.7.10)

values of k, as a function of the available E_b/N_0 of the link. The improvement over BPSK binary signaling by increasing the block length is apparent.

The disadvantage of the orthogonal sequence waveform system (besides the decoder complexity) is that 2^k chips must be sent to represent the k bits. This requires that the chips must be sent $2^k/k$ times faster if a particular data rate is to be maintained (requiring a bandwidth $2k/k$ times larger than the bit-rate bandwidth). Conversely, the transmitted bit rate will be $k/2^k$ times slower than the chip rate permitted by the link. Hence, coded waveform PE improvement is achieved at the expense of channel bandwidth or reduced data bit rate.

Multiple Phase Shift Keying

Another common waveform encoding procedure is *multiple phase shift keying* (MPSK), where the k data bits are transmitted at one time by encoding into one of $M = 2^k$ phase shifts of the same carrier. The system still uses constant-envelope carriers with phase modulation, but the phase modulation can be any one of M phases instead of two phases as in BPSK. During a word transmission time T_w, the carrier again has the form

$$c(t) = A\cos[\omega_c t + \theta(t) + \psi] \tag{2.7.2}$$

where $\theta(t)$ is one of the M phases $i(2\pi/M)$ radians, $i = 0, \ldots, M - 1$, depending on the sequence of k bits forming the word. The carrier then switches phase for the next k-bit word, and so on. The spectrum is again that of a carrier phase shifted at rate T_w s, except that T_w is the word time (time to transmit k bits). The carrier bandwidth, however, does not increase as long as T_w is held constant.

Decoding of MPSK requires phase measurements on the received carrier to determine which phase θ is being received during each word time. This again requires a phase-coherent perfectly timed reference in order to distinguish the modulation phase shifts. Phase-referencing subsystems for MPSK however are more elaborate than for BPSK (see Appendix B), and generally they perform more poorly for the same carrier power levels. Word-error probabilities for MPSK are given approximately by

$$\text{PWE} = 2Q\left[\left(\frac{2E_w}{N_0}\right)^{1/2} \sin\frac{\pi}{M}\right] \tag{2.7.3}$$

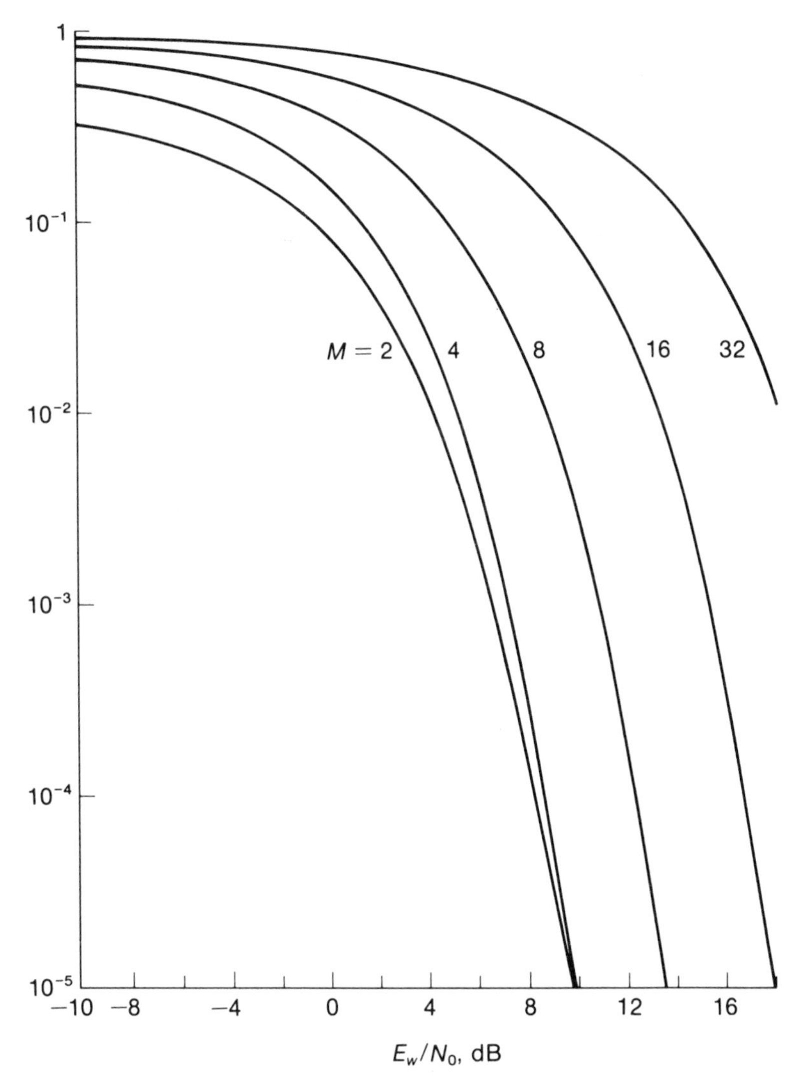

FIGURE 2.31 MPSK word-error probabilities versus word energy $E_w = P_c T_w$. M = number of carrier phases; log M = number of bits in T_w s.

where PWE is the probability of decoding a block word in error, and E_w is the word energy, $E_w = P_c T_w$ (see Section A.4, Appendix A). Plots of PWE are shown in Figure 2.31. Although MPSK is transmitting $\log_2 M$

bits during each word time, Eq. (2.7.3) shows the serious disadvantage if M gets too large. The argument decreases as M increases, producing a rapidly increasing PWE. Alternatively, we need to increase E_w approximately as $M^2 = (2^k)^2$ to maintain PWE with increasing word size k. This degradation is caused since the possible signaling phases θ become closer together as M increases and are harder to detect in the presence of noise. For this reason MPSK systems are usually limited to low M values—usually not much larger than 8 (three-bit words). Note that MPSK with

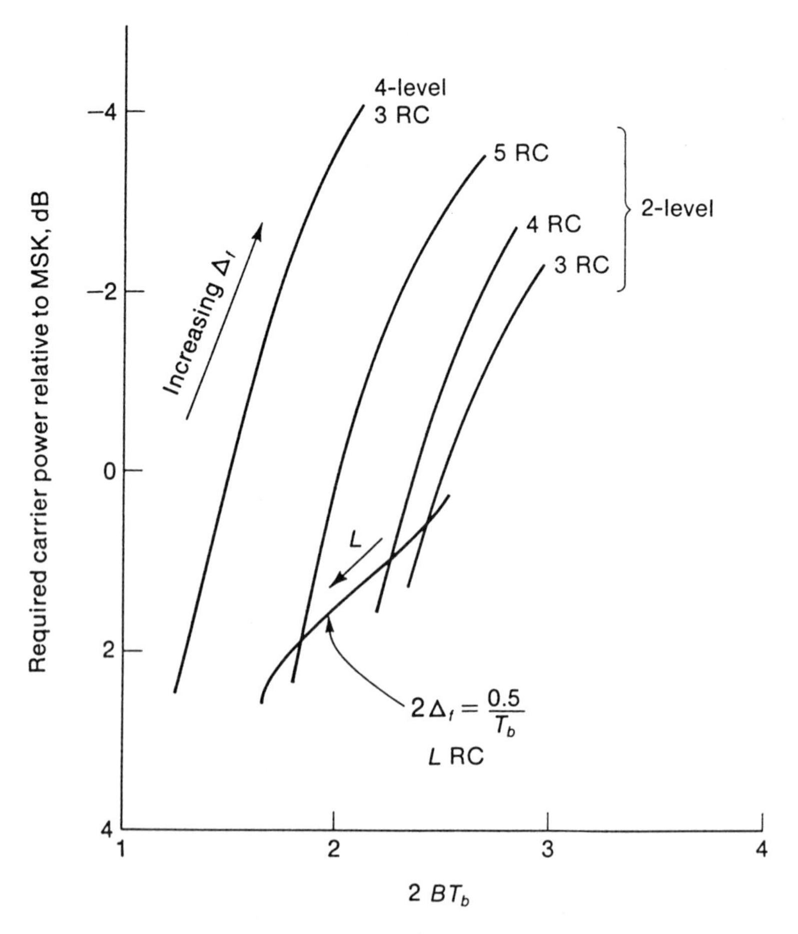

FIGURE 2.32 Multilevel CPFSK E_b performance relative to MSK versus 99% carrier bandwidth B (from Aulin et al., 1981).

$M = 4$ corresponds to carrier phase shifts of $2\pi/4 = \pi/2$ rad $= 90°$, which, as we pointed out earlier, corresponds to QPSK. Thus, QPSK can be alternatively viewed as word coding of two bits at one time onto one of four phases of an RF carrier.

Multilevel CPFSK Encoding

Block waveform encoding with constant amplitude carriers can also be achieved by multilevel CPFSK (Benedetto et al., 1987; Mazur and Taylor, 1981; Proakis, 1988). In this case, each data word is encoded into one of M frequency deviations that are used to pulse-frequency-modulate the carrier. If the FM waveform is selected as an extended raised-cosine pulse, as in Figure 2.10, the M-ary word encoding can be coupled with the spectral shaping of this format. Phase-tracking demodulation can be used to decode the multilevel phase trajectories over the memory (number of extended RC bit times) of the encoding. Decoding performance is again estimated by PE bounds, as listed in Table 2.6. These bounds are useful for evaluating the basic bandwidth-error probability trade-off of these types of carriers.

Figure 2.32 displays the raised-cosine CPFSK performance directly in terms of the required carrier bandwidth, which is defined as the bandwidth needed to pass 99% of the carrier power. Performance is shown in terms of the E_b required relative to that of MSK with the same PE. At low values of Δ_f ($\Delta_f \cong 0.5/T_b$), the bandwidth reduction is achieved at the expense of a slight (several dB) power degradation. By trading back bandwidth with increased deviations, or by increasing the number of pulse levels with phase-tracking algorithms, performance can be made to surpass MSK. Increased decoding processing capability will continue to foster future development of these systems.

Multiple Frequency Shift Keying

A word of k bits can be transmitted without the need for phase-coherent decoding by the use of *multiple frequency shift keying* (MFSK). The k-bit word is transmitted as a carrier burst at one of $M = 2^k$ different frequencies. The frequencies must be separated by $1/T_w$ Hz to avoid carrier energy from one frequency appearing as energy at another frequency. MFSK can be decoded noncoherently by a bank of tuned bandpass filters at each frequency, followed by an energy comparison (i.e., the M-ary extension of the binary FSK decoder in Figure 2.24). The resulting MFSK bit-error probability is computed in Appendix A and is listed in Table 2.6. It is closely bounded by

$$\mathrm{PE} \leq \frac{M}{4} e^{-E_b k/N_0} \tag{2.7.4}$$

where E_b is the carrier energy per bit. Figure 2.33 plots PE for several values of k. Note that unlike MPSK, MFSK performance improves with increasing word size, but the required number of frequencies increases as $M = 2^k$. Since the spacing between frequencies is fixed, the required RF bandwidth also increases with word size. Also, since each frequency requires a separate bandpass filter and energy detector, the complexity of the decoding structure also increases exponentially with block size.

Multiple Amplitude Shift Keying

Block waveform encoding can also be achieved by using a fixed-frequency carrier burst with multiple amplitude levels. This is referred to as *multiple amplitude shift keying* (MASK). In this format, each k-bit data word is transmitted as a T_w s burst of a carrier with one of $M = 2^k$ amplitude levels. Thus, MASK is a form of amplitude modulation with the encoded data word determining the carrier amplitude every T_w s. This means the carrier power varies with the encoding word, and an average carrier power can be determined by averaging over all amplitude levels. Decoding is achieved by using a phase-coherent reference multiplier followed by a T_w s integrator, as in the BPSK decoder in Figure 2.16, but comparing the integrator output to each of the M possible integrator levels. MASK decoding is improved by increasing the separation between amplitude levels, which in turn increases the overall average carrier power.

Since MASK involves amplitude modulation, it is directly affected by any amplitude distortion in the carrier link. For this reason, MASK encoding is generally not used in satellite communications, unless the link can be guaranteed to have little or no amplitude nonlinearities. On the other hand, MASK has the distinct advantage that it can transmit any number of amplitude levels M using a carrier bandwidth that is always about $2/T_w$ Hz, just as in MPSK.

Word-error and bit-error probabilities for MASK signaling can be determined by computing the probability that the integrated decoder noise causes the incorrect carrier amplitude to be decided, as shown in Section A.10. The results are listed in Table 2.6, in terms of M and the average word energy E_w (product of average carrier power and T_w). We again see that as M increases (more levels are used with fixed spacing)

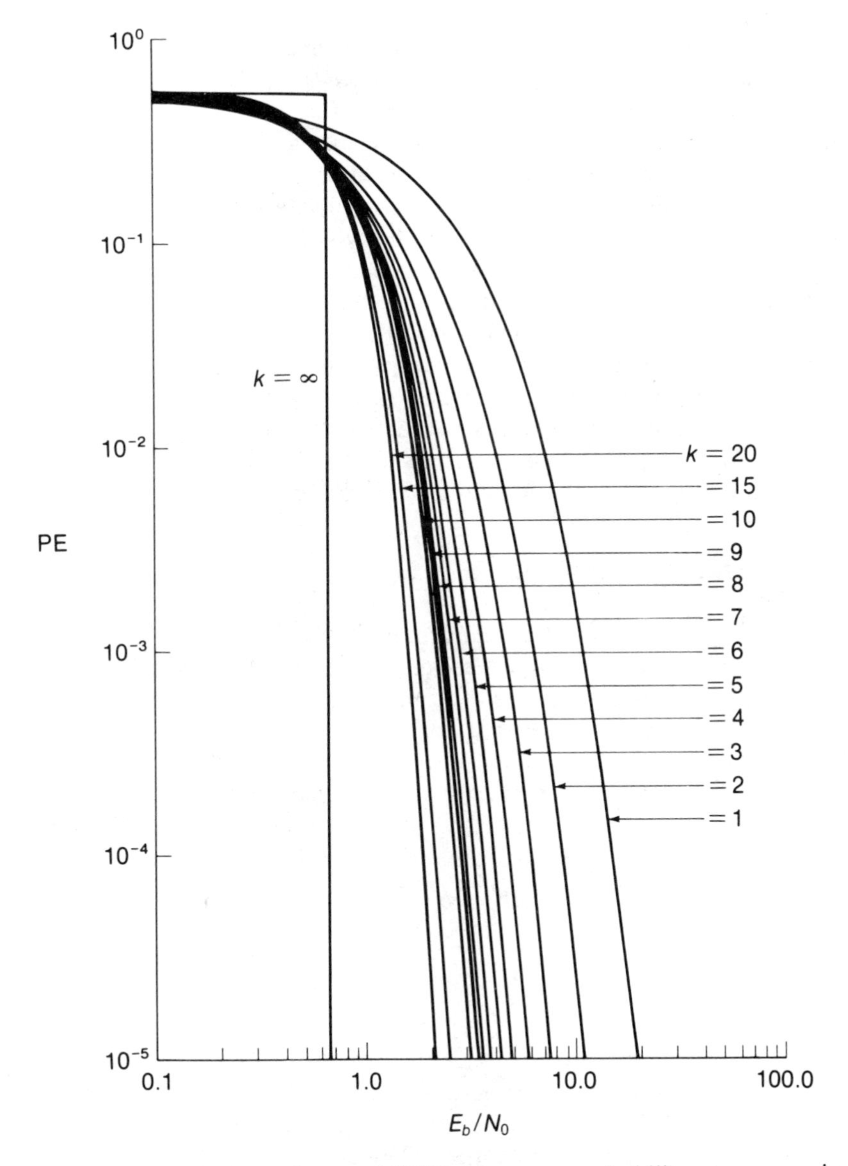

FIGURE 2.33 Noncoherent MFSK bit-error probability versus equivalent bit energy.

the average word energy must necessarily increase as M^2 in order to maintain the same PWE performance. This is similar to the MPSK performance, which also uses carrier bursts but has constant amplitudes and slightly more complicated decoder processing.

MASK encoding can be simultaneously applied to quadrature components of a common carrier. Each component is separately encoded in amplitude by a different data word, and the two components are combined into a single carrier. This is referred to as *quadrature amplitude shift keying* (QASK). QASK is equivalent to the encoding of a $2k$-bit word (or two k-bit words) into one of 2^{2k} joint amplitude values of the two quadrature components. Since each quadrature amplitude pair defines a combined carrier with a specific amplitude and phase, QASK is equivalent to simultaneous amplitude and phase encoding of the carrier. QASK carriers can be decoded by a phase-coherent quadrature correlator, as in Figure 2.16b, followed by separate amplitude decoding in each arm.

2.8 DIGITAL THROUGHPUT

In satellite systems, excessive demand for limited satellite bandwidth forces the communication link to be extremely efficient in the use of this bandwidth. For this reason satellite signaling formats tend toward modulation and coding schemes that combine both power efficiency with bandwidth advantages. In digital systems this bandwidth can be somewhat controlled by proper selection of the waveshapes used for the encoding, as we previously described. If a digital link transmits R bits/s (bps) while requiring a carrier bandwidth of B_c Hz, the bandwidth efficiency is measured by the so-called *channel throughput*, defined as

$$\eta = \frac{R}{B_c} \text{ bps/Hz} \tag{2.8.1}$$

This parameter indicates how well the digital link makes use of the available bandwidth B_c. The larger the throughput, the more bits can be transmitted for a given carrier bandwidth. Since the achievable bit rate is related to the decoding E_b/N_0 via Eq. (2.5.1),

$$\frac{E_b}{N_0} = \frac{P_c T_b}{N_0} = \frac{P_c}{N_0 R} \tag{2.8.2}$$

and since the value of E_b/N_0 determines error-probability performance, PE, we see that throughput implicitly depends on the design PE. If we boldly increase bit rate R by simply sending bits at a faster rate through the same carrier bandwidth, we will eventually distort waveshape, and a higher E_b/N_0 will be needed to produce the same PE. This means increased carrier power P_c must be provided to maintain the higher rate at the same PE. We therefore expect that throughput could be increased at the expense of carrier power. In satellite systems operating under bandwidth constraints, this trade-off may in fact be both desirable and cost efficient. The communication design task is then to find signaling formats that provide the best trade-off (greater throughput increase as a function of carrier power).

Consider a standard BPSK system sending bits at a rate $R = 1/T_b$ bps. From Figure 2.7 we require a carrier bandwidth of approximately $B_c = 2/T_b$ Hz, and a carrier power $P_c = \gamma N_0 B_c/2$, where γ is the required value of E_b/N_0 to achieve the desired PE, obtained by reading from Figure 2.17. The throughput is then

$$\eta_{\text{BPSK}} = \frac{R}{B_c} = \frac{1/T_b}{2/T_b} = \frac{1}{2}\frac{\text{bps}}{\text{Hz}} \tag{2.8.3}$$

Thus, for example, if 1 Mbps is to be transmitted, a carrier bandwidth of 2 MHz is required. If we use QPSK, or OQPSK, the bit rate is doubled for the same bandwidth and PE. Hence,

$$\eta_{\text{QPSK}} = \frac{2/T_b}{2/T_b} = 1\,\frac{\text{bps}}{\text{Hz}} \tag{2.8.4}$$

One megabit per second can now be transmitted with 1 MHz of carrier bandwidth.

To further increase the throughput while maintaining the constant-envelope property, it is necessary either to increase the number of bits encoded onto the carrier burst without increasing bandwidth or to encode the same number of bits with a reduced bandwidth. The former can be achieved by inserting additional phase states over the QPSK carrier; that is, MPSK. The latter, bandwidth reduction, can be achieved by phase modulation with pulse waveforms extended in time or by the CPFSK waveforms previously described to reduce bandwidth. In MPSK the throughput becomes

$$\eta_{\text{MPSK}} = \frac{\log_2 M}{B_c T_w} = \frac{\log_2 M}{2} \tag{2.8.5}$$

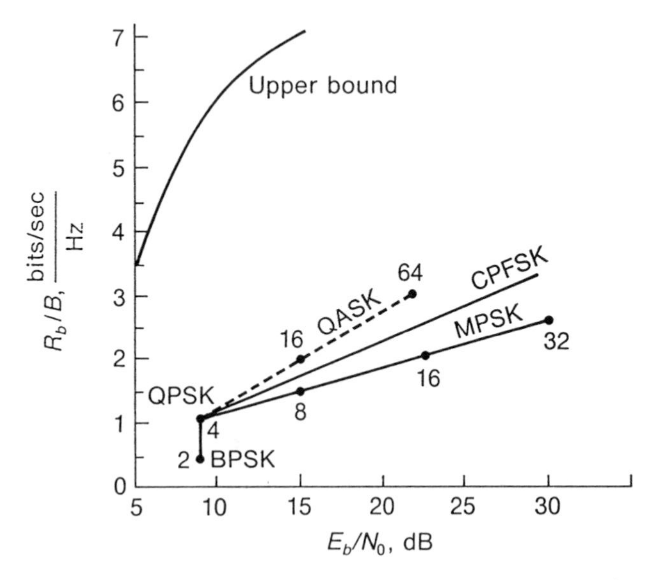

FIGURE 2.34 Throughput versus E_b/N_0. PE $= 10^{-5}$.

However, the carrier power must be correspondingly increased, as evidenced by Eq. (2.7.2), to maintain performance. Figure 2.34 shows how the throughput increase in MPSK requires an increasing carrier power to sustain a PE $= 10^{-5}$ as the number of phase states increases. Points for BPSK ($M = 2$) and QPSK ($M = 4$) are included.

MFSK also transmits $\log_2 M$ bits with each T_w s frequency burst, but requires a bandwidth that increases as $M(2/T_w)$. Hence,

$$\eta_{\text{MFSK}} = \frac{\log_2 M}{2M} \tag{2.8.6}$$

and the MFSK throughput actually decreases with increasing M, having a maximum value of $\eta_{\text{MFSK}} = \frac{1}{4}$ at $M = 2$ and 4.

Bandwidth reduction by pulse shaping in frequency-modulation encoding increases throughput, but requires complex decoding and increased carrier power to overcome the resulting intersymbol interference. Points for raised-cosine phase encoding for various symbol extensions are included in Figure 2.34. Note that an improved trade-off of throughput for power is achievable by these methods.

If error-correction encoding is used to reduce the required carrier power, the coded chips will require a *wider* bandwidth to maintain a prescribed bit rate. With a code rate r, $r < 1$, the chip time must be rT_b, and the error-corrected throughput becomes

$$\eta_{ec} = \frac{1/T_b}{2/rT_b} = \frac{r}{2} \tag{2.8.7}$$

While error correction improves PE, it reduces the BPSK throughput. In essence, error correction purposely increases bandwidth to lower the required carrier power for a given PE, which is directly contrary to desired throughput improvement.

Throughput can also be increased if the constant-envelope condition is not required. In this case we can increase the number of bits encoded onto a carrier by allowing more than two amplitude levels, as we do in MASK. If we use M levels of amplitude, then $\log_2 M$ bits can be encoded onto each carrier burst, producing a throughput equal to Eq. (2.8.5). However, the carrier power (defined as an average overall amplitude level) must be increased to allow sufficient separation of the M levels for adequate decoding performance. (See Table 2.6.) Points for various levels of M-ary QASK are also shown in Figure 2.34. We see that removal of the constant-envelope condition allows even further throughput improvement. Unfortunately, nonconstant amplitude carriers are susceptible to other forms of distortion in satellite links that may preclude the throughput advantage.

Theoretical analysis in information theory has shown that a rigorous upper bound to the data rate exists in any digital communication system. In operating over an additive Gaussian white noise channel (level N_0) with carrier power P_c and bandwidth B_c, the maximal bit rate that can be achieved is (Wozencraft and Jacobs, 1965)

$$R_b = B_c \log\left(1 + \frac{P_c}{N_0 B_c}\right) \tag{2.8.8}$$

This implies that the maximal throughput R_b/B_c must satisfy the equation

$$\frac{R_b}{B_c} = \log\left[1 + \left(\frac{P_c}{N_0 R_b}\right)\left(\frac{R_b}{B_c}\right)\right] \tag{2.8.9}$$

or

$$\eta = \log\left(1 + \frac{E_b \eta}{N_0}\right) \tag{2.8.10}$$

The value of η satisfying this equation at each value of E_b/N_0 is plotted as the upper bound curve in Figure 2.34. This shows the maximum possible throughput achievable by *any* type of encoding and decoding operation. We see that practical encoding methods are generally well below this bound, and approach it rather slowly. However, our discussion has shown that any attempts to approach rapidly the maximal throughput will necessarily be accompanied by increasingly higher levels of decoder processing and the implementation of rather complex decoding equipment.

REFERENCES

Aulin, T., Rydbeck, N., and Sundberg, C. (1981), "Continuous Phase Modulation—Part I and Part II," *IEEE Trans. Comm.*, **Com-29** (March), 196–225.

Bell Telephone Laboratories (1970), *Transmission Systems for Communications*, 4th ed., Bell Telephone Laboratories, New Jersey.

Benedetto, S., Biglieri, E. and Castellani, V. (1987), *Digital Transmission Theory* Prentice-Hall, Englewood Cliffs, NJ.

Cahn, C. R. (1974), "Phase Tracking and Demodulation with Delay," *IEEE Trans. Information Theory*, **IT-20** (January), 50–58.

Carlson, B. (1986), *Communication Systems*, 3rd ed., McGraw-Hill, New York.

Deshpande, G., and Wittle, P. (1981), "Correlative Encoded Digital FM," *IEEE Trans. Comm.*, **Com-29** (February).

Divsalar, D. and Simon, M. (1990) "Multiple-Symbol Differential Detection of MPSK", *IEEE Trans. on Comm.* **Com-29** (March), 300–308.

Forney, D. (1973), "The Viterbi Algorithm," *Proc. IEEE*, **61** (March) 268–273.

Freeman, R. (1975), *Telecommunication Transmission Handbook*, Wiley, New York.

Gagliardi, R. (1988), *Introduction to Communication Engineering*, 2nd ed., Wiley, New York.

Gregg, W. (1977), *Analog and Digital Communications*, Wiley, New York.

Gronemeyer, S., and McBride, A. (1976), "MSK and OQPSK Modulation," *IEEE Trans. Comm.*, **Com-24** (August).

Mazur, B., and Taylor, D., (1981), "Demodulation and Carrier Synchronization of Multi-*h* Phase Codes," *IEEE Trans. Comm.*, **Com-29** (March), 257–266.

McLane, P. (1983), "Viterbi Receiver for Correlation Encoded MSK Signals," *IEEE Trans. Comm.*, **Com-31** (February), 290–295.

Proakis, J. (1988), *Digital Communications*, 2nd ed., McGraw-Hill, New York.

Rodin, M. (1985), *Analog and Digital Communication Systems*, 2nd ed., Prentice-Hall, Englewood Cliffs, NJ.

Scharf, L. L., Cox, D. D., and Masreliez, C. J. (1980), "Modulo-2π Phase Sequence Estimation," *IEEE Trans. Information Theory*, **IT-26** (September), 615–620.

Stremler, F. (1982), *Introduction to Communication Systems*, 2nd ed., Addison-Wesley, Reading, MA.

Taub, H., and Schilling, D. (1986), *Principles of Communication Systems*, 2nd ed., McGraw-Hill, New York.

Ungerboeck, G. (1974), "New Applications for the Viterbi Algorithm: Carrier Phase Tracking in Synchronous Data-Transmission Systems," *NTC'74 Conference Record*, San Diego, CA, (December), pp. 734–738.

Viterbi, A. J. (1967), "Error Bounds for Convolutional Codes and an Asymptotically Optimum Decoding Algorithm," *IEEE Transactions Information Theory*, **IT-13**, 250–269.

Viterbi, A. J., and Omura, J. K. (1979), *Principles of Digital Communication and Coding*, McGraw-Hill, New York.

Wozencraft, J. and Jacobs, I. (1965), *Principle of Communication Engineering*, Wiley, New York.

Ziemer, R., and Tranter, W. (1976), *Principles of Communications*, Houghton Mifflin, Boston, MA.

PROBLEMS

2.1 Convert the following numbers to decibels:
(a) 200 (b) 0.01 (c) 1/8 (d) 27 (e) 10^4
Convert the following to numbers:
(a) 37 dB (b) 6 dB (c) -6 dB (d) 20 dB (e) -190 dB.

2.2 A communication link is to transmit a 6-MHz TV channel over a 36 MHz RF bandwidth. A demodulation input threshold of 12 dB is required. The noise level is -150 dBw/Hz.
(a) How much received carrier power is needed, and what SNR_d will be achieved, if AM is used? (b) Repeat for FM, assuming the maximum possible frequency deviation is used.

2.3 An FM system has a receiver threshold of 15 dB. How much received carrier power and RF bandwidth is needed to transmit a 4-KHz baseband signal with a demodulated SNR_d of 40 dB? ($N_0 = 10^{-10}$ W/Hz.)

2.4 An FM system transmits a 1-MHz bandwidth waveform with a 10-MHz deviation. (a) If the demodulator requires a threshold of

15 dB, compute the demodulated SNR_d. (b) If the deviation is limited to 1% of the carrier frequency, what is the maximum SNR_d attainable with a 100-MHz RF carrier?

2.5 Let $x(t)$ be a random voltage waveform with probability density $p(x)$ at any t. Determine the compander gain function required to produce an output $y(t)$ that is uniformly distributed over $(-1,1)$ V. [*Hint*: If $y = g(x)$, then $p(y) = (p(x)/|dg/dx|$ evaluated at $x = g^{-1}(y)$.]

2.6 A voltage waveform is sampled, and each sample is quantized into one of q different levels. If the sampling rate is f_s samples per second, show that the resulting bit rate that will be generated is $R = f_s\,(\log q)$ bps.

2.7 Derive the spectrum of the NRZ and Manchester waveforms in Eqs. (2.3.3) and (2.3.4) by formally transforming each waveshape.

2.8 Derive the spectrum of the raised-cosine pulse in Eq. (2.4.1). [*Hint*: $p(t)$ in Eq. (2.4.1) is also the result of passing a square pulse of width $T_b/2$ into a filter having impulse response $h(t) = (\pi/T_b)\sin(2\pi t/T_b)$, $0 \le t \le T_b/2$.]

2.9 Let a signal $s(t), 0 \le t \le T$, be expandable into the orthonormal set

$$s(t) = \sum_{i=1}^{v} s_i \Phi_i(t), \qquad 0 \le t \le T$$

where

$$\int_0^T \Phi_i(t)\Phi_j(T)dt = 1, \qquad i = j$$
$$= 0, \qquad i \ne j$$

We then plot $\mathbf{s} = (s_1, s_2, \ldots, s_v)$ as a point in a vector space of v dimensions. (a) Show that $|\mathbf{s}|^2$ = energy in $s(t)$. (b) Show that for two such expansions, $y(t)$ and $s(t)$, each with energy E, $\int_0^T y(t)s(t)\,dt = E(\cos\psi)$, where ψ is the angle between the vectors $\mathbf{s}$ and $\mathbf{y}$. (c) Show that any carrier $\cos(\omega t + \theta)$ can be plotted as a unit phasor at angle θ in a two-dimensional space. (d) Use (c) to plot vector points for a BPSK signal pair, a QPSK signal set, and an MPSK signal set.

2.10 An M-ary pulse position modulated (MPPM) system communicates by placing a τ s carrier pulse in one of M τ s slots to represent a data word of $\log_2 M$ bits. Use the main hump bandwidth of the carrier to determine the MPPM throughput.

CHAPTER 3
The Satellite Channel

Communication between points is achieved by analog or digital modulation of information onto carriers, and by transmission of the carriers as an electromagnetic field from one point to the other. As we discussed in Chapter 2, the amount of received carrier power invariably determines the ability of the receiver to demodulate or decode the information. In satellite systems it is extremely important to know the key parameters that directly determine this received power so that proper trade-off in system design between spacecraft and earth stations can be achieved. In this chapter we examine the basic power flow equations associated with satellite channels.

3.1 ELECTROMAGNETIC FIELD PROPAGATION

A basic communication link is shown in Figure 3.1. The transmitter field is characterized by its *effective isotropic radiated power* (EIRP) defined by

$$\text{EIRP} = P_T g_t(\phi_z, \phi_l) \tag{3.1.1}$$

where P_T is the available antenna input carrier power from the transmitter power amplifier, including circuit coupling losses and antenna radiation losses, and $g_t(\phi_z, \phi_l)$ is the transmitting antenna gain function in the

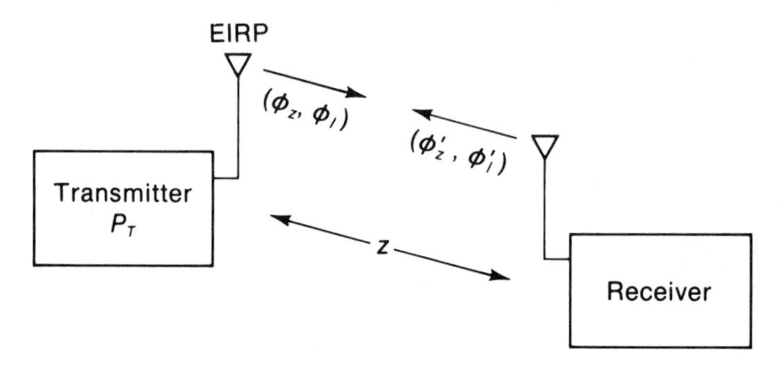

FIGURE 3.1 Communication link.

angular direction (ϕ_z, ϕ_l) of the receiver. Here (ϕ_z, ϕ_l) refer to an azimuth and elevation angle, respectively, measured from a coordinate system centered at the transmitting antenna. The *flux density*, or *field intensity* of the electromagnetic field at the receiver due to the transmitter field is then

$$I(z) = \frac{(\text{EIRP})L_a}{4\pi z^2} \tag{3.1.2}$$

where z is the propagation distance to the receiver and L_a accounts for the atmospheric losses during propagation. The received carrier power collected by the receiving antenna having area A_r normal to the direction of the transmitter is then

$$P_r = I(z)A_r = \frac{(\text{EIRP})L_a}{4\pi z^2} A_r \tag{3.1.3}$$

The receiving area A_r can be written in terms of the receiving antenna gain function g_r in the direction of the transmitter:

$$A_r = \left(\frac{\lambda^2}{4\pi}\right) g_r(\phi_z', \phi_l') \tag{3.1.4}$$

where λ is the carrier wavelength and (ϕ_z', ϕ_l') are the azimuth and elevation angles of the transmitter relative to the receiver coordinate system. Combining this equation with Eq. (3.1.3) allows us to rewrite

$$P_r = (\text{EIRP})L_a L_p g_r(\phi_z', \phi_l') \tag{3.1.5}$$

where we have defined

$$L_p = \left(\frac{\lambda}{4\pi z}\right)^2 \tag{3.1.6}$$

The parameter L_p appears as an effective loss occurring during transmission and is referred to as the *propagation loss* of the link. Note that L_p depends on both the carrier frequency through the wavelength λ and on the distance z, and its loss is always present, even if there are no atmospheric losses (i.e., if there is free space transmission outside the Earth's atmosphere). We often state P_r in terms of decibels, and write

$$(P_r)_{\text{dB}} = (\text{EIRP})_{\text{dB}} + (L_p)_{\text{dB}} + (L_a)_{\text{dB}} + (g_r)_{\text{dB}} \tag{3.1.7}$$

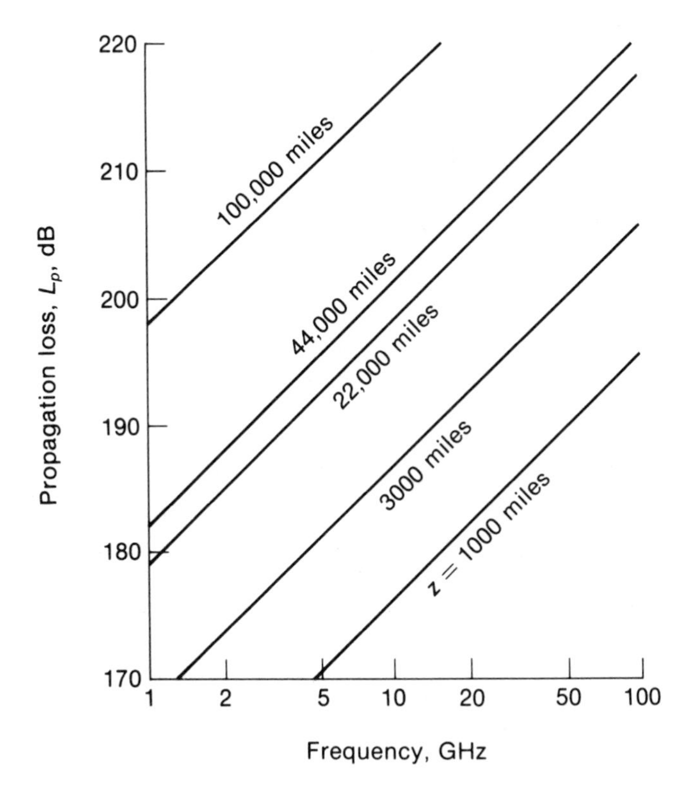

FIGURE 3.2 Satellite propagation loss.

where each term is computed in decibels. Note that gain values (greater than 1) are always positive (add) decibels, whereas attenuation losses (less than 1) are always negative (subtract) decibels. The propagation loss, L_p, when converted to frequency has the decibel value

$$(L_p)_{\text{dB}} = -36.6 - 20 \log_{10}[z(\text{miles}) f(\text{MHz})] \tag{3.1.8}$$

A plot of $(L_p)_{\text{dB}}$ for typical satellite distances and satellite frequencies is shown in Figure 3.2. Note that about several hundred decibels of propagation loss will generally occur in satellite communication paths for geostationary orbits. The exact value depends on the actual satellite slant range to the earth station, as discussed in Section 1.3.

Another parameter often specified in satellite link analysis is the *received isotropic power* (RIP). This is obtained from Eq. (3.1.5) as

$$\text{RIP} = (\text{EIRP}) L_a L_p \tag{3.1.9}$$

or in decibels as

$$(\text{RIP})_{\text{dB}} = (\text{EIRP})_{\text{dB}} + (L_a L_p)_{\text{dB}} \tag{3.1.10}$$

The RIP represents the transmitted power that would be collected by the receiver antenna if it were ideally isotropic. Thus, whereas EIRP indicates the ability of a transmitter to radiate power, the RIP represents the available field power at the receiver.

In addition to its power content, an electromagnetic field also has a designated *polarization* (orientation in space). This polarization is determined by the manner in which the electromagnetic field is excited at the antenna feeds prior to propagation. An additional receiver power loss will occur if the receiving antenna subsystem is not properly aligned with the received wave polarization. This is referred to as a *polarization loss*, and should be included in the L_a loss term in Eqs. (3.1.2) and (3.1.7). The common polarizations in satellite links are *linear* and *circular*. In *linear polarization* the electromagnetic field is aligned in one planar direction throughout the entire propagation, as shown in Figure 3.3a. These directions are usually designated as horizontal or vertical polarizations (relative to the receiving antenna coordinates). The receiving antenna system must have a matching planar receptor in order to maximize the collected power. For example, terrestrial commercial television is transmitted as a horizontally polarized field, and our rooftop antennas utilize horizontal dipole rods for reception.

In *circular polarization* (CP) the field is excited and transmitted with

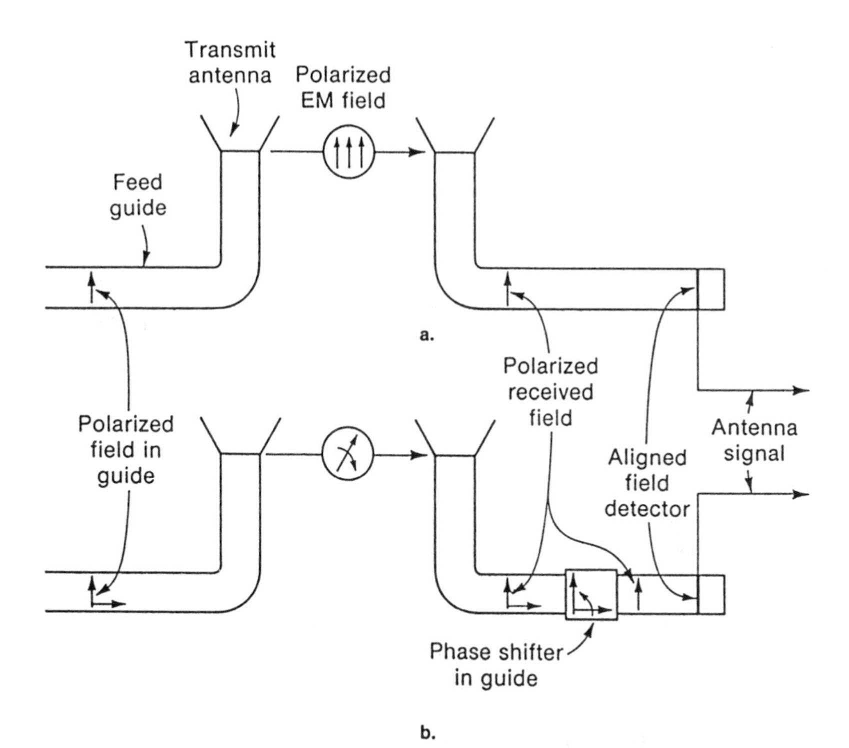

FIGURE 3.3 Transmitted and detected field polarizations: (a) linear; (b) circular.

components in two orthogonal coordinates (one horizontal and one vertical) that are phased so that the combination of the two produces a resultant field polarization that appears to rotate circularly as the wave propagates (Figure 3.3b). CP reception is achieved by an antenna that feeds both components into an antenna waveguide and that uses internal phase shifters to reorient one orthogonal polarization onto the other, thereby collecting the total power available in both components. CP has the advantage that any extraneous rotation of the polarization axis caused by the atmosphere will not affect CP reception, whereas such rotation will produce a polarization loss in a linearly polarized system.

3.2 ANTENNAS

The antenna converts electronic carrier signals to polarized electromagnetic fields and vice versa. A transmitting antenna is composed of a feed assembly that illuminates an aperture or reflecting surface, from which

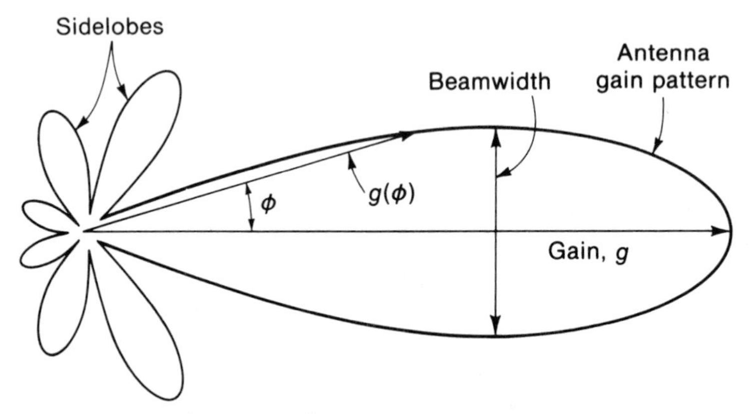

FIGURE 3.4 Antenna gain pattern.

the electromagnetic field then radiates. A receiving antenna has an aperture or surface focusing an incident radiation field to a collecting feed, producing an electronic signal proportional to the incident radiation. Antennas are one of the key elements in a communication system, since their gain values directly determine the amount of received power. Their size and structure are perhaps the easiest to modify from a hardware point of view.

Antennas are described by their gain pattern $g(\phi_z, \phi_l)$, which indicates how the gain of the antenna is distributed spatially with respect to the antenna coordinate system. Although a function of the two-dimensional angles, gain patterns are often displayed in terms of a single planar angle ϕ, as shown in Figure 3.4. For nonsymmetrical patterns, a separate planar pattern must be shown for both azimuth and elevation.

The most important parameters of an antenna pattern are its *gain* (maximum value of the gain pattern), its *beamwidth* (a measure of the angle over which most of the gain occurs), and its *sidelobes* (amount of gain in off-axis directions). For most communication purposes, desirable antenna patterns should be highly directional, with high maximum gain concentrated over a narrow beamwidth with negligibly small sidelobe transmission. As a rule of thumb for all antennas, the maximum gain g and the half-power beamwidth ϕ_b in radians are given by

$$g = \rho_a\left(\frac{4\pi}{\lambda^2}\right)A \tag{3.2.1}$$

$$\phi_b \cong \frac{\lambda}{d\sqrt{\rho_a}} \tag{3.2.2}$$

where A is the antenna aperture cross-sectional area, d is the cross-sectional diameter, and ρ_a is the antenna efficiency factor. The latter is dependent on the type of antenna and on how the antenna aperture is electromagnetically illuminated. Note that the antenna gain is always proportional to the square of the carrier frequency and the antenna size, whereas the beamwidth varies inversely with frequency and size. Hence, the larger the antenna or the higher the frequency, the larger is the gain and the narrower the beamwidth. Thus, a given antenna has an increasingly more directional pattern as higher frequencies are used. At a fixed frequency, the pattern becomes more directional as the antenna is made larger.

In satellite systems the common antenna types are the linear dipole, the helix, the horn, the antenna array, and the parabolic reflector. These are shown in Table 3.1, along with a sketch of their gain patterns and parameter values. The parabolic reflector is the most common, giving a highly directional, symmetric pattern. The dipole has a pattern that is hemispherical and produces a propagating field polarized in the direction of the dipole. Helix and horns are smaller antennas with reasonable directivity, but higher sidelobes than parabolic reflectors. A helix antenna produces a circularly polarized field, while a horn is generally used to produce linearly polarized fields. A phased array is a group of small antennas (dipoles, horns, helices) that individually radiate phase-shifted fields that combine in space to produce a resulting field pattern having directivity and narrow beamwidths. The more elements in the array, the larger the pattern gain and the narrower the beamwidth. By properly selecting the phase shifts between the array elements, the direction of the beam can be controlled.

The most common antenna for satellites and earth stations is the parabolic reflector, or *dish*. Figure 3.5 shows two different feed diagrams for a parabolic dish. The field to be transmitted is excited in the feed waveguide with the desired polarization and is then radiated to the reflector. The feed may be located in front at the focus of the parabolic dish, or it may be fed from behind using a subreflector. If the dish is uniformly illuminated by the feed, the reflected transmitted gain pattern is circularly symmetric, as given in Table 3.1. The gain, half-power beamwidth ϕ_b, and the sidelobe gain depend on the manner in which the feed radially distributes the field intensity over the dish. Table 3.2 lists the relation among type of illumination, the resulting aperture efficiency ρ_a, beamwidth ϕ_b, and the sidelobe peak values. Note that reduction of sidelobes is accompanied by a spreading of the radiated beam.

For small angles off boresight, such that the pattern skirt is within about 6 dB of the peak value, the parabolic antenna pattern is adequately

TABLE 3.1 Antenna patterns

Antenna Type	*Pattern*	*Gain* (g)	*Half-power Beamwidth*
Short dipole $l \ll \lambda$	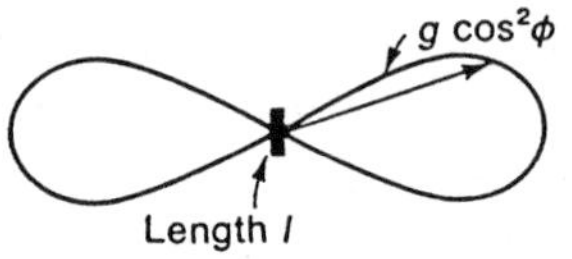	1.5	90°
Long dipole $l \gg \lambda$	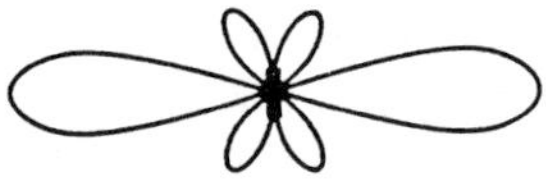	1.5	47°
$l = \lambda/2$		1.64	78°
Phased array	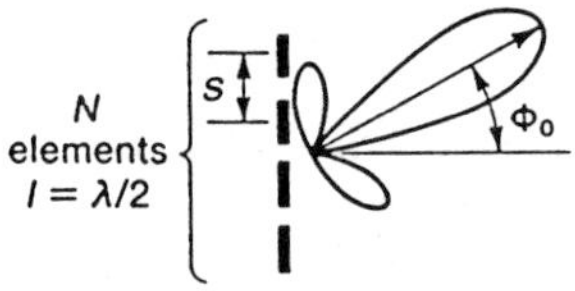	$\left\{\frac{N\pi s}{1.4\lambda}\right\}^2$	$50°\left\{\frac{\lambda}{Ns}\right\}$ $\theta_0 = 0$
Helix	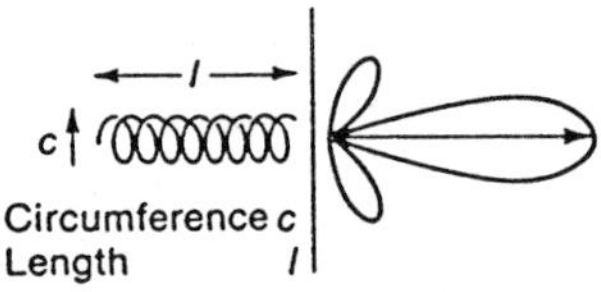	$15\left\{\frac{cl}{\lambda^2}\right\}$	$52°\left\{\frac{\lambda}{c\sqrt{l}}\right\}$
Square horn, dimension d 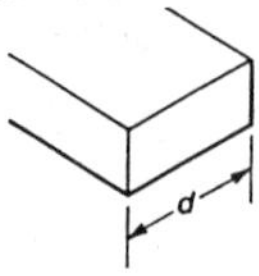	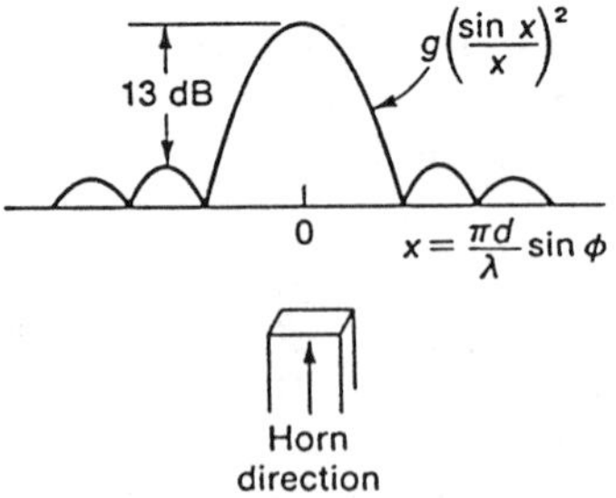	$\frac{4\pi d^2}{\lambda^2}$	$\frac{0.88\lambda}{d}$ rad
Circulator reflector	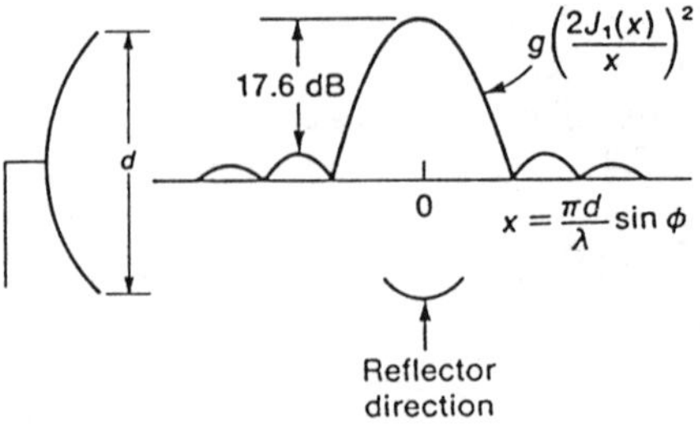	$\left\{\frac{\pi d}{\lambda}\right\}^2$	$\frac{1.02\lambda}{d}$ rad

TABLE 3.2 Parameter values of parabolic reflector

Aperture field intensity distribution over dish:

$$\left[l - \left(\frac{2r}{d}\right)^2\right]^{\gamma}, \qquad r \leq d/2$$

γ	$\rho_a(\%)$	ϕ_b *(rad)*	*Sidelobe (dB)*	*Gain (g)*
0	100	$1.02\ \lambda/d$	-17.6	$9.86\ (d/\lambda)^2$
1	75	$1.27\ \lambda/d$	-24.6	$7.39\ (d/\lambda)^2$
2	55	$1.47\ \lambda/d$	-30.6	$5.4\ (d/\lambda)^2$
3	44	$1.65\ \lambda/d$	-40.6	$4.83\ (d/\lambda)^2$

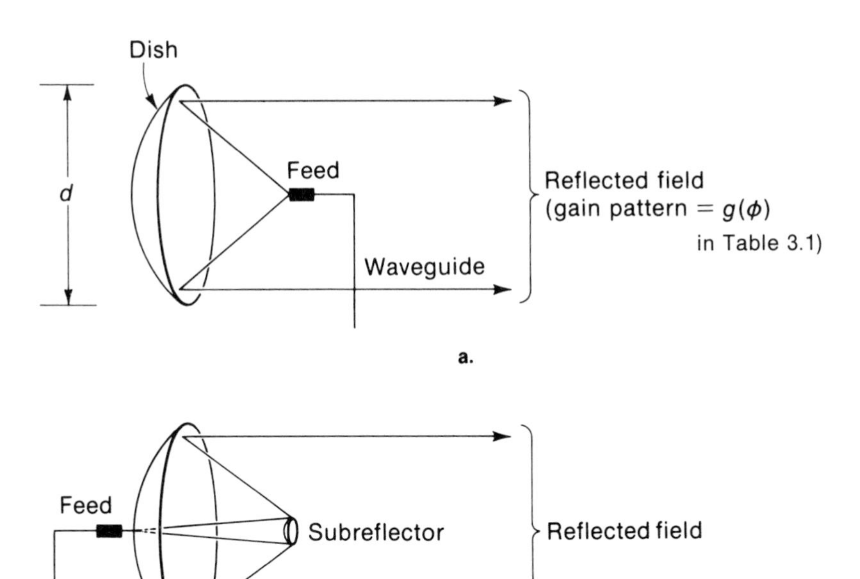

FIGURE 3.5 Parabolic reflector: (a) front feed; (b) rear (Cassegrainian) feed.

represented by the simpler relation

$$g(\phi) \approx \rho_a \left(\frac{\pi d}{\lambda}\right)^2 e^{-2.76(\phi/\phi_b)^2} \tag{3.2.3}$$

Equation (3.2.3) states that parabolic gain falls off approximately exponentially with angle at small angles, and serves as a more convenient expression for evaluating off-axis gains.

In addition to efficiency loss, the roughness of an antenna dish surface can cause radiation scattering and loss of gain in the desired direction. This surface roughness loss is typically given as

$$L_r = e^{-(4\pi\sigma/\lambda)^2} \tag{3.2.4}$$

where σ is the rms roughness in wavelength dimensions (Ruze, 1966). Figure 3.6 plots L_r for several roughness values as a function of frequency. Note that as higher carrier frequencies are used, the loss due to the roughness of the antenna surface eventually becomes important, and overcomes the theoretical increase in gain with frequency. Thus, while Eq. (3.2.1) predicts that an arbitrarily large gain is theoretically feasible by continuing to increase frequency, Figure 3.6 shows that eventually surface effects will begin to degrade that gain, and beyond that point, scattering will actually produce poorer performance. Both cost and weight of reflecting surfaces generally increase as higher surface tolerance is required.

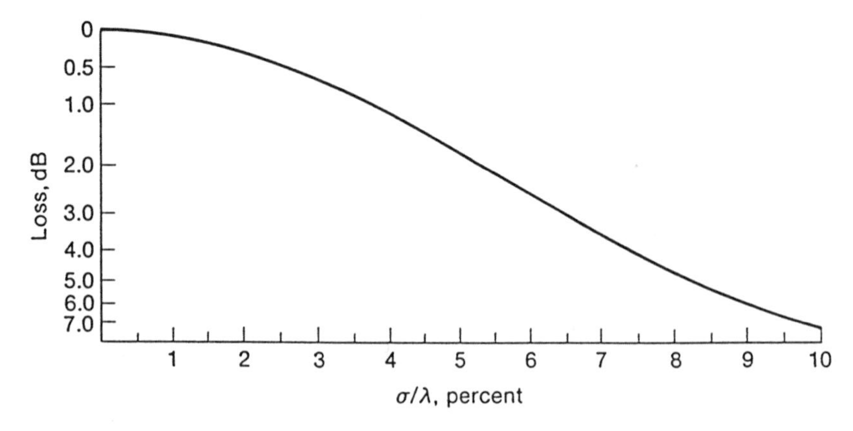

FIGURE 3.6 Surface roughness loss in reflector antennas (σ = rms surface roughness).

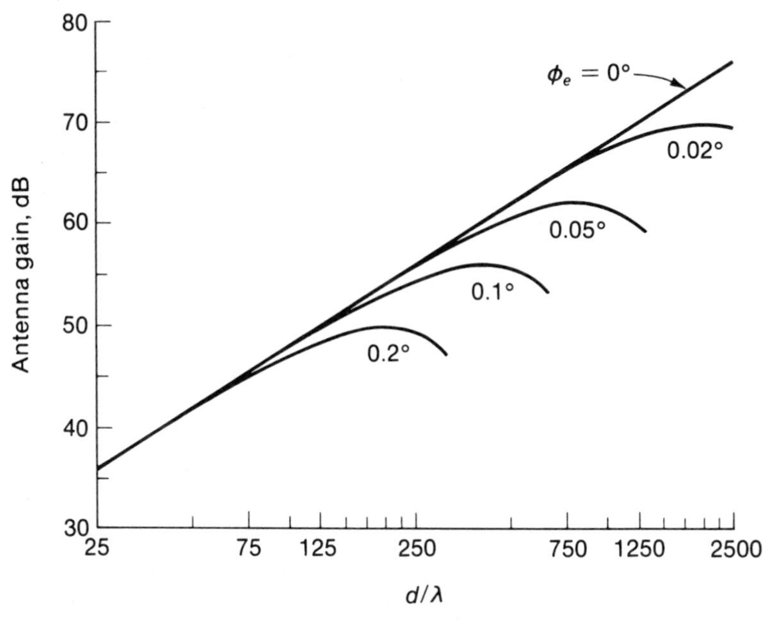

FIGURE 3.7 Pointing-error loss in antennas (ϕ_e = pointing error). Antenna efficiency = 55%.

While the gain g represents the peak gain of the antenna, the actual gain used in the power calculation depends on the angular direction from transmitter to receiver. That is, it depends on how accurately the antennas are pointed at the transmitter and receiver. This accuracy depends, in turn, on the pointing error, which arises from both the inability to aim an antenna in exactly the right direction and the inaccurate knowledge of the target location. Pointing errors to geostationary satellites are usually about 0.1°–0.5°, depending on the antenna pointing and tracking capabilities, and are slightly higher for orbiting satellites. The presence of a pointing error means that the power to the receiver is determined by the antenna gain on the skirts of the pattern, rather than by the peak gain g. For a parabolic antenna with pointing error ϕ_e, Eq. (3.2.3) shows that

$$g(\phi_e) = ge^{-2.76(\phi_e/\phi_b)^2} \tag{3.2.5}$$

where g is the peak gain and ϕ_b the half-power beamwidth. Equation (3.2.5) indicates that an exponential decrease in gain occurs for small pointing errors, as the pointing error increases relative to beamwidth. When narrow beams are used, a given pointing error therefore becomes more critical. Since beamwidth depends directly on the d/λ (diameter/wavelength) ratio, pointing-error losses increase exponentially with this product. Figure 3.7 plots $g(\phi_e)$ as a function of d/λ for several values of pointing errors. This pointing-error loss is added directly to surface roughness loss L_r in Eq. (3.2.4). Note that while roughness losses prevent use of high frequencies, pointing errors prevent use of narrow beams, and therefore constrain both frequency and antenna size. For example, if no more than 1 dB of pointing loss can be tolerated, it is necessary that $(\phi_e/\phi_b) \leq 0.28$. From Eq. (3.2.2) this requires

$$d/\lambda \leq 0.28/\phi_e \tag{3.2.6}$$

where ϕ_e is in radians. Hence, link pointing error limits the usable d/λ ratio of the antenna.

There is also an increasing interest in the use of phased-array antennas in satellite communications. As shown in Figure 3.8a a pointable, focused beam can be formed by simultaneously phase shifting an electromagnetic carrier field in N separate branches, and radiating each shifted component as the output of an individual radiator (dipole, horn, etc.). The resulting radiated pattern will reinforce (phase-combine) in some directions and interfere in others, the net result being a combined beam that points in a direction determined by the phase shifts used. By properly selecting the phase shifts, the directivity of the beam can be oriented in a given direction. Hence, a phased array can theoretically produce beams in arbitrary off-axis directions. The array gain increases with the square of the number of array elements, and thus high gain is achieved with large arrays. The beamwidth is inverse to the number of elements, but changes with the beam direction, being narrowest for broadside beams, and becoming increasingly larger for off-axis beams.

Recent advances in ferrite phase shifters that can be installed directly inside microwave waveguides have made the phased-array antenna an easily implementable, lightweight assembly for satellite use. The phase shifters operate as shown in Figure 3.8b. A ferrite material surrounds a dielectric inserted directly inside the waveguide. An external current magnetizes the ferrite, which acts to vary the index of refraction of the dielectric. This varies the retardation of the field, producing an adjustable phase shift relative to an unretarded field. The properties of the dielectric determine the amount of phase shift per unit length, so that the overall length of the dielectric can be adjusted to a desired phase range. Lengths on the order of only inches are generally needed at satellite frequencies to obtain a full 360° phase shift. Because of the proportional relation between current and magnetism in the ferrite, a fairly linear phase–

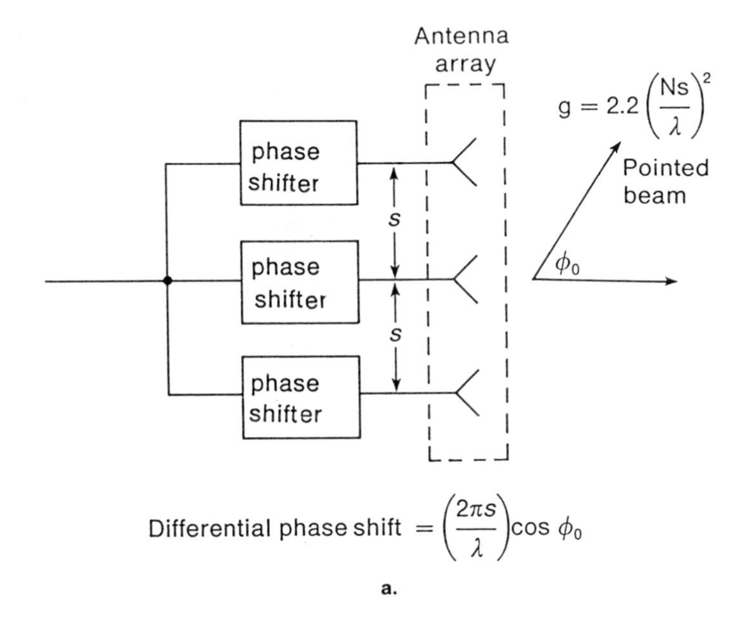

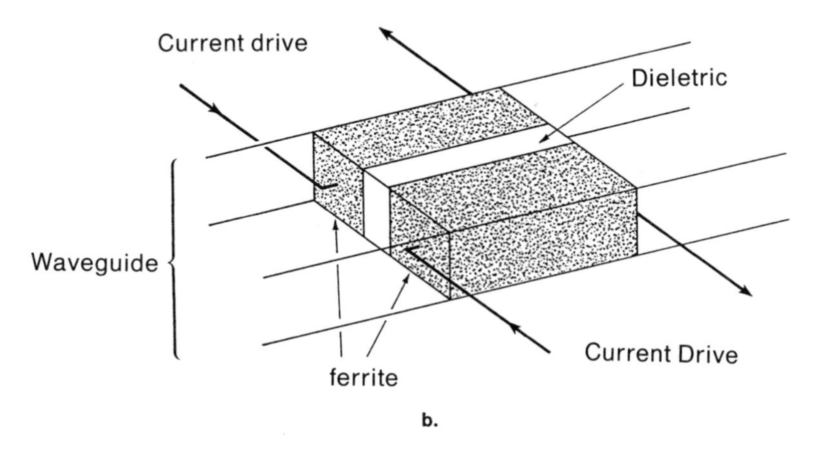

FIGURE 3.8 Phased-array antenna design: (a) block diagram; (b) ferrite phase shifter inserted in waveguide.

current characteristic occurs. By properly biasing the drive current, an extremely accurate phase shift can be set for each phase shifter. The antenna array is obtained by placing such phase shifters in each radiator guide and adjusting the phase shifts to the proper value, as shown in Figure 3.8a.

Phased-array antennas on satellites represent a rapidly advancing technology that will continue to undergo substantial development. Phased arrays allow beam steering without physically moving an antenna by simply readjusting the phase shifters in the array feed. This electronic steering can be used to advantage to point toward moving transmitters or to reduce the receiving gain in the direction of noise sources (null steering).

3.3 ATMOSPHERIC LOSSES

The Earth is surrounded by a collection of gases, atoms, water droplets, pollutants, and so on captured by the Earth's gravity field and extending to an altitude of about 400 miles. This constitutes the Earth's atmosphere. The heaviest concentration of these particulates is near the Earth, with particle density decreasing with altitude. The particles in the upper part of the atmosphere (*ionosphere*) absorb and reflect large quantities of radiated energy from the sun. The absorbed energy is then reradiated in all directions by the ionosphere. In addition, the absorbed energy ionizes the ionospheric atoms, producing bands of upper-atmospheric free electrons that surround the Earth. These electrons directly interact with any electromagnetic field passing through them.

A propagating electromagnetic field undergoes a basic power loss (L_p) that increases inversely with the square of the distance propagated, even in free space. When propagating in or through the Earth's atmosphere, additional losses occur that further degrade power flow. These losses are caused by absorption and scattering of the field by the atmospheric particulates. These effects become more severe as the field carrier frequency is increased to a point where the wavelength begins to approach the size of the particulates. Figure 3.9a shows the average decibel loss that can be expected as a function of frequency for the elevation angle shown. Below about 10 GHz, atmospheric attenuation is nominal (less than 2 dB); at frequencies above that, however, it increases rapidly. Higher amounts of attenuation occur at those particular frequencies having wavelengths corresponding to specific gas molecules in the atmosphere. For example, severe absorption due to water vapor molecules occur at about 22 and 180 GHz, whereas oxygen absorption occurs at about 60 and 118 GHz. This attenuation increase at the higher frequencies is the primary dis-

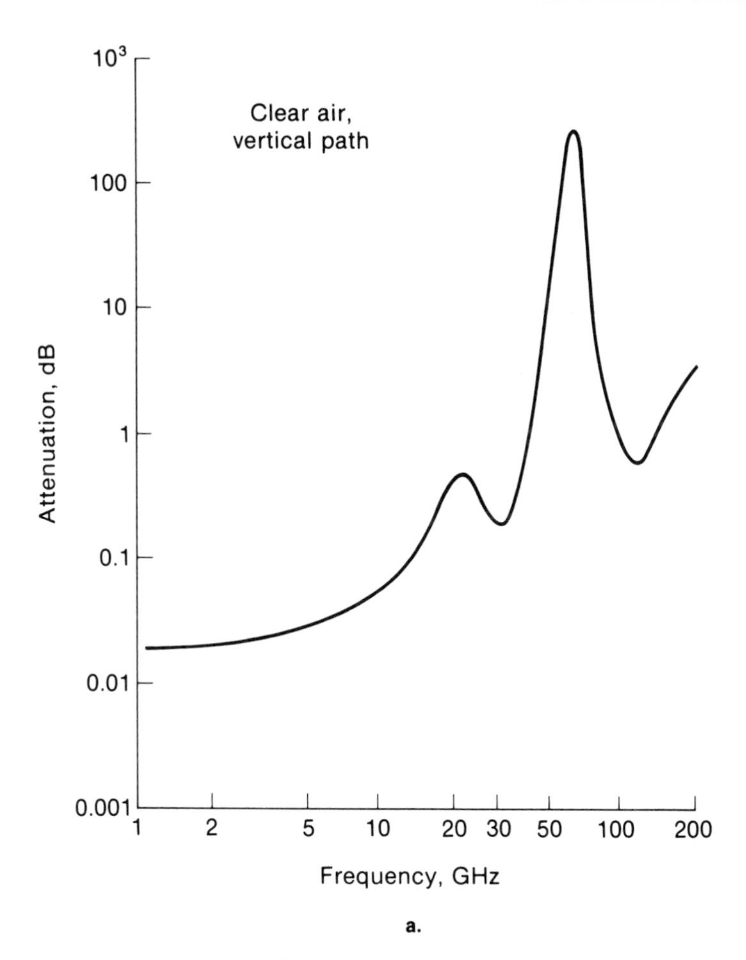

FIGURE 3.9 Atmosphere attenuation: (a) versus carrier frequency; (b) versus elevation angle.

advantage of shifting operation to the higher-frequency bands in satellite systems.

Since the density of particulates in the atmosphere decreases with distance above Earth, atmospheric attenuation is lower at the higher altitudes. This means that the total attenuation expected in a satellite link will depend on both the link elevation angle and the slant range. Most severe attenuation occurs at angles close to the horizon, and decreases for vertical paths directly overhead. Figure 3.9b plots the atmospheric attenuation at various satellite frequency bands as a function of elevation angle measured from ground. From these curves, it is clear that it is

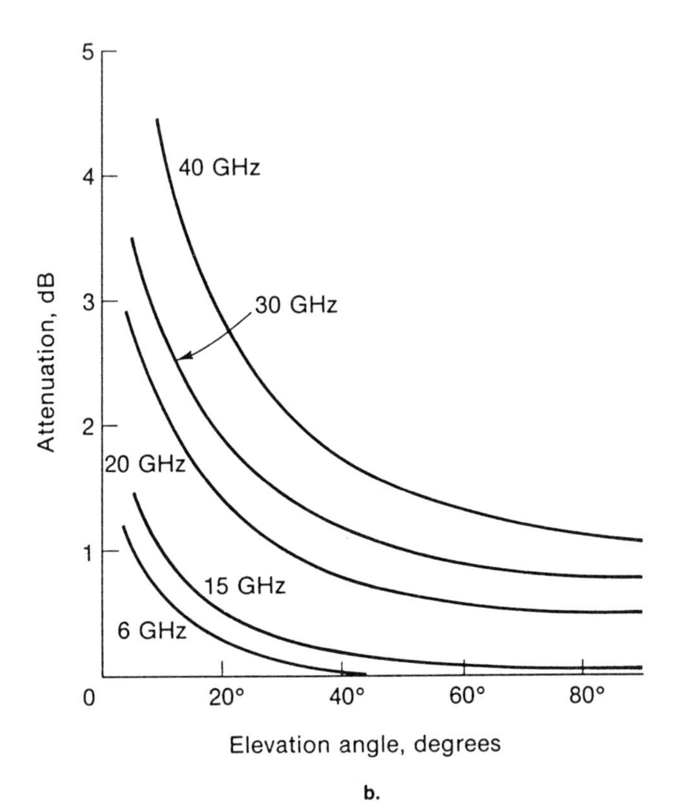

FIGURE 3.9 (Continued).

preferable that satellites be viewed above a minimum elevation angle. Domestic satellite systems serving particular areas are usually designed to ensure that the satellite will be higher than some acceptable viewing angle.

The most serious atmospheric effect in a satellite link is rainfall. Water droplets scatter and absorb impinging radiation, causing attenuation many times the clear air losses shown in Figure 3.9. Rain effects become most severe at wavelengths approaching the water drop size, which is dependent on the type of rainfall. In heavy rains raindrop size may approach a centimeter, and severe absorption may occur at frequencies as low as 10 GHz. Thus rainfall effects can become extremely severe at satellite frequencies at X-band and above.

If a satellite link is to be maintained during a rainfall, it is necessary that enough extra power (called *power margin*) be transmitted to overcome the maximum additional attenuation induced by the rain. Hence it is necessary to have an accurate assessment of expected rain loss when evaluating link parameters. The expected additional rain loss depends on the operating frequency, the amount of rainfall, and the path length of

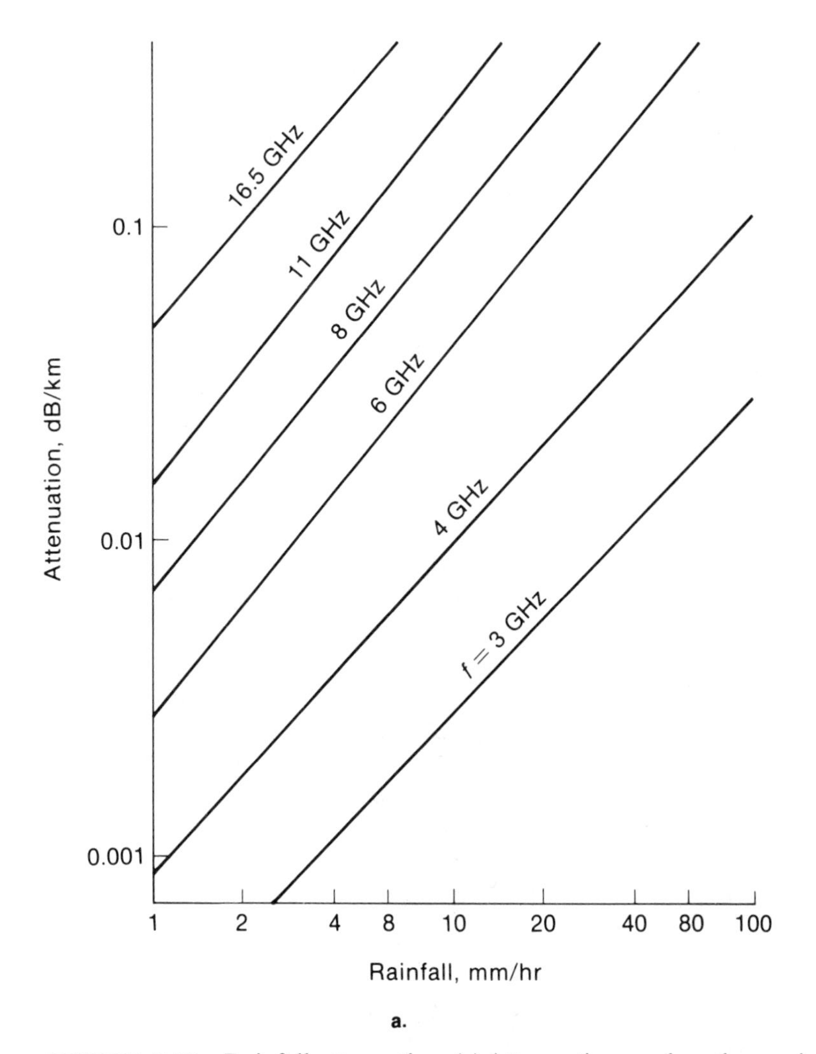

FIGURE 3.10 Rainfall attenuation. (a) Attenuation per length vs. rainfall rate. (b) Approximate average rainstorm path length vs. elevated angle and rainfall rate. Designations: 0.25 mm/hr = drizzle, 1.25–12.5 = light rain, 12.5–25 = medium rain, 25–100 = heavy rain, >100 = tropical downpour.

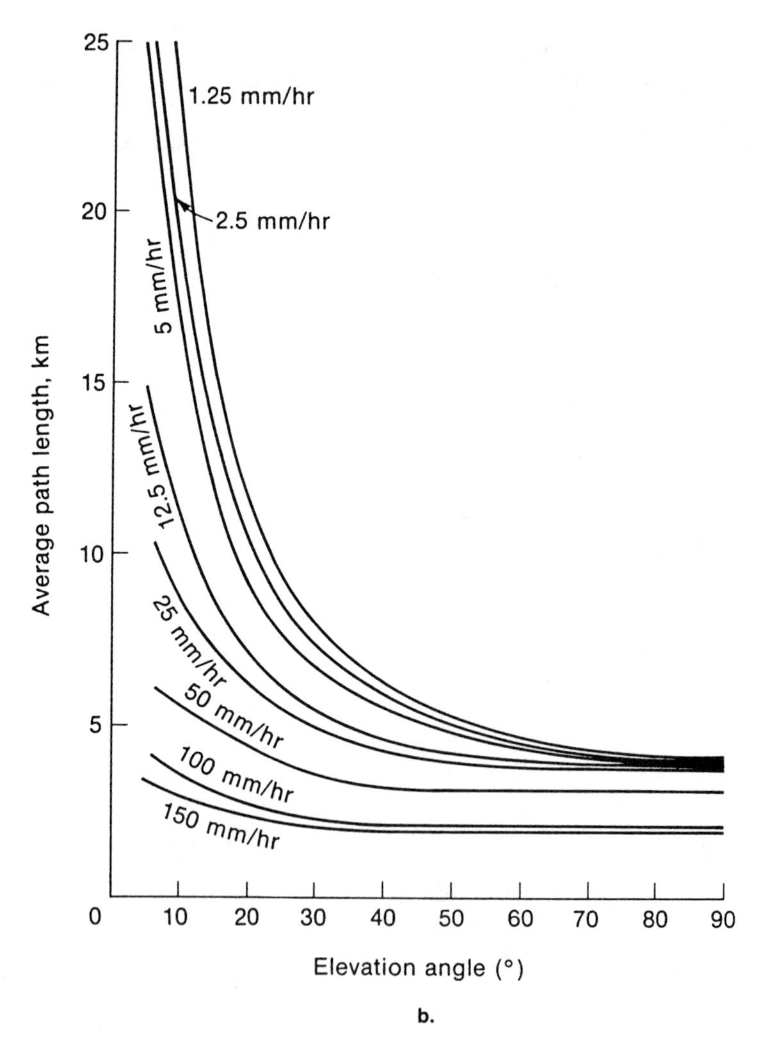

FIGURE 3.10 (Continued).

the propagation through the rain (Angelakos and Everhart, 1968; Crane, 1977; Ippolito, 1986; Morrison, et al., 1973; Oguchi, 1973). To evaluate this additional rain loss we first obtain the expected rainfall rate in millimeters per hour for the region of the communication link. We then use curves such as those shown in Figure 3.10a to read off decibel loss per path length at the operating frequency. These curves are generated from combinations of empirical data and mathematical models that fit the data. The rainfall attenuation model commonly used (Ippolito, 1986) is of the form

$$\text{dB loss/length} = ar^b \tag{3.3.1}$$

where r is the rainfall rate, and a and b are frequency-dependent coefficients obtained via curve-fitting and extrapolation (e.g., see Problem 3.7). The mean path length of the rain is then determined for the given elevation angle. This length also depends on rainfall rate, as shown in Figure 3.10b, since rainclouds with heavy rain are, in general, closer to the ground. These path-length curves are also obtained by fitting mathematical models to measured data (Crane, 1977; Ippolito, 1986). With the

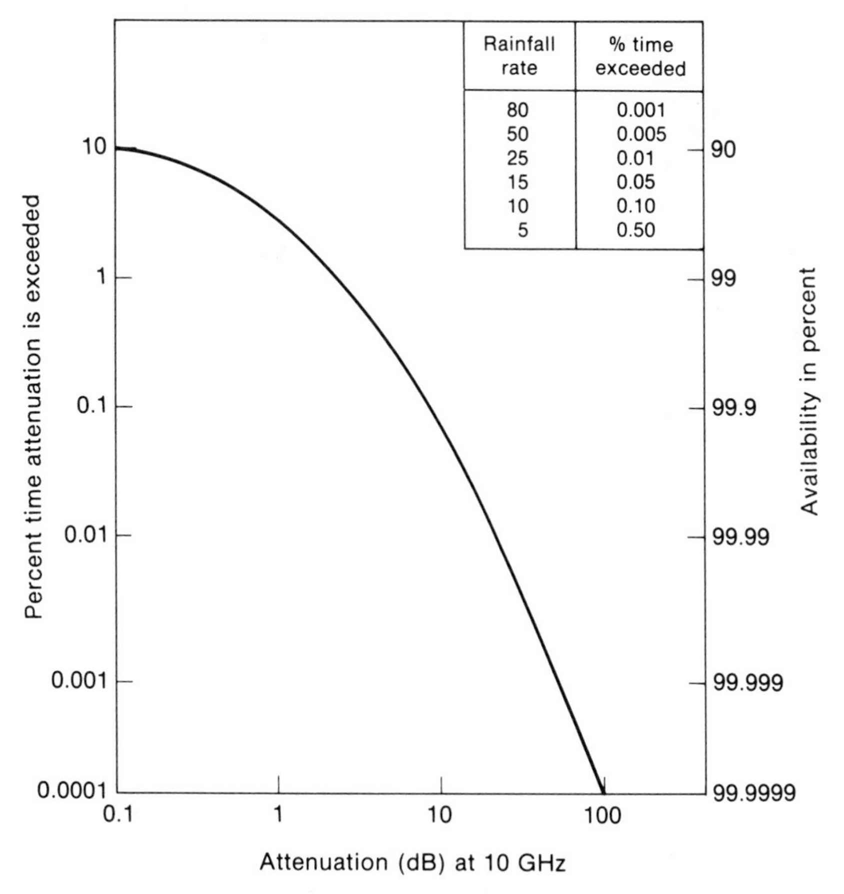

FIGURE 3.11 Attenuation distribution at 10 GHz and 30° elevation angle.

mean path distance estimated, expected rainfall attenuation is obtained by multiplying the rain dB/length by the mean path length. When this rainfall computation is carried out for the data in Figures 3.10, we see that even at C-band frequencies a heavy rainfall can add an additional 5 dB or more of attenuation to the path loss.

Rainfall attenuation computed as just described is actually a conditional attenuation dependent on the rainfall rate selected. Since the latter is itself a statistical phenomenon, it is often more meaningful to take into account the probability that a given rainfall rate will be exceeded. Data on such probabilities are usually available for most regions of the world. When these probabilities are taken into account, we compute rain attenuation statistics similar to that shown in Figure 3.11. This shows, for a particular region, the expected percentage of time that a given rainfall attenuation level will be exceeded. Since Figure 3.10 allows us to convert rainfall rate to expected attenuation at specific frequencies, these rainfall probabilities allow us to associate a probability with a given attenuation being exceeded at that frequency and elevation angle.

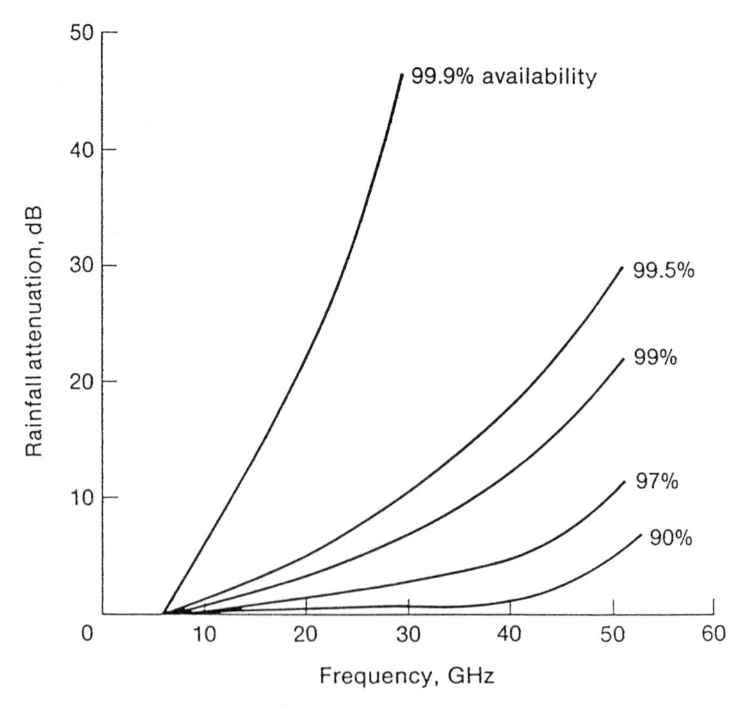

FIGURE 3.12 Availability data for attenuation and frequency (elevation angle = 30°).

Curves of this type are much more meaningful for allowing safe design margins and preventing an inordinate amount of link overdesign. If $P(\alpha)$ is the percentage of time an attenuation α is exceeded, then $1 - P(\alpha)$ is the percent *availability* of the link with an attenuation less than α. The value $P(\alpha)$ is often called the percent *outage* time of the link, with an attenuation exceeding α. Repeated conversions of this type will generate a crossplot of link availability with respect to a given rain attenuation level and frequency. Such a curve is shown in Figure 3.12, using the data of Figures 3.10 and 3.11. This shows us, for example, that if a 30-GHz link at an elevation angle of 30° can withstand a rain attenuation of 10 dB, it will be available 99.5% of the time, or, equivalently, it will have an outage 0.5% of the time (about 45 hr a year). If the elevation angle decreases, the link will be available for less time.

3.4 RECEIVER NOISE

In addition to receiving the impinging field from a transmitting source, a receiving antenna also collects other forms of electromagnetic energy that may be present. This other radiation appears as interference, or noise, that tends to mask the desired field propagating from the transmitter. The primary cause of these unwanted fields is background radiation from galactic and cosmic sources in the sky and atmospheric reradiation, and *radio frequency interference* (RFI) from other transmitters and emitters. In addition to this noise collected by the receiving antenna, noise can be generated directly within the electronics of the receiver immediately following the antenna. The contributions from all these sources of interference combine to define a total noise level for the receiving system. This noise sets the minimal required power from the desired transmitter for achieving reliable communications. That is, even before any field processing is performed, it is necessary for the receiving antenna to collect enough transmitter power (P_r in Section 3.1) to overcome the total noise occurring at the receiver.

The amount of receiver noise present is defined by a receiver noise temperature T°_{eq}. The parameter T°_{eq} is an effective equivalent temperature that an external noise source would have to have to produce the same amount of receiver noise. This equivalent temperature is written as

$$T^\circ_{\text{eq}} = T^\circ_b + (F - 1)290^\circ \tag{3.4.1}$$

where T°_b is the background noise temperature accounting for contributions collected by the antenna (from galactic and sky noise) and F is the *noise figure* of the receiver. This latter accounts for the contribution due to receiver electronics. A noise source with a temperature T°_{eq} produces an effective one-sided noise spectrum level of

$$N_0 = kT^{\circ}_{eq} \text{ W/Hz} \tag{3.4.2}$$

at the antenna input, where k is Boltzmann's constant:

$$\begin{aligned} k &= 1.379 \times 10^{-23} \text{ W/K-Hz} \\ &= -228.6 \text{ dB W/K-Hz} \end{aligned} \tag{3.4.3}$$

From the spectral level in Eq. (3.4.2), the total noise power entering the receiver over a bandwidth B_{RF} is then

$$P_n = kT^{\circ}_{eq} B_{RF} \tag{3.4.4}$$

Here, B_{RF} typically represents the RF bandwidth of the receiving system. Hence, a computation of total receiver noise can be obtained from knowledge of the T°_{eq} of the receiver.

The value of T°_{b} depends on the specific characteristics of the actual noise background observed by the receiving antenna. For example, an earth-based antenna facing skyward would see the galactic noise sources (stars, planets, moon, etc.) in the antenna beamwidth, along with the reradiation effects of the sun's noise energy by the atmosphere. On the other hand, a satellite-based antenna, looking back toward Earth, would see primarily the reradiation noise emissions of the Earth itself. These latter emissions would appear as a uniform bright spot encompassing the Earth, set against the blackness of the outer universe. When viewed from outer space, the Earth has a reradiation temperature of about 300 K. Thus, the background noise effects may be significantly different for the two cases.

Background noise temperatures for earth-based antennas also depend indirectly on frequency, since they are primarily a combination of reflected galactic noise and reradiated atmospheric noise, both of which are frequency-dependent. Figure 3.13 shows a typical sky background noise temperature T°_{b}, showing the galactic noise decreasing with frequency, and the atmospheric reradiated noise increasing with frequency at fixed elevation angle. The combination of the two produces a minimum noise window in the range of about 1–10 GHz. This is a primary reason why the bulk of microwave terrestrial and satellite communication systems (specifically C-band and X-band satellite links) have evolved in this range.

It should be pointed out that rainfall may alter the effective sky background temperature slightly, since it introduces more scatterers (raindrops) to the antenna input. In general, rainfall rates below about 10 mm/hr cause no appreciable increase in T_b°, but may add an additional 10–50° in severe rainstorms.

Internal circuit noise is accounted for in T_{eq}° by specifying the receiver noise figure F, which depends on the specific electronic circuitry following the antenna. The noise figure for an electronic system is defined as

$$F = \frac{\text{Total system output noise power}}{\text{Output system noise power due to system input noise at 290 K}} \tag{3.4.5}$$

When defined in this way, $(F - 1)290°$ is the temperature of an equivalent noise source at the system input that will produce the same contribution to the system output noise as the internal noise of the system itself. For an antenna system, this equivalent circuit temperature adds directly to the background noise collected by the antenna, producing the combined equivalent noise T_{eq}° in Eq. (3.4.1). Explicit formulas for the noise figure

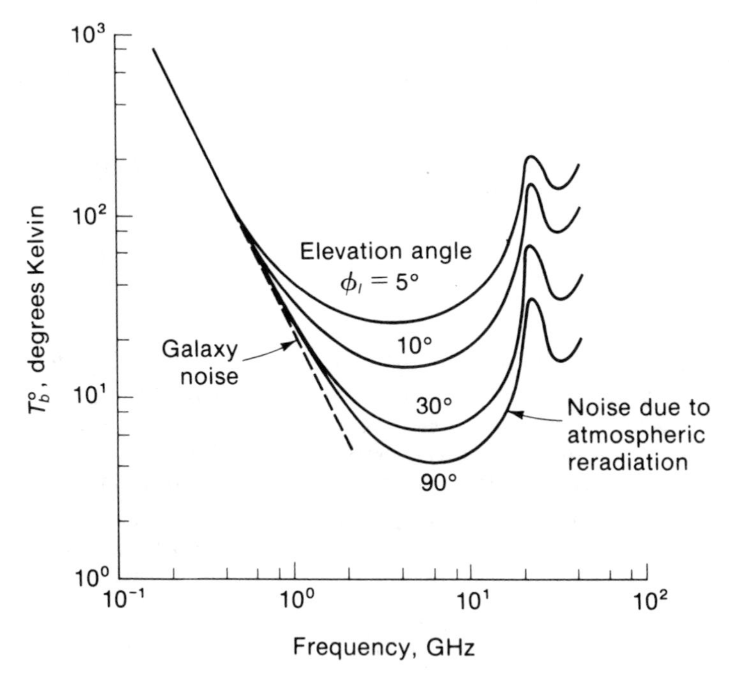

FIGURE 3.13 Background noise temperature for earth-based receivers.

for typical receiver front ends will be given in Chapter 4, when specific satellite circuits are considered.

3.5 CARRIER-TO-NOISE RATIOS

One of the key parameters characterizing the performance of a communication receiver is the ratio of the received carrier power to the total noise power of the receiver. This RF carrier-to-noise ratio (CNR) is defined by

$$\text{CNR} = \frac{P_r}{P_n} \tag{3.5.1}$$

where P_r and P_n are defined in Eqs. (3.1.5) and (3.4.4). This ratio indicates the relative strength of the desired transmitter-received power and the total interference. Typically, communication links require P_r to be at least 10 times (10 dB greater than) the noise power P_n (i.e., $\text{CNR} \geq 10$) for adequate receiver processing.

Substituting for P_r and P_n yields

$$\text{CNR} = \frac{(\text{EIRP})L_p L_a g_r}{kT^\circ_{\text{eq}} B_{\text{RF}}} \tag{3.5.2}$$

This shows how the specific link parameters directly affect receiver CNR. When stated in decibels, this becomes

$$(\text{CNR}) = (\text{EIRP})_{\text{dB}} + (L_p)_{\text{dB}} + (L_a)_{\text{dB}} + (g_r)_{\text{dB}} - (k)_{\text{dB}} - (T^\circ_{\text{eq}})_{\text{dB}} - (B_{\text{RF}})_{\text{dB}} \tag{3.5.3}$$

The resulting CNR in decibels is therefore obtained by adding and subtracting the preceding decibel values. Note that Boltzmann's constant in Eq. (3.5.2), when converted to decibels, has the value $(k)_{\text{dB}} = -228.6$, which enters Eq. (3.5.3) as a positive number in the link budget.

Since the receiver bandwidth B_{RF} is often dependent on the modulation format, we often isolate the RF link power parameters by normalizing out bandwidth dependence. We therefore define instead RF carrier-to-noise level ratio

$$\left(\frac{C}{N_0}\right) = \frac{(\text{EIRP})L_p L_a g_r}{kT^\circ_{\text{eq}}} \tag{3.5.4}$$

which does not depend on B_{RF}. Equation (3.5.4) can also be stated in terms of the received isotropic power (RIP) in Eq. (3.1.9) as

$$\left(\frac{C}{N_0}\right) = \frac{\mathrm{RIP}}{k}\left(\frac{g_r}{T^{\circ}_{\mathrm{eq}}}\right) \tag{3.5.5}$$

In digital systems the C/N_0 ratio allows us to compute directly the receiver bit energy-to-noise level ratio as

$$\frac{E_b}{N_0} = \left(\frac{C}{N_0}\right)T_b \tag{3.5.6}$$

where T_b is the bit time. Thus, from knowledge of the link C/N_0, which depends only on the RF link parameters, we can directly compute either analog CNR by dividing in bandwidth, or digital E_b/N_0 by multiplying in bit times.

It is often convenient to factor Eq. (3.5.4) as

$$\left(\frac{C}{N_0}\right) = \left[\frac{\mathrm{EIRP}}{k}\right][L_p L_a]\left[\frac{g_r}{T^{\circ}_{\mathrm{eq}}}\right] \tag{3.5.7}$$

The first bracket contains only transmitter parameters, the second bracket contains propagation parameters, and the last receiver parameters. Thus, Eq. (3.5.7) separates out the contribution of each subsystem to the total C/N_0. This interpretation is interesting in that it shows, for example, that in terms of C/N_0 the only effect of the receiving system is through the ratio $g_r/T^{\circ}_{\mathrm{eq}}$ (i.e., the ratio of the receiver antenna gain to its equivalent noise temperature). It in fact shows that as far as the receiver is concerned, performance can be maintained with a lower receiver gain (smaller antenna) if T°_{eq} can be suitably reduced. Hence, there is a direct trade-off of receiver antenna size and receiver noise temperature in achieving a desired performance. Receiver temperature can be controlled by careful control of the antenna background (orbit selection, elevation angles, etc.) and receiver electronics (lower noise figure). Since antenna size directly impacts overall receiver cost and construction, this type of receiver-noise-quality–antenna-size trade-off is often desirable. This is especially true when many receivers (earth stations) are to be used. Figure 3.14 plots the value of $g/T^{\circ}_{\mathrm{eq}}$ for a given frequency–antenna size product at various values of noise temperature T°_{eq}.

Conversely, use of a high-gain antenna allows a noisier receiver to be employed; such a receiver will generally dominate the noise contribution from the background. This makes the performance of the system less

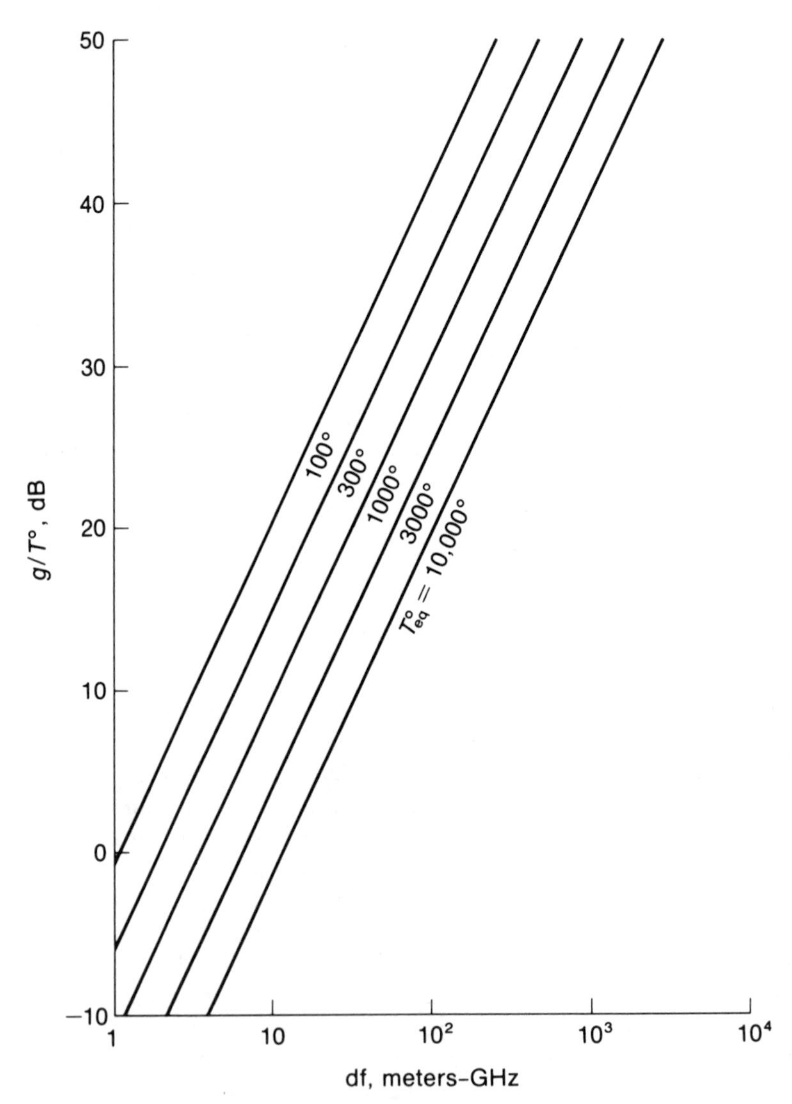

FIGURE 3.14 Receiver g/T° values (d = meters, f = GHz, T°_{eq} = degrees kelvin).

sensitive to parameters influencing the background contribution, such as field of view, rainfall, and so on. That is, small changes in T°_b will not affect T°_{eq}, and therefore C/N_0, if the second term dominates in Eq. (3.4.1).

One must be careful, however, of overemphasizing the g/T° parameter in attempting to improve satellite performance, since it often leads to serious misconceptions. For example, at a glance Figure 3.14 implies that g/T° can be improved by using a higher carrier frequency with a given antenna size, since this directly increases receiving antenna gain. However, we recall from Eq. (3.1.6) that propagation loss also increases as f^2, so that with a fixed-transmitter EIRP in Eq. (3.5.4), C/N_0 does not directly depend on carrier frequency. (In fact, it may be reduced since the atmospheric losses L_a generally increase with f.) This can be made more obvious by reinserting Eqs. (3.1.6) and (3.2.1) to rewrite Eq. (3.5.7):

$$\left(\frac{C}{N_0}\right) = \left[\frac{(\mathrm{EIRP})L_a}{4\pi k T^\circ_{\mathrm{eq}}}\right]\left[\frac{A_r}{z^2}\right] \tag{3.5.8}$$

For a fixed transmitting EIRP and T°_{eq}, C/N_0 is independent of frequency, and depends only on propagation path length and receiving area. The apparent dependence on frequency appears in L_p only if we choose to write A_r in terms of receiving gain g_r. Thus, selecting a higher-frequency band will not directly aid the receiving system. Of course, increasing frequency does aid the transmitter, since a given EIRP can be achieved with a smaller transmitting antenna.

3.6 SATELLITE LINK ANALYSIS

The CNR link budgets of the preceding section can now be directly applied to the analysis of specific satellite links.

Satellite Uplink

Figure 3.15 sketches a simplified earth-station–satellite uplink. Transmitter power for earth stations is generally provided by high-powered amplifiers, such as TWTs and klystrons. Since the amplifier and transmitting antenna are located on the ground, size and weight are not prime considerations, and fairly high transmitter EIRP levels can be achieved. Earth-based power outputs of 40–60 dBw are readily available at frequency bands up through K-band, using cavity-coupled TWTA or klystrons (Angelakos and Everhart, 1968). These power levels, together with the transmitting antenna gains, determine the available EIRP for uplink communications.

In the design of satellite uplinks, the beam pattern may often be of more concern than the actual uplink EIRP. Whereas the latter determines the power to the desired satellite, the shape of the pattern determines the amount of off-axis (sidelobe) interference power impinging on nearby

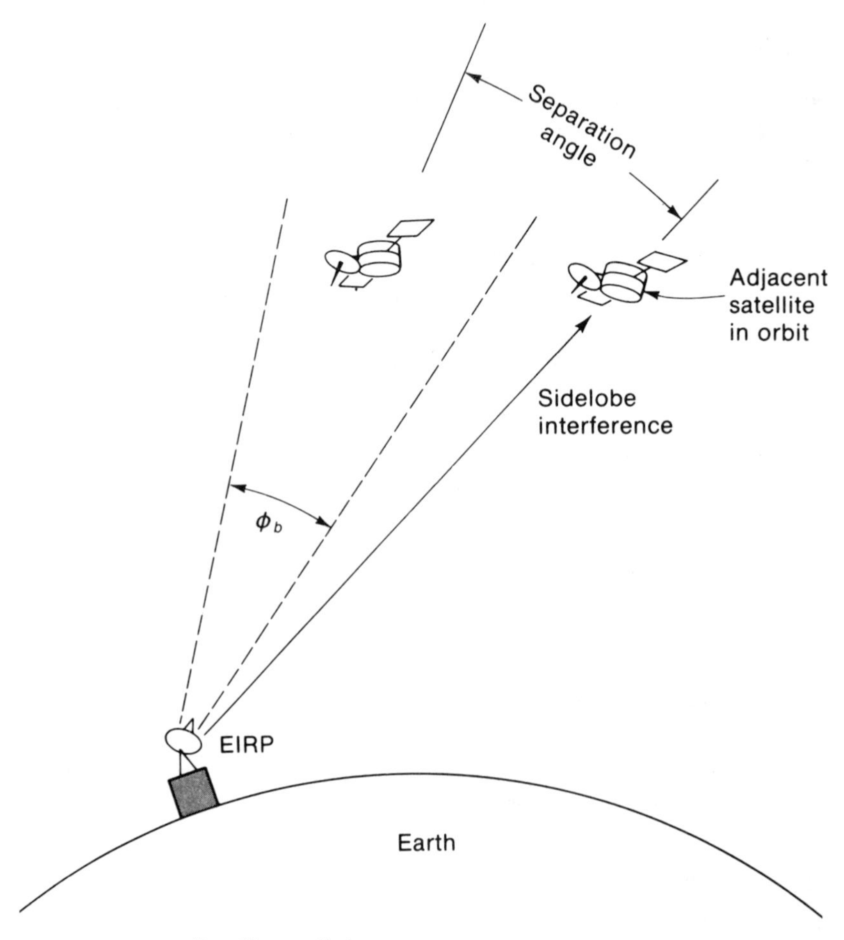

FIGURE 3.15 Satellite uplink.

satellites. The beam pattern therefore establishes an acceptable satellite spacing, and thus the number of satellites that can simultaneously be placed in a given orbit with a specified amount of communication interference. The narrower the earth-station beam, the closer an adjacent satellite can be placed without receiving significant interference. On the other hand, an extremely narrow beam may incur significant pointing losses due to uncertainties in exact satellite location. For example, if a satellite location is known only to within $\pm 0.2^\circ$ (see Section 1.4), a minimum earth-station half-power beamwidth of about 0.6° is necessary. This sets the transmit antenna gain at about 55 dB. For the parabolic ground antennas in Table 3.2, this produces the off-axis gain

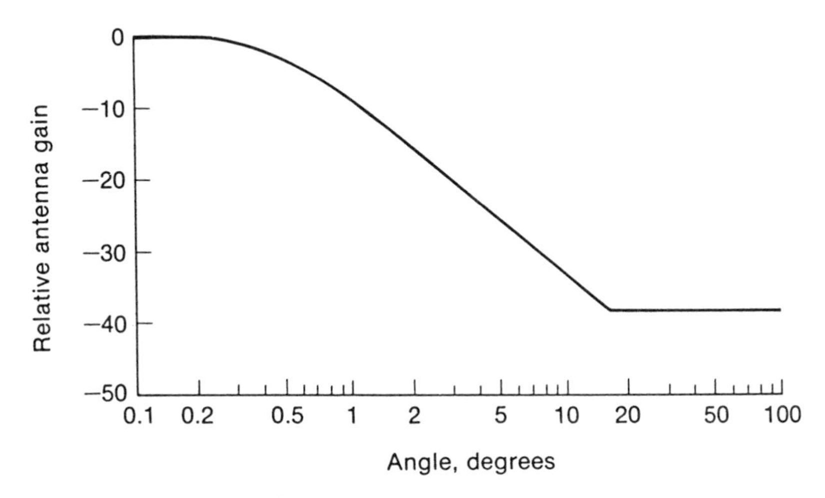

FIGURE 3.16 Uplink earth-station antenna pattern. Angle measured from boresight.

curve shown in Figure 3.16. For a 20 dB reduction in adjacent satellite interference, we see that the nearest satellite must be at least 3° away. That is, when observed from Earth, two satellites in the same orbit must be separated by about 3° in Figure 3.15. Thus, the uplink beamwidth is set by the pointing accuracy of the earth station, whereas satellite orbit separation is determined by the acceptable sidelobe interference. If satellite pointing is improved, the uplink beamwidth can be narrowed, allowing closer satellite spacing in the same orbit. This would increase the total number of satellites placed in a common orbit, such as the synchronous orbit.

With the half-power beamwidth set, a higher carrier frequency will permit smaller earth stations. Figure 3.17 shows the relation between earth-station antenna diameter and frequency in producing a given uplink beamwidth and gain, using Eqs. (3.2.1) and (3.2.2). Note that while increase of carrier frequency does not directly aid receiver power, we see that an advantage does accrue in reducing earth-station size and, possibly, in improving satellite trafficking (allowing more satellites in orbit).

With a 0.6° uplink beamwidth (gain $\approx$ 55 dB) earth-station EIRP values of about 80–90 dBw are readily available. Table 3.3 lists an example uplink budget for computing the CNR at the satellite in a 10-MHz bandwidth, showing the way in which the entries of Eq. (3.5.3) are individually computed and combined. Figure 3.18 generalizes this budget to show how CNR will vary with earth-station EIRP and satellite receiver

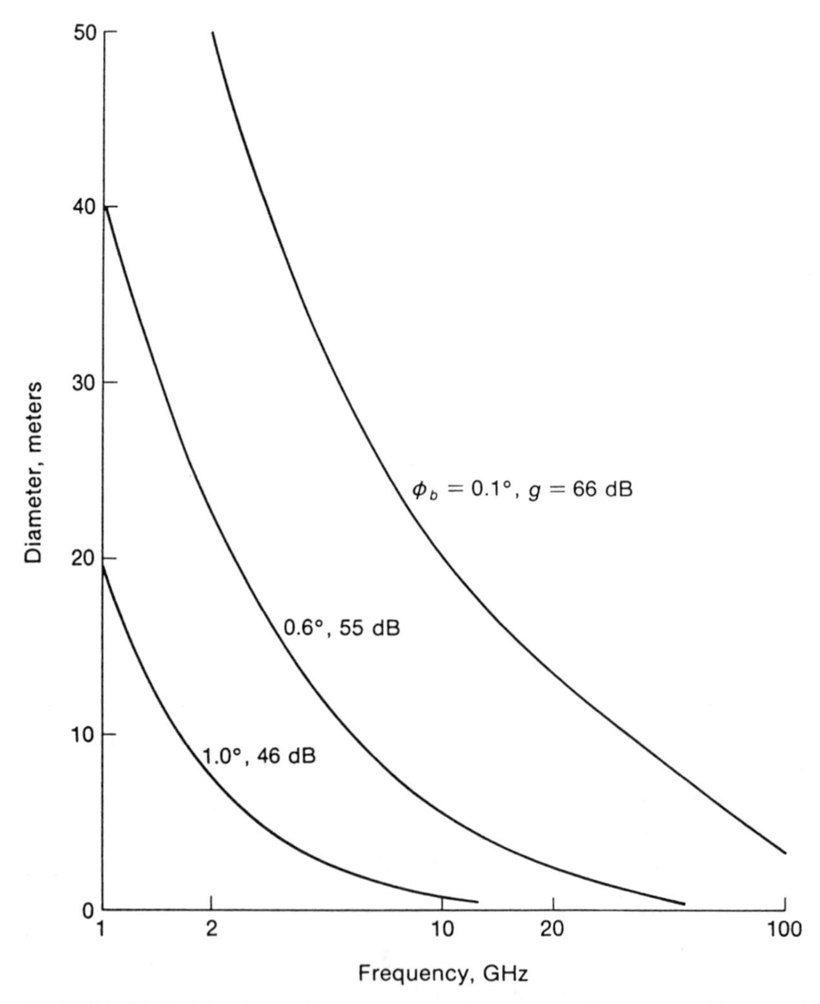

FIGURE 3.17 Earth-station antenna size versus frequency (ϕ_b = half-power beamwidth, g = gain, $\rho_a = 1$).

g/T°_{eq}. Even with significant range losses ($\approx$200 dB) and relatively low g/T° values, an acceptable uplink communication link can usually be established.

Satellite downlink

A satellite downlink (Figure 3.19a) is constrained by the fact that the power amplifier and transmitting antenna must be spaceborne. This limits

TABLE 3.3 C-band uplink power budget

Frequency, 6 GHz		
Transmitter power, $P_T = 30$ W	15 dBW	
Transmitter antenna gain, g	55 dB	
EIRP		70 dBW
Path length, 23,000 miles		
Propagation loss, L_P		−199 dB
Atmospheric loss (rain)		−4 dB
Polarization loss		−1 dB
Pointing loss		−0.6 dB
Satellite receive antenna, 1.5 ft		
beamwidth 8°		
gain (efficiency = 55%)	26 dB	
Background temperature $T_b^\circ = 100$ K		
Receiver noise figure $F = 7.86$ dB		
Receiver noise temperature $T_{eq}^\circ = 1584$ K	32 dB	
Receiver g/T°		−6 dB
Boltzmann constant		228.6 dB
Bandwidth, 10 MHz		−70 dB
CNR		18 dB

the power amplifiers to the efficient, lightweight devices discussed in Section 1.4, with limited output power capabilities that are dependent on the carrier frequency (see Figure 3.19b). The spacecraft antenna, while similarly limited in size, must use beam patterns that provide the required coverage area on Earth. Recall that the coverage area for a specified minimal viewing elevation angle depends only on the satellite altitude. Hence, the satellite downlink beamwidth for a given coverage area is automatically selected as soon as the satellite orbit altitude is selected. This also means the corresponding downlink antenna gain is established by the orbit altitude. Table 3.4 extends the results of Table 1.2 by showing the beamwidth, resulting gain, and required antenna d/λ factor. By using higher-frequency bands (smaller λ), this required downlink beamwidth can be achieved with smaller satellite antenna sizes, as stated before. Spacecraft antennas that provide the maximal coverage area are referred to as *global antennas.*

With satellite power level and antenna gain established, the carrier power collected at the earth station depends only on the g/T factor, just as for the uplink. Figure 3.20 shows an example of a downlink power

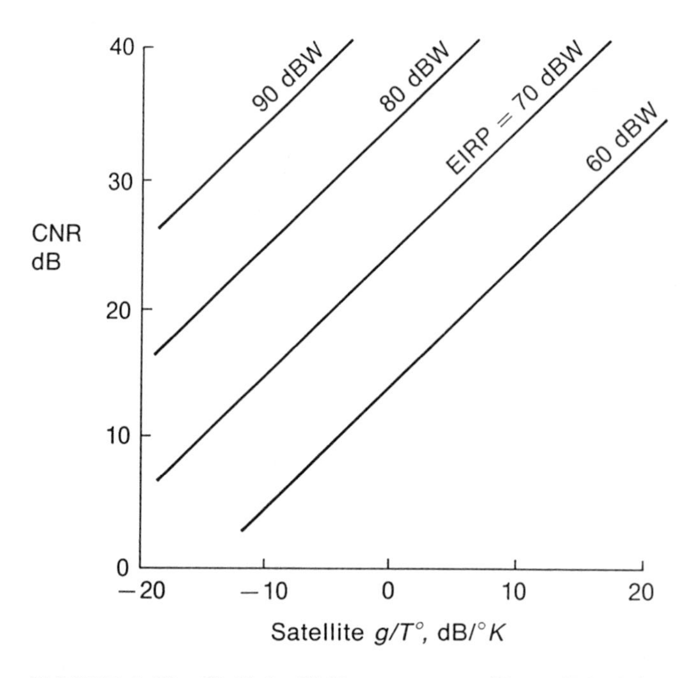

FIGURE 3.18 Uplink CNR versus satellite g/T° C-band link, bandwidth = 10 MHz; link parameters are given in Table 3.3.

budget for a 10-MHz bandwidth and global satellite antennas, and again generalizes to a CNR plot in terms of satellite power and receiver g/T°_{eq}. It is evident that relatively large earth-station g/T°_{eq} is needed to overcome the smaller EIRP of the satellite. This means small earth stations will be severely limited in their ability to receive wide bandwidth carriers.

TABLE 3.4 Spacecraft antenna parameters

Beamwidth (deg)	*Gain (dB)* ($\rho_a = 1$)	d/λ
17.4	22.2	4.18
9.38	27.58	7.75
6.42	29.12	11.33
3.94	35.12	18.46
2.0	41.01	36.38
0.2	61.05	363.8

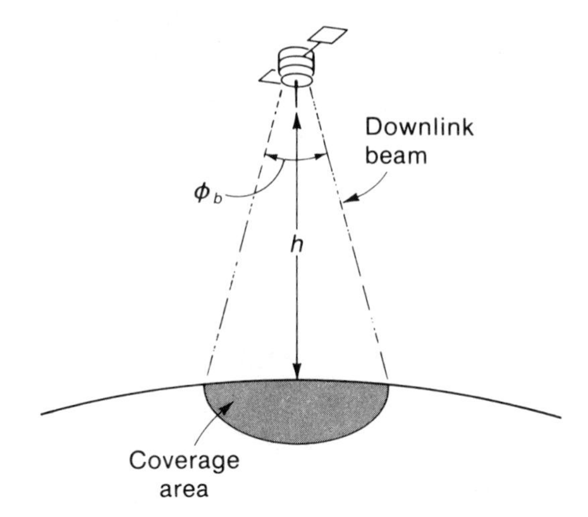

a.

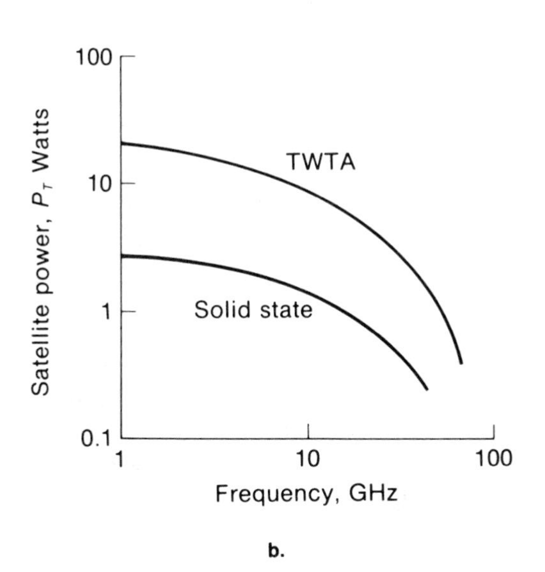

b.

FIGURE 3.19 Satellite downlink: (a) model; (b) spacecraft power sources.

Downlink C-band link budget example		
Frequency, 4 GHz		
Satellite power, 158 W	22 dBW	
Satellite antenna 1.5 ft gain (55% eff.)	23 dB	
Satellite EIRP		45 dBW
Path loss L_p		−196 dB
Atmospheric rain loss		−4 dB
Polarization loss		−1 dB
Receive antenna, 10 ft gain (55% eff.)	40 dB	
Noise figure, 4 dB		
Noise temperature, 435°K	26 dB	
Receiver $g/T°$		14 dB
Boltzmann constant		228.6 dB
Bandwidth, 10 MHz		−70 dB
CNR		16.6 dB

a.

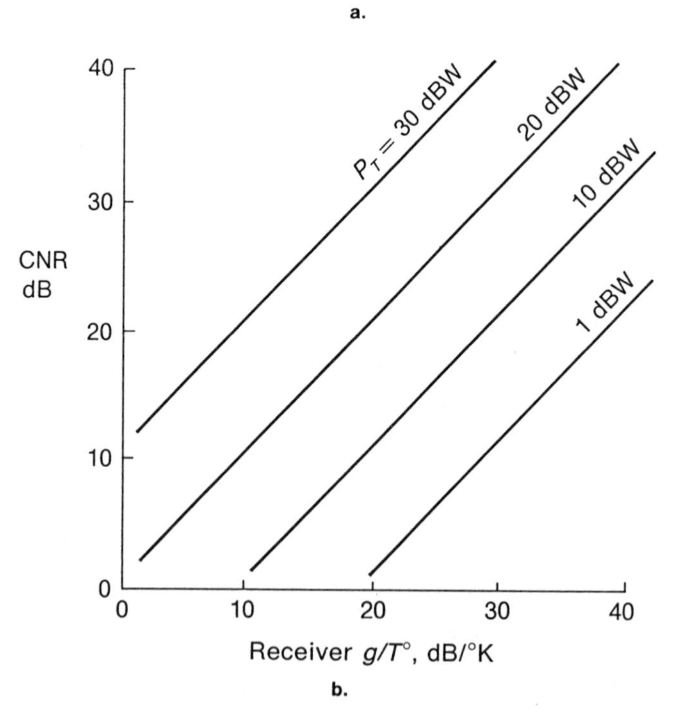

b.

FIGURE 3.20 (a) Downlink C-band link budget. (b) Downlink CNR versus receiver $g/T°$. P_T = satellite power, global antenna, bandwidth = 10 MHz, link parameters are given in Figure 3.20a.

Although use of higher carrier frequencies allows smaller satellite antennas, care must again be used in accounting for its effect in downlink analysis, as discussed in Section 3.5. It will produce higher earth-station g/T values, but it will not increase CNR owing to the increased downlink space loss. To emphasize this point, let us rewrite the CNR formula in Eq. (3.5.8), with satellite beamwidth $\phi_b^2 \cong (4\pi/g_t)$ inserted, as

$$\mathrm{CNR}_d = \frac{P_T A_r L_a}{\phi_b^2 z^2 k T^\circ_{\mathrm{eq}} B_{\mathrm{RF}}} \tag{3.6.1}$$

With the terms in the denominator fixed by the link, we see that downlink CNR depends only on available satellite power P_T and on receiver collecting area A_r. Note that neither satellite EIRP nor receiver gain directly affects downlink quality. The choice of frequency band is, of course, important in determining available P_T (Figure 3.19b), and in determining atmospheric losses. A secondary consideration in frequency-band selection is the possible advantage that may be attained by allowing wider RF bandwidths.

Repeater Link Analysis

In transponding satellites, the primary function of the spacecraft is to relay the uplink carrier into a downlink. The circuit details of such relay satellites will be considered in Chapter 4. In this section we analyze a transponder channel as a combined uplink–downlink, where we model the satellite simply as an ideal linear power amplifier. We neglect the frequency translation between uplink and downlink, and simply convert the former to the latter through an amplifier with power gain G (Figure 3.21). This represents the most basic, idealized, repeater link that can be constructed. The uplink power is composed of a signal term from the uplink earth station, P_{us}, and the noise power collected at the satellite front end, P_{un}. The downlink power P_T is composed of an amplified signal and noise power term

$$P_T = GP_{us} + GP_{un} \tag{3.6.2}$$

Let L represent the combined total power gain (or loss) in the downlink, including antenna gains and channel losses. From Eq. (3.1.5)

$$L = g_t L_a L_p g_r \tag{3.6.3}$$

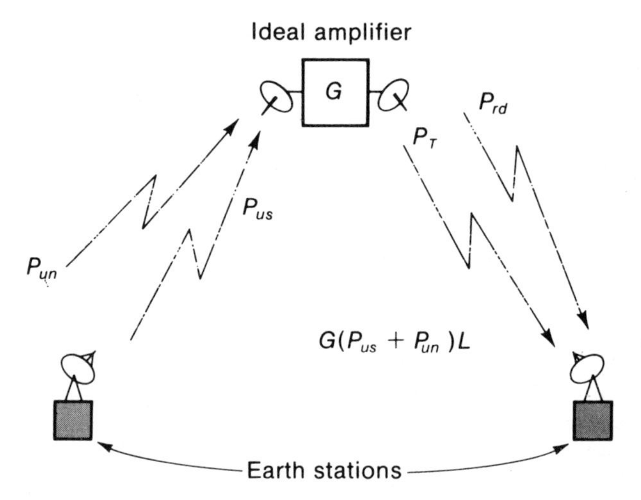

FIGURE 3.21 Combined up–down link repeater. (L = combined downlink losses.)

The received downlink carrier power is then

$$P_{rs} = GP_{us}L \tag{3.6.4}$$

The retransmitted uplink noise appearing at the downlink receiver is

$$P_{rn} = GP_{un}L \tag{3.6.5}$$

In addition, a noise power $P_{rd} = kT^{\circ}_{d}B_{d}$ appears at the downlink receiver due to its noise temperature T°_{d} and bandwidth B_{d}. Hence, the combined CNR at the downlink receiver is

$$\begin{aligned}\mathrm{CNR}_d &= \frac{P_{rs}}{P_{rn} + P_{rd}} \\ &= \frac{P_{us}}{P_{un} + (P_{rd}/GL)}\end{aligned} \tag{3.6.6}$$

Dividing by P_{un}, we have

$$\begin{aligned}\mathrm{CNR}_d &= \frac{P_{us}/P_{us}}{1 + (P_{rd}/P_{rn})} \\ &= \frac{(\mathrm{CNR}_u)(\mathrm{CNR}_r)}{\mathrm{CNR}_u + \mathrm{CNR}_r}\end{aligned} \tag{3.6.7}$$

where we have denoted

$$\text{CNR}_u \triangleq \frac{P_{us}}{P_{un}} \tag{3.6.8}$$

as the uplink CNR at the satellite, and

$$\text{CNR}_r \triangleq \frac{P_{us}GL}{P_{rd}} \tag{3.6.9}$$

as the downlink receiver CNR. This last CNR equation is based on satellite-transmitted carrier power and receiver noise only, that is, as if there were no uplink noise. Thus, even with a relatively simple and ideal satellite repeater model, we establish a basic property of repeater systems. The downlink CNR depends on *both* the uplink CNR and the receiver CNR, and can never exceed either one. Thus, the weakest of the uplink and downlink channels will determine the performance level of a repeater system. Even with perfect repeater amplifiers, design of the uplink, as well as the satellite downlink, must be taken into account. In Chapter 4 we shall see that more practical, nonideal satellite models produce similar conclusions, but with additional forms of degradations occurring. We point out that by inverting CNR_d we can rewrite Eq. (3.6.7) as

$$(\text{CNR}_d)^{-1} = (\text{CNR}_u)^{-1} + (\text{CNR}_r)^{-1} \tag{3.6.10}$$

This is sometimes more convenient to use in computing CNR_d in repeater analyses.

For a transponded digital link, the downlink CNR_d can be converted to E_b/N_0 to determine bit-error probability for the linear amplifier satellite. This requires replacing B_d by $1/T_b$, and writing

$$\left(\frac{E_b}{N_0}\right)_d = \frac{(E_b/N_0)_u(E_b/N_0)_r}{(E_b/N_0)_u + (E_b/N_0)_r} \tag{3.6.11}$$

where $(E_b/N_0)_u$ and $(E_b/N_0)_r$, are obtained from Eqs. (3.6.8) and (3.6.9) by taking the noise bandwidth as $1/T_b$ Hz.

The resulting bit-error probability, PE, for a phase coherent BPSK link in Table 2.3 is then

$$\text{PE} = Q\left[\left(\frac{2E_b}{N_0}\right)_d^{1/2}\right] \tag{3.6.12}$$

where $Q(x)$ is the Gaussian tail integration in Table 2.3. Note again that digital performance depends on both uplink and downlink CNR. Figure 3.22 plots PE in Eq. (3.6.12) for both these parameters, showing how the weaker of the two links determines the overall PE performance.

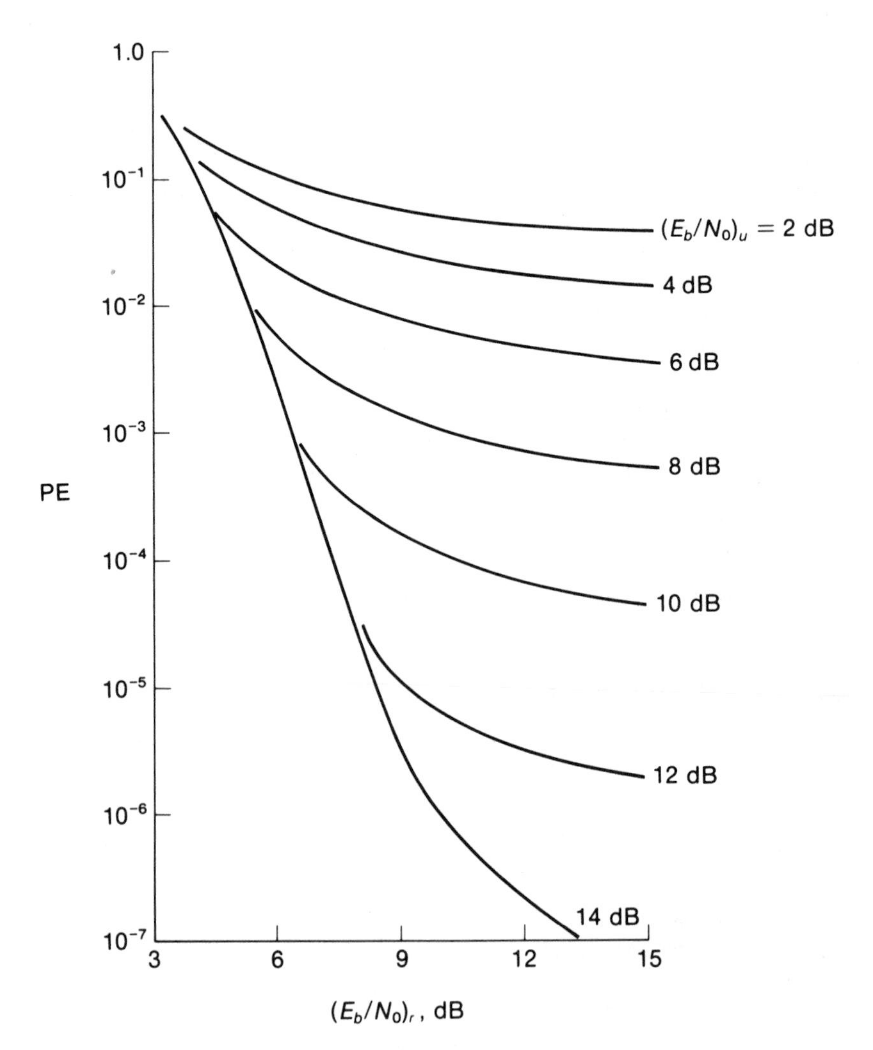

FIGURE 3.22 Bit-error probability, PE, for a BPSK up-down link. $(E_b/N_0)_u$ refers to uplink, $(E_b/N_0)_r$ to downlink.

Satellite Crosslinks

Satellite systems often require communications between two satellites via a crosslink. A crosslink can be established between synchronous satellites, low-earth-orbiting satellites, or deep-space satellites. A crosslink between two orbiting satellites is referred to as an *intersatellite link* (ISL). As a communication link, an intersatellite link has the disadvantage that both transmitter and receiver are spaceborne, limiting operation to both low P_T and low g/T° values. To compensate in long links, it is necessary to increase EIRP by resorting to narrow transmit beams for higher-power concentration. With satellite antenna size constrained, the narrow beams are usually achieved by resorting to higher carrier frequencies. Hence, satellite crosslinks are typically designed for K-band (20–30 GHz) or EHF (60 GHz) frequencies. The emerging possibilities of optical (laser) crosslinks are considered in Chapter 10.

Consider the crosslink model in Figure 3.23. Two satellites at altitude h are separated by angle ϕ_s as shown. The transmitting satellite has transmission power P_T available, and we assume both satellites use

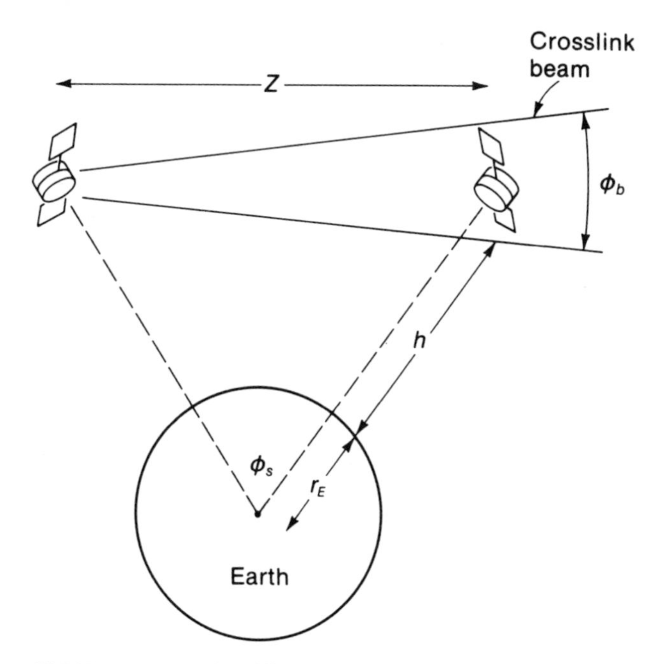

FIGURE 3.23 Satellite crosslink model.

antennas of diameter d. The receiving satellite has noise temperature T°_{eq}. The propagation distance between the satellites is given by

$$z = 2(h + r_E)\sin(\phi_s/2) \tag{3.6.13}$$

where r_E is the Earth's radius. The maximum line-of-sight distance occurs when

$$z_{max} = 2[(h + r_E)^2 - r_E^2]^{1/2} \tag{3.6.14}$$

which, for $h \gg r_E$, is approximately 2 h. We assume first that both satellite locations are known exactly by each, and each is perfectly stabilized, so that each satellite uses antenna beamwidths ϕ_b pointed exactly at each other. The CNR delivered to the receiving satellite over the crosslink, using Eq. (3.6.1) with $L_a = 1$, is then

$$\text{CNR} = \frac{P_T(\pi d^2/4)}{\phi_b^2 z^2 k T^\circ_{eq} B_{RF}} \tag{3.6.15}$$

The crosslink transmitting beamwidth ϕ_b is related to the carrier frequency through Eq. (3.2.2). Assuming $\rho_a = 1$, we substitute $\phi_b \cong \lambda/d$, and Eq. (3.6.15) can be simplified to

$$\text{CNR} = 8.7 P_T\left[\frac{f^2 d^4}{z^2 T^\circ_{eq} B_{RF}}\right] \tag{3.6.16}$$

where P_T is in watts, f is in GHz, and d and z are in meters. We see that CNR increases as the fourth power of the antenna diameters. Hence, the most efficient way to improve a crosslink is to increase antenna size. With the antenna selected, CNR can be increased only by lowering front end temperature, increasing transmitter power, or operating at the higher EHF crosslink frequencies.

The increase in CNR in Eq. (3.6.16) with antenna size and frequency is a direct result of a more concentrated beamwidth. The limitation of extremely narrow crosslink beams, however, is the pointing error that exists due to the relative uncertainty of the location of each satellite with respect to each other and to the satellite attitude error (the inability of a satellite to properly orient itself so as to point exactly in a desired direction). These errors are invariably much larger than those encountered in earth-based links, where ground control of tracking and pointing is feasible. If r is the relative location uncertainty distance when observed from each satellite, and if the attitude error is ϕ_a, the total uncertainty angle is

$$\phi_e = \left(\frac{r}{z} + \phi_a\right) \text{rad} \tag{3.6.17}$$

The transmitted beamwidth must be wide enough to encompass these pointing errors. This shows that a key element in crosslink systems is the ability to point between satellites. It is evident that a trade-off exists between reducing the crosslink beam (more concentrated power) and improving the pointing accuracy. A detailed discussion of the overall crosslink design and the associated trade-offs of these parameters is considered in Chapter 10.

Deep-Space Links

A special type of satellite link is the deep-space link in which image and sensor data from unmanned spacecraft at planetary distances are transmitted to Earth. While the satellite links considered previously involved transmissions over distances on the order of synchronous altitudes, the deep-space link involves distances on the order of astronomical units (1 AU = 1.5×10^8 km). The combination of the extremely long propagation paths, along with constraints on spacecraft size, weight, and power, severely limits the available receiver power that can be collected at an earth station from an interplanetary probe. To compensate, a network of extremely large receiving dishes ($\cong$60 m) are located worldwide to provide sensitive earth-station reception with minimal ground-based interference. In addition, significant levels of coding and error correction are inserted to improve the link performance. Planetary probes have successfully been completed since the 1960s. Data rates from deep space have increased from a few bits per second with the early Moon and Venus probes, to hundreds of kilobits per second for the Jupiter probes of the 1980s.

The frequency bands allocated specifically for deep-space research and communications were listed in Table 1.5. Because of the weak signal strengths of the return link, and the high-power transmissions needed for the forward command and ranging links to the vehicle, these bands are carefully monitored and reserved to ensure complete restriction to deep-space operations. These particular bands were selected for optimal communication performance, minimal propagation effects, available efficient hardware, and compatibility with existing terrestrial systems. The deep-space bands are subdivided into individual channels that are assigned specifically for each space mission. Primary use is made of the 2-GHz and

8-GHz bands, with future applications planned for the K-band frequencies.

Space vehicles invariably use turnaround ranging systems continually to monitor location and velocity throughout the mission. This requires that the forward carrier frequency must be translated by a fixed, precise factor at the spacecraft to generate the coherent return carrier. Hence, the forward and return frequencies selected for any deep-space mission must be an exact ratio. For example, the return carrier in the 2-GHz band is always selected to be exactly 240/221 times the forward carrier.

Modulation on the transmitted carriers involves the combination of the ranging waveform (see Appendix C) and the data or command subcarriers, which are then phase-modulated on the carrier. In the forward link the ranging waveform and command subcarriers are separated after phase demodulation at the vehicle. The recovered ranging signal is then instantaneously returned by adding to the data subcarrier and remodulating onto the return carrier. At the ground-tracking station the returned ranging waveform is processed for two-way ranging and Doppler estimation, whereas the data subcarrier is separated and decoded to recover the spacecraft probe, sensor, or video information.

Table 3.5 summarizes a typical deep-space return link budget for a Jupiter mission, transmitting 130 kbps from a 10-W power source to a

TABLE 3.5 Deep-space return link budget (Jupiter mission)

Frequency, 8.42 GHz	
Vehicle amplifier power, 10 W	40 dBm
Vehicle antenna, 5 m	
beamwidth 0.48°	
gain (55% eff)	51 dB
Propagation loss, 6.37 AU ($=9.5 \times 10^8$ km)	−290.5 dB
Pointing, atmospheric loss	−1.2 dB
Receive antenna (64 m), gain	71.7 dB
$N_0(=kT^\circ_{eq})$, $T^\circ_{eq} = 26°$	−184.4 dBm/Hz
Received C/N_0	55.41 dB-Hz
Receiver implementation loss	−1.55 dB
Data rate, 130 kbps	−51 dB
E_b/N_0	2.76 dB
Required E_b/N_0 (with coding, for PE $= 10^{-3}$)	2.31 dB
Margin, with single receiver	0.45 dB

200-ft earth-based antenna. Excessive coding is inserted to aid the link performance, but the resulting link margin is still somewhat low, requiring accurate and precise control of link losses to complete the telemetry link. Often, diversity reception using carrier reinforcement from other ground stations can be inserted to help improve the link margin.

3.7 DUAL POLARIZATION

The satellite links considered in Section 3.6 were assumed to transmit a fixed carrier bandwidth as an electromagnetic field having the prescribed power level with either a linear or a circular field polarization, as described in Figure 3.3. It is possible, however, to transmit two separate fields simultaneously over the same link, each utilizing the same carrier frequency and bandwidth but with different field polarizations. If the two polarizations are spatially orthogonal, then the two fields can be separated completely at the receiver, with no interference from each other, even though they occupy the same bandwidth. This format is called *dual polarization.*

In dual polarization two modulated carriers are transmitted by placing each on orthogonal polarizations generated at the transmitter. The two fields can be generated from separate antennas or from a single antenna with separate feeds. As long as the polarizations are maintained as truly orthogonal during propagation and are individually received by antenna subsystems aligned with each polarization, the carrier bands can be uniquely separated with no interference from each other. Hence, the same bandwidth can be transmitted with two different data modulations at the same time, using two different polarizations of the same field. Therefore, this doubles the information sent from the transmitter.

This concept of using the same frequency band to transmit separate carriers is referred to as *frequency reuse*. Frequency reuse increases the information capacity of the link without increasing link bandwidth. Dual polarization achieves frequency reuse by using the spatial orthogonality of the field alignments during propagation.

In a linearly polarized system, dual polarization is obtained by using two orthogonal planar directions (e.g., one horizontal and one vertical). By properly aligning the receiving antenna or by tapping off the desired orientation in a common antenna waveguide, the fields can be separated adequately. Dual polarization can also be obtained by using two circular polarizations (CP) of the same field, with each phased to rotate in opposite directions (called *right-hand* CP and *left-hand* CP). Separation at the receiver is achieved by controlling spatial phase shifters inserted into the antenna waveguide. The phase shifter is designed so that one field is rotated in the guide in such a way that its field components cancel, whereas the other field components add, similar to that shown in Figure 3.3.

A problem with dual-polarized satellite downlinks is that the atmosphere, in addition to producing attenuation effects, may also affect the orthogonality of the two polarizations. This would cause an inherent cross-coupling of one channel onto the other at the same frequency band, producing interference and crosstalk in that channel, even with perfectly aligned receiving systems. This latter effect is referred to as *depolarization*, and is strongly dependent on the nature of the atmosphere and on the field wavelength (Arnold et al., 1981; Cox and Arnold, 1982; Gale and Mon, 1982; Ippolito, 1986; Morrison et al., 1973; Oguchi, 1973; Semplak, 1974). Atmospheric depolarization is basically negligible below about 10 GHz, and dual polarization is therefore commonly used in C-band satellite systems. However, depolarization significantly increases at higher carrier frequencies, especially when a high water content is present in the atmosphere. This makes the feasibility of using dual polarization for frequency reuse in the 10–30-GHz range somewhat questionable.

Depolarization is caused primarily by nonspherical water droplets that affect the orthogonal polarizations of the impinging field by different amounts (Figure 3.24). The result is differential attenuation and differential phase shifts between the orthogonal field components in space, thereby

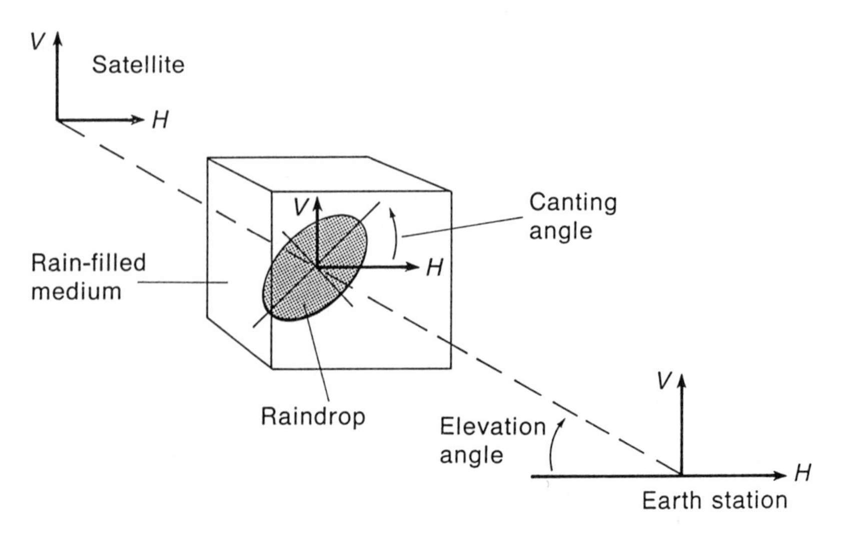

FIGURE 3.24 Rain depolarization vector model.

altering the spatial alignments of these components. The amount of these differential effects depends on the amount of water and ice in the atmosphere, and therefore becomes most severe during rain and snow. In particular, depolarization depends on rainfall rate and extent (path length), raindrop size, raindrop orientation (rain *canting* angle), and carrier wavelength. Since falling rain tends to have a fixed average canting angle, depolarization will depend on the field polarization orientation relative to this angle, and therefore will be more severe for some orientations than for others. A linearly polarized field has minimum depolarization if it is horizontal or vertical relative to the canting angle. A CP field has a depolarization corresponding to the worst case angle of a linearly polarized field.

A formal analytical approach to depolarization can be developed by treating the dual polarized fields as a two-dimensional vector waveform, and representing the rainfall effect in Figure 3.24 as an effective matrix transformation on that vector. Let the dual polarized waveforms be denoted by the complex fields $c_1(t)$ and $c_2(t)$, and let the atmospheric matrix be T. The received complex fields in the same polarizations then become

$$\begin{bmatrix} r_1(t) \\ r_2(t) \end{bmatrix} = T \begin{bmatrix} c_1(t) \\ c_2(t) \end{bmatrix} \tag{3.7.1}$$

The received fields are then the real parts of $r_1(t)$ and $r_2(t)$. The depolarization matrix T has been shown (Ippolito, 1986) to be of the form

$$T = \begin{bmatrix} a_{11}e^{j\psi_{11}} & a_{12}e^{j\psi_{12}} \\ a_{21}e^{j\psi_{21}} & a_{22}e^{j\psi_{22}} \end{bmatrix} \tag{3.7.2}$$

where $\{a_{ij}\}$ represent amplitude attenuations, and $\{\psi_{ij}\}$ are carrier phase shifts (field delays). The a_{ij} and ψ_{ij} for $i = j$ are called *coplanar* parameters, whereas those for $i \neq j$ are *cross-planar* parameters. The coplanar attenuation is the same atmospheric losses that would occur on either polarization alone, as described in Section 3.3. For most atmospheric conditions $a_{12} \approx a_{21}$ and $\psi_{12} \approx \psi_{21}$, so that the cross-planar effects can be assumed symmetrical. Equations (3.7.1) and (3.7.2) account for the fact that the dual polarized fields are each individually attenuated and phase shifted, and a depolarizing complex cross-coupling can occur between the polarizations owing to the atmosphere. The values of the a_{ij} and ψ_{ij} parameters in Eq. (3.7.2) depend on the inherent characteristics of the atmosphere and, as stated earlier, are primarily dependent on the rainfall content of the propagation path.

To account for the depolarization matrix on the communication waveforms, the field components can be written as the general angle modulated carriers.

$$c_1(t) = Ae^{j[\omega_c t + \theta_1(t)]}$$
$$c_2(t) = Ae^{j[\omega_c t + \theta_2(t)]} \tag{3.7.3}$$

After the atmospheric propagation in Figure 3.24, the received fields in the receiver polarizations in Eq. (3.7.1) become

$$\text{Real}\, r_1(t) = a_{11}A\cos[\omega_c t + \theta_1(t) - \psi_{11}] + a_{12}A\cos[\omega_c t + \theta_2(t) - \psi_{12}]$$
$$\text{Real}\, r_2(t) = a_{21}A\cos[\omega_c t + \theta_1(t) - \psi_{21}] + a_{22}A\cos[\omega_c t + \theta_2(t) - \psi_{22}] \tag{3.7.4}$$

An antenna oriented to receive one particular polarization will therefore receive the corresponding waveform above. Note that each such waveform contains both the coplanar and the cross-planar components. It is convenient to normalize Eq. (3.7.4) by rewriting it as

$$\text{Real}\, r_1(t) = Aa_{11}\{\cos(\omega_c t + \theta_1 - \psi_{11}) + (a_{12}/a_{11})\cos(\omega_c t + \theta_2 - \psi_{12})\}$$
$$\text{Real}\, r_2(t) = Aa_{22}\{\cos(\omega_c t + \theta_2 - \psi_{22}) + (a_{21}/a_{22})\cos(\omega_c t + \theta_1 - \psi_{21})\} \tag{3.7.5}$$

The *cross-polarization discriminants* are defined as

$$\text{XPD}_1 = (a_{12}/a_{11})^2$$
$$\text{XPD}_2 = (a_{21}/a_{22})^2 \tag{3.7.6}$$

and are generally stated in decibels. The XPD for either polarization indicates the fraction of the field power cross-coupled during dual polarization. The XPD is therefore a number less than 1, and will be expressed in negative decibels. The more negative the decibel value, the less is the cross-coupling. The reciprocal of XPD is called the *cross-polarization isolation* (XPI), and is a measure of the isolation between the two polarizations at the receiver. The *XPI* has the decibel value of XPD (without the negative sign) and, therefore, the higher the isolation, the less the cross-coupled power.

Over the years much effort has been devoted to attempting to characterize the XPD parameters of the atmosphere from both analytical

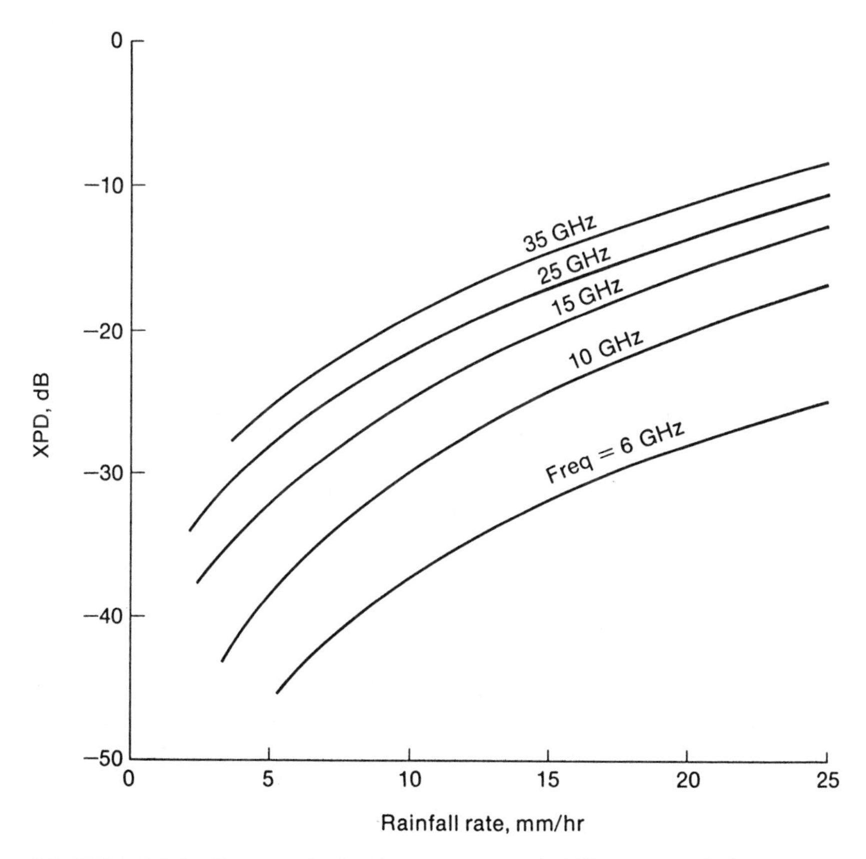

FIGURE 3.25 Cross-polarization versus rainfall rate and frequency. Worst case polarization, 5 km rain path.

models and curve-fitting with measured data. Mathematical models were developed by assuming propagation through rain-filled media of given dimensions and specified raindrop shape and canting angle (Morrison et al., 1973; Oguchi, 1973). This model has been improved over the years to include the effects of randomness in canting angle and raindrop shape. From these models empirical formulas have been generated for describing the XPD as a function of rain rate, rain path length, elevation angles, and canting angle statistics. Figure 3.25 shows a typical plot of average XPD as a function of rainfall rate and carrier frequency. In general, XPD increases (more cross-coupling) as rainfall rate and frequency increase, but it must be remembered that XPD is a random parameter that in fact changes over time.

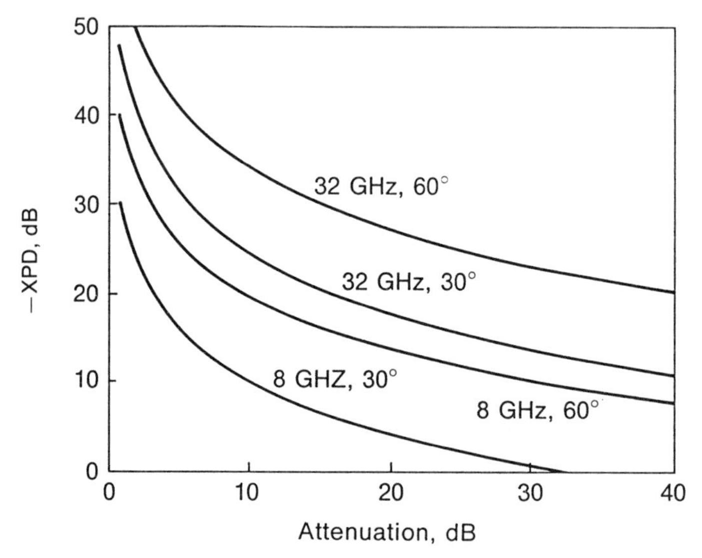

FIGURE 3.26 Cross-polarization discriminant versus path attenuation at various frequencies and elevation angles.

One result that has been noted is a direct correlation between XPD and coplanar attenuation. The atmospheric conditions at which the attenuation is most severe are also those for which the XPD is the greatest. This means that XPD can often be predicted from the amount of signal loss that occurs. Figure 3.26 shows the result of an accepted CCIR formula (Ippolito, 1986) relating XPD and attenuation loss during depolarization. The curves are dependent on frequency and elevation angle, and follow the general formula

$$-\text{XPD} = U - V(\log A) \tag{3.7.7}$$

where U and V are parameters estimated from measured data, and depend on the atmospheric state, frequency, and elevation angles. Since C-band systems have a smaller probability of having a large attenuation, the XPD is therefore highly likely to be low. This again shows why dual polarization links are less affected at these frequency bands.

3.8 EFFECT OF DEPOLARIZATION ON DUAL POLARIZED COMMUNICATIONS

In dual polarized communications, atmospheric depolarization causes a cross-coupled interference onto the coplanar waveforms, as evident in Eq.

(3.7.5). For analog communications, such as an FM carrier, the interference appears as added in-band noise, and the simplest procedure to determine the depolarization effect is to compute the receiver CNR. The CNR can be computed by extending the results of the previous sections. From Eq. (3.5.1) the CNR and depolarization becomes

$$\mathrm{CNR_{dp}} = \frac{P_r}{P_n + P_r(\mathrm{XPD})} \tag{3.8.1}$$

where P_r is received power per polarization mode. Defining $\mathrm{CNR} = P_r/P_n$ as the CNR without depolarization,

$$\mathrm{CNR_{dp}} = \mathrm{CNR}\left[\frac{1}{1 + \mathrm{CNR(XPD)}}\right] \tag{3.8.2}$$

The bracket appears as a suppression effect on CNR, dependent on the value of XPD. Figure 3.27 plots this XPD suppression as a function of rainfall rate and frequency, using the data in Figure 3.25. This suppression represents degradation due purely to cross-coupling and is independent of the attenuation of the same channel. Whereas the latter can be overcome by increasing transmitter power, XPD represents a residual suppression present in any dual polarization system. Since rainfall rates of a given amount occur only a fixed percentage of the time, Figure 3.27 can be relabeled in terms of fractional time or occurrence probability. It can then also be interpreted as a depolarization distribution curve, similar to the attenuation effect in Figure 3.11. It should be pointed out that attempts to reduce XPD by using receiver circuitry to restore the field orthogonality have been reported (Chu, 1971/1972; DiFonza et al., 1976).

In digital dual-polarized links the effect of cross-polarization must be assessed in terms of decoded bit-error probability in either polarization. The latter depends on the actual waveshape that is cross-coupled and not simply its power level. Consider a dual polarized system using BPSK modulation in each polarization. We assume two independent bit streams with equal bit rates and synchronized bits in each coordinate. A polarized receiver therefore receives the waveform

$$\text{Real } r(t) = Aa_{11}d_1(t)\cos(\omega_c t - \psi_{11}) + Aa_{12}d_2(t)\cos(\omega_c t - \psi_{12}) \tag{3.8.3}$$

in additive white noise, where $d_1(t)$ and $d_2(t)$ denote the respective bit modulations. A coherent decoder for the $d_1(t)$ data will therefore phase reference to ψ_{11}, and decode with an integrate-and-dump decoder over

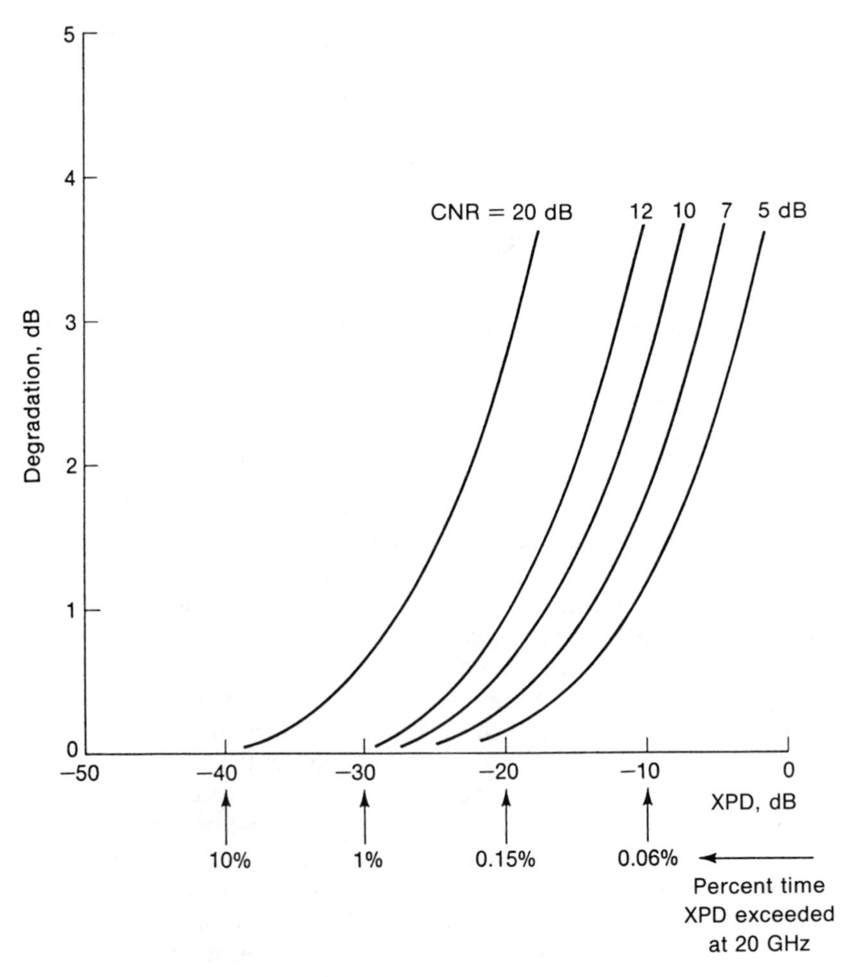

FIGURE 3.27 Depolarization degradation in CNR. Percent time based on rain rate occurrence and XPD at 20 GHz.

the bit time T_b. Since the cross-coupled bit, d_2, may be equal or opposite to d_1, the resulting PE follows as

$$\text{PE} = \tfrac{1}{2}Q\left[\sqrt{\frac{2E_b}{N_0}}\cos\psi_e + \sqrt{\frac{2E_b(\text{XPD})}{N_0}}\sin(\psi_e + \delta)\right] + \tfrac{1}{2}Q\left[\sqrt{\frac{2E_b}{N_0}}\cos\psi_e - \sqrt{\frac{2E_b(\text{XPD})}{N_0}}\sin(\psi_e + \delta)\right] \quad (3.8.4)$$

where ψ_e is the phase reference error of the coplanar bit, $\delta = (\psi_{11} - \psi_{12})$ is the difference phase between the coplanar and cross-planar carrier, and $E_b/N_0 = A^2 a_{11}^2 T_b/2N_0$ is the attenuated bit energy to noise level ratio. We see that PE depends on both the XPD and the differential phase shift δ due to depolarization. If δ is assumed to be uniformly distributed over $(0, 2\pi)$, then the cross-planar term is independent of the phase reference error ψ_e. Figure 3.28 shows a plot of PE in Eq. (3.8.4) for a perfectly referenced ($\psi_e = 0$) BPSK dual polarized system with a uniform δ, as a function of E_b/N_0 and the XPD coefficient. Note that even with perfect referencing a significant PE degradation occurs if the cross-polarization becomes excessive.

A similar result can be derived for a dual polarized QPSK system. In this case

$$\mathrm{PE} = \tfrac{1}{8} Q\left[\sqrt{\frac{2E_b}{N_0}} [\cos \psi_e \pm \sin \psi_e \pm \sqrt{\mathrm{XPD}} \cos(\psi_e + \delta) \pm \sqrt{\mathrm{XPD}} \sin(\psi_e + \delta)] \right] \qquad (3.8.5)$$

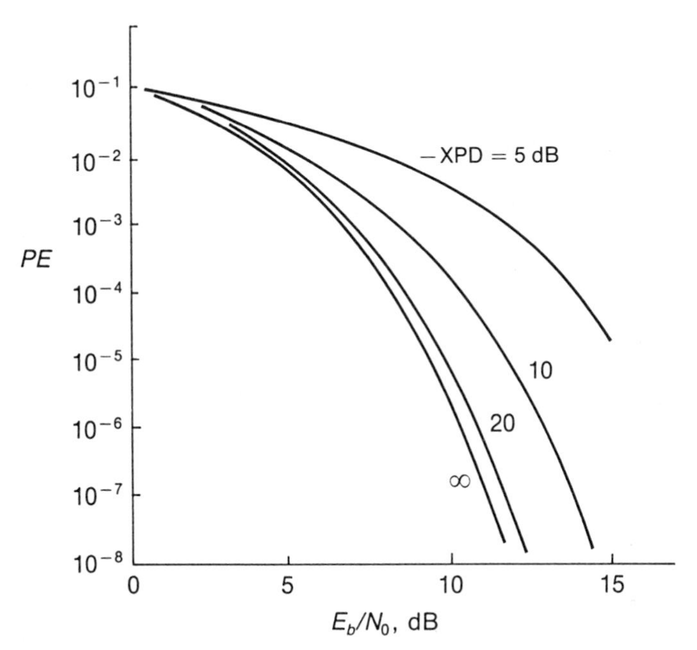

FIGURE 3.28 BPSK bit-error probability with cross-polarization XPD. Uniformly distributed cross-polar phase angle.

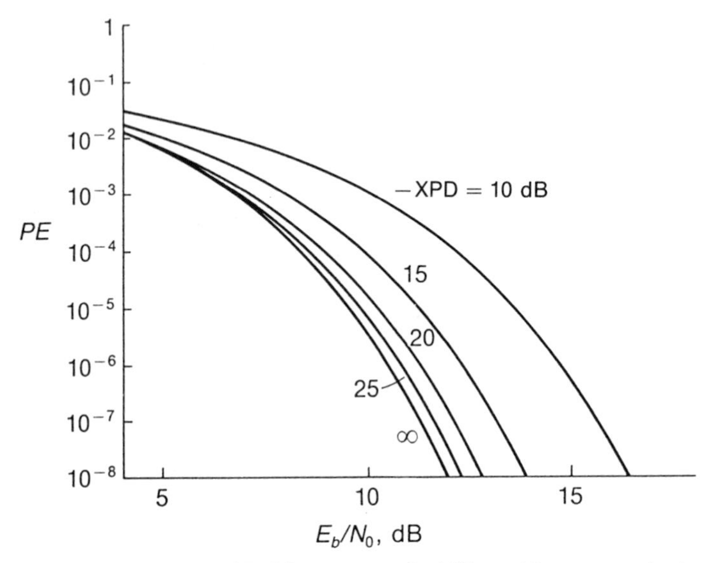

FIGURE 3.29 QPSK bit-error probability with cross-polarization XPD. Uniform cross-planar phase.

where the $\pm$ indicates either bit sign can occur, and Eq. (3.8.5) must be averaged over all the possible bit signs. Figure 3.29 includes these results for the perfectly referenced QPSK case with uniform differential angle δ, assuming equal likely bits in all coordinates. The degradation is slightly higher for QPSK than for BPSK at a specified level of XPD. It must be remembered that the degradation depicted in Figures 3.28 and 3.29 is due to depolarization only, and the additional loss due to coplanar attenuation must be included. The use of data similar to Figure 3.26 is useful here for relating the expected attenuation loss with the corresponding XPD coefficient. Further studies for evaluating depolarization effects, and possible corrective measures have been reported (Aguirre and Gagliardi, 1987).

3.9 SPOT BEAMS IN SATELLITE DOWNLINKS

In satellite downlinks the field intensity on the ground is limited by the available satellite EIRP. This means that small earth stations with low g/T° values will have difficulty receiving wideband information. One way to improve reception is to increase satellite EIRP by using narrower beams [decreasing ϕ_b in Eq. (3.6.1)] so as to concentrate power into areas

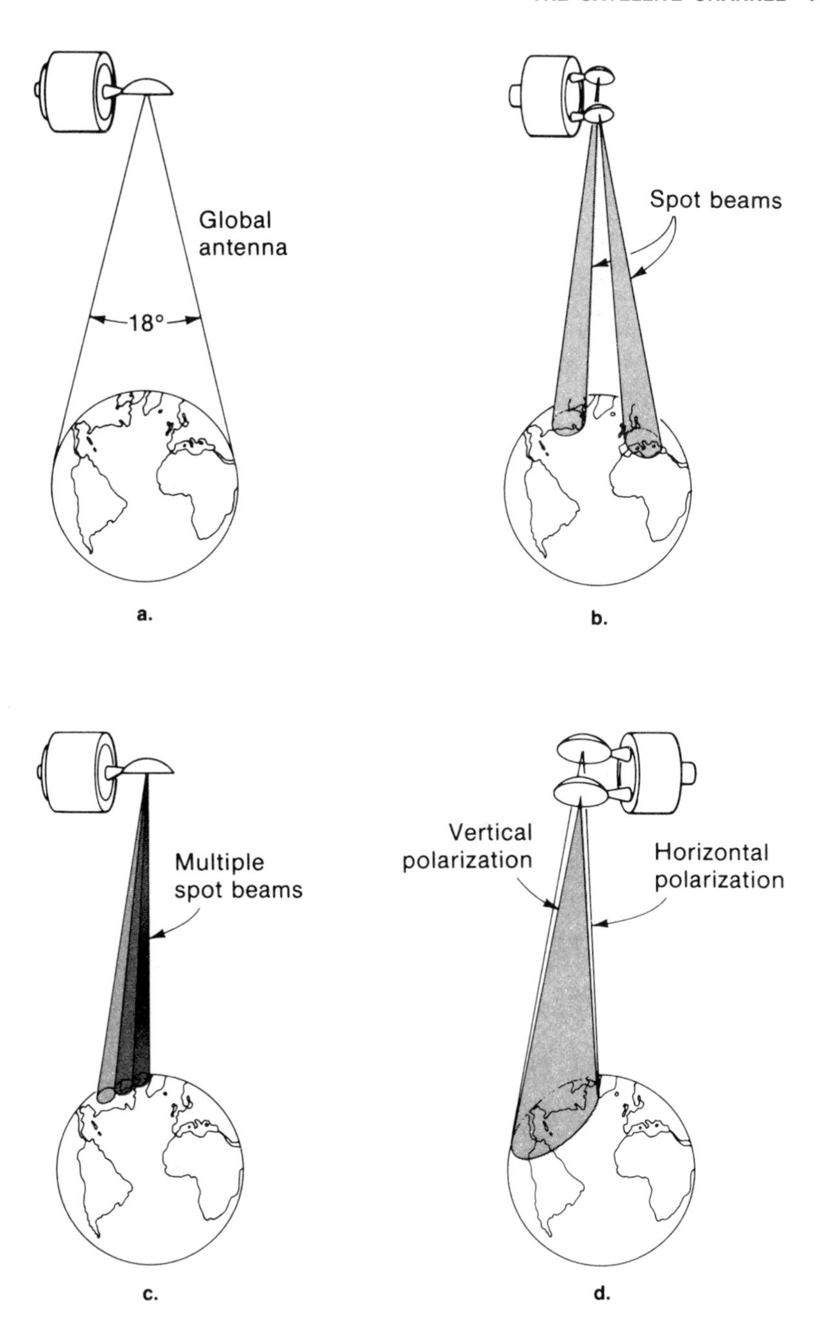

FIGURE 3.30 Spot beaming from satellites: (a) simple global beam; (b) two spot beams; (c) multiple spots; and (d) dual polarized spots.

TABLE 3.6 Spot beam parameters

Spot Beamwidth	*Earth Coverage Diameter (miles)*	*EIRP Increase over Global Beam (dB)*
10°	3921	5.1
5.7°	2235	9.9
2.8°	1117	16.1
1.0°	392	25.0
0.57°	223	29.9

smaller than those covered by global beams. These beams are referred to as *spot beams* (Figures 3.30a and 3.30b). Spot beams increase the CNR within the beam, but reduce the coverage area. To cover the original maximum area, it may be necessary to use multiple spots, the sum of which spans the desired coverage area (Figure 3.30c). Smaller earth stations anywhere in the combined area will receive with the higher EIRP. Often, transmissions within a given spot area can be further separated by the use of dual polarization within a spot beam (Figure 3.30d). That is, some earth stations would use only horizontal polarizations, say, whereas others would use only vertical polarizations.

Table 3.6 lists the beamwidth, earth-coverage diameter of the beam footprint, and the effective EIRP increase relative to a global (18°) beam. Note that the EIRP per beam increases as the spot is reduced, but the number of spots needed to cover a fixed area also increases. The disadvantage is that multiple antenna beams must be provided from the satellite, increasing the complexity and structure of the spacecraft. In addition, beam pointing is now more critical with the narrower beams.

Multiple beams from a satellite can be produced in one of three basic ways (Figure 3.31). The simplest involves the use of separate antennas for each beam, where each is pointed to an appropriate area. This is the simplest feed assembly, with maximum isolation between feeds and little beam interference at the satellite. An uplink transmission to be relayed to a particular earth station must be routed to the proper antenna and transmitted. Even with the same frequency band, the physical separation of the feeds aids the isolation. Multiple beams can also be generated from a single reflector or microwave lens using multiple feeds (Figure 3.31b). The feeds simultaneously illuminate a

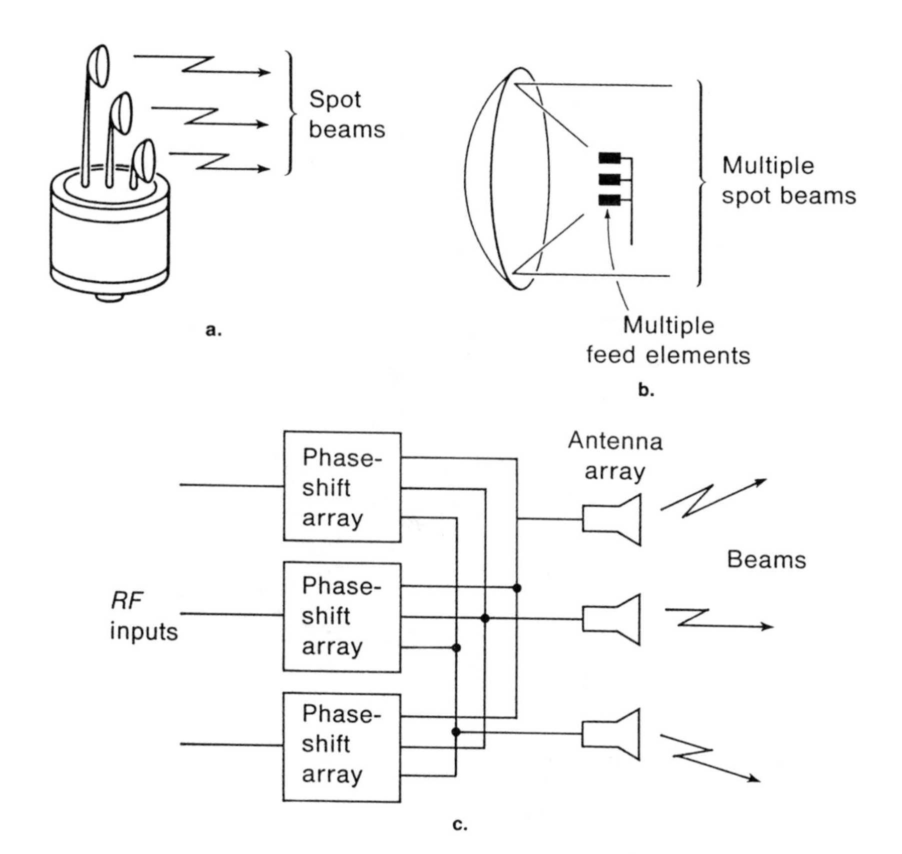

FIGURE 3.31 Multiple beam antennas: (a) separate antennas; (b) multiple feed, single reflector; and (c) arrays.

common parabolic dish, which focuses the entire feed field in a given direction. By offsetting the feeds and positioning each to point to a different section of the dish, the reflected fields can be spatially separated, producing the desired spot beams. Each beam will have the polarization of the corresponding feed. By properly positioning a group of feeds so as to overlap their spots slightly, downlink earth patterns can be formed that match the contours of specific countries or continents.

A third method for producing multiple beams is by phased antenna arrays. By properly phasing a modulated carrier the field can be transmitted in a given direction. Spot patterns can be formed by phasing each carrier of a given band into a single spot, as was shown in Figure 3.8. Each individual carrier can be phased separately to form another spot. A set of M spots is formed by M sets of phase-shifter banks feeding a

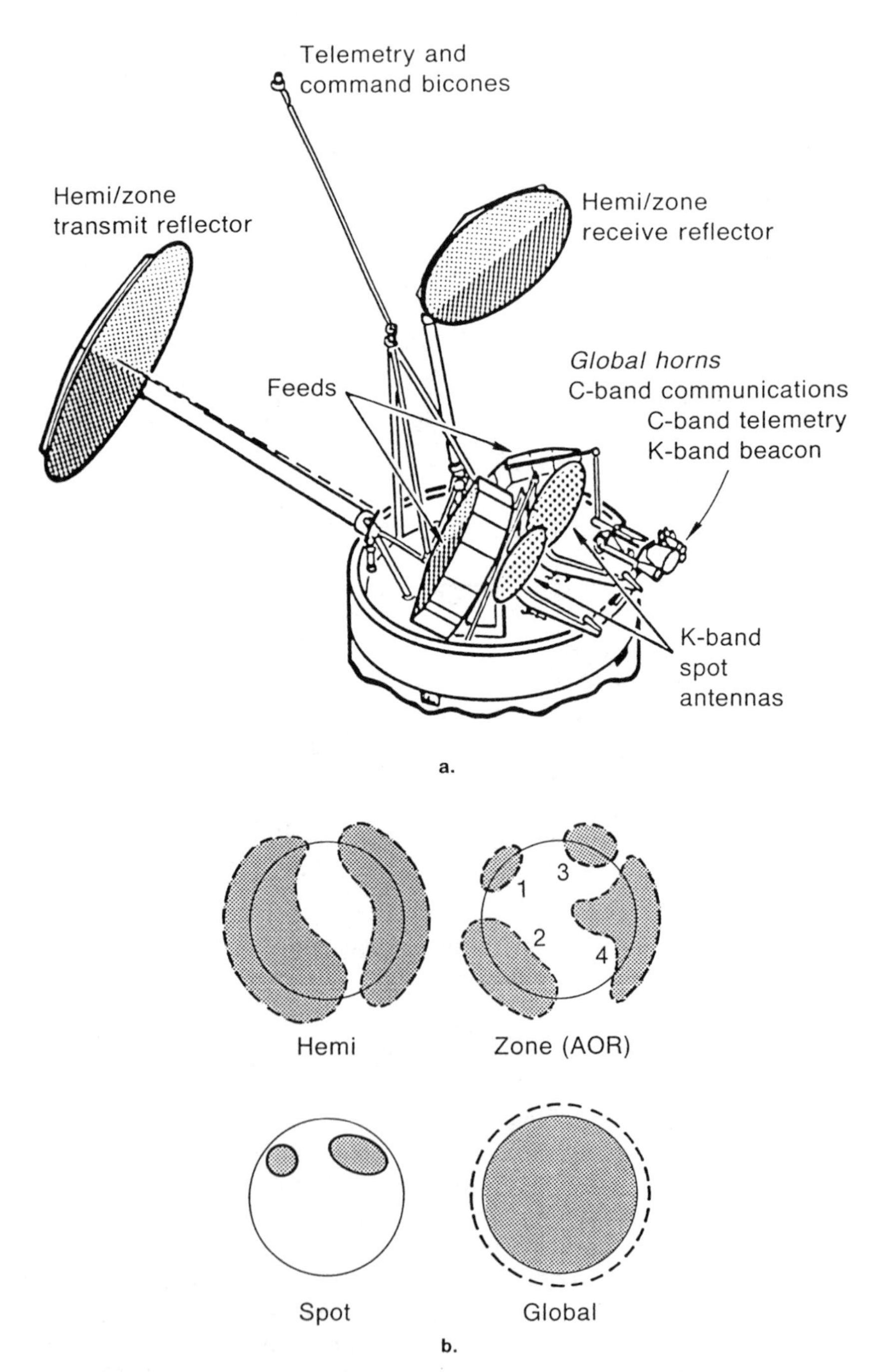

FIGURE 3.32 Intelsat VI antenna: (a) antenna assembly; (b) earth beam footprints (courtesy of Intelsat Corp.).

common antenna (horn, dipole, helix) array, with each bank producing the proper array shift for that bank (Figure 3.31c). Only a single antenna array is used, but the feed (phase-shifters) mechanism becomes complicated. The disadvantage in forming spot beams with arrays is that the size of the array (number of elements) becomes large when it is forming a small spot. Recall from Table 3.1 that beamwidth decreases with the number of elements of the array. This means narrow spot beam arrays require an increase in both the array size and the number of phase-shift feeder elements at the satellite.

Figure 3.32a shows the sophisticated antenna assembly of the Intelsat VI satellite (1989), containing separate large and small reflectors with multiple feeds, and wide angle horns. These operate to produce combinations of two hemispherical beams, four zonal beams, two spot beams, and a wide angle global beam (Figure 3.32b).

A way to achieve spot beaming advantages, while still obtaining wide coverage areas, is by the use of hopping beams. Rather than simultaneously generating multiple spots for the desired coverage, as in Figure 3.30c, a single spot can be moved in sequence (hopped) from one point to another, as sketched in Figure 3.33. The wide coverage is achieved with a single beam, but only on a shared-time basis. Only one spot antenna assembly is needed, but it must have the capability of repositioning sequentially. This can be achieved by gimbal control, in which the antenna assembly is mechanically repointed in each desired direction. A more promising way is via the phase-array system in Figure 3.8. By periodically readjusting the bias current of ferrite phase shifters, the phase shifts of a

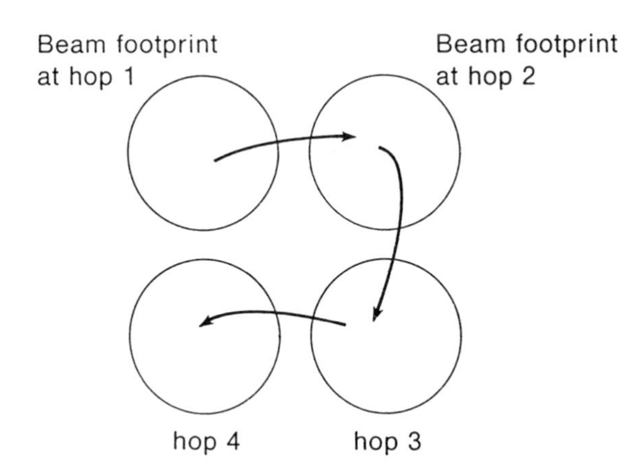

FIGURE 3.33 Hopping beam patterns.

single array feed can be converted on command to repoint the array beam in Figure 3.8. The ability to achieve highly accurate current control via precise digital-to-analog conversion circuits, along with the fairly linear phase–current characteristics of ferrite, has made this approach to beam hopping of considerable interest.

Spot beaming also provides an alternate form of frequency reuse. Suppose different information is to be sent to different points within the coverage area, and each point is within a different spot beam area. A global beam will send the same carrier to all points. However, spot beams can be used to send the separate information simultaneously to each point. One modulated carrier is sent in one beam and another modulated carrier, at the same frequency, is sent in the other. Thus, the total amount of information relayed can be increased. Roughly, if we have M beams, then theoretically M times the amount of information of one beam can be achieved. This corresponds to another form of frequency reuse, since the same frequency band is used in each beam. The operation does, however, require some method of allowing earth-station connectivity, that is, of connecting together stations within different beams. This connectivity can be accomplished through on-board processing directly at the satellite. Methods for accomplishing it depend on the multiple-accessing format, and will be considered in the following chapters.

The main problem with frequency reuse on multiple beams is avoiding interference of one beam into an adjacent beam. This requires careful separation and shaping of each spot beam pattern so as to reduce the beam spillover. One way to further reduce this interference is to use separate frequency bands on adjacent beams. Receiver front-end filtering will aid in reducing adjacent beam spillover, but the total required frequency range now increases. Some degree of frequency reuse is still possible, since only adjacent beams need have band separation. By increasing the number of available bands for a given number of spots, better isolation can be achieved but the information per total bandwidth decreases. This point is considered in more detail in Chapter 9.

When only a few spots are involved, beam separation can also be aided by orthogonal beam polarization. One field polarization is used on one beam, and an orthogonal polarization is used on the adjacent beam. Receivers in each spot receive only transmissions of the correct polarization. This again allows complete frequency reuse within a single band. Depolarization caused by downlink atmospheric effects produces interference of the two fields, and in fact may become the limitation to true polarization separability.

REFERENCES

Aguirre, S., and Gagliardi, R. (1987), "Signal Design For Dual Polarization Digital Space Communications," *USCEE Report*, Contract No. 957645 (December).

Angelakos, D., and Everhart, T. (1968), *Microwave Communications*, McGraw-Hill, New York.

Arnold, D., et al. (1981), "Measurements of Prediction of the Polarization Dependent Properties of Rain and Ice Depolarization," *IEEE Trans. Comm.*, **COM-29** (May), 710–716 and 716–721.

Chu, T. (1971/1972), "Restoring the Orthogonality of Two Polarizations in Radio Communications," Part 1, *BSTJ*, **50** (November); Part II, *BSTJ*, **52** (March).

Cox, D., and Arnold, H. (1982), "Results from 19–28 GHz Comstar Satellite Propagation Experiment at Crawford Hill," *Proceedings of the IEEE*, **70**(5), (May), 458–488.

Crane, R. K. (1977), "Prediction of the Effect of Rain on Satellite Communication Systems," *Proceedings of the IEEE*, **65** (March), 456–474.

Difonza, D., Trachtman, W., and Williams, A. (1976), "Adaptive Polarization Control for Satellite Frequency Re-Use Systems," *Comsat Technical Review*, **6**(2) (Fall).

Gale, P., and Mon, J. (1982), "Effect of Ice Induced Cross-Polarization on Earth-Space Links," *Inter. Comm. Conference (ICC) Proceedings*, Philadelphia, PA, (June), Paper 6H.2.1.

Ippolito, Jr., L. (1986), *Radiowave Propagation in Satellite Communications*, Van Nostrand Reinhold, New York.

Morrison, J., Cross, M., and Chu, T. (1973), "Rain Induced DA and DPS at Microwave Frequencies," *BSTJ*, **52** (4) (April), 579–604.

Oguchi, T. (1973), "Attenuation and Phase Rotation of Radio Waves Due to Rain," *Journal of Radio Science*, **8** (1) (January), pp. 31–38.

Ruze, R. (1966), "Effects of Surface Error Tolerance on Efficiency and Gain of Parabolic Antennas," *Proceedings of the IEEE*, **54** (April).

Semplak, R. (1974), "Measurements of Depolarization by Rain for LP and CP at 18 GHz," *BSTJ*, **53** (2), (March).

PROBLEMS

3.1 A transmitter transmits with power P_T and beamwidth ϕ_b over a distance z. Show that the received field power collected over a receiving area A is proportional to the ratio of the receiving area to the beam front area of the radiated field arriving at the receiver.

3.2 A transmitter radiates 10 W at carrier frequency 1.5 GHz with the following gain pattern:

$$g(\phi_l, \phi_z) = (100)e^{-(\phi_l^2+\phi_z^2)/2(15^\circ)^2}$$

A receiver is located 30° off antenna boresight, a distance 100 km away, and has a 0.1 rad receiving beamwidth aimed at the transmitter. Neglect aperture, coupling, and atmospheric losses, and determine the carrier power at the receiving antenna.

3.3 Two parabolic antennas are pointed as shown in Figure P3.3. If the diameter of each is increased (separately), will the received power increase or decrease? Explain.

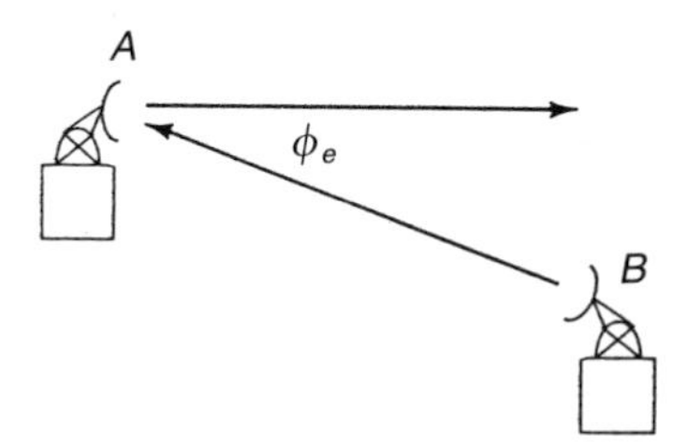

FIGURE P3.3

3.4 The gain function $g(\phi_l, \phi_z)$ of an antenna is also defined by

$$g(\phi_l, \phi_z) = \frac{\text{Power radiated per unit solid angle in direction } (\phi_l, \phi_z)}{P_T/4\pi}$$

where P_T is the total radiated power. (a) Show that g always integrates over a sphere to a constant for any pattern. (b) The noise temperature T_b° of an antenna is defined as

$$T_b^\circ = \frac{1}{4\pi}\int_{\text{sphere}} T(\phi_l, \phi_z)g(\phi_l, \phi_z)d\Omega$$

where $T(\phi_l, \phi_z)$ is the noise temperature of the background in the direction (ϕ_l, ϕ_z). Show that if the background has constant temperature T_s°, then $T_b^\circ = T_s^\circ$.

3.5 An FM system is operated with the following RF parameters:

Transmitter power = 10 kW	Rain attenuation $\doteq$ 0.2 dB/mile
Range 500 miles	Free space loss = 150 dB
Tr. ant. gain = 60 dB	Sky background = 50 K
RF bandwidth = 0.2 MHz	Rec. noise fig. = 3 dB
Basebandwidth = 10 kHz	Rec. ant. gain = 10 dB

What is the achievable baseband SNR_d at the receiver after FM demodulation?

3.6 A communication station on Earth is to transmit to the moon at 5 GHz. (a) Design a transmitting antenna so that the transmission from Earth just covers the moon. Assume an antenna efficiency of 50%. The distance to the moon is 200,000 miles and its diameter is 0.27 that of the Earth's. (b) What is the resulting antenna gain?

3.7 Given Table P3.7, showing values of *a* and *b* coefficients in Eq. (3.3.1) as a function of frequency [from (Ippolito, 1986)]. Plot dB/km curves as a function of rain rate for frequencies 1 GHz, 10 GHz, and 30 GHz, and compare to Figure 3.9.

TABLE P3.7

Frequency (GHz)	*a(f)* *(dB/mm/hr)*	*b(f)* *(exponent)*
1	0.00015	0.95
4	0.00080	1.17
6	0.00250	1.28
10	0.0125	1.18
15	0.0357	1.12
20	0.0699	1.10
30	0.170	1.075
40	0.325	0.99
60	0.650	0.84
80	0.875	0.753

3.8 The attenuation distribution curve in Figure 3.10 is convex. In some models the curve is shown concave. Explain how the shape of the curve will influence conclusions concerning link-availability–link-margin trade-offs.

3.9 Show that if a device has a noise figure F, defined in Eq. (3.4.5), then $(F - 1)290°$ is the temperature of an equivalent noise source that must be added at the input of the device to produce the same output noise power. [*Hint*: Divide the output noise into a part due to input noise and a part due to internal noise.]

3.10 Show that the number of satellites that can be placed in geosynchronous orbit with no more than 20 dB sidelobe interference from the ground is $n \approx (20) fd$, where f is frequency in GHz and d is antenna diameter in meters.

3.11 Sketch a curve of required antenna size needed from geosynchronous orbit at frequency 10 GHz to provide a given spot size on Earth.

3.12 Given the uplink–downlink satellite system in Figure 3.21. The carrier is transmitted to the satellite, amplified, and retransmitted. Determine the received CNR at the ground receiver, using the following parameters:

Uplink EIRP = 90 dB	Satellite $g/T^{\circ} = -40$ dB
Uplink loss = -200 dB	Satellite bandwidth = 1 MHz
Satellite gain, $Gg_t = 100$ dB	Downlink loss = -190 dB
Satellite NF, $F = 5$ dB	Receiver $g/T^{\circ} = 50$ dB
Uplink sky temp = 300°	Receiver bandwidth = 1 MHz

3.13 The satellites in Figure P3.13 use square horn antennas, each having a dimension of 0.3 m, and operates at $f = 10$ GHz. Neglect all system noise. Determine the carrier to interference power ratio at the receiver when both satellites are transmitting.

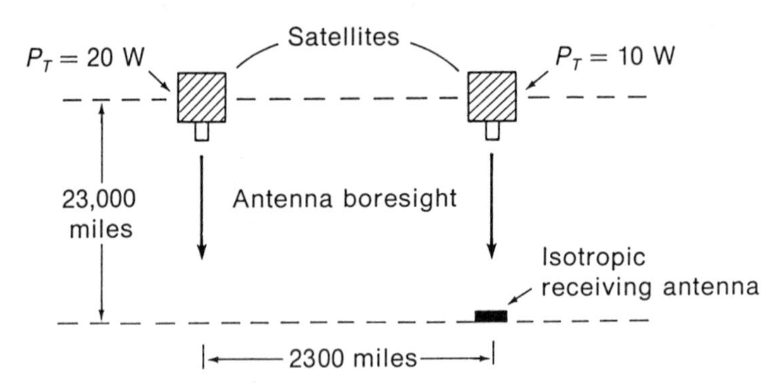

FIGURE P3.13

CHAPTER 4
Satellite Electronics

In Chapter 3 we concentrated on the basic power flow and power budget equations for various links of a satellite system. In this and the next chapters we focus on the detailed electronics of a satellite payload, and examine the specific waveshapes and noise processes generated by such electronics. Waveform characteristics become important when investigating distortion and suppression effects imposed by the satellite hardware, or when seeking methods to further improve an existing system. The entire question of satellite system optimization requires an accurate waveform model in order to understand the exact manner by which system anomalies are introduced. In this chapter we concentrate on the basic single-channel transponder, in which a single uplink carrier is processed for downlink transmission.

4.1 THE TRANSPONDER MODEL

A satellite transponder receives and retransmits the RF carrier. A simplified diagram was shown in Figure 1.5 to indicate the basic block structure. A more detailed diagram is shown here in Figure 4.1. The RF front end receives and amplifies the uplink carrier, while filtering off as much receiver noise as possible. The received carrier is then processed so as to prepare the retransmitted waveform for the return link. Carrier

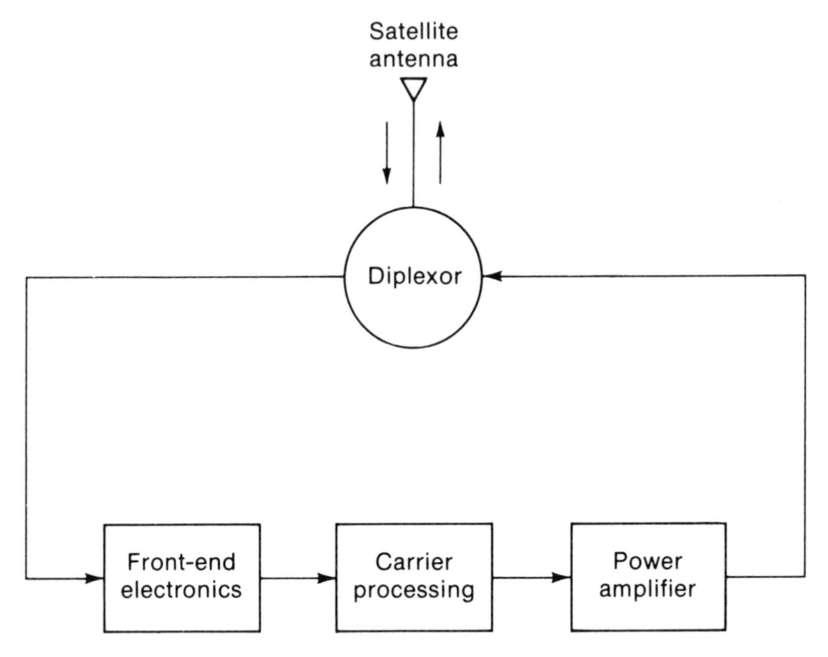

FIGURE 4.1 Transponder block diagram.

processing involves either some form of direct spectral translation or some form of remodulation. In spectral translation, the entire uplink spectrum is simply shifted in frequency to the desired downlink frequency. In remodulation processors, the uplink waveforms are demodulated at the satellite, and then remodulated onto the downlink carrier. Remodulation processors involve more complex circuitry, but provide for changeover of the modulation format between uplink and downlink. This restructuring of the downlink can provide advantages in decoding, power concentration (e.g., spot beams), and interference rejection. While the earlier transponder diagram showed a separate antenna for uplink receiving and downlink transmission, the same antenna can actually be used for both. This is possible since the uplink and downlink frequency bands are separated. A *diplexor* is used in the front end to allow simultaneous transmission and reception. The diplexor is a two-way microwave gate that permits received carrier signals from the antenna and transmitted carrier signals to the antenna to be independently coupled into and out of the antenna cabling. The carriers, being at different frequency bands, can flow in the same cabling and antenna feeds without interfering.

After front-end filtering and signal processing, the downlink carrier is power-amplified to provide the required power level for the downlink receiver. As discussed in Section 1.3 this power amplification is generally provided by traveling-wave-tube amplifiers (TWTA). These amplifiers may be preceded by stages of preamplification that set operating points for the TWTA. If we interpret the transponder as an ideal "amplifier in the sky," the entire transponder must provide the required overall gain needed to multiply the uplink power level to that of the downlink. Since uplink carrier power levels will be on the order of fractions of a microwatt (see Section 3.6), and since downlink power values of watts are needed, a typical transponder must provide a composite power gain of around 80–100 dB. Taking into account the power losses of filters and cabling, the required gain is generally far above that which can be provided by a TWTA alone. Hence, transponders require intermediate amplification to achieve the power levels suitably matched to the capability of the high-power downlink amplifier. These intermediate gain stages are usually provided by low-weight semiconductor amplifiers. Note that a high-power amplifier (TWTA) requires both high-gain and high-output power levels.

Transponder output power levels are directly related to available primary power. If the satellite has a power conversion efficiency of ϵ (ϵ is the ratio of usable downlink carrier power to primary power provided by solar panels and batteries), then a given transponder power requires $1/\epsilon$ times as much primary power. The lower the efficiency, the less the carrier power for a given amount of primary power, or conversely, the more primary power to achieve a desired downlink carrier power. In addition, the unused primary power [$(1 - \epsilon)$ times the primary power] must be dissipated as heat through the use of heat sinks if the satellite temperature is to be maintained. Thus higher transponder output levels may not only require larger TWTA, but also increased sizes of primary power sources and heat sinks, adding to the weight of the satellite. If the primary power of the satellite is limited by weight or cost, the efficiency parameter ϵ then becomes a critical design parameter, often dictating the transponder downlink capabilities.

4.2 THE SATELLITE FRONT END

Front-end electronics in satellite transponders are usually implemented as shown in Figure 4.2. The receiving antenna is coupled through cabling, usually to a diplexor or power splitter and followed by stages of RF filtering and amplifiers. The cabling, diplexors, and power splitters are lossy elements, while the filters and amplifiers determine bandwidth and power gain values.

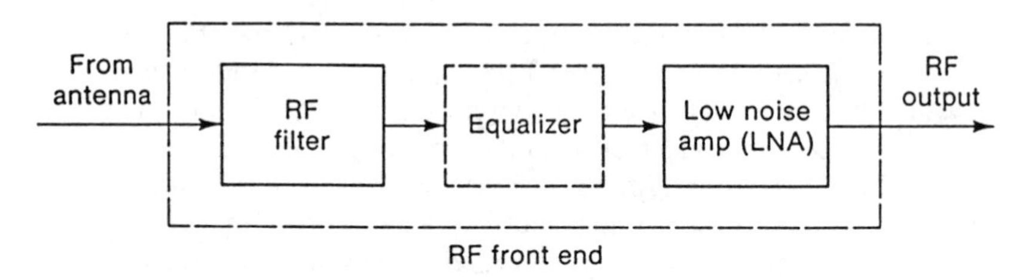

FIGURE 4.2 Front-end model.

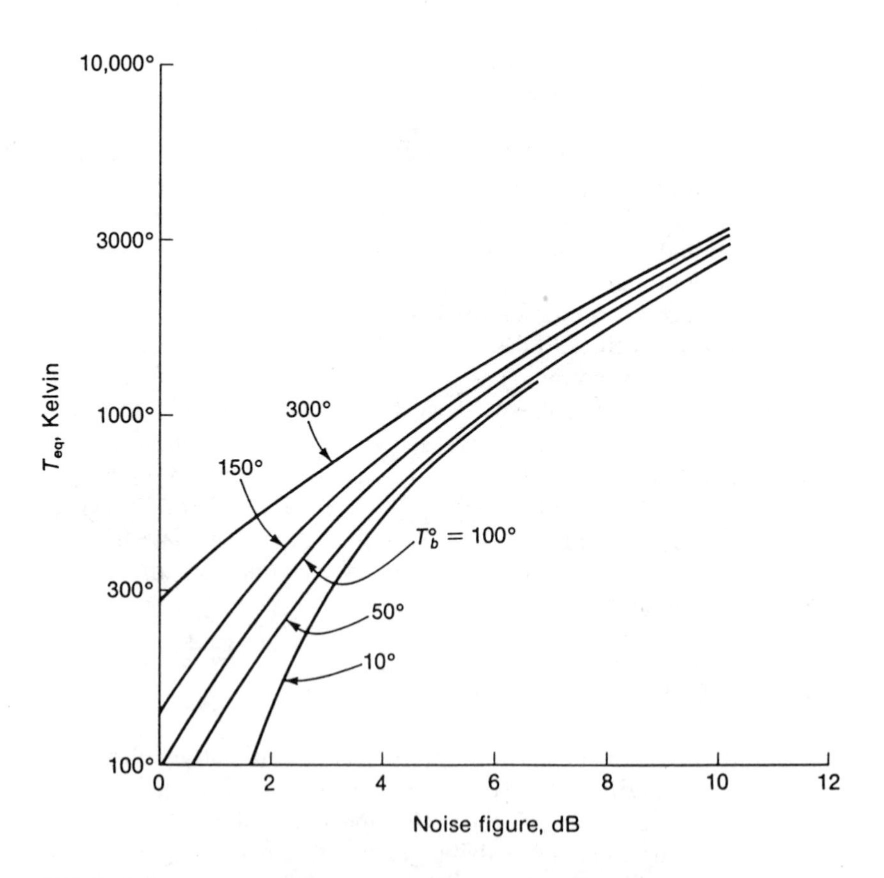

FIGURE 4.3 Equivalent receiver front-end noise temperature (T_b° = background temperature).

The satellite front end establishes the uplink carrier signal and noise levels for the transponder. As stated in Section 3.4 front-end noise is contributed from two basic sources: antenna noise, determined by the background temperature, T_b°, and the front-end circuit noise, determined by the noise figure F. These parameters set the front-end spectral level as

$$N_0 = kT^\circ_{\mathrm{eq}} \tag{4.2.1}$$

where k is Boltzmann's constant and

$$T^\circ_{\mathrm{eq}} = T^\circ_b + (F - 1)290^\circ \tag{4.2.2}$$

Background noise temperature T°_b was discussed in Section 3.4. An earth-based antenna sees the sky and galactic background from all directions, producing the approximate T°_b values shown in Figure 3.13. A satellite looking down at the Earth sees the illuminated Earth at a uniform reradiation temperature of about 300 K, surrounded by the galactic noise of outer space. A global spacecraft antenna will see the Earth in its mainlobe, and the lower-level galactic noise in its sidelobes. The antenna averaging will then tend to lower the combined background temperature. A satellite spot beam sees the illuminated Earth in its entire pattern, and therefore will have a background temperature closer to 300 K. When these background models are inserted into Eq. (4.2.2), the equivalent receiver temperature T°_{eq} plots shown in Figure 4.3 are derived, as a function of the receiver front-end noise figure. The higher the noise figure, the higher the receiver noise temperature, for a specific background model.

Receiver noise figure F depends on the elements forming the front end. The noise figure of each element is first determined, then all are combined into the total noise figure F. Table 4.1 lists the noise figure equations of various front-end elements. If F_i is the noise figure of the ith element and G_i is its power gain (or loss), the total noise figure F in Eq. (4.2.2) is given by

$$F = F_1 + \frac{F_2 - 1}{G_1} + \frac{F_3 - 1}{G_1 G_2} \cdots \frac{F_N - 1}{G_1 G_2 \cdots G_{N-1}} \tag{4.2.3}$$

Thus, the noise figure of a front end composed of cascaded circuit elements is computed by combining the noise figure of each as in Eq. (4.2.3). Figure 4.4 shows several front-end models, and the corresponding equivalent noise figure, obtained by computing F in Eq. (4.2.3). Table 4.1 and Figure 4.4 indicate the following points: When the front end uses an RF preamplifier immediately after the antenna, its noise figure determines

TABLE 4.1 Noise figure of front-end elements

Element	Noise figure
Lossy cable Length L, temperature T_g° α dB/L $L_g = 10^{-0.1(\alpha L)}$	$1+\left(\frac{1-L_g}{L_g}\right)\frac{T_g^\circ}{290^\circ}$
Lossy network Gain = L_g, temperature = T_g°	$1+\left(\frac{1-L_g}{L_g}\right)\frac{T_g^\circ}{290^\circ}$
Power split, 290° Power out/power in = α	$1/\alpha$
Amplifier (rated noise figure F_a) Lossy network (L_g) + amplifier at 290°	F_a/L_g

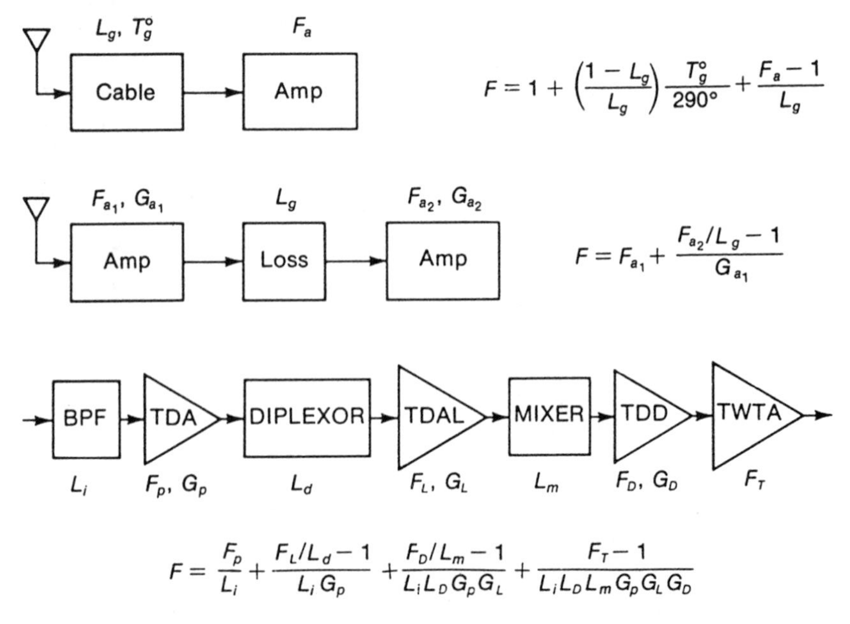

FIGURE 4.4 Front-end models and corresponding noise figures, F.

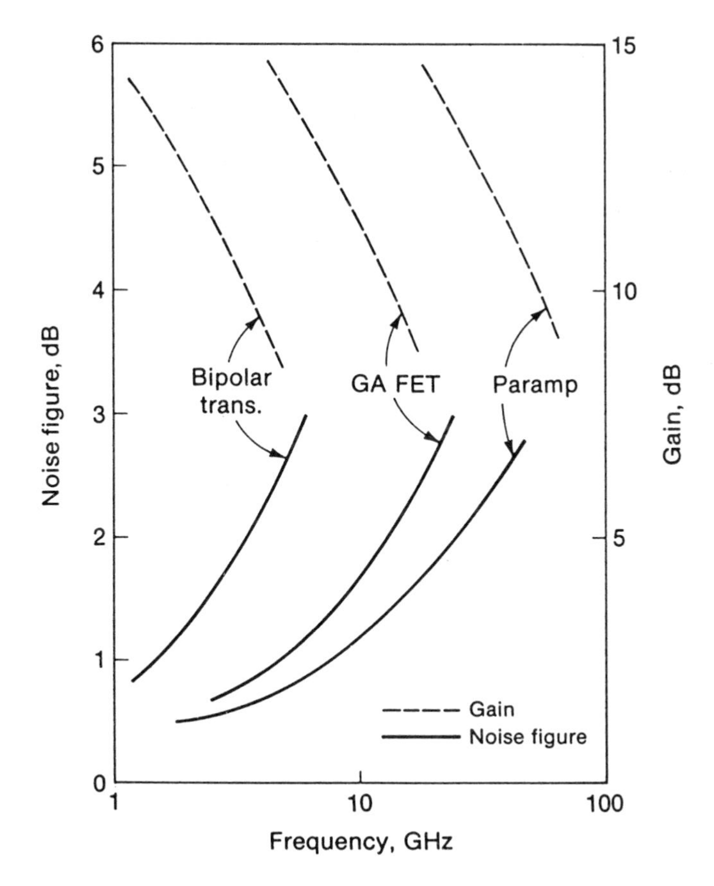

FIGURE 4.5 Noise figures and gain for typical front-end satellite amplifiers.

the entire front-end noise figure if its gain is suitably high. Thus, a basic requirement in preamplifier design is for high-gain, *low-noise amplifiers* (LNA) for this purpose. Figure 4.5 plots some preamplifier noise figures and gain values as a function of carrier frequency. Solid-state microwave amplifiers, such as *tunnel diode amplifiers* (TDA), with their inherent low-noise contributions and moderately high power gains are almost universally used for C-band and K-band satellite front ends. Unfortunately, TDA have decreasing gain values at the higher microwave frequencies (>10 GHz), and it is expected that TDA will be rapidly replaced by newer technological advances in LNA design, such as *field-effect transistors* (FET), for K-band satellites.

Equation (4.2.3) also shows that front-end elements further down the cascade contribute less to overall noise figure since they are divided down by the gain up to that point. Hence, noisy, high-gain preamplifiers are placed further back in the front-end stages. Also, lossy elements (devices with power gains less than 1) preceding an amplifier effectively increase the amplifier noise figure by the reciprocal of the element loss. Hence, an LNA should be placed as close to the antenna terminals as possible, prior to any significant coupling losses. Cable losses leading from antenna to amplifier increase the effective receiver noise and should be minimized. Typically, spacecraft front-end noise figures are about 5–10 dB, producing receiver noise temperatures in Figure 4.3 of about 1000–3000 K. An earth-based station can generally use extensive cooling to lower its front-end noise figure to the 2–5 dB range. Hence, earth-station noise temperatures are more typically about 50–800 K.

While background and noise figure set the spectral level of the receiver noise, the amount of noise power entering the front end depends on the noise bandwidth of the front-end filtering. This noise bandwidth depends on the type of filtering used, its spectral width, and its actual spectral shape. If $H(\omega)$ is the combined filter transfer function of the front end, its noise bandwidth is given by

$$B_c = \frac{1}{2\pi} \int_{-\infty}^{\infty} |H(\omega)|^2 \, d\omega \tag{4.2.4}$$

Figure 4.6 shows the amplitude plot $|H(\omega)|$ of common RF and IF filter functions of various orders. The plots show the amplitude characteristic for frequencies on one side of the carrier center frequency, with the abscissa normalized to the filter 3-dB frequency. Butterworth filters have flat in-band response with most in-band distortion occurring at bandedges. Chebyshev filters have a rippling in-band response with better response at the bandedges. Both filters have similar out-of-band attenuation, with Chebyshev filters having the better rejection at the higher orders. Bessel and Legendre filters have smoother amplitude functions, with slightly poorer out-of-band rejection. Table 4.2 lists the values of the noise bandwidths B_c in Eq. (4.2.4) for the class of Butterworth and Chebyshev bandpass filters.

The noise bandwidth of the RF filter must be balanced against carrier distortion. If the front-end filter is made too narrow, the carrier-signal spectrum will be distorted as it passes through. Distortion is caused by both the nonflat amplitude responses over the carrier bandwidth and by group-delay distortion (different frequencies being delayed by different amounts). Figure 4.7 shows the delay plot corresponding to the same

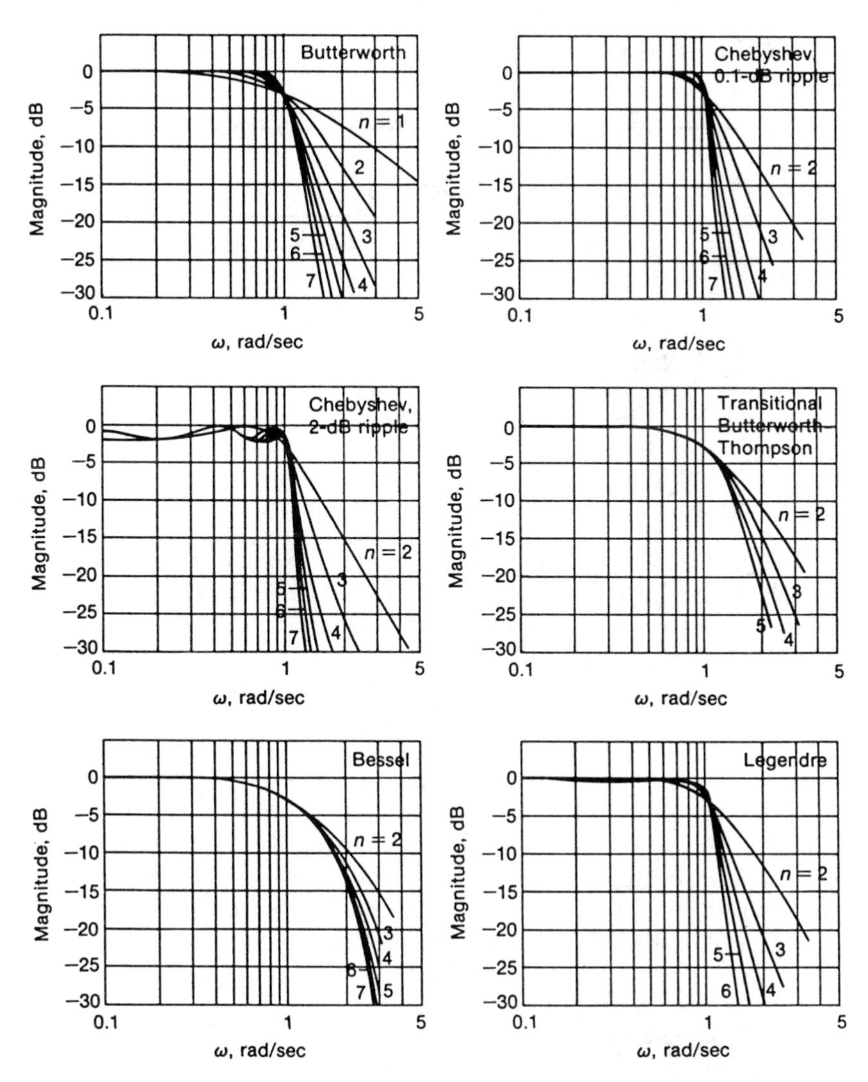

FIGURE 4.6 Amplitude characteristics of front-end bandpass filters (n = order of filter).

filters in Figure 4.6. Delay variation occurs primarily at bandedge, where the amplitude characteristic is rapidly decreasing. Bessel and Legendre filters have the better delay characteristics and are the preferable filter when delay distortion is critical (usually where in-band response is more important than out-of-band noise rejection). The amount of tolerable distortion caused by amplitude and group delay depends on the properties

TABLE 4.2 Bandpass-filter noise-bandwidths relative to 3-dB frequency for Butterworth and Chebyshev filters (ϵ = dB ripple)

Filter	*Order*	$\frac{B_c}{B_{3dB}}$	*Filter*	*Order*	$\frac{B_c}{B_{3dB}}$	*Filter*	*Order*	$\frac{B_c}{B_{3dB}}$
Butter-	1	1.570	Chebyshev	1	1.570	Chebyshev	1	1.57
worth	2	1.220	($\epsilon = 0.1$)	2	1.15	($\epsilon = 0.158$)	2	1.33
	3	1.045		3	0.99		3	0.86
	4	1.025		4	1.07		4	1.27
	5	1.015		5	0.96		5	0.81
	6	1.010		6	1.06		6	1.26

of the signal spectrum. If the uplink spectrum is that of a single wideband modulated carrier, with center frequency in the middle of the RF bandwidth, bandedge effects may not be that significant and some degree of band limiting may be tolerable. However, when the uplink is an FDMA format, with multiple carriers spread over the bandwidth, bandedge effects become extremely important to the outer carriers.

RF carrier spectra in satellite systems often have designated masks inside of which the downlink spectrum must be contained (Figure 4.8). These masks control the out-of-band spectral content of the carrier. If the uplink carrier spectrum does not satisfy the mask at the satellite, RF filtering must be applied either at the transmitter or at the front end to accomplish the desired filtering. As the order of the filter is increased for a given 3-dB bandwidth, the out-of-band attenuation increases, along with the in-band bandedge group delay. This means that if the mask attenuation is to be held constant, the 3-dB frequency of the filter can be increased, providing a reduction in group-delay distortion. Hence, implementing higher-order RF filters can be traded directly for reductions in delay distortion. Group-delay distortion can also be partially compensated with delay *equalization* networks. These networks are designed to have no attenuation distortion (flat gain curves across the RF bandwidth) but have delay variations that tend to cancel the delay distortion of the RF filters. Usually a filter is first constructed to satisfy the mask, then an equalizer is added to correct the delay distortion.

Satellite filters must be designed to be lightweight while achieving the desired mask, noise rejection, and equalization. With the high satellite frequencies involved, RF filters are typically constructed as microwave waveguide filters (Atia, 1976; Atia et al., 1974; Kurzrok, 1966; Kudsia and

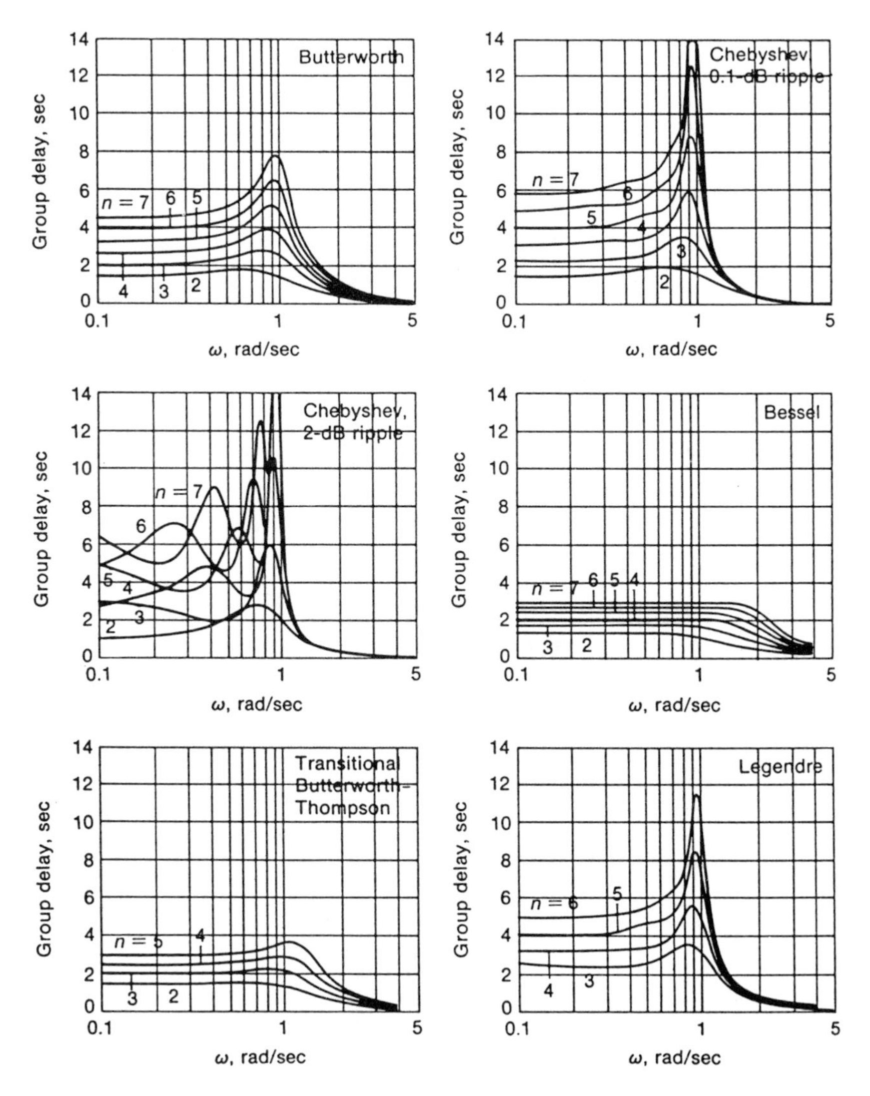

FIGURE 4.7 Group-delay characteristics of filters in Figure 4.6.

O'Donovan, 1974; Matthaei et al., 1964; Rhodes, 1969; Williams and Atia, 1977). Such integrated circuits allow better packaging and lower weight, and require less power. Increased use is also being made of dual-mode filters using cavity coupling. Such filters have center frequency-to-bandwidth ratios of about 10^4 at C-band and about 10^3 at K-bands. For example, satellites at C-band typically have carrier filters with RF bandwidths of about 36 MHz.

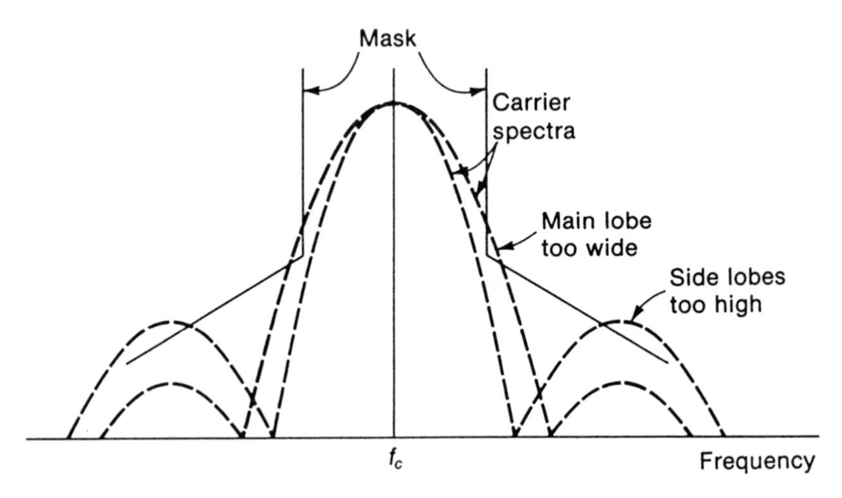

FIGURE 4.8 Front-end mask on satellite carrier spectrum.

Bandpass Gaussian noise appearing in the RF front end can be represented by the mathematical expression

$$n_{\mathrm{RF}}(t) = \tilde{n}_c(t)\cos(\omega_c t + \psi) - \tilde{n}_s(t)\sin(\omega_c t + \psi) \tag{4.2.5}$$

where ψ is an arbitrary phase angle and $\tilde{n}_c(t)$ and $\tilde{n}_s(t)$ are random quadrature noise components. These quadrature components are each low-pass noise processes whose power spectrum is obtained by shifting the one-sided bandpass spectrum to the origin, as shown in Figure 4.9.

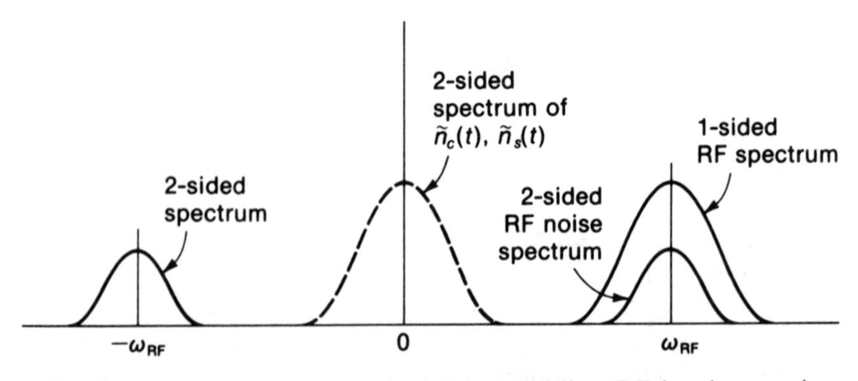

FIGURE 4.9 Front-end spectral models (solid line, RF bandpass noise spectrum; dashed line, low-pass noise component spectra).

Note that the one-sided spectral level of the bandpass noise becomes the two-sided spectral level of the noise components. When the spectrum of the bandpass RF noise is symmetric about the center frequency, ω_c, the noise components $\tilde{n}_c(t)$ and $\tilde{n}_s(t)$ are statistically independent processes, and their cross-correlation is everywhere zero. Also, if $n_{RF}(t)$ is a Gaussian noise process, $\tilde{n}_c(t)$ and $\tilde{n}_s(t)$ are each also Gaussian processes, and Eq. (4.2.5) is an exact statistical model for the RF noise.

The effect of additive bandpass noise on a carrier waveform in the satellite front end is to modify the waveform. Consider a general carrier

$$c(t) = a(t)\cos[\omega_c t + \theta(t) + \psi] \tag{4.2.6}$$

similar to the carrier forms discussed in Section 2.1. The addition of the noise waveform in Eq. (4.2.5) produces the combined waveform

$$\begin{aligned} x(t) &= c(t) + n(t) \\ &= a(t)\cos[\omega_c t + \theta(t) + \psi] + \tilde{n}_c(t)\cos(\omega_c t + \psi) \\ &\quad - \tilde{n}_s(t)\sin(\omega_c t + \psi) \end{aligned} \tag{4.2.7}$$

Trigonometrically expanding produces

$$\begin{aligned} x(t) = {} & [a(t) + n_c(t)]\cos[\omega_c t + \theta(t) + \psi] \\ & -n_s(t)\sin[\omega_c t + \theta(t) + \psi] \end{aligned} \tag{4.2.8}$$

where

$$\begin{aligned} n_c(t) &= \tilde{n}_c(t)\cos\theta(t) + \tilde{\eta}_s(t)\sin\theta(t) \\ n_s(t) &= \tilde{n}_c(t)\sin\theta(t) - \tilde{n}_s(t)\cos\theta(t) \end{aligned} \tag{4.2.9}$$

The noise components $n_c(t)$ and $n_s(t)$ are formed from combinations of the quadrature components $\tilde{n}_c(t)$ and $\tilde{n}_s(t)$. It can be shown (Gagliardi, 1988, Chap. 4) that $n_c(t)$ and $n_s(t)$ are also Gaussian for *any* $\theta(t)$, and will essentially have the same spectra as $\tilde{n}_c(t)$ and $\tilde{n}_s(t)$ as long as the RF bandwidth is larger than that of the modulation $\theta(t)$. Equation (4.2.8) can be rewritten as

$$x(t) = \alpha(t)\cos[\omega_c t + \theta(t) + \psi + v(t)] \tag{4.2.10}$$

where

$$\alpha(t) = \{[a(t) + n_c(t)]^2 + n_s^2(t)\}^{1/2} \tag{4.2.11a}$$

$$v(t) = \tan^{-1}\left(\frac{n_s(t)}{a(t) + n_c(t)}\right) \tag{4.2.11b}$$

Here $v(t)$ is referred to as *phase noise*. Hence, the effect of adding bandpass RF noise to the uplink carrier in the satellite front end is to convert any uplink amplitude modulation to $\alpha(t)$ in Eq. (4.2.11a) and to insert the phase noise $v(t)$ in Eq. (4.2.11b) onto the carrier phase. Since satellite carriers are usually constant envelope $[a(t) = A]$, and since uplink carrier levels exceed noise levels, we often approximate in Eq. (4.2.11) by

$$\alpha(t) \approx A + n_c(t) \tag{4.2.12a}$$

$$v(t) \approx \frac{n_s(t)}{A} \tag{4.2.12b}$$

The amplitude is therefore corrupted primarily by $n_c(t)$, which is often referred to as *amplitude noise*, or *in-phase* noise. The phase is affected primarily by $n_s(t)$, which is referred to as *quadrature noise*. Having somewhat precise mathematical forms for the front-end waveforms will be extremely beneficial in subsequent processing analysis.

4.3 RF FILTERING OF DIGITAL CARRIERS

The effect of excessive RF filtering on carriers modulated with digital data is extremely important in maintaining digital performance. This filtering may occur either at the transmitter or at the satellite. To examine this effect in detail, let a digital carrier at the input to an RF filter be the generalized constant envelope quadrature carrier

$$c(t) = A[m_c(t)\cos(\omega_c t + \psi) + m_s(t)\sin(\omega_c t + \psi)] \tag{4.3.1}$$

introduced in Chapter 2. Here A is the carrier amplitude, and $m_c(t)$ and $m_s(t)$ are the quadrature data-modulated waveforms. From Eq. (2.3.12) this carrier has the spectrum

$$S_c(\omega) = \tfrac{1}{4}A^2[S_{m_c}(\omega) + S_{m_s}(\omega)]_{\omega \pm \omega_c} \tag{4.3.2}$$

depending on the individual spectral densities of the quadrature modulation. The effect of a bandpass RF filter $H_{RF}(\omega)$ is to produce a filtered spectrum

$$S_y(\omega) = S_c(\omega)|H_{RF}(\omega)|^2 \tag{4.3.3}$$

at the filter output. The output carrier waveform can be determined by first writing

$$c(t) = A\,\mathrm{Re}\{m_c(t)e^{j(\omega_c t+\psi)}\} + A\,\mathrm{Im}\{m_s(t)e^{j(\omega_c t+\psi)}\} \tag{4.3.4}$$

where Re[·] and Im[·] refer to real and imaginary parts. The filter output waveform is then

$$\begin{aligned} y(t) = A\,\mathrm{Re}&\left\{\int_{-\infty}^{\infty} m_c(t-z)e^{j[\omega_c(t-z)+\psi]}h_{RF}(z)dz\right\} \\ &+ A\,\mathrm{Im}\left\{\int_{-\infty}^{\infty} m_s(t-z)e^{j[\omega_c(t-z)+\psi)}h_{RF}(z)dz\right\} \end{aligned} \tag{4.3.5}$$

where $h_{RF}(t)$ is the bandpass-filter impulse response. This means

$$\begin{aligned} y(t) = A\,\mathrm{Re}&\left\{e^{j(\omega_c t+\psi)}\int_{-\infty}^{\infty} m_c(t-z)e^{-j\omega_c z}h_{RF}(z)dz\right\} \\ &+ A\,\mathrm{Im}\left\{e^{j(\omega_c t+\psi)}\int_{-\infty}^{\infty} m_s(t-z)e^{-j\omega_c z}h_{RF}(z)dz\right\} \end{aligned} \tag{4.3.6}$$

The integrals are complex and correspond to the filtering of each $m(t)$ by a filter whose impulse response is $\tilde{h}_{RF}(t) = e^{-j\omega_c t}h_{RF}(t)$. By Fourier transforming, $\tilde{h}_{RF}(t)$ corresponds to an equivalent filter function

$$\begin{aligned} \tilde{H}_{RF}(\omega) &= \int_{-\infty}^{\infty} h_{RF}(t)e^{-j\omega_c t}e^{-j\omega t}\,dt \\ &= H_{RF}(\omega+\omega_c) \end{aligned} \tag{4.3.7}$$

The latter is simply the low-pass filter function obtained by shifting the bandpass RF filter function centered at ω_c to $\omega = 0$. Hence, Eq. (4.3.6) becomes

$$\begin{aligned} y(t) = A\,\mathrm{Re}&\{\tilde{m}_c(t)e^{j(\omega_c t+\psi)}\} \\ &+ A\,\mathrm{Im}\{\tilde{m}_s(t)e^{j(\omega_c t+\psi)}\} \end{aligned} \tag{4.3.8}$$

where $\tilde{m}_c(t)$ and $\tilde{m}_s(t)$ are the filtered versions of $m_c(t)$ and $m_s(t)$, each having transform

$$\tilde{M}_c(\omega) = M_c(\omega)\tilde{H}_{\mathrm{RF}}(\omega)$$
$$\tilde{M}_s(\omega) = M_s(\omega)\tilde{H}_{\mathrm{RF}}(\omega) \tag{4.3.9}$$

The filtered carrier spectrum is then

$$S_y(\omega) = \tfrac{1}{4}A^2[|M_c(\omega)\tilde{H}_{\mathrm{RF}}(\omega)|^2 + |M_s(\omega)\tilde{H}_{\mathrm{RF}}(\omega)|^2]_{\omega \pm \omega_c} \tag{4.3.10}$$

Equation (4.3.10) shows that the filtered carrier spectrum in Eq. (4.3.3) is equivalent to the filtering of the data modulation by a shifted RF filter. Hence, the details of RF carrier spectra can be determined by examining the low-pass data filtering. Figure 4.10a shows filtered QPSK spectra corresponding to Chebyshev filtering of a QPSK carrier for several values of filter bandwidth and order. The reduction in out-of-band carrier power is apparent, although main-hump spectral reduction can also occur if the filter is too narrow.

Excessive filtering in trying to reduce spectral tails and noise, however, may produce distortions on the quadrature data. Let $m(t)$, prior to the filtering, be written as the NRZ waveform

$$m(t) = \sum_{n=-\infty}^{\infty} a_n p(t - nT) \tag{4.3.11}$$

where $p(t) = 1$, $0 \leq t \leq T$, $a_n = \pm 1$, and T is a bit time for BPSK and a symbol time for QPSK. Equation (4.3.9) shows that the RF-filtered carrier will have data waveforms

$$\tilde{m}(t) = \sum_{n=-\infty}^{\infty} a_n \tilde{p}(t - nT) \tag{4.3.12}$$

where $\tilde{p}(t)$ is now the filtered pulse having transform

$$\tilde{P}(\omega) = P(\omega)\tilde{H}_{\mathrm{RF}}(\omega) \tag{4.3.13}$$

and $P(\omega)$ is the bit transform. Noticeable waveform distortion begins to appear when $B_{\mathrm{RF}}T \gtrsim 2$, where B_{RF} is the RF filter 3-dB bandwidth (that is, when the RF filter bandwidth becomes less than the main spectral hump of the modulated carrier). The effect is to begin rounding off the bit waveforms $p(t)$ and begin introducing temporal tails that extend over

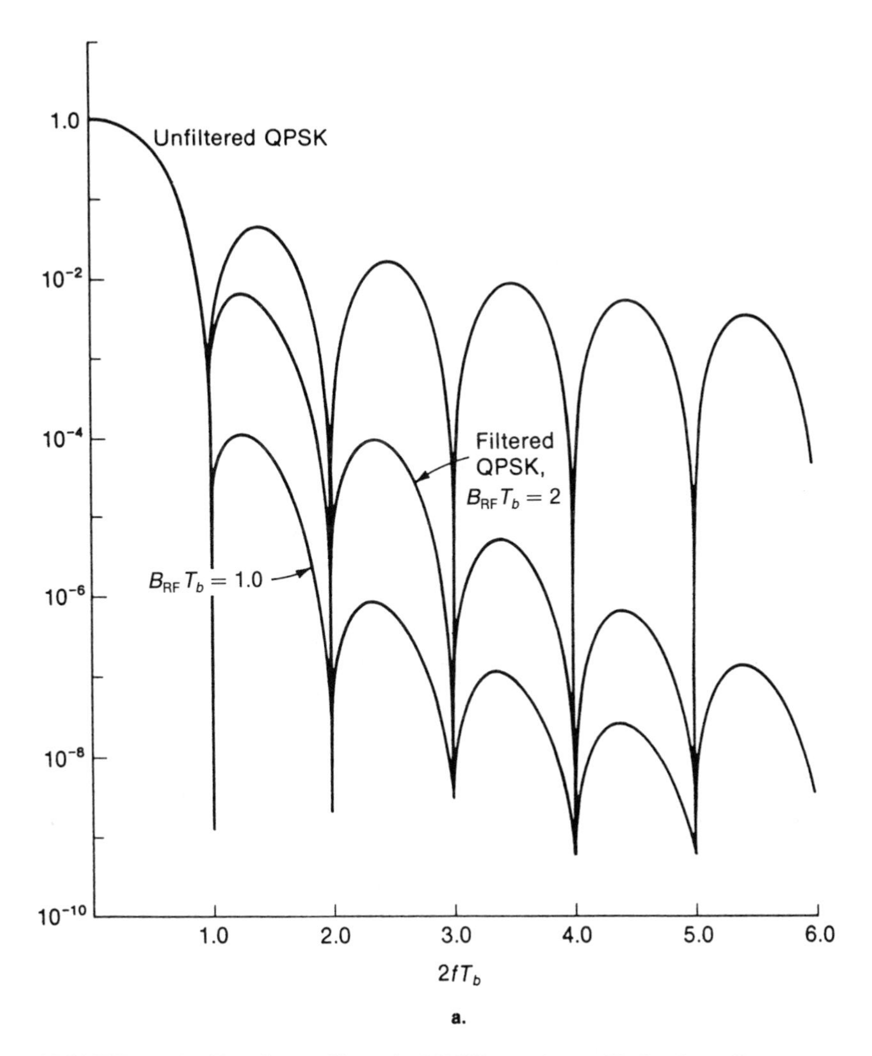

FIGURE 4.10 Bandpass-filtered QPSK carriers, Chebyshev filters: (a) spectrum; (b) pulse time response. (n = order of filter, $B_{RF}T_b = 1.0$.)

a bit time. Figure 4.10b, for example, shows the response of a single-bit pulse to the Chebyshev filter of various orders and $B_{RF}T$ products. Excessive filtering of the pulse is seen to cause a loss of bit energy during the bit time, whereas the tails create *intersymbol interference* in the form of bit spillover onto adjacent bits. This means the value of $\tilde{m}(t)$ at any t no longer depends on the present a_n but rather exhibits the intersymbol effect of other $\{a_n\}$ through the overlaps of $\tilde{p}(t)$.

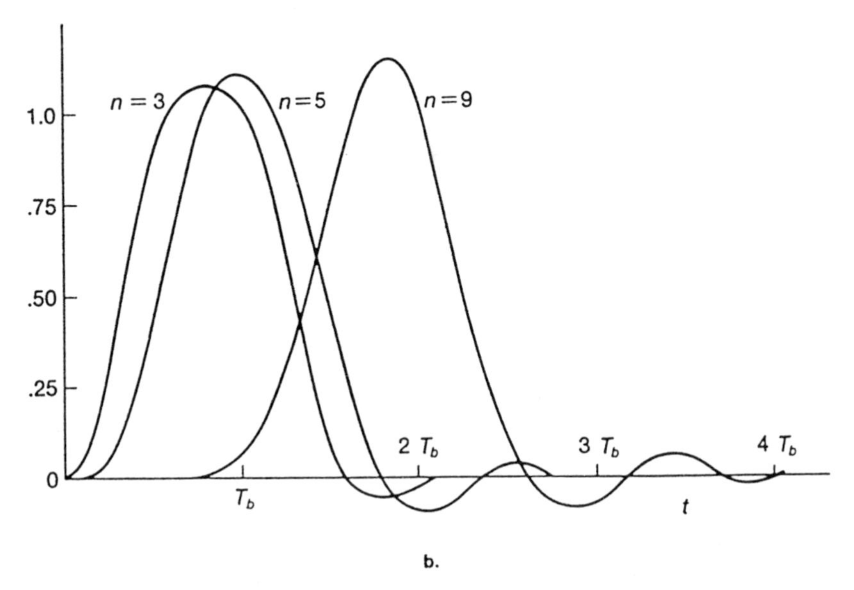

FIGURE 4.10 (continued) Bandpass-filtered QPSK carriers, Chebyshev filters: (a) spectrum; (b) pulse time response. (n = order of filter, $B_{\mathrm{RF}}T_b = 1.0$.)

Bit distortion means that even if the transponder retransmitted the filtered carrier directly to the decoder without further distortion, the effect of waveshape on bit decoding must still be considered. This topic is examined in Section A.2 of Appendix A. Figure 4.11a exhibits results derived in the section, showing the resulting degradation to PE caused by decreasing values of $B_{\mathrm{RF}}T_b$. Figure 4.11b replots the same data by showing the effective increase in carrier power needed to overcome the combined bit energy loss and intersymbol interference in order to maintain a bit-error probability of $\mathrm{PE} = 10^{-6}$, as a function of the $B_{\mathrm{RF}}T_b$ product. Thus, whereas the role of bandpass RF filtering is to reduce noise and spectral tails, excessive filtering will require increases in carrier power levels to maintain desired performance. Thus, the communication engineer must carefully evaluate the trade-off in reduced front-end bandwidth (increased throughput in bits per second per hertz) for decreased decoding performance.

Another important effect of narrow filtering is the possibility of creating a nonconstant envelope carrier from one that was originally constant envelope. At any t the envelope of $c(t)$ in Eq. (4.3.1), using Eq. (4.3.11) is

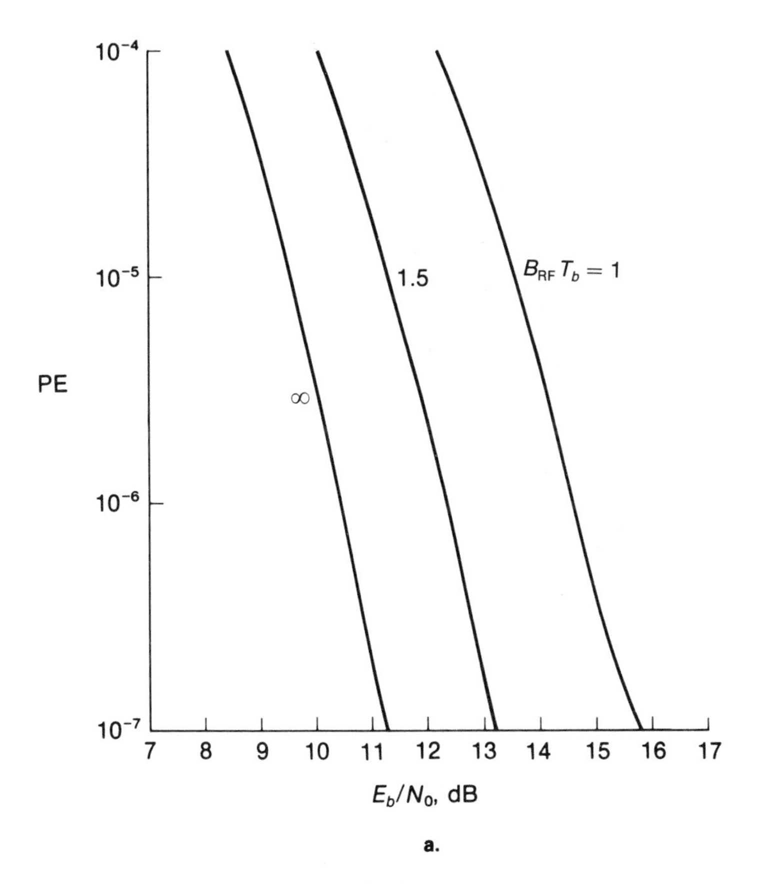

FIGURE 4.11 BPSK PE degradation due to bandpass filtering, B_{RF} = 3dB bandwidth: (a) PE vs E_b/N_0; (b) required increase in E_b to maintain PE $= 10^{-6}$. Integrate and dump decoder, Chebyshev filter, order n.

$A[p_c^2(t) + p_s^2(t)]^{1/2}$, designed to be a constant at all t. However, $y(t)$ in Eq. (4.3.8) will have the envelope

$$|y(t)| = A[\tilde{m}_c^2(t) + \tilde{m}_s^2(t)]^{1/2}$$
$$= A\left[\left(\sum_{n=-\infty}^{\infty} a_n \tilde{p}(t - nT)\right)^2 + \left(\sum_{n=-\infty}^{\infty} b_n \tilde{p}(t - nT)\right)^2\right]^{1/2} \quad (4.3.14)$$

This envelope is no longer constant, owing to both the distortion in $\tilde{p}(t)$ and the intersymbol effect. Hence, excessive RF filtering can be detri-

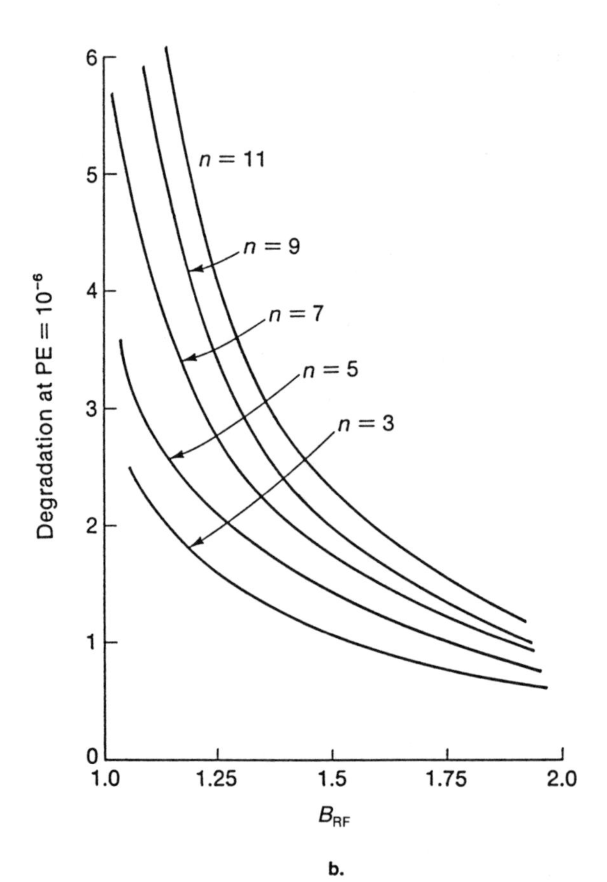

FIGURE 4.11 (continued) BPSK PE degradation due to bandpass filtering, $B_{\mathrm{RF}} =$ 3dB bandwidth: (a) PE vs E_b/N_0; (b) required increase in E_b to maintain PE $= 10^{-6}$. Integrate and dump decoder, Chebyshev filter, order n.

mental to subsequent system processing designed for constant-envelope carriers.

The effect of RF filtering on constant-envelope phase-modulated carriers is more difficult to determine. If we write the general PM carrier as

$$c(t) = \mathrm{Re}\{Ae^{j\theta(t)}e^{j(\omega_c t + \psi)}\} \tag{4.3.15}$$

where $\theta(t)$ represents the phase modulation, then the RF-filtered version, corresponding to Eq. (4.3.6), becomes

$$y(t) = A\ \mathrm{Re}\left\{e^{j(\omega_c t + \psi)} \int_{-\infty}^{\infty} e^{j\theta(t-z)}\tilde{h}_{\mathrm{RF}}(z)dz\right\} \tag{4.3.16}$$

This contains in the integral the effective filtering by the low-pass filter $\tilde{H}_{RF}(\omega)$ on the complex modulation

$$m(t) = e^{j\theta(t)} \tag{4.3.17}$$

The waveform $m(t)$ is known to have a correlation function related to the second characteristic function of the process $\theta(t)$ (Gagliardi, 1988, chap. 3) and a power spectrum related to sums of repeated convolutions of the spectra of $\theta(t)$. Since convolutions spread the spectra of the modulation, the effect of filtering is to tend to reduce the significance of the higher-order convolutions. The spectrum of the filtered carrier in Eq. (4.3.16) is therefore dominated by the first few terms of the convolved spectrum of $\theta(t)$. Hence, we often approximate the filtered phase-modulation carrier spectrum as

$$S_y(\omega) = \tfrac{1}{4}A^2[[S_\theta(\omega) + S_\theta(\omega) \otimes S_\theta(\omega)]|\tilde{H}_{RF}(\omega)|^2]_{\omega \pm \omega_c} \tag{4.3.18}$$

where $\otimes$ denotes spectral convolution. Thus, to a first-order approximation, $S_y(\omega)$ appears as a filtered version of the modulation spectra, plus the result of filtering the convolved modulation spectrum, all shifted to the carrier frequency.

4.4 SATELLITE SIGNAL PROCESSING

Satellite signal processing generates the microwave carrier for the return link. This carrier can be achieved by direct frequency translation or by carrier remodulation. Subsystems that accomplish these operations are shown in Figure 4.12. Figure 4.12a shows a direct RF-to-RF conversion using a single mixer system. Figure 4.12b shows a double conversion from RF to an intermediate lower IF, then back up to RF using a single mixing oscillator. This system has the advantage of allowing carrier filtering and amplification to be performed at the lower IF frequency band rather than at the higher RF band. In addition, RF–IF–RF systems allow uplink command carriers to be more easily removed (or new telemetry carriers inserted) at the satellite before return retransmission. Note that frequency translators of either type always cause all uplink noise to be similarly translated, and, therefore, directly superimposed onto the return spectra, as we assumed in Chapter 3. When the transponder merely frequency translates the uplink carrier to the downlink, the satellite appears to be simply "bending" the earth-station uplink into the receiving downlink. For this reason, frequency-translating satellite links are often referred to as *bent pipes*.

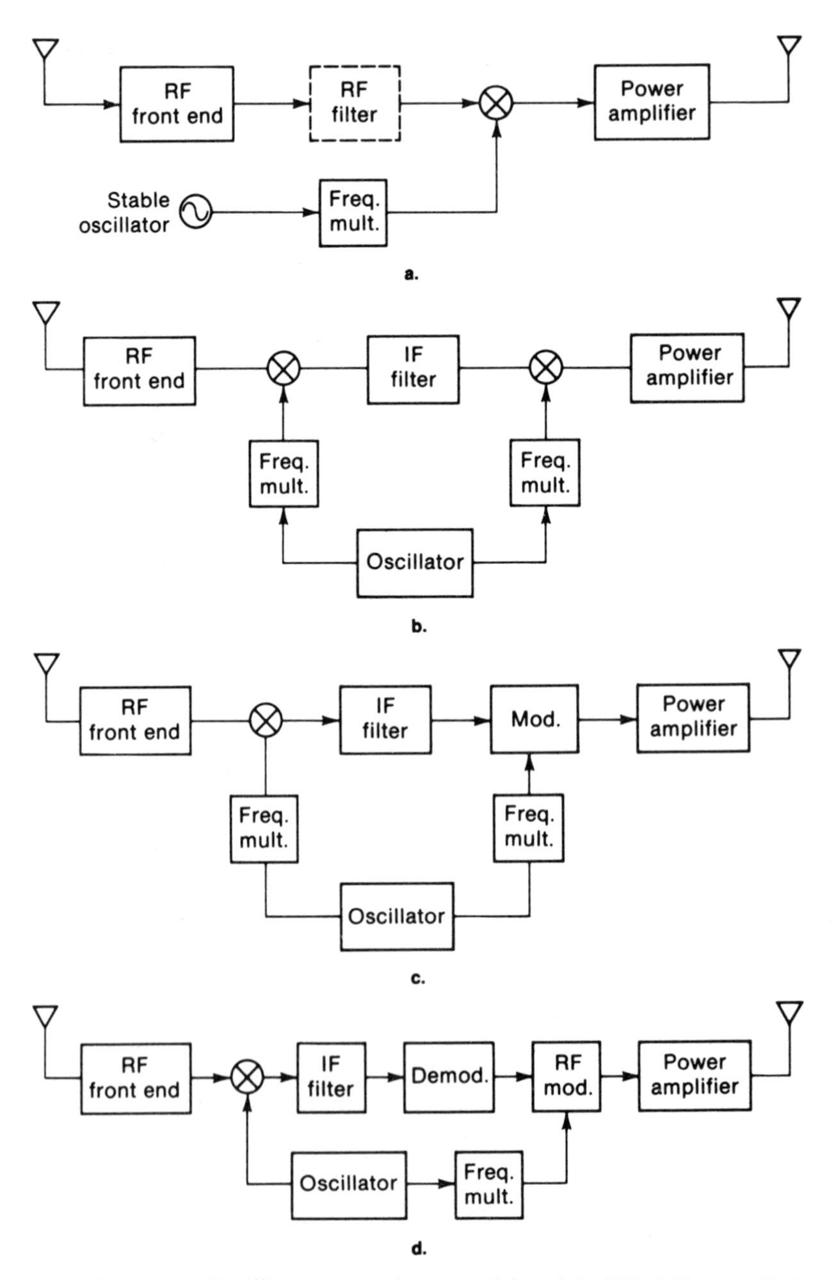

FIGURE 4.12 Satellite processing models: (a) RF–RF translation; (b) RF–IF–RF translation; (c) IF remodulation; (d) demodulation–remodulation.

Remodulating processors involve a stage of modulation that generates the return carrier. This is accomplished by (1) translating the uplink RF spectrum down to a low IF band, and then modulating the entire IF onto the return RF, as shown in Figure 4.12c; or (2) directly demodulating the uplink carrier to baseband, and remodulating the baseband onto the return carrier, as in Figure 4.12d. With digitally modulated carriers, baseband demodulation can involve baseband decoding of the information bits, followed by digital remodulation. Remodulation removes uplink noise and interference from the downlink modulation, while facilitating uplink on-board digital processing and downlink bit insertion. In the following sections we analyze the effect of a single uplink carrier processed in the transponder by each of these processing systems.

RF–RF TRANSLATION

Let the uplink front-end waveform be written as

$$x(t) = c_u(t) + n_u(t) \tag{4.4.1}$$

where $c_u(t)$ is an uplink carrier of the form

$$c_u(t) = \sqrt{2P_c}\cos[\omega_u t + \theta_u(t) + \psi] \tag{4.4.2}$$

The uplink noise is bandpass-filtered Gaussian noise, having the form of Eq. (4.2.5). The uplink noise has the spectral level N_0 in Eq. (4.2.1). In RF–RF translation, the RF spectrum of $x(t)$ is merely shifted to the return carrier frequency. If the mixer local oscillator is written as

$$c_l(t) = \sqrt{2}\cos(\omega_l t + \psi_l) \tag{4.4.3}$$

then the mixer output produces

$$\begin{aligned} x_2 &= k_m x(t) c_l(t) \\ &= k_m\sqrt{P_c}\cos[(\omega_u - \omega_l)t + \theta(t) + \psi - \psi_l] \\ &\quad + k_m\sqrt{P_c}\cos[(\omega_u + \omega_l)t + \theta(t) + \psi + \psi_l] \\ &\quad + n_2(t) \end{aligned} \tag{4.4.4}$$

where k_m is the mixer gain and $n_2(t)$ is the translated mixer output noise. The waveform $x_2(t)$ has its uplink RF spectrum shifted to the two center frequencies $(\omega_u \pm \omega_l)$. The bandpass filter following the mixer can be tuned

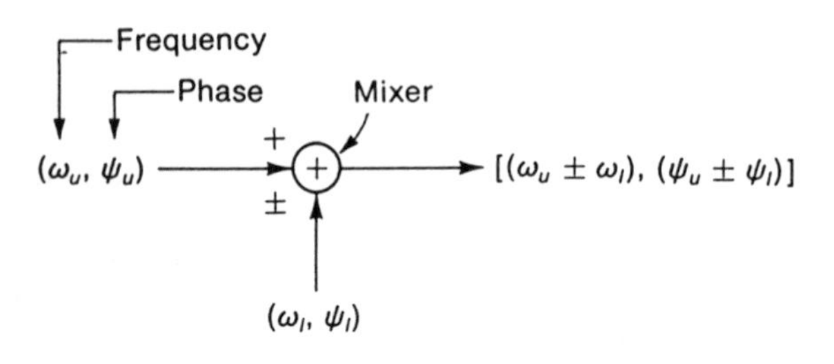

a.

(ω_u, ψ_u) $(\omega_u \pm K_1\omega_0, \psi_u \pm K_1\psi_0)$ $(\omega_u \pm (K_1 \pm K_2)\omega_0, \psi_u \pm (K_1 \pm K_2)\psi_0)$

$(K_1\omega_0, K_1\psi_0)$ $(K_2\omega_0, K_2\psi_0)$

$\times K_1$ $\times K_2$

(ω_0, ψ_0)

b.

FIGURE 4.13 Frequency phase model for mixing: (a) single mixer; (b) multiple mixing.

to select either the sum or difference frequency of the mixer output. Hence, the uplink spectrum can be either up-converted to the sum frequency or down-converted to the difference frequency. For the commercial C-band satellite communication, the 6-GHz uplink is down-converted to the 4-GHz downlink.

Note from Eq. (4.4.4) that mixing scales the carrier amplitude while generating the difference frequency (for the down-converter) and the difference phase of the uplink and local carriers. This frequency and phase subtraction (addition for up-conversion) allows us to establish the carrier frequency–phase conversion diagram shown in Figure 4.13. Such a diagram allows us to keep track of carrier frequency and phase, especially if we pass through several stages of mixing prior to arriving at the desired RF frequency.

At the input to the mixer the carrier power in Eq. (4.4.2) is P_c whereas the noise power is $N_0 B_{\mathrm{RF}}$. Hence, the mixer input CNR is

$$\mathrm{CNR}_u = \frac{P_c}{N_0 B_{\mathrm{RF}}} \tag{4.4.5}$$

After mixing, the carrier and noise are both scaled identically by the mixer gain k_m in frequency conversion. Hence, the CNR remains the same as Eq. (4.4.5) and ideal frequency mixing (either up-conversion or down-conversion) does not degrade CNR. This means the analysis of Section 3.6, where earth-station CNR_d was computed for an ideal amplifier as

$$(\mathrm{CNR}_d)^{-1} = (\mathrm{CNR}_u)^{-1} + (\mathrm{CNR}_r)^{-1} \tag{4.4.6}$$

is still valid, and CNR_u can be referred to either before or after frequency translation. In other words, ideal transponder frequency conversion does not affect downlink receiver CNR.

It is important to recognize the importance of the stability of the satellite master oscillator in the frequency and phase shifting of the RF translation operation. Offsets in the oscillator frequency produce offsets in the downlink carrier frequency. Phase variations on the local oscillator, such as phase noise variations, are transferred directly to the translated RF carrier. These effects become subtly important, since they can cause phase and frequency errors to permeate the entire system. For this reason an extremely stable master oscillator is usually selected for transponder translation.

To aid the frequency stability of the satellite oscillator, a pilot carrier is often used. The pilot is an unmodulated RF carrier transmitted from a control earth station to the satellite at a fixed frequency outside the modulation spectral band. The transponder oscillator then frequency and phase locks to the pilot frequency. If the phase lock is nearly perfect (i.e., the transponder master oscillator is operating exactly at the pilot frequency or a fixed multiple of it), the transponder mixing frequency is directly connected to the earth segment. Its frequency stability can therefore be controlled from the ground and can be adjusted for the proper return RF frequency. When the pilot frequency lock is not ideal, the local oscillator will appear to have additional phase noise components that become superimposed on the translated RF carrier.

For intersatellite links (ISL), RF–RF up-conversion is a simple procedure that can be used for generating the crosslink carrier. The uplink carrier (e.g., at 6 GHz) can be directly frequency translated to the K-band, Q-band, or V-band crosslink frequency for TWT amplification and transmission. At the ISL receiving satellite, the received carrier is merely down-converted to the desired downlink frequency (e.g., 4 GHz), and retransmitted to the downlink.

RF–IF–RF Translation

With double-conversion frequency translation, an intermediate conversion frequency is used. The mixing is accomplished by a double mixer, usually driven by a common master oscillator, as was shown in Figure 4.12b. The oscillator frequency is simultaneously multiplied to each mixing frequency, the first to down-convert to IF and the second to up-convert back to RF. The translation can be analyzed by applying the results of Figure 4.13a to both stages, as in Figure 4.13b. We assume that the frequency conversions from the master frequency f_0 to the desired conversion frequencies is obtained by a frequency multiplier g, which simultaneously multiplies up (or divides down) by the same factor both the oscillator frequency and phase. Examination of Figure 4.13b shows that if the translator input signal is that given in Eq. (4.4.2), the RF signal after double conversion has the form

$$x_3(t) = \sqrt{P_c}k_{m_1}k_{m_2}\cos[\omega_3 t + \theta(t) + \psi_u - (g_1 + g_2)\psi_0] + n_3(t) \quad (4.4.7a)$$

where $\omega_3 = \omega_u - (g_1 + g_2)\omega_0$ is the desired return RF carrier frequency. The noise $n_3(t)$ has the RF front-end noise spectrum filtered by the IF bandpass filter and shifted to ω_3, with its spectral level scaled by $(k_{m_1}k_{m_2})^2$. Just as in RF–RF mixing, the sequence of mixing operations ideally preserves CNR during its operation. Note that the carrier portion of Eq. (4.4.7a) can be written with the phase

$$\omega_3 t + \theta(t) + \psi_u + [(\omega_3 - \omega_u)/\omega_0]\psi_0]. \quad (4.4.7b)$$

Thus, the phase inserted by the local oscillator appears to have been multiplied up by the frequency-conversion difference normalized by the oscillator frequency. Hence, it is only the total frequency *difference* of the translation that is important to the phase insertion, and not the number of translation stages.

IF Remodulation

Satellite processing using IF remodulation was shown in Figure 4.12c. The uplink is translated to IF, and the entire IF is modulated onto a downlink carrier. Since the IF bandwidth will be expanded into the RF bandwidth by the modulation, this conversion technique has application only if the uplink IF bandwidth is much smaller than that available in the downlink. The system, however, has the advantage that the noise spectrum in the satellite uplink is not retransmitted directly into the downlink. Instead, the uplink noise is effectively modulated onto the downlink carrier along with the desired signal. To analyze the processing, consider again the single uplink carrier in Eq. (4.4.2) containing the phase modulation $\theta(t) = \Delta_u m(t)$. The uplink RF is translated to IF, generating the IF signal

$$x_{\text{IF}}(t) = A_{\text{IF}} \sin[\omega_{\text{IF}} t + \theta(t) + \psi_{\text{IF}}] + n_{\text{IF}}(t) \tag{4.4.8}$$

The preceding waveform has a CNR of

$$\text{CNR}_u = \frac{A_{\text{IF}}^2/2}{N_{0u} B_{\text{IF}}} \tag{4.4.9}$$

where B_{IF} is the IF noise bandwidth. The IF signal has its amplitude adjusted to generate the desired phase index Δ_d for the downlink. This is equivalent to multiplying $x_{\text{IF}}(t)$ by Δ_d/A_{IF}. The downlink-modulating waveform is then the bandpass signal

$$\begin{aligned} x_d(t) &= \left(\frac{\Delta_d}{A_{\text{IF}}}\right) x_{\text{IF}}(t) \\ &= \Delta_d \sin[\omega_{\text{IF}} t + \Delta_u m(t) + \psi_{\text{IF}}] + \left(\frac{\Delta_d}{A_{\text{IF}}}\right) n_{\text{IF}}(t) \end{aligned} \tag{4.4.10}$$

After transponder amplification the downlink carrier then becomes

$$c_d(t) = \sqrt{2P_T} \sin[\omega_d t + x_d(t)] \tag{4.4.11}$$

where P_T is the available satellite downlink power. Note that Eq. (4.4.11) contains no noise added to the downlink carrier. Instead, the IF noise is contained within the modulation and will appear only following downlink phase demodulation. The carrier in Eq. (4.4.11) will be received on the ground with power $P_T L$ with L given in Eq. (3.6.3). The downlink CNR is then

$$\text{CNR}_d = \frac{P_T L}{N_{0d} B_c} \tag{4.4.12}$$

where N_{0d} is the receiver noise level of the downlink receiver. If this threshold is high enough to satisfy the receiver, the downlink carrier can be phase demodulated, producing the recovered baseband

$$x_d(t) = \Delta_d \sin[\omega_{\text{IF}} t + \theta(t)] + \left(\frac{\Delta_d}{A_{\text{IF}}}\right) n_{\text{IF}}(t) + \frac{n_d(t)}{\sqrt{2P_T L}} \tag{4.4.13}$$

where $n_d(t)$ is the downlink noise, having power $N_{0d}B_{\text{IF}}$. The total modulation interfering noise in Eq. (4.4.13) can be interpreted as having been caused by an equivalent downlink receiver noise of spectral level

$$N_{0d}\left[1 + \Delta_d^2 \frac{2N_{0u}P_T L}{A_{\text{IF}}^2 N_{0d}}\right] = N_{0d}\left[1 + \Delta_d^2\left(\frac{\text{CNR}_d}{\text{CNR}_u}\right)\right] \tag{4.4.14}$$

where the CNR are given in Eqs. (4.4.9) and (4.4.12). Hence, the receiver demodulated noise level is effectively increased by the uplink IF noise during demodulation, with the increase dependent on the ratio of the uplink CNR to the downlink CNR. We note, however, that the uplink noise does not alter the CNR in Eq. (4.4.12) required for establishing the receiver demodulation threshold. Contrast this result with the case of RF–RF conversion in Eq. (4.4.6) in which uplink noise directly reduced CNR_d.

IF remodulation methods have also been suggested for intersatellite links (ISL). Instead of simply up-converting the uplink carrier directly to the crosslink frequency, the uplink carrier is instead down-converted to an IF frequency (MHz), remodulated back onto the uplink carrier via FM, and up-converted to the crosslink frequency (20–60 GHz). At the ISL receiving satellite, the crosslink carrier is down-converted to RF, demodulated to regenerate the IF carrier, and the latter is up-converted to the downlink frequency. This method provides a higher downlink CNR than simple RF–RF ISL conversion with the same ISL CNR, since it has the advantage of the SNR_d improvement during FM demodulation. Conversely, the ISL can be operated with a smaller CNR (smaller antennas and TWT power) for the same downlink CNR, at the expense of inserting the additional mod–demod equipment. This reduction in crosslink CNR is particularly significant in considering antenna size, since we recall from Eq. (3.6.16) that crosslink CNR varies as the *fourth power* of antenna diameter.

Demodulation–Modulation Conversion

In satellite processing using demodulation and remodulation, the uplink is demodulated to baseband, and the entire baseband is remodulated onto a downlink carrier. Since demodulation is used, this conversion has primary application to the case when a single uplink carrier is involved (unless a separate satellite demodulator was provided for each uplink carrier). This processing format allows for (1) uplink commands transmitted with the carrier modulation to be recovered during the demodulation, and (2) satellite telemetry to be inserted into the baseband for downlink modulation. As with IF modulation, the system also has the advantage that the uplink noise spectrum is not retransmitted directly in the downlink.

Analysis of baseband remodulation is carried out in a manner similar to that of IF remodulation, except that demodulated waveforms are

$$x(t) = \Delta_u m(t) + \frac{n_u(t)}{A_u} \tag{4.4.15}$$

Here $n_u(t)$ is the uplink quadrature noise in the modulation bandwidth of $m(t)$. The baseband signal $x(t)$ now has its amplitude adjusted to generate a new phase index, Δ_d, for the downlink. The downlink carrier is then formed as in Eq. (4.4.11) with

$$x_d(t) = \Delta_d m(t) + \left(\frac{\Delta_d}{\Delta_u}\right)\frac{n_u(t)}{A_u} \tag{4.4.16}$$

After downlink phase demodulation, the equivalent receiver noise spectral level then becomes

$$N_{0d}\left[1 + \left(\frac{\Delta_d}{\Delta_u}\right)^2 \frac{\text{CNR}_d}{\text{CNR}_u}\right] \tag{4.4.17}$$

which is similar to Eq. (4.4.14).

When digital modulation is used, the transponder demodulation can be reinterpreted as bit decoding. Remodulation corresponds to encoding these decoded bits back onto the RF return carrier. This means a given source bit, in traveling through the transponder to the specific earth station, undergoes two stages of decoding in cascade. This may be diagrammed as shown in Figure 4.14. A given bit will be decoded correctly at the end if two correct decodings occurred, or if two incorrect decodings occurred. Let PE_1 denote the probability of a bit error during the uplink decoding and let PE_2 be that for the downlink decoding. The average bit-error probability is then

$$\begin{aligned}\text{PE} &= (1 - \text{PE}_1)\text{PE}_2 + \text{PE}_1(1 - \text{PE}_2) \\ &= \text{PE}_1 + \text{PE}_2 - 2\text{PE}_1\text{PE}_2 \end{aligned} \tag{4.4.18}$$

If both links have been designed for small PE ($\text{PE}_1, \text{PE}_2 \leq 10^{-1}$), the last term will be negligible compared to the sum. This means the overall PE for the transponder link is the sum of the individual link error probabili-

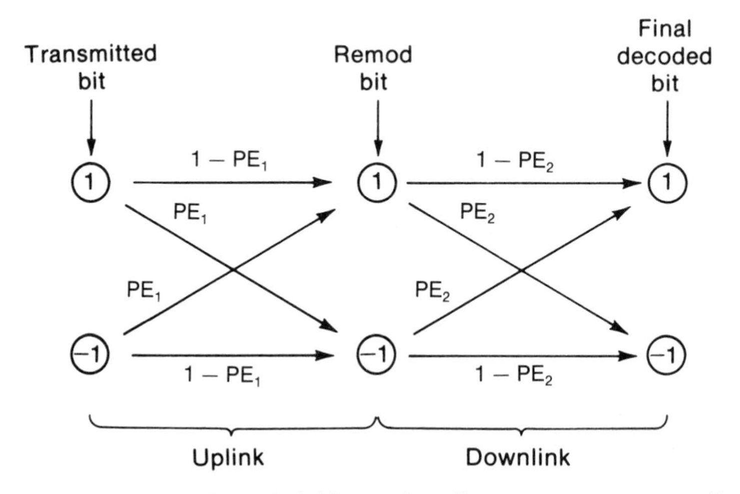

FIGURE 4.14 Cascaded binary-decoding stages, corresponding to demod–remod digital transponders.

ties. The weakest link (largest PE_i) will therefore determine the overall PE, and there is no advantage in having one digital link significantly better than the other in terms of error probability.

As an example, consider a BPSK uplink and downlink, operating with an uplink $(E_b/N_0)_u$ and a downlink $(E_b/N_0)_r$, each calculated by the methods of Section 3.5. We wish to operate the overall transponder remodulation link with an error probability PE. Figure 4.15 shows the uplink–downlink E_b/N_0 combinations needed to produce a specified PE of 10^{-6}–10^{-3} for the cascade demod–remod link. Also included is the required E_b/N_0 values for the same PE using a linear amplifying up–down link, obtained from Figure 3.22. The result shows clearly the advantage of inserting decoding–encoding hardware at the satellite. Curves of this type are useful for performing initial system design and sizing the hardware for satellite and earth station.

4.5 FREQUENCY GENERATORS

The signal processing in Figure 4.12 requires carrier frequency tones to be generated in the satellite at all the proper frequencies to accomplish the frequency mixing, translation, and modulation. In satellite payloads these required frequency tones are produced in subsystems referred to as

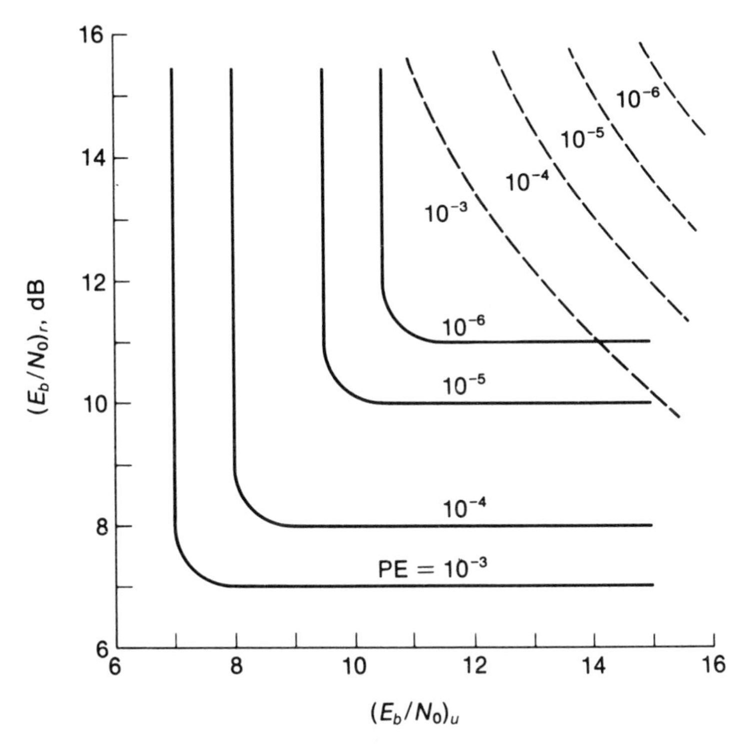

FIGURE 4.15 Required values of uplink E_b/N_0 and downlink E_b/N_0 to achieve a given PE (solid line, demod–remod system; dash line, ideal frequency translation).

frequency generators. Frequency generators are constructed by taking a pure oscillator carrier tone and multiplying or dividing or both its output frequency to all the desired frequencies needed. Oscillators are simple electronic amplifier devices coupled to tuning mechanisms via some type of feedback (Smith, 1986), as shown in Figure 4.16a. Resonances of the tuning circuits allows a sustained feedback oscillation to occur, producing an output tone at the resonant frequency. The oscillator tuning circuits commonly used are the resistance, inductance, capacitance (*RLC*) electronic or microwave circuit, the crystal quartz resonator, and the atomic resonators. *RLC* circuits are the simplest and easiest to construct and therefore are the more frequently used oscillator. However component imperfections and aging often makes it difficult to set and maintain a precise tone frequency over long time intervals. Crystal oscillators use the crystal structure itself as a resonant circuit (or a component of a resonant

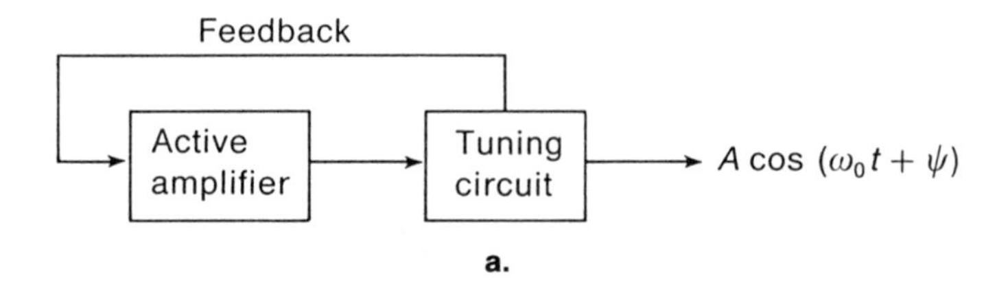

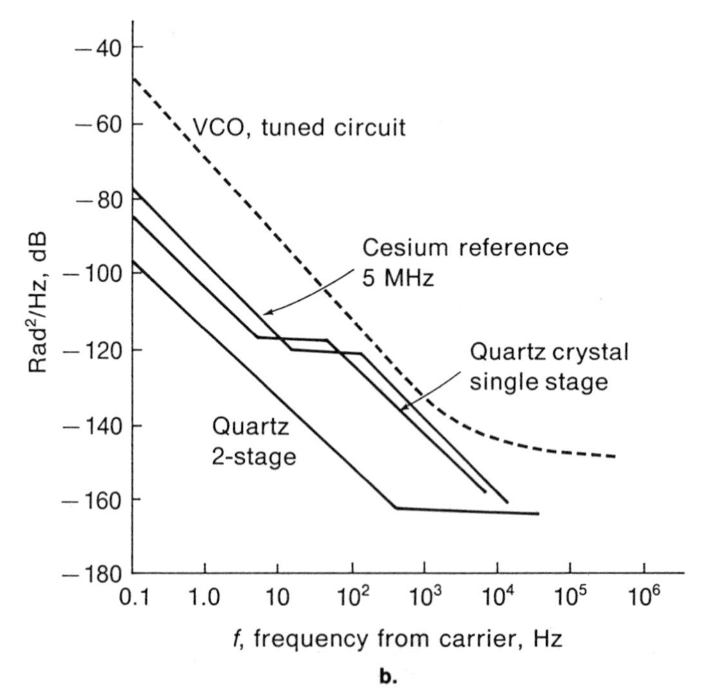

FIGURE 4.16 Oscillators: (a) circuit diagram; (b) typical phase-noise spectra.

circuit) to produce sharply tuned resonances and relatively stable output tones. A cut quartz crystal with properly attached electrodes acts like a high-Q series–parallel electronic resonant circuit (Frerking, 1978; Heising, 1946; Matthys, 1983; Parzen, 1983). The frequency of the oscillation can be set by either the series resonance or the parallel resonances of the crystal. These resonant frequencies depend on the way the crystal is cut and formed and the crystal thickness. The parallel resonant frequency is usually used for oscillator frequencies below about 20 MHz, since high-frequency oscillation with parallel resonance requires extremely thin crystal cuts. Hence, the series resonance of the crystal is usually used for the higher frequencies. Quartz crystal has a sharply tuned resonance, producing highly stable oscillators. Crystal oscillators at 5 and 10 MHz are common communication elements in modern RF systems.

The common atomic resonator is the cesium beam, which uses a stream of cesium atoms to interact with a magnetic field so as to produce an almost perfect oscillator at the specific frequency 9.152 GHz. Rubidium resonators, using light beams interacting with rubidium vapor, produce a fixed oscillation at 6.8 GHz. Atomic oscillators are often inserted as frequency measurement standards and are used primarily as reference tones for systems requiring extreme frequency accuracy, such as time measurement and ranging systems.

Often there is need to vary the oscillator output frequency according to an external control signal. This is achieved by using the control voltage to vary or adjust the tuning circuit parameter, resulting in an output frequency directly proportional to the control. Such devices are referred to as *voltage-controlled oscillators* (VCOs). *RLC* oscillators are most often used as a VCO, since electronic elements such as resistors and capacitors are easily adjusted, and voltage control over a relatively wide frequency range is usually feasible. Voltage-controlled crystal oscillators (VCXOs) can also be used, but frequency range is often limited, because of the sharp tuning of the crystal structures. Hence, a VCXO is only used if the "pulling range" of the oscillator is relatively small.

An ideal oscillator produces a pure sinusoidal carrier with fixed amplitude, frequency, and phase. Practical oscillators, however, produce carrier waveforms with parameters that may vary in time, owing to temperature changes, component aging, and inherent tuning circuit noise. Amplitude variations are somewhat tolerable since they can be easily controlled with electronic clipper circuits and limiting amplifiers. More important to a communication system are the variations in frequency and phase that may appear on an oscillator output. Although preliminary system design may be based on the supposition of ideal carriers, the possibility of imperfect oscillators and the degradation they may produce must eventually be considered.

Frequency offsets in oscillators are usually specified as a fraction of the oscillator design frequency. This fraction is generally normalized by a 10^{-6} factor and stated in units of parts per million (ppm). An offset of Δf Hz in an oscillator designed for f_0 Hz output frequency will therefore be stated as having an offset of $(\Delta f/f_0)10^6$ ppm. Thus, for example, a 5-MHz oscillator, specified as having a stability of ± 2 ppm, will be expected to produce an output frequency that is within $\pm 2 \times 10^{-6} \times 5 \times 10^6 = \pm 10$ Hz of the desired 5-MHz output.

Oscillator frequency offsets are contributed primarily by frequency uncertainty (inability to exactly set the desired frequency), frequency drift (long-term variations due to component changes, stated in ppm/s), and short-term, random frequency variations. The latter appears as a phase jitter on the oscillator, which effectively converts the fixed carrier phase of an ideal oscillator to a randomly varying phase noise process. This phase noise typically has a spectrum that is predominantly low frequency, extending out to about several kilohertz. Figure 4.16b shows some typical phase-noise spectra for some common oscillators operating at 5 MHz. In general *RLC* oscillators tend to have higher phase noise than crystal oscillators, whereas atomic resonators have the lowest phase noise. Since oscillator phase noise will add directly to any phase or frequency modulation placed on that carrier, phase noise will always be of primary concern in angle-modulated systems and phase-tracking subsystems. Some specific phase noise effects will be considered in Chapter 9 and Appendix B.

Frequency generators are constructed from oscillators by multiplying a reference oscillator carrier to the desired frequencies needed throughout the transmitter. Multiplication of an oscillator carrier frequency can be achieved by straightforward generation of its harmonics via an electronic nonlinearity. The most common is a simple squaring device followed by a bandpass filter, as shown in Figure 4.17a. If the oscillator produces the reference tone $A\cos(\omega_0 t + \psi)$, its square is

$$[A\cos(\omega_0 t + \psi)]^2 = \tfrac{1}{2}A^2 + \tfrac{1}{2}A^2\cos(2\omega_0 t + 2\psi) \qquad (4.5.1)$$

This corresponds to a constant (dc) signal plus the second harmonic tone at $2\omega_0$. The latter is extracted by the bandpass filter centered at $2\omega_0$, producing a direct frequency doubling at the output. Higher-order nonlinearities can be used, but the higher harmonics tend to be reduced progressively in amplitude. Additional amplification is then necessary to restore the desired carrier amplitude. Note that the phase is also multiplied by the same frequency multiplication factor. This means that any phase noise on the input carrier will have its spectrum scaled up during frequency multiplication.

By cascading N sequential stages of a "times two" multiplier, the reference can be multiplied to $2^N\omega_0$ (Figure 4.17b). Multiplier chains of this type are commonly used in modern systems to generate multiple carrier frequencies from a single reference oscillator, as shown. Subsystems of this type, producing all the required carriers from a common source, are referred to as *master frequency generators* (MFGs).

Frequency multiplication can also be achieved by frequency synthesizers (Rohde, 1983), as shown in Figure 4.18a. These are feedback systems

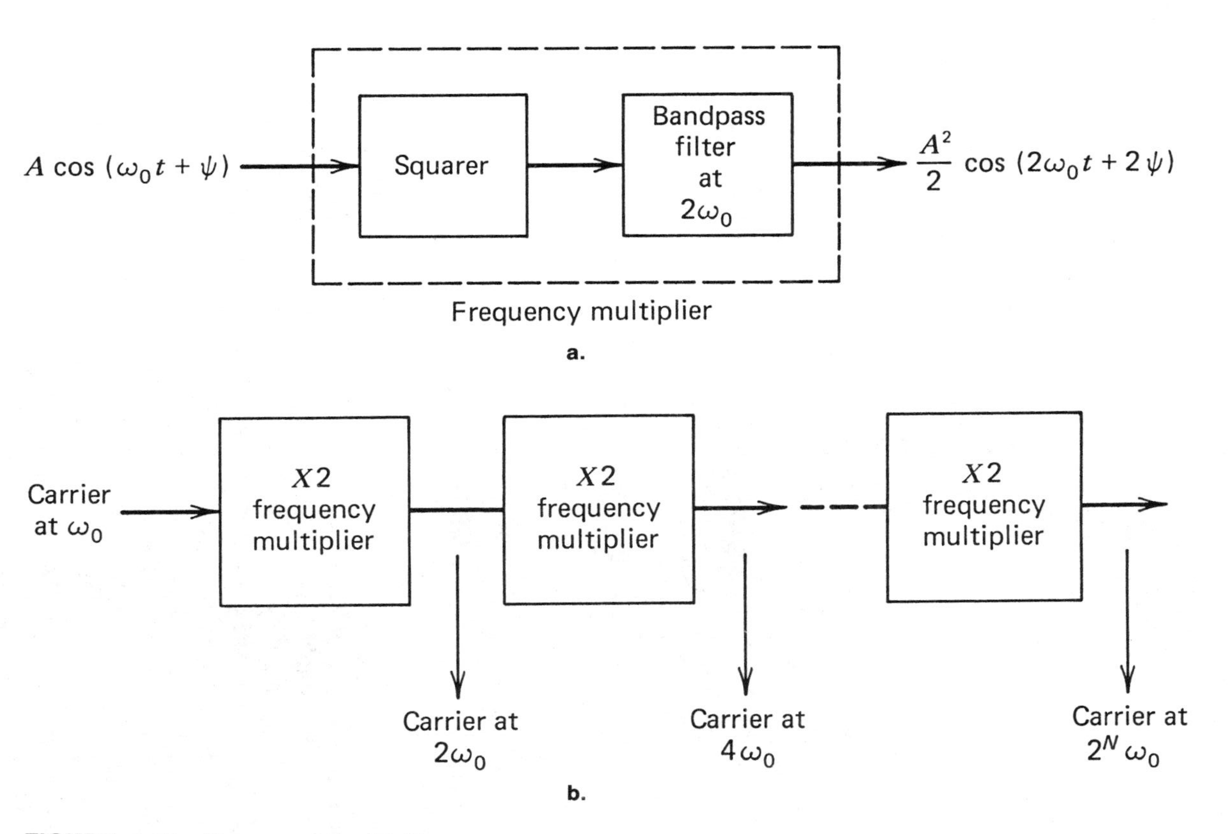

FIGURE 4.17 Frequency multipliers.

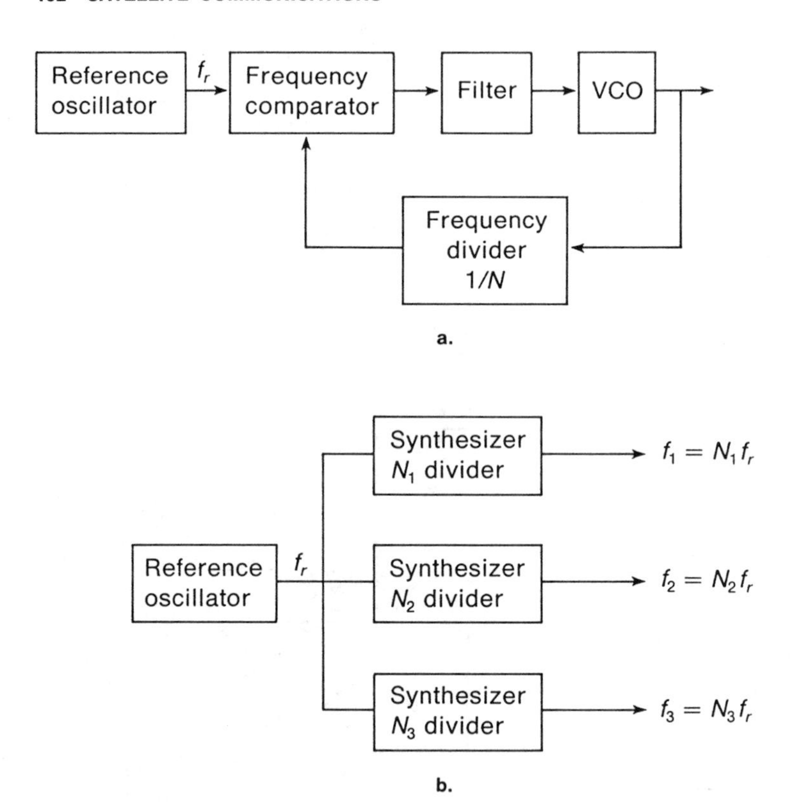

FIGURE 4.18 Frequency synthesizers: (a) block diagram; (b) frequency generator using multiple synthesizers.

in which the reference oscillator is used to control the frequency of a separate VCO, via a frequency-divider circuit. Divide-down counters are obtained by simple cycle counting subsystems that produce an output pulse every N input cycle zero crossings. A low-pass filter then passes the first harmonic of the pulse sequence, having a frequency of $1/N$th of the divider input frequency. If the VCO output frequency in Figure 4.18a is at f_0 Hz, then the divide-down frequency is at f_0/N Hz. A frequency comparison circuit produces an output error voltage proportional to the frequency difference between the reference carrier frequency f_r and the divider output carrier. This error voltage is then used to control the VCO frequency so as to null out the frequency difference. The loop therefore stabilizes such that $f_0/N - f_r = 0$, or at the VCO frequency

$$f_0 = Nf_r \tag{4.5.2}$$

Hence, the synthesizer (VCO) output frequency is maintained at a frequency that is a direct multiplication of the input reference oscillator frequency by the divided factor N. Since divider circuits can be easily constructed with relatively high (2000–4000) factors, synthesizers can be achieved with similar multiplication factors. Furthermore, since the reference oscillator is generally selected as an extremely stable reference, the

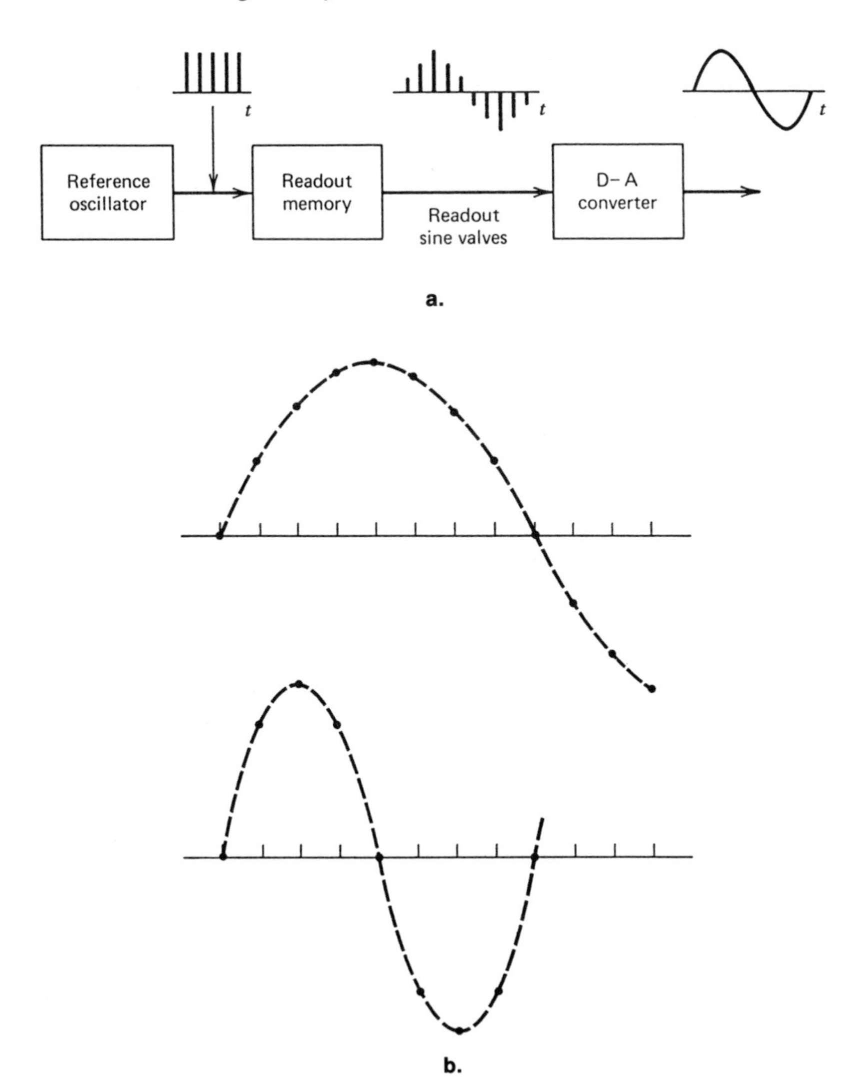

FIGURE 4.19 (a) Block diagram. (b) Read-out samples.

output frequency of the VCO will be maintained with a similar stability. Thus, the feedback operation effectively shifts the frequency stability of the reference to the higher-frequency VCO. Frequency synthesizers are becoming increasingly important in modern high-frequency systems, and entire frequency generators can be constructed by using a single reference to control a multiple set of VCO frequencies via parallel divide-down feedback loops (Figure 4.18b).

There has also been rapid advancement in the development of digital frequency generators (DFG). These devices use digital software in conjunction with a stable reference oscillator to produce a subharmonic output carrier. A DFG operates by storing in memory a sequence of numerical values of one period of a sine wave. These numbers are then continually read out in sequence at a prescribed clock rate driven by the reference oscillator. This produces in series the discrete numerical values of a sine wave, which are then converted to an analog waveform by digital–analog (D–A) conversion, forming the output carrier, as shown in Figure 4.19a.

Since the sine wave values (including the zero crossings) are clocked by the reference oscillator, the output frequency contains the periodic stability of the reference, with the voltage accuracy of the stored samples. If there are M stored sine words in memory, the output frequency in Figure 4.19a will be $f_1 = f_0/M$, where f_0 is the reference frequency.

Suppose that we now alter the readout from memory by reading out every other value at the same clock rate (Figure 4.19b). We now produce a higher output frequency at $f_2 = 2f_1 = 2f_0/M$. By reading out every fourth sine wave value, the output frequency is again doubled. Continuing in this manner, the output frequency can increase to $f_0/4$ (at least four values, including zero crossings, are needed to define a sine wave). Hence, a DFG is primarily used for generating low-frequency, fine-resolution harmonics, with the lower limit depending only on the storage capability of the memory.

4.6 TRANSPONDER LIMITING

Often transponder signal processing includes hard limiting following frequency conversion. The ideal hard limiter is a nonlinear gain device having the input–output characteristics shown in Figure 4.20. Hard limiters are usually followed by bandpass filters tuned to the input carrier frequency, and the combined limiter–filter is referred to as a *bandpass*

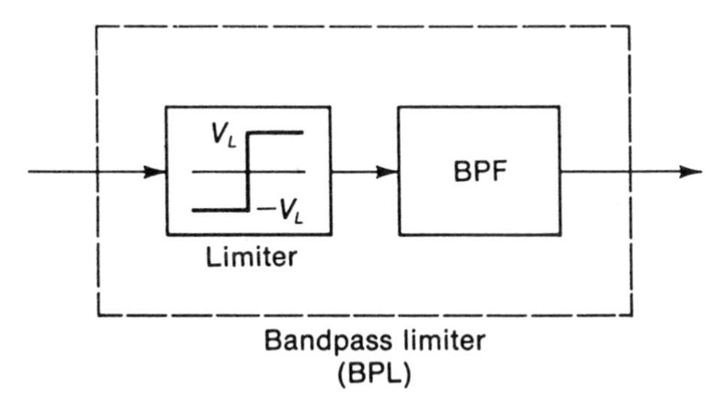

FIGURE 4.20 Bandpass limiter model.

limiter (BPL). Let us again write the combined RF carrier and noise at the BPL input as

$$x(t) = \alpha(t) \cos[\omega_c t + \theta(t)] \tag{4.6.1}$$

The output of the BPL is then

$$y(t) = \left(\frac{4V_L}{\pi}\right) \cos[\omega_c t + \theta(t)] \tag{4.6.2}$$

The output produces an RF carrier with the same phase and frequency modulation as the input but with a constant amplitude level. The BPL eliminates amplitude modulation while preserving angle modulation. It is obvious that transponders with BPL are designed for constant-envelope carriers. Note that the power at the output of a bandpass limiter is always given by

$$P_L = \frac{1}{2}\left(\frac{4V_L}{\pi}\right)^2 = \frac{8V_L^2}{\pi^2} \tag{4.6.3}$$

Since the limiter level V_L can be preset, the power of the BPL output can be accurately adjusted. Hence, a BPL serves as a simple power and amplitude control device. They are usually placed in transponders to prevent large amplitude swings and to set power levels for the remaining circuitry.

When the input to the BPL is the sum of an angle-modulated carrier plus additive receiver noise, it is often desirable to know the extent by which the carrier waveform has been preserved in passing through the BPL. This is difficult to determine from Eq. (4.6.3), since the limiter output noise is incorporated entirely into the phase noise of the carrier. However, it is known (Gagliardi, 1988, Chap. 4) that the carrier power at the BPL output is given by

$$P_{co} = \left(\frac{2V_L^2}{\pi}\right)\mathrm{CNR}_i e^{-\mathrm{CNR}_i}\left[I_0\left(\frac{\mathrm{CNR}_i}{2}\right) + I_1\left(\frac{\mathrm{CNR}_i}{2}\right)\right]^2 \tag{4.6.4}$$

where $I_0(x)$ and $I_1(x)$ are imaginary Bessel functions of order 0 and 1, respectively, and CNR_i is the input CNR of the BPL,

$$\mathrm{CNR}_i = \frac{P_c}{N_0 B_{\mathrm{RF}}} \tag{4.6.5}$$

Since the total power at the output of the BPL is P_L in Eq. (4.6.3), it follows that the noise must constitute the difference. Hence, the BPL output noise power is

$$P_{no} = P_L - P_{co} \tag{4.6.6}$$

The resulting bandpass limiter output CNR is then

$$\begin{aligned}\mathrm{CNR}_{\mathrm{BL}} &= \frac{P_{co}}{P_{no}} \\ &= \frac{P_{co}/P_L}{1 - (P_{co}/P_L)}\end{aligned} \tag{4.6.7}$$

From Eq. (4.6.4)

$$\frac{P_{co}}{P_L} = \left(\frac{\pi}{4}\right)(\mathrm{CNR}_i)e^{-\mathrm{CNR}_i}\left[I_0\left(\frac{\mathrm{CNR}_i}{2}\right) + I_1\left(\frac{\mathrm{CNR}_i}{2}\right)\right]^2 \tag{4.6.8}$$

A plot of the normalized ratio

$$\Gamma = \frac{\mathrm{CNR}_{\mathrm{BL}}}{\mathrm{CNR}_i} \tag{4.6.9}$$

is shown in Figure 4.21 as a function of CNR_i. The result shows the way in which the CNR is altered in passing through a BPL. Note that the effect of the BPL is to cause an increase in the CNR if the ratio is large but to cause a slight degradation (by about 2 dB) if the input CNR

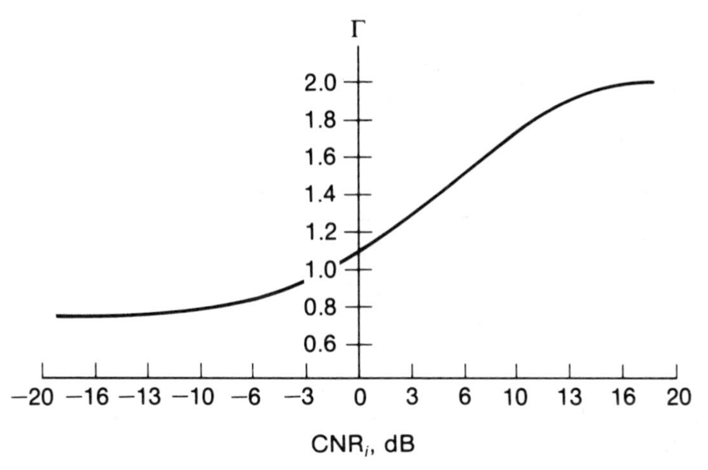

FIGURE 4.21 Output/Input CNR ratio versus input CNR for BPL ($\Gamma = \mathrm{CNR_{BL}}/\mathrm{CNR}_i$).

is low. Whereas pure frequency translation did not alter the CNR, we find that bandpass limiting does. This means that in retransmitting the limiter output by a linear power amplifier, the ground receiver CNR_d in Eq. (4.4.6) is instead

$$\begin{aligned}\mathrm{CNR}_d &= \frac{(\Gamma\mathrm{CNR}_u)(\mathrm{CNR}_r)}{\Gamma\mathrm{CNR}_u + \mathrm{CNR}_r} \\ &= [(\Gamma\mathrm{CNR}_u)^{-1} + (\mathrm{CNR}_r)^{-1}]^{-1}\end{aligned} \tag{4.6.10}$$

with Γ dependent on CNR through Figure 4.21. Hence, limiting modifies the effect of uplink CNR on the downlink receiver.

Since the CNR is altered in passing through the limiter, we can interpret limiting as effectively modifying the input carrier and noise powers at the output. Hence, we denote α_s^2 and α_n^2 as the limiter carrier and noise *suppression factors*, defined by

$$\begin{aligned}P_{co} &= \alpha_s^2 P_c \\ P_{no} &= \alpha_n^2 (N_0 B_{\mathrm{RF}})\end{aligned} \tag{4.6.11}$$

Since $\mathrm{CNR_{BL}} = P_{co}/P_{no}$, it follows that

$$\mathrm{CNR_{BL}} = \left(\frac{\alpha_s^2}{\alpha_n^2}\right)\mathrm{CNR}_i \tag{4.6.12}$$

This therefore defines the suppression factor ratio as

$$\frac{\alpha_s^2}{\alpha_n^2} = \Gamma \tag{4.6.13}$$

Thus, Figure 4.21 actually plots the ratio of the carrier and noise factors given in Eq. (4.6.11). This means the suppression factors themselves depend on the input CNR_i. To determine the individual suppression values, we use Eqs. (4.6.6) and (4.6.11) to write

$$\alpha_s^2 P_c + \alpha_n^2 (N_0 B_{\text{RF}}) = P_L \tag{4.6.14}$$

Substituting from Eq. (4.6.13) and solving for the limiter output carrier power $\alpha_s^2 P_c$ defines the suppression factor as

$$\alpha_s^2 P_c = P_L \left[\frac{\text{CNR}_{\text{BL}}}{1 + \text{CNR}_{\text{BL}}} \right] \tag{4.6.15a}$$

Similarly, we obtain the noise suppression factor with

$$\alpha_n^2 (N_0 B_{\text{RF}}) = P_L \left[\frac{1}{1 + \text{CNR}_{\text{BL}}} \right] \tag{4.6.15b}$$

Thus, the α_s^2 and α_n^2 power suppression factors in passing through a limiter are such that they divide the available BPL output power P_L in accordance with the ratios in Eq. (4.6.15).

An equally important consideration with BPL is the effect of the limiting on the spectrum of the carrier. Let us consider a filtered quadrature carrier as in Eq. (4.3.8) at the BPL input,

$$\tilde{c}(t) = A\tilde{m}_c(t) \cos(\omega_c t + \psi) + A\tilde{m}_s(t) \sin(\omega_c t + \psi) \tag{4.6.16}$$

The output of the BPL is given by Eq. (4.6.2), where

$$\theta(t) = \tan^{-1} \left[\frac{\tilde{m}_s(t)}{\tilde{m}_c(t)} \right] \tag{4.6.17}$$

If we write Eq. (4.6.2) as a quadrature carrier

$$y(t) = f_c(t) \cos(\omega_c t + \psi) + f_s(t) \sin(\omega_c t + \psi) \tag{4.6.18}$$

it is evident that

$$[f_c^2(t) + f_s^2(t)]^{1/2} = 4V_L/\pi \tag{4.6.19}$$

and

$$\frac{\tilde{m}_s(t)}{\tilde{m}_c(t)} = \frac{f_s(t)}{f_c(t)} \tag{4.6.20}$$

Solution of these equations requires

$$f_c(t) \triangleq \frac{\tilde{m}_c(t)}{[\tilde{m}_c^2(t) + \tilde{m}_s^2(t)]^{1/2}} \left[\frac{4V_L}{\pi}\right] \tag{4.5.21a}$$

$$f_s(t) = \frac{\tilde{m}_s(t)}{[\tilde{m}_c^2(t) + \tilde{m}_s^2(t)]^{1/2}} \left[\frac{4V_L}{\pi}\right] \tag{4.6.21b}$$

We see that the BPL has converted the filtered quadrature carrier with components $\tilde{m}_c(t)$ and $\tilde{m}_s(t)$ into the new components in Eq. (4.6.21). In particular, the limiting has introduced coupling between the QPSK components, since both $f_c(t)$ and $f_s(t)$ depend on both data bit sequences. For example, in an offset QPSK format, if $\tilde{m}_s(t)$ changes bit sign during a bit time of $\tilde{m}_c(t)$, it is clear that $f_s(t)$ changes bit sign also. However, since $\tilde{m}_s(t)$ passes through zero during the bit change, $f_c(t)$ in Eq. (4.6.21a) must undergo a temporary transition to the value 1 during the bit change. Hence, $f_c(t)$ develops blips in its waveform at the transitions of the quadrature bit changes, that is, during the middle of its own bit patterns (Figure 4.22). In QPSK, bit changes occur simultaneously when they occur, and the blips are effectively superimposed on each bit transition. We see that hard limiting has caused independent bit changes in the quadrature channels to be coupled effectively into other channels as waveform perturbations. These perturbations tend to increase the bandwidth of the limited carrier.

The spectrum of the limited carrier $y(t)$ in Eq. (4.6.18) can be computed as in Eq. (4.3.2). This leads to

$$S_y(\omega) = \frac{1}{T}[|F_c(\omega)|^2 + |F_s(\omega)|^2]_{\omega \pm \omega_c} \tag{4.6.22}$$

where $F_c(\omega)$ and $F_s(\omega)$ are the spectra of $f_c(t)$ and $f_s(t)$, respectively, in Eq. (4.6.21). In general, these spectra are difficult to compute since they

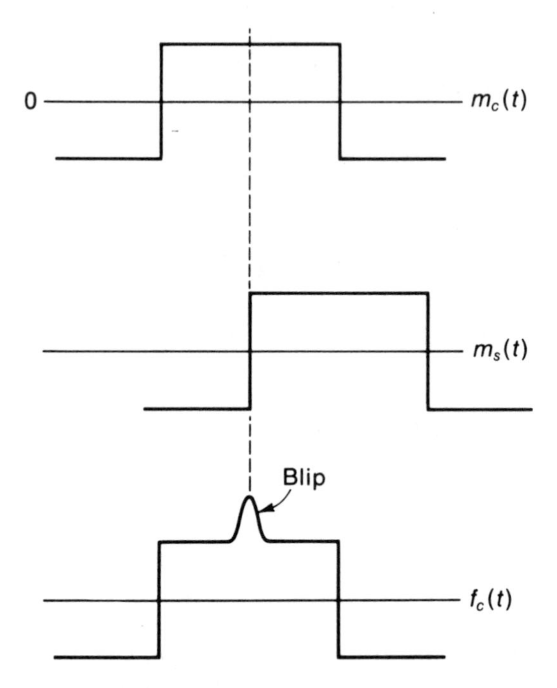

FIGURE 4.22 Blip development in QPSK with limiting.

involve ratios of time functions and two different data sequences, and are usually evaluated by simulation or numerical transform techniques. Figure 4.23 shows published (Morais and Feher, 1980) spectral results indicating the effect of hard limiting on the filtered forms of QPSK, OQPSK, and MSK carrier waveforms. We see that while filtering successfully reduces the spreading of the tails (as we discussed in Section 4.3), the hard limiting tends to restore the tails to prefiltered values. The result is that the limited carrier in the transponder may no longer satisfy spectral masks for which the filtered carrier was designed.

From Figure 4.23, we see that standard QPSK has the higher degree of tail restoration, whereas offsetting tends to reduce the amount of this regeneration, as exhibited by the OQPSK and MSK systems. This can possibly be attributed to the fact that filtered offset carriers, such as MSK, tend to retain a more constant envelope [the denominators in Eq. (4.6.21) are nearly constant] so that $f_c(t)$ and $f_s(t)$ are simply scaled versions of the corresponding prelimited components. In this case the hard-limited carrier retains the spectra of the filtered offset carrier in spite of the limiting. The reduced tail regeneration of offset carriers has been found to occur also in other types of bit waveforms, such as raised-cosine pulses (Divsalar and Simon, 1981, 1982).

Note that MSK, which has a slightly wider main hump spectrum than QPSK (see Figure 2.14), may often be filtered to reduce this main lobe in attempting to satisfy a satellite mask. We see from Figure 4.23 that while limiting only partly restores the tails, it restores almost fully the main lobe. This may now violate the required mask, even though the tail reduction may be satisfactory.

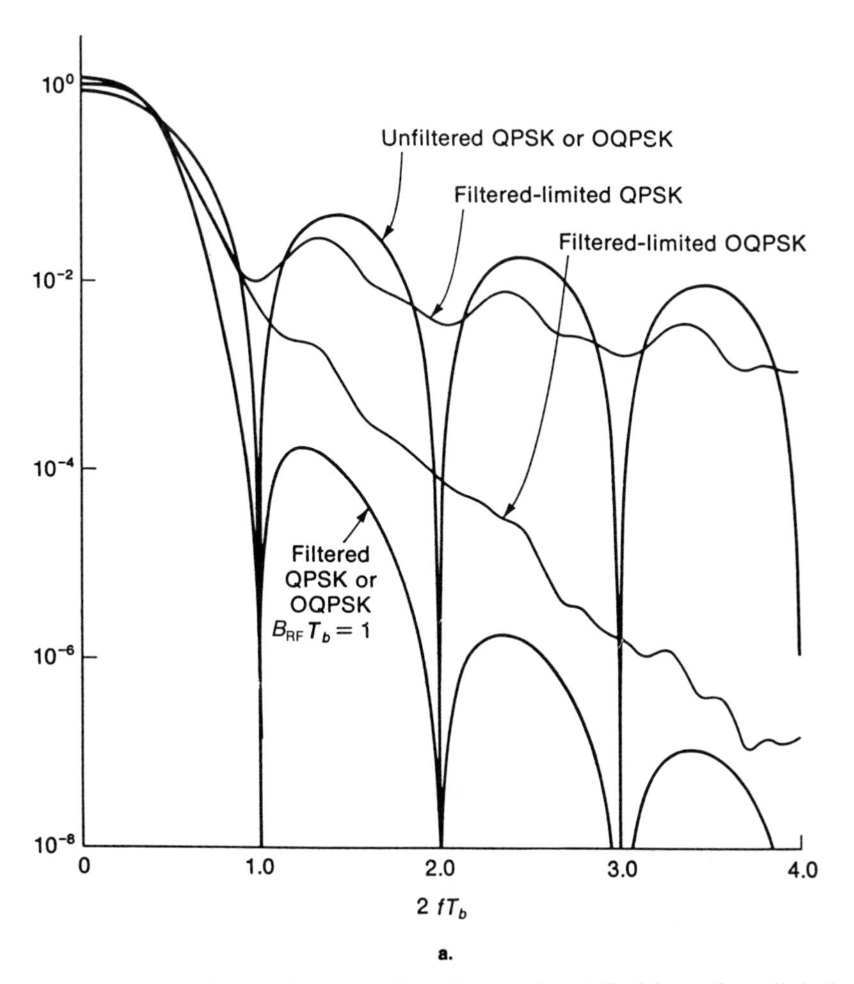

FIGURE 4.23 Spectral restoration due to hard limiting after digital carrier bandpass filtering: (a) QPSK and OQPSK; (b) MSK.

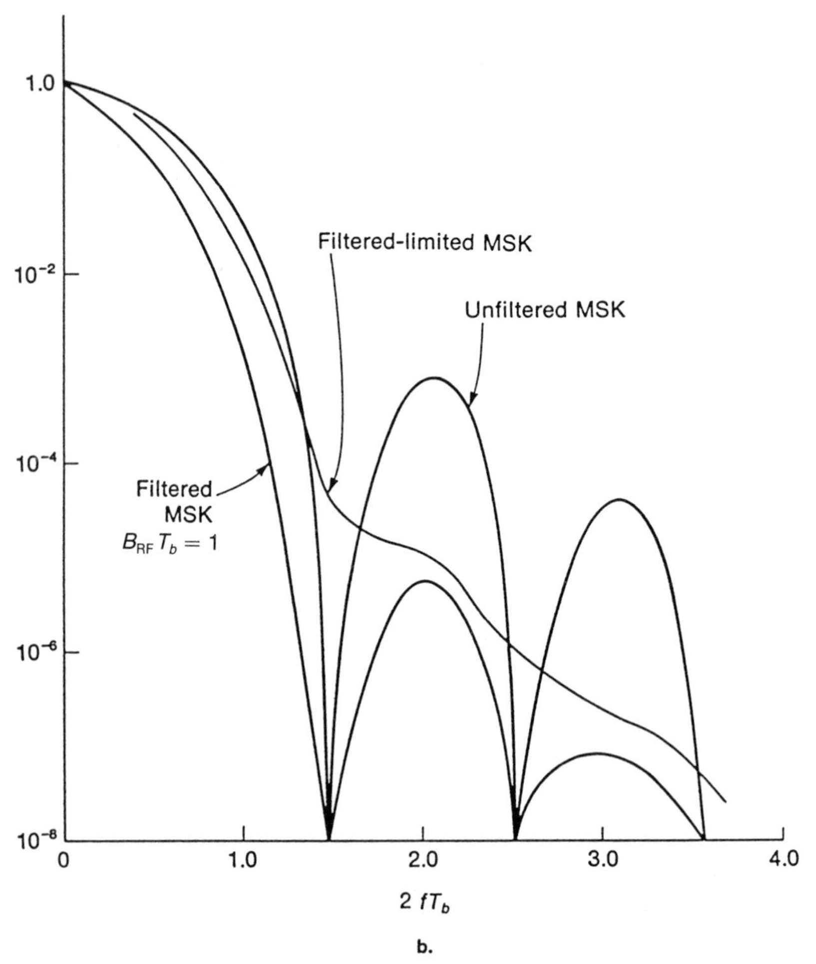

FIGURE 4.23 (continued) Spectral restoration due to hard limiting after digital carrier bandpass filtering: (a) QPSK and OQPSK; (b) MSK.

4.7 NONLINEAR SATELLITE AMPLIFIERS

The traveling-wave-tube amplifier (TWTA) is the commonly used power amplifier for satellites. TWTAs achieve their power gain by using the input microwave carrier to phase-control resonant waves in a cavity so as to produce wave reinforcement. An output from the resonant cavity or tube, then, is an amplified, phase replica of the input carrier. Such mechanisms achieve the significantly large power gains needed for the satellite trans-

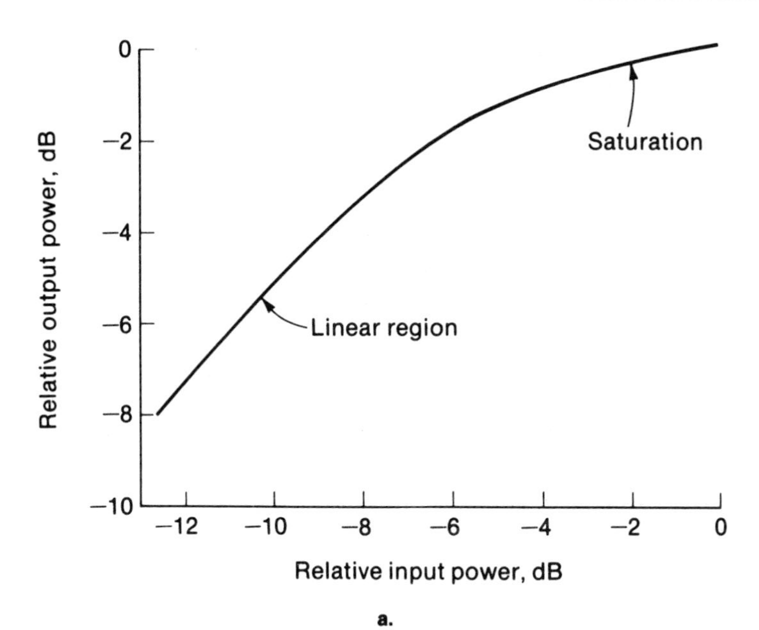

a.

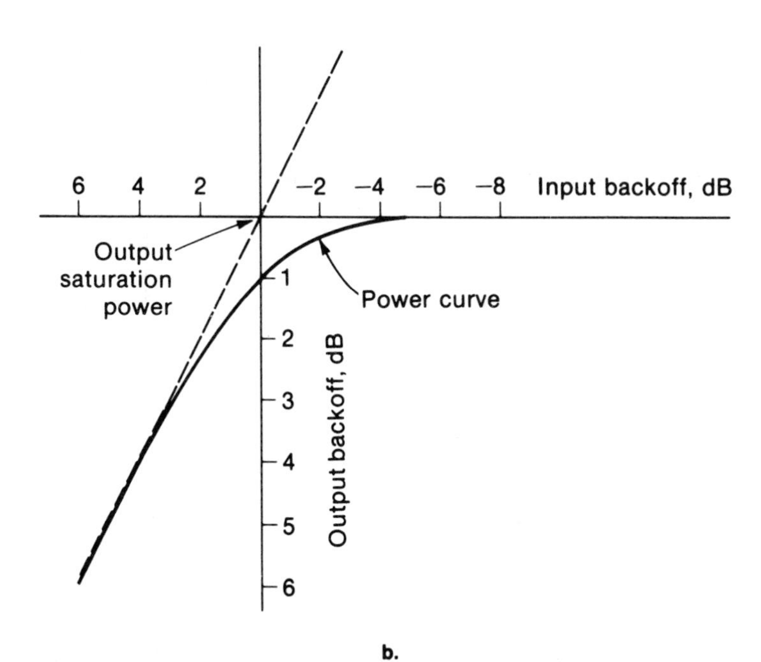

b.

FIGURE 4.24 TWTA power-gain characteristic: (a) normalized power; (b) standardized for backoff definitions.

ponder links. Since TWTAs amplify phase-modulated carriers, they are intended primarily for constant-amplitude carriers. Variations in input carrier amplitudes during amplification produce an additional unintentional phase modulation that appears as phase interference on the amplified carrier. This effect will be examined later in the section.

For constant-amplitude input carriers there is a direct gain conversion to the output amplitude. As the amplitude of the input carrier is increased, the output amplitude is also increased until a saturation effect occurs within the cavity. This is exhibited by the input–output power curve of the TWTA, which is typically of the form shown in Figure 4.24a. As the input power is increased, a direct linear gain occurs in output power until the output power saturates, as shown, and further increase in input power no longer produces larger outputs. Achievement of this maximum output power is therefore accompanied by a nonlinear amplification within the amplifier as the saturation condition is approached. When only a single input carrier is involved, this saturation causes no carrier distortion, but only a limitation to its output power. When the TWTA input corresponds to multiple carriers, however, as in an FDMA format, this nonlinearity of the saturation effect becomes extremely important, as will be discussed in Chapter 5.

The drive power at which the output power saturation occurs is called the *input saturation power*. The ratio of input saturation power to desired drive power is called the amplifier *input backoff*. Increasing input backoff (decreasing input drive power) produces less output power but improves the linearity of the device, since the degree of nonlinearity is reduced. The *output saturation power* of the amplifier is the maximum total power available from the amplifier. *Output backoff* is the ratio of the maximum output (saturation) power to actual output power. Output backoff obviously depends on input backoff, that is, where the drive power is operated. Increasing input backoff lowers the output power and increases the output backoff.

Since the actual input power at which saturation occurs may be difficult to specify exactly, backoff definitions are sometimes defined relative to a standardized power curve as shown in Figure 4.24b. The 0-dB value of input backoff is specified at the point where the output power is 1 dB below saturation. This means that full output saturation will often occur at a negative decibel value of input backoff.

Proper input control of the operating drive power of an amplifier is important in TWTA operation. Power control for the TWTA is often obtained by a BPL–amplifier combination, similar to that shown in Figure 4.25. Since the limiter produces fixed-output power levels, proper gain adjustment of the drive amplifier can carefully set the TWTA input to

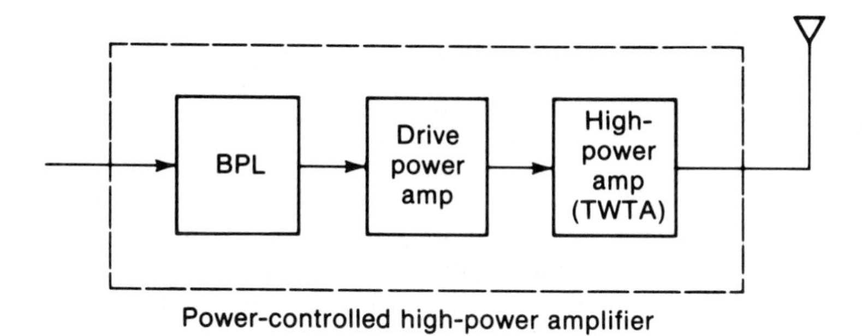

FIGURE 4.25 Power-controlled TWTA subsystem.

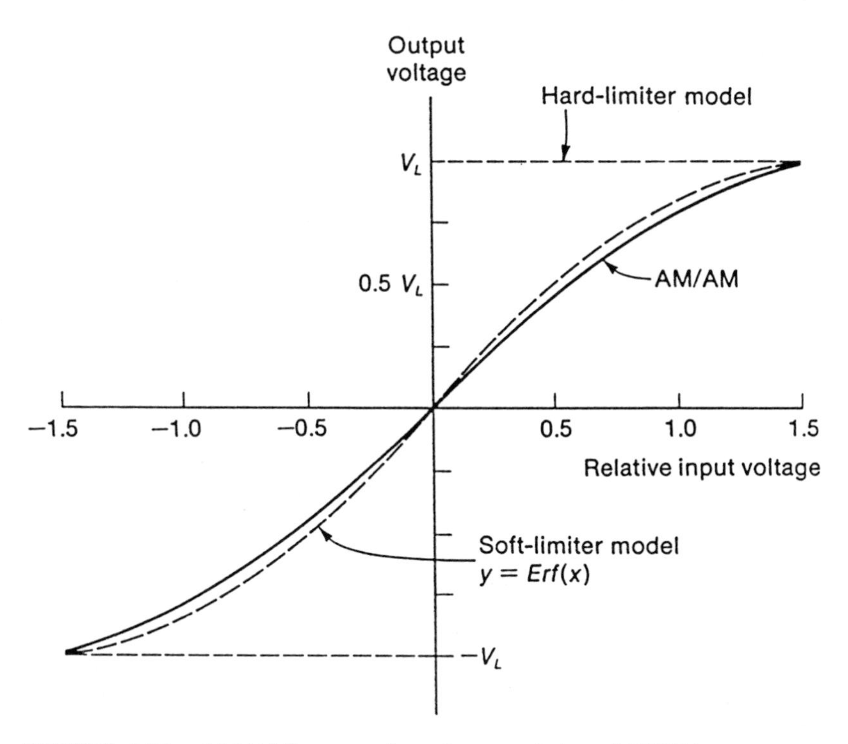

FIGURE 4.26 AM/AM conversion characteristics of TWTA.

desired input backoff. This backoff setting can be extremely critical in satellite operation for achieving satisfactory downlink performance.

AM/AM Conversion

For constant amplitude carriers, carrier power is simply one-half the square of the amplitude. Hence, a TWTA power curve can be simply square-rooted to obtain the corresponding input–output amplitude curve. Figure 4.26 shows amplitude characteristics corresponding to TWTA power curves. This amplitude curve is often called the *AM/AM conversion* characteristics of the TWTA. When the input backoff is high, operation occurs on the linear part of the amplitude characteristic. For less backoff, the TWTA operation enters into the nonlinear region. For strong input drive powers, the amplifier is almost always in the saturated region, and the AM/AM conversion can be considered to have a hard-limiter characteristic, as shown. For intermediate operation, the characteristic is referred to as a *soft-limiter* curve, which accounts for the conversion between linear and hard-limiting operation.

Several mathematical forms for the TWTA AM/AM curve have been proposed. The most convenient is

$$\begin{aligned} g(\alpha) &= V_L \operatorname{Erf}\left(\frac{\alpha}{b}\right), \qquad \alpha \geq 0 \\ &= -V_L \operatorname{Erf}\left(\frac{|\alpha|}{b}\right), \qquad \alpha \leq 0 \end{aligned} \tag{4.7.1}$$

where α is the input carrier amplitude and Erf(x) is the function

$$\operatorname{Erf}(x) = \frac{2}{\sqrt{\pi}} \int_0^x e^{-u^2}\, du \tag{4.7.2}$$

The parameter b defines the input saturation voltage, and the corresponding input saturation power is then $b^2/2$. The input backoff is then

$$\beta_i = \frac{b^2/2}{\alpha^2/2} = \left(\frac{b}{\alpha}\right)^2 \tag{4.7.3}$$

Other functional models for AM/AM conversion have been suggested for modeling tunnel diode amplifiers (Forsey et al., 1978). Although slightly more accurate in representing TDA AM/AM conversion, these functions tend to be analytically unwieldy.

Analysis of the nonlinear gain in Eq. (4.7.1) is carried out in detail in Appendix D. It is shown that if a single phase-modulated carrier with amplitude A is amplified in the nonlinear gain device in Eq. (4.7.1), the power in the output carrier is given by

$$P_T = P_{\text{sat}}\left[\frac{1}{\pi}\int_0^\infty J_1(Az)e^{-b^2z^2/2}\,dz/z\right]^2 \tag{4.7.4}$$

where P_{sat} is the output saturation power and $J_1(x)$ is the first-order Bessel function. Increasing input backoff (reducing A relative to b) reduces the output power, since the amplifier is operated further from saturation. This can be seen by substituting $u = Az$, and rewriting Eq. (4.7.4) in terms of β_i in Eq. (4.7.3) as

$$P_T = P_{\text{sat}}\left[\frac{1}{\pi}\int_0^\infty J_1(u)e^{-\beta_i^2u^2/2}\,\text{du/u}\right]^2 \tag{4.7.5}$$

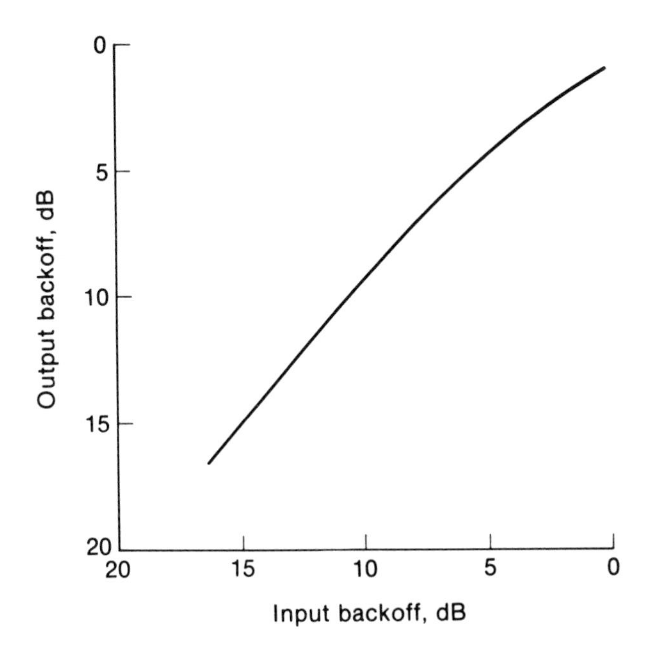

FIGURE 4.27 Output backoff versus input backoff for Erf(x) soft-limiting characteristics.

The output backoff is then

$$\beta_o = \frac{P_{sat}}{P_T} \tag{4.7.6}$$

Equation (4.7.6) is plotted in Figure 4.27 in terms of input backoff β_i for a single carrier input. This figure demonstrates how increasing input backoff to improve satellite linearity also reduces the available downlink carrier power for satellite retransmission.

AM/PM Conversion

TWTAs are designed to operate with constant-amplitude input carriers to achieve cavity gain. Variations in input amplitude produce an unintentional phase modulation on the amplified carrier, referred to as *AM/PM conversion.* AM/PM conversion is a form of carrier distortion in which envelope variations on the total multiple-carrier waveform being amplified are converted into phase variations on each individual carrier. These phase variations appear as additive waveform interference in angle-modulated carriers. Although the possibility of this conversion exists theoretically in any nonlinear device, its presence in cavity amplifiers is physically caused by the amplification mechanism of the cavity device itself. Envelope variations on the cavity field being amplified cause a variable retardation on the field in the cavity. This time-varying retardation appears as a phase-delay variation, or a phase modulation—in synchronism with the envelope variations—on the cavity field. The resulting amplifier output waveform then has an additive phase modulation proportional to the envelope variation.

When the envelope variations are due to thermal noise, the additive phase variation appears as added random phase-noise interference. When the envelope variation is due to baseband modulation (either intentional or not), the AM/PM of the nonlinear power amplifier may cause the modulated information to be coupled into the carrier as a form of intelligible crosstalk.

To examine AM/PM conversion analytically, again consider the amplifier input to be the general carrier waveform

$$x(t) = [A + \Delta(t)] \cos[\omega_c t + \theta(t)] \tag{4.7.7}$$

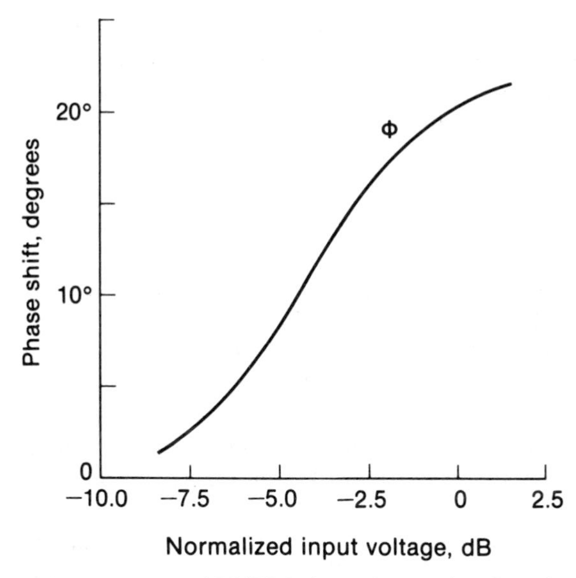

FIGURE 4.28 AM/PM phase conversion function versus input backoff.

where $\Delta(t)$ represents amplitude variations around a fixed level A. AM/PM conversion of the amplifier causes the amplifier output carrier to have phase

$$\Omega(t) = \omega_c t + \theta(t) + \Phi[\Delta(t)] \tag{4.7.8}$$

where $\Phi(\Delta)$ is called the AM/PM conversion function. A typical plot of $\Phi(\Delta)$ is shown in Figure 4.28, showing the manner in which the phase is coupled in. The actual form of this function depends on the particular characteristics of the amplifier tube itself.

A simple model, valid for small levels of input amplitude variation, is to assume $\Phi(\Delta)$ is linear with respect to amplitude variations around the selected bias amplitude A in Eq. (4.7.7). We can then write

$$\Phi[\Delta(t)] = \eta\Delta(t) \tag{4.7.9}$$

The coefficient η is often called the AM/PM *conversion coefficient* of the amplifier. Its value depends on the degree of nonlinearity of the amplifier operation, which in turn depends on the input backoff through the amplitude A in Eq. (4.7.7). When receiver noise is added to the uplink carrier, as in Eq. (4.2.10), and the envelope is written as in Eq. (4.2.12), we can see from Eq. (4.7.8) that an additional phase noise is coupled into the downlink carrier by the amplifier nonlinearity. Thus, AM/PM noise is due to the uplink in-phase noise, and produces a phase noise of power $\eta^2 P_{nu}$, where P_{nu} is the satellite uplink noise power.

REFERENCES

Atia, A. E. (1976), "Computer-Aided Design of Waveguide Multiplexers," *IEEE Trans. MTT*, **MTT-22** (March) 332–336.

Atia, A. E., Williams, A. E., and Newcomb, R. W. (1974), "Narrow-Band Multiple-Coupled Cavity Synthesis," *IEEE Trans. Circuits Syst.*, **CAS-21** (Sept.), 649–654.

Divsalar, D., and Simon, M. (1981), "Performance of Overlapped Raised Cosine Modulation over Nonlinear Channels," *Proceedings of the ICC*, Denver, CO.

Divsalar, D., and Simon, M., (1982), "The Power Spectral Density of Digital Modulations Transmitted over Nonlinear Channels," *IEEE Trans. Comm.*, **COM-26.**

Forsey, R., Gooding, V., McLane, P., and Campbell, L. (1978), "M-ary PSK Transmission Via a Coherent 2-link Channel with AM-AM and AM-PM Nonlinearities," *IEEE Trans. Comm.*, **Com-26** (January),116–123.

Frerking, M. 1978, *Crystal Oscillator Design*, Van Nostrand-Reinhold, New York.

Gagliardi, R. (1988), *Introduction to Communication Engineering*, 2nd ed., Wiley, New York.

Heising, R. (1946), *Quartz Crystals and Electronic Circuits*, Van Nostrand-Reinhold, New York.

Kurzrok, R. M. (1966), "General Four-Resonator Filters in Waveguide," *IEEE Trans. MTT*, **MTT-16** (January), 46–47.

Kudsia, C. M., and O'Donovan, V., (1974), *Microwave Filters for Communications Systems*, Artech House, Dedham, MA.

Matthaei, G. L., Young, L., and Jones, E. M. T. (1964), *Microwave Filters, Impedance Matching Networks and Coupling Structures*, McGraw-Hill, New York.

Matthys, R. (1983), *Crystal Oscillator Circuits*, Wiley, New York.

Morais, D., and Feher, K. (1980), "The Effects of Filtering and Limiting on the Performance of QPSK and MSK Systems," *IEEE Trans. Comm.*, **COM-28** (December), 2152–2163.

Parzen, B. (1983), *Design of Crystal and Other Harmonic Oscillators*, Wiley, New York.

Rhodes, J. D. (1969), "The Design and Synthesis of a Class of Microwave Band Pass Linear Phase Filters," *IEEE Trans. MTT*, **MTT-17** (April), 386–399.

Rohde, U. (1983), *Digital PLL Frequency Synthesizers*, Prentice-Hall, Englewood Cliffs, NJ.

Smith, J. (1986), *Modern Communication Circuits*, McGraw-Hill, New York.

Williams, A. E., and Atia, A. E. (1977), "Dual Mode Canonical Waveguide Filters," *IEEE Trans. MTT*, **MTT-25** (Dec.), 1021–1026.

Problems

4.1. The front end of an RF receiver is shown in Figure P4.1. How much power gain G must the tunnel diode have to produce a front-end noise figure of 10 dB?

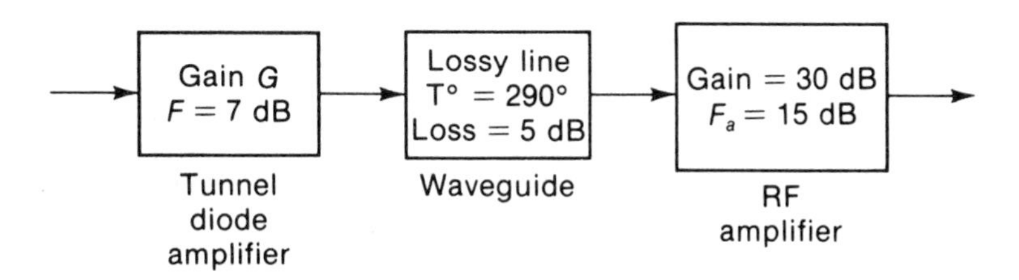

FIGURE P4.1

4.2. Given the parallel combination of two devices with noise figures F_1, F_2 and power gains G_1, G_2. Determine the overall noise figure, using the definition in Eq. (3.4.5).

4.3. A low-pass Butterworth filter has the transfer function

$$H(\omega) = \frac{1}{1 + j\left(\dfrac{\omega}{2\pi f_3}\right)^n}$$

where f_3 is the half-power frequency. Compute its noise bandwidth.

[*Hint*: $\int_0^\infty (1 + x^q)^{-1}\, dx = (\pi/q)/\sin(\pi/q)$.]

4.4. (a) Show that the correlation function of $n_{RF}(t)$ in Eq. (4.2.5) is given by

$$R_{n_{RF}}(\tau) = R_{cc}(\tau) \cos \omega\tau - R_{cs}(\tau) \sin \omega\tau$$

where $R_{cc}(\tau) = \overline{\tilde{n}_c(t)\tilde{n}_c(t+\tau)} = R_{ss}(\tau)$, and $R_{cs}(\tau) = \overline{\tilde{n}_c(t)\tilde{n}_s(t+\tau)} = -R_{cs}(-\tau)$. (b) Compute the spectral density of $n_{RF}(t)$, using the spectrum of $\tilde{n}_c(t)$ and $\tilde{n}_s(t)$ in Figure 4.9 and assuming $R_{cs}(\tau) = 0$.

4.5. *Frequency noise* is defined as the derivative of phase noise $v(t)$ in Eq. (4.2.12b). Show that the power spectral density of frequency noise is quadratic in frequency when the phase noise has a flat spectrum. [*Hint*: Use the fact that dv/dt is the output of a filter whose input is $v(t)$ and whose transfer function is $H(s) = s$.]

4.6. Show that the *rms offset frequency* of an oscillator is the square root of the mean squared value of its phase-noise spectrum.

4.7. Given a sequence of phase values θ_i, $i = 1, 2, 3, \ldots, N$, of an oscillator. The *Allan variance* (AV) is defined as

$$\mathrm{AV} = \frac{1}{N-3} \sum_{n=3}^{N} (\theta_n - \theta_{n-1})^2 - (\theta_{n-1} - \theta_{n-2})^2$$

Show that the AV is an estimate of the difference frequency stability of the oscillator.

4.8. The frequency synthesizer shown in Figure 4.18a is modified to that in Figure P4.8. (a) Determine the frequency at the output at which the system will stabilize. (b) Explain how this design can improve the multiplication factor or the resolution or both of the synthesizer. (*Hint*: Q can be an integer and/or fractional multiplier.)

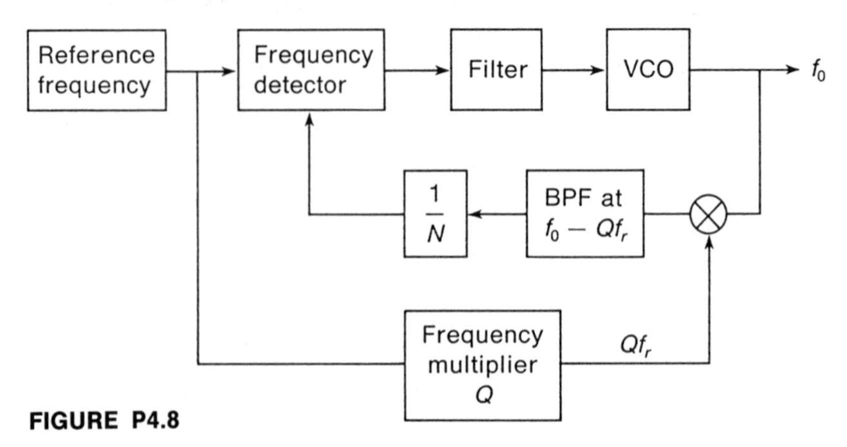

FIGURE P4.8

4.9. Let N sequential values of a sine wave (taken at $360/N$ degrees apart) be stored in memory as k-bit words. These sample words are then read out at a clock rate f_s words/s. (a) Plot the sequence of sample values, carefully labeling the time axis, and connect the points to form a continuous sine wave. (b) Use the same readout rate, but assume that every other sample value is read out, and plot the sequence waveform on the same time axis. (c) Repeat for every fourth sample value. Show that the maximum and minimum frequency of the output sine wave is

$$\text{Minimum frequency} = \frac{f_s}{2^k}$$

$$\text{Maximum frequency} = \frac{f_s}{4}$$

4.10. A *fractional divider* can be obtained as follows. After each N input zero crossings, produce an output crossing. Do this for $M - 1$ output crossings. For the Mth, count one extra input crossing before producing the output crossing. Show that this will fractionally divide the input frequency.

4.11. A carrier squaring device is used to obtain frequency doubling of a unit power carrier. Determine the carrier power loss in decibels for the doubled frequency carrier. Repeat for the tripled frequency carrier when a cubic device is used.

4.12. The carrier-plus-noise waveform into a hard-limiter amplifier has CNR = 3 dB. The amplifier output power is 10 W. Estimate how much power is in the output carrier and in the output noise at the amplifier output. Repeat for CNR = 10 dB and -3 dB.

4.13. An ISL is operated with an FM carrier and has a total carrier bandwidth of 2 GHz. The downlink requires a CNR of 40 dB. (a) What receiver CNR must the ISL provide in an RF–RF conversion system? (b) An IF remodulation ISL uses a 450-MHz carrier with a bandwidth of 100 MHz. What receiver CNR must this ISL have? (c) Compare the results in (a) and (b) in terms of antenna sizes.

4.14. A TWTA has the AM/AM gain function $g_1(\alpha)$ and AM/PM function $\phi_1(\alpha)$. Show that the required corrective nonlinearity to be inserted after the TWTA to linearize its gain and phase is given by $g_2(\alpha) = kg_1^{-1}(\alpha)$ and $\phi_2(\alpha) = -\phi_1[g_1^{-1}(\alpha)] + \psi$. (b) Is the same nonlinearity used before the TWTA for linear operation?

CHAPTER 5
Frequency-Division Multiple Access

In Chapter 4 we analyzed a single-channel transponder with a single uplink carrier. We now extend the discussion to transponders designed for multiple carriers. Recall that when multiple carriers are used in satellite communications, it is necessary that a multiple-accessing format be established over the system. This format allows distinct separation of the uplink transmissions in passing through the satellite processor. In Section 1.6 three of the most common multiple-access formats were described. Each format has its own specific characteristics, advantages, and disadvantages. Satellite anomalies, such as nonlinear amplification and power division, will therefore have widely different effects on system performance for each method. Hence, it is necessary to carry out separate analysis procedures to assess analytically (and therefore design) satellite systems of each type. In this and the next three chapters we present comparisons of the accessing formats described in that section. Our basic objective is to describe the relationship, analytically and graphically, between key system parameters of the link and the established performance criterion, such as SNR and bit-error probability, of the system receivers. In this chapter we concentrate on *frequency-division multiple accessing* (FDMA) systems.

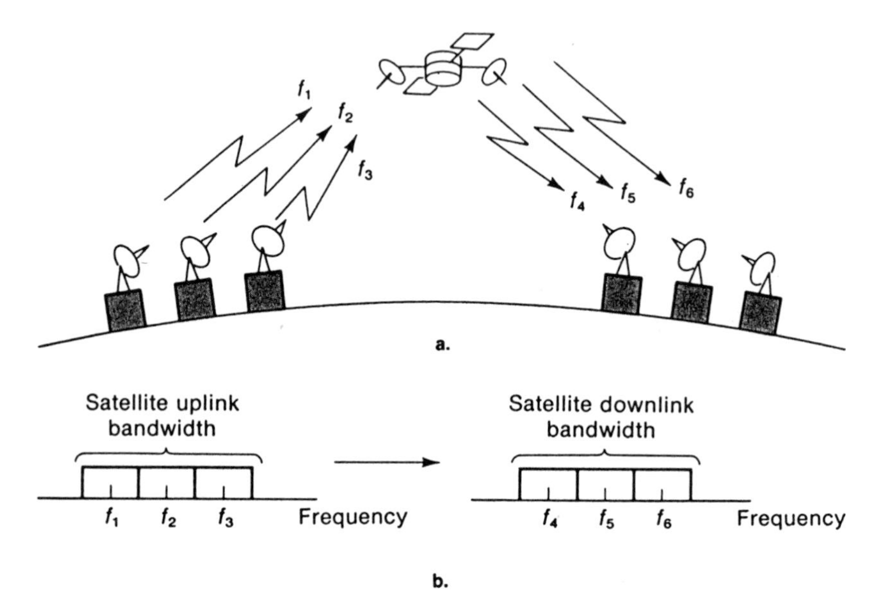

FIGURE 5.1 FDMA system model: (a) block diagram; (b) frequency plan.

5.1 THE FDMA SYSTEM

The basic model of an FDMA satellite system is shown in Figure 5.1a. A set of earth stations transmit uplink carriers to be relayed simultaneously by the satellite to various downlink earth stations. Each uplink carrier is assigned a frequency band within the available RF bandwidth of the satellite (Figure 5.1b). In the basic satellite transponder, the entire RF frequency spectrum appearing at the satellite input is frequency-translated to form the downlink. A receiving station receives a particular uplink station by tuning to, and filtering off, the proper band in the downlink spectrum. Each carrier can independently be modulated, either analog or digital, from all others. With digital carriers, only synchronization between the desired carrier and the receiving station must be established without regard to other carriers at other frequency bands. FDMA represents the simplest form of multiple accessing, and the required system technology and hardware are almost all readily available in today's communication market.

Each uplink carrier may originate from a separate earth station, or several carriers may be transmitted from a particular station. Frequency

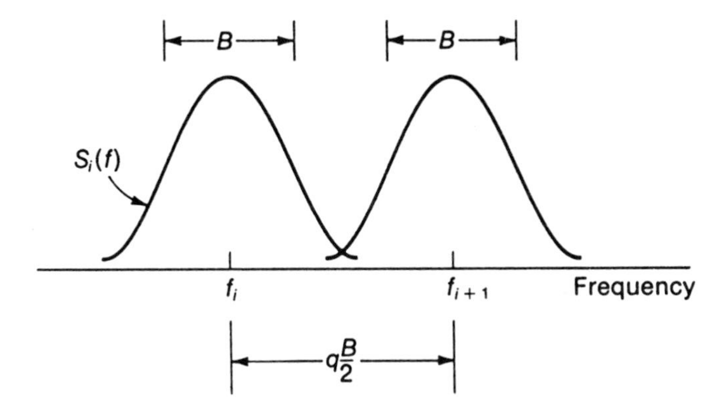

FIGURE 5.2 FDMA adjacent spectra.

band selection may be fixed or assigned. In *fixed-frequency* operation each carrier is assigned a dedicated frequency band for the uplink, and no other carrier utilizes that band. In *demand-assignment multiple access* (DAMA) frequency bands are shared by several carriers, with a particular band assigned at time of need, depending on availability. DAMA systems can serve a larger number of carriers if the usage time of each is relatively low, but may require more complex ground routing hardware (Van Trees, 1979).

Individual carrier spectra in an FDMA system must be sufficiently separated from each other both to allow filtering off of the carriers at the downlink stations and to prevent carrier crosstalk (frequencies of one carrier spectrum falling into the band of another carrier, as shown in Figure 5.2). This is why spectral tails associated with digital carriers was discussed in detail in Sections 2.4 and 4.3. However, excessive separation causes needless waste of satellite bandwidth. To determine the proper spacing between FDMA carrier spectra, crosstalk power must be calculated. Spacings can then be selected for any acceptable crosstalk level desired. Let the ith carrier have a center frequency ω_i, and a one-sided power spectrum given by $S_i(\omega)$. The power of the carrier, referred to the satellite uplink receiver, is then

$$P_i = \frac{1}{2\pi}\int_0^\infty S_i(\omega)dx \qquad (5.1.1)$$

Note that P_i includes the uplink station EIRP, uplink space losses, and satellite antenna and front-end gains. The spectrum $S_i(\omega)$ is assumed to occupy a 3-dB carrier spectral bandwidth of B Hz about ω_i. With the aid of Figure 5.2, we see that the fractional crosstalk power of the ith carrier falling into the adjacent carrier bandwidth is then

$$C_i = \frac{1}{2\pi P_i} \int_{\pi B(q-1)}^{\pi B(q+1)} S_i(\omega_i + x)dx \tag{5.1.2}$$

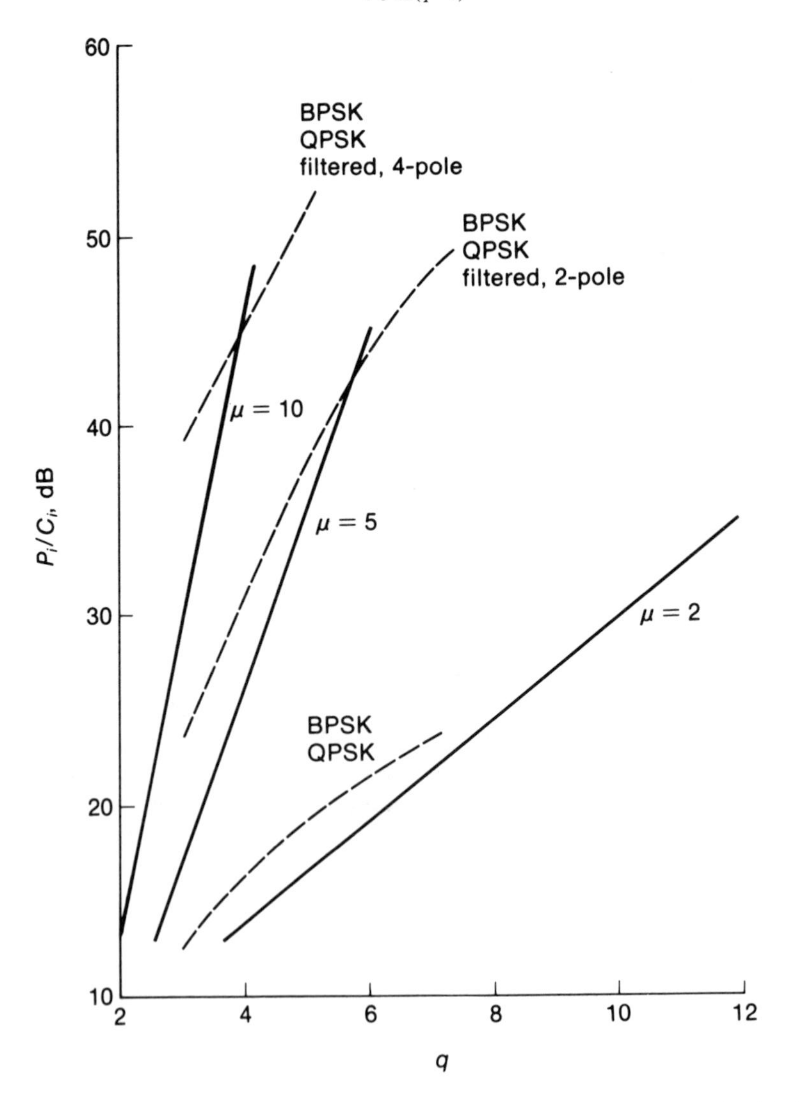

FIGURE 5.3 Signal to crosstalk ratio for adjacent carriers (solid lines, Butterworth spectra, fall-off rate μ; dashed lines, digital carriers).

where the carrier center frequency spacing is denoted $qB/2$. Equation (5.1.2) is plotted in Figure 5.3 for the case of a Butterworth-shaped carrier spectrum

$$S_i(\omega) = \frac{P_i(\mu/\pi B)\sin(\pi/2\mu)}{1 + [(\omega - \omega_i)/2\pi B]^{2\mu}} \tag{5.1.3}$$

where the bandwidths and power of both adjacent carriers are assumed equal. Also included is the corresponding result for several types of digital carrier using the spectral information from Figure 2.7. The result shows the carrier-to-crosstalk power ratio, as a function of the spacing between center frequencies. Note that a spacing of approximately three bandwidths is needed to obtain a crosstalk ratio of more than 20 dB with a $\mu = 2$ Butterworth spectrum, but little more than one bandwidth is needed when $\mu \geq 10$. This latter case corresponds to only a slight separation between carrier spectra. The crosstalk power level should be small enough to produce an interference level significantly smaller than the acceptable receiver noise level.

For digital carriers, separation is plotted in terms of mainlobe bandwidths. Crosstalk is computed based on peak spectral levels falling into adjacent bandwidths. If crosstalk is too high, either carrier separation must be increased, or spectral filtering must be applied (usually at the transmitting earth station) to reduce spectral tails. The latter leads to possible carrier distortion and decoding degradation, as we discussed in section 4.3, and is susceptible to the tail-regeneration problem associated with nonlinear amplification.

When an FDMA format is used with a linear transponding channel, the downlink performance can be determined by extending the single-channel analysis of Section 3.6. Let an RF satellite bandwidth B_{RF} be available to the earth-station carriers, each carrier using an individual bandwidth of B Hz (including spectral spacing). The number of FDMA carriers allowed by the satellite bandwidth is then

$$K = \frac{B_{\mathrm{RF}}}{B} \tag{5.1.4}$$

Let P_{ui} be the ith carrier uplink power at the satellite amplifier input, and let P_{un} be the corresponding uplink noise in the same carrier bandwidth. The total amplifier input power is then

$$P_u = \sum_{i=1}^{K} P_{ui} + KP_{un} \tag{5.1.5}$$

For a linear amplifier transponder, the satellite RF bandwidth is frequency-translated and amplified by the power gain, as in Eq. (3.6.2),

$$G = \frac{P_T}{P_u} = \frac{P_T}{\sum_{i=1}^{K} P_{ui} + KP_{un}} \tag{5.1.6}$$

The downlink receiver power of the ith carrier after amplifying and downlink transmission is then

$$P_{di} = LGP_{ui} = LP_T\left[\frac{P_{ui}}{\sum_{i=1}^{K} P_{ui} + KP_{un}}\right] \tag{5.1.7}$$

where L is again the combined downlink power losses and gains from satellite amplifier output to earth-station receiver input. Note that the power robbing on the downlink carrier that was caused in the single carrier case only by the noise (Section 3.6) is now increased by the additional power robbing of the other carriers.

The total receiver noise is the sum of the transponded uplink noise and the receiver noise. Hence, the downlink CNR of a single carrier in its own bandwidth B is then

$$\text{CNR}_d = \frac{P_{di}}{LGP_{un} + N_{0d}B} \tag{5.1.8}$$

where N_{0d} is the receiver noise spectral level. We again rewrite this as

$$(\text{CNR}_d)^{-1} = (\text{CNR}_u)^{-1} + (\text{CNR}_r)^{-1} \tag{5.1.9}$$

where

$$\text{CNR}_u = \frac{P_{di}}{LGP_{un}} = \frac{P_{ui}}{P_{un}} \tag{5.1.10}$$

$$\text{CNR}_r = \frac{P_{di}}{N_{0d}B} = \frac{P_T L}{N_{0d}B}\left(\frac{P_{ui}}{P_u}\right) \tag{5.1.11}$$

Here CNR_u is the uplink CNR of a single carrier. The receiver CNR_r is that which the satellite power P_T can produce at the receiver, reduced by the power-robbing loss of the uplink. If the system is power-balanced (all carriers have some uplink power) and if the uplink CNR_u is high ($\text{CNR}_u \gg 1$), then $\text{CNR}_r \approx P_T L/KN_{0d}B$. That is, the available satellite P_T

is equally divided among the FDMA downlink carriers. If we solve Eq. (5.1.9) for P_T, we can compute the required satellite power to support an FDMA system with K carriers. This yields

$$P_T = \frac{KN_{0d}B}{L}\left[\frac{\mathrm{CNR}_d}{1-(\mathrm{CNR}_d/\mathrm{CNR}_u)}\right] \tag{5.1.12}$$

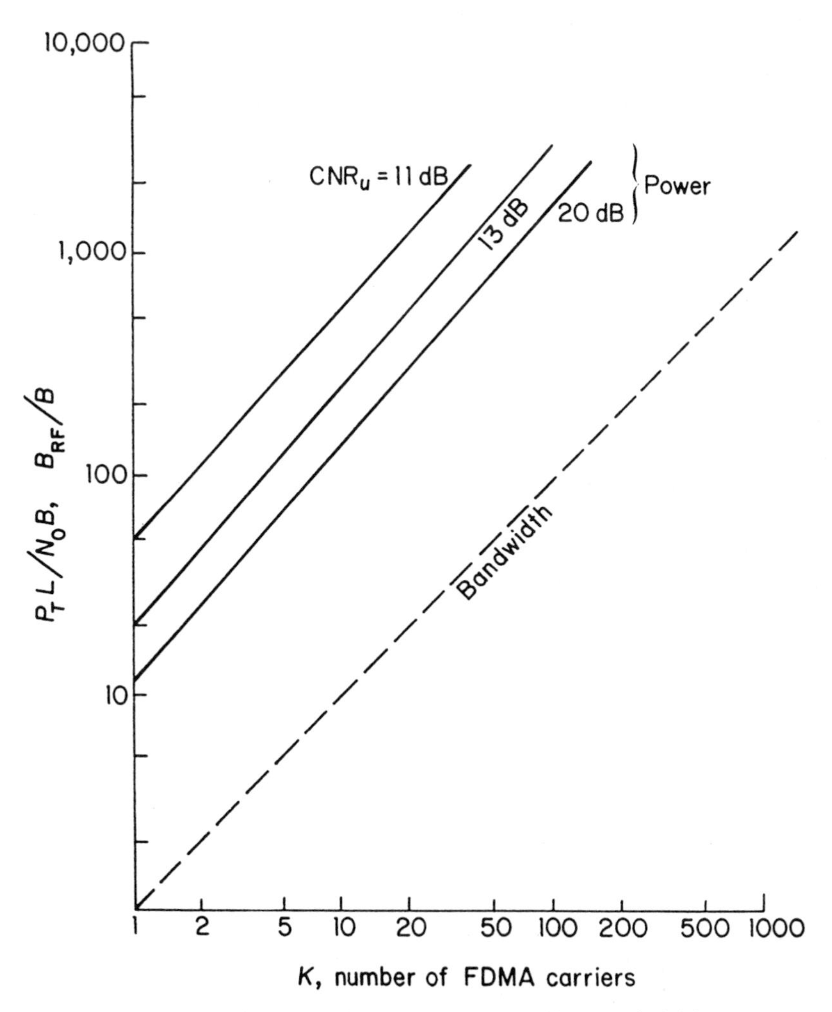

FIGURE 5.4 Required satellite power and satellite bandwidth to support K FDMA carriers; linear channel. $\mathrm{CNR}_d = 10$ dB.

With a specified value of CNR_d and CNR_u we see that P_T increases linearly with the number of carriers. Equation (5.1.12) is plotted in Figure 5.4. Also superimposed is the relation between the satellite bandwidth and the number of carriers in Eq. (5.1.4). This figure can also be used to balance the number of carriers allowed by the satellite power against that allowed by the satellite bandwidth. By entering the ordinate at the proper value of P_T and B_{RF}, we can read off the number of FDMA carriers that can be separately supported by each. The smaller of the two then determines the available FDMA system capacity. Hence, an FDMA system may be either power-limited or bandwidth-limited, in terms of the number of carriers that can be simultaneously supported. In any FDMA design tradeoff, it is always important to know which is the limiting factor, since any techniques to increase power levels or bandwidth will not necessarily increase multiple access capacity if the other parameter determines performance.

5.2 NONLINEAR AMPLIFICATION WITH MULTIPLE FDMA CARRIERS

Nonlinear transponder effects, discussed in Section 4.6 for a single carrier, become even more important when an FDMA format is used. As multiple carriers pass through a transponder, the nonlinear effects of limiters and power amplifiers cause intermodulation products (beats) among these carriers. These beat terms produce additional frequency components at the nonlinear output that can interfere with the desired carriers. This can be seen from the example depicted in Figure 5.5. Suppose two frequency tones (unmodulated carriers) are to be passed simultaneously into the general nonlinear device shown. The inherent nonlinearity of the device will cause the two carriers to beat together, producing cross-products that generate new frequency components at all the multiple combinations of the two input frequencies. The output of the device will therefore have the output frequency spectrum similar to that shown. The amplitude of the individual output carrier terms will depend on the degree of the nonlinearity. If the device is highly nonlinear, there will be many such terms of significant amplitude, possibly dominating the two desired frequencies. For a weak nonlinearity (i.e., a device that is nearly linear), the amplitude of the additional beat frequency terms will be reduced, and only the two input carriers will dominate. When the input tones correspond to modulated carriers, the output cross-product terms involve convolutions of the carrier spectra of the input, centered at all the beat frequencies previously shown. The combined beat terms now appear as

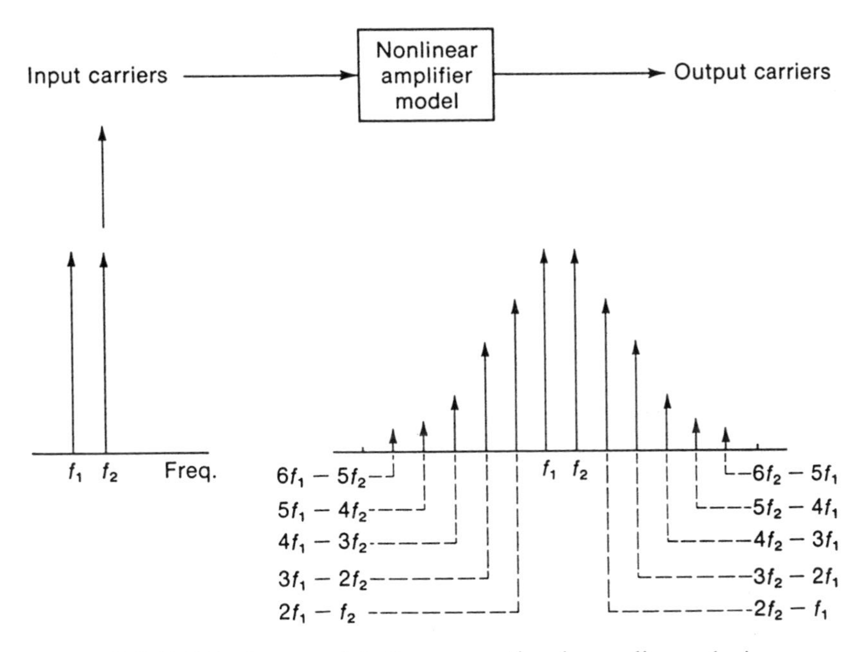

FIGURE 5.5 Intermodulation generation in nonlinear devices.

intermodulation noise distributed over the entire frequency band. This intermodulation noise must be included in assessing FDMA performance. In order to do this, however, it is first necessary to derive a somewhat rigorous nonlinearity model that will analytically account for the intermodulation terms.

A mathematical analysis for the nonlinear amplification of FDMA waveforms can be formulated by extending the analysis of Section 4.7 for the single-carrier case. The voltage nonlinearity in Figure 5.5 is modeled as the function $g(x)$, so that the amplifier output can be written

$$y(t) = g[x(t)] \tag{5.2.1}$$

where, now, the $x(t)$ is the FDMA waveform

$$x(t) = \sum_{i=1}^{K} A_i \sin[\omega_i t + \theta_i(t)] \tag{5.2.2}$$

This represents the sum of K angle-modulated carriers, each with amplitude A_i, frequency ω_i, and phase modulation $\theta_i(t)$. For the nonlinear saturating power amplifier in Eq. (4.7.1) it is shown in Appendix D that the FDMA output can always be written as

$$y(t) = \sum_{m_K=0}^{\infty} \cdots \sum_{m_1=0}^{\infty} h(m_1, \ldots, m_K) \sin\left[\sum_{i=1}^{K} [m_i \omega_i t + m_i \theta_i(t)] \right] \quad (5.2.3)$$

where

$$h(m_1, \ldots, m_K) = \frac{2V_L}{\pi} \int_0^{\infty} \left[\prod_{i=1}^{K} J_{m_i}(A_i u) \right] e^{-b^2u^2/2} \frac{du}{u} \quad (5.2.4)$$

Here V_L is the saturation power of the amplifier and b is the saturation voltage in Figure 4.26. Equation (5.2.3) is the extension of the single-carrier case to the FDMA multiple-carrier case. The output is composed of a multiple summation of carrier sine wave terms. Each term of the sum corresponds to a particular integer vector $(m_1, m_2, \ldots, m_K)$ that generates a specific frequency and phase function for the sine wave. These frequencies involve all the cross-product frequencies of the input carriers, obtained by taking all positive and negative integer combinations of the input frequencies. Each such term contains a phase modulation composed of a similar combination of the input phases. The summation over all such integer vectors defines the total output produced by the nonlinear amplification. The coefficients $h(m_1, \ldots, m_K)$ are the amplitudes of each such term, and $h^2\,(m_1, \ldots, m_K)/2$ is its power. The terms corresponding to the particular vector $[m_i = 0, i \neq j, (m_j) = 1]$ produce the jth carrier at the output, and this will have power

$$P_{cj} = \frac{8V_L^2}{\pi} \left[\int_0^{\infty} u^{-1} J_1(A_j u) \prod_{q=1, q \neq j}^{K} J_0(A_q u) e^{-b^2u^2/2}\, du \right]^2 \quad (5.2.5)$$

The preceding can be evaluated for each particular carrier. Contributions from each cross-product term can be evaluated from Eq. (5.2.4) in a similar manner. In the following we examine such terms for specific power-amplifier models.

Consider first the case where the amplifier is operated in saturation so that a hard-limited model can be used. This can be accounted for by letting $b = 0$ in the previous equations. The output power contributions can be determined by evaluating Eq. (5.2.5) under this condition. For a single input carrier, $K = 1$, the output power is that of a single-bandpass hard-limiter, with level V_L, and represents the maximum output carrier power of the amplifier. In Section 4.7, this was called the saturated output power, and was given as

$$P_{\text{sat}} = \frac{8V_L^2}{\pi} \tag{5.2.6}$$

For the case of two carriers ($K = 2$), with amplitudes A_1 and A_2, respectively, the power in each output carrier, normalized to the available saturation power in Eq. (5.2.6) is then

$$\frac{P_{c_1}}{P_{\text{sat}}} = h^2(1, 0) = \left[\int_0^\infty J_1(A_1 u) J_0(A_2 u) \frac{du}{u}\right]^2$$

$$\frac{P_{c_2}}{P_{\text{sat}}} = h^2(0, 1) = \left[\int_0^\infty J_1(A_2 u) J_0(A_1 u) \frac{du}{u}\right]^2 \tag{5.2.7}$$

The ratio of these output carrier powers is then

$$\frac{P_{c_1}}{P_{c_2}} = \frac{h^2(1, 0)}{h^2(0, 1)} \tag{5.2.8}$$

Equation (5.2.8) is plotted in Figure 5.6 as a function of the input carrier power ratio $(A_1/A_2)^2$. The result exhibits the interesting fact that the output carrier power ratio is not equal to that of the input, but, instead, has the stronger input carrier even stronger at the output. That is, the stronger carrier tends to suppress the weaker carrier during hard-limiting amplification. We see that this suppression approaches a factor of 4 (i.e., the output ratio is 6 dB larger than the input ratio). This effect is referred to as *carrier suppression*, and illustrates the disadvantage of having strong and weak carriers simultaneously amplified in a saturating amplifier.

Similar studies for larger numbers of carriers have been reported, making use of computerized versions of Eq. (5.2.7) (Sevy, 1966; Shaft, 1965, Spoor, 1967). The results for four carriers are shown in Figure 5.7. The curves exhibit the carrier-suppression effect of the stronger carriers on the weaker carriers, for various combinations of input power distributions, in the noiseless case. With one strong carrier a maximum of about 5.5 dB suppression of a weaker carrier occurs, whereas only about 1 dB occurs for three strong carriers. In the case of two strong carriers, the stronger signals are actually suppressed relative to the weaker ones during amplification. (Apparently, the two strong signals destructively interfere with each other, allowing the weaker signal to obtain a larger portion of the available output power.) Note that the 1-dB suppression with multiple

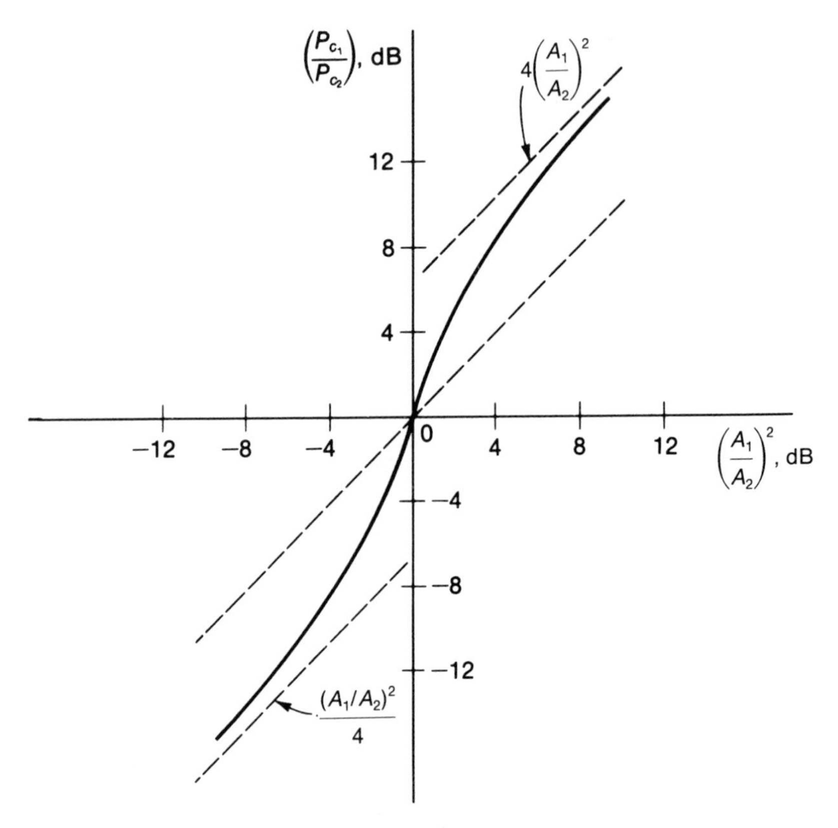

FIGURE 5.6 Two-carrier power suppression with hard-limiting nonlinearities [(A_1/A_2) = input carrier amplitudes; P_{c_1}, P_{c_2} = output carrier power].

carriers is similar to the effect of strong noise on a single carrier for the BPL, as discussed in Section 4.6. Hence, the combination of many carriers appears as an additive noise to a single carrier, as far as hard-limiting suppression is concerned.

The total normalized output carrier power for K equal amplitude carriers is obtained from Eq. (5.2.5) as

$$\frac{P_T(K)}{P_{\text{sat}}} = Kh^2(1, 0, \ldots, 0) = K\left[\int_0^{\infty} J_1(Au)J_0^{K-1}(Au)\frac{du}{u}\right]^2 \tag{5.2.9}$$

The result is plotted in Figure 5.8 as a function of K. The curve shows the hard-limited total power loss as more carriers are simultaneously

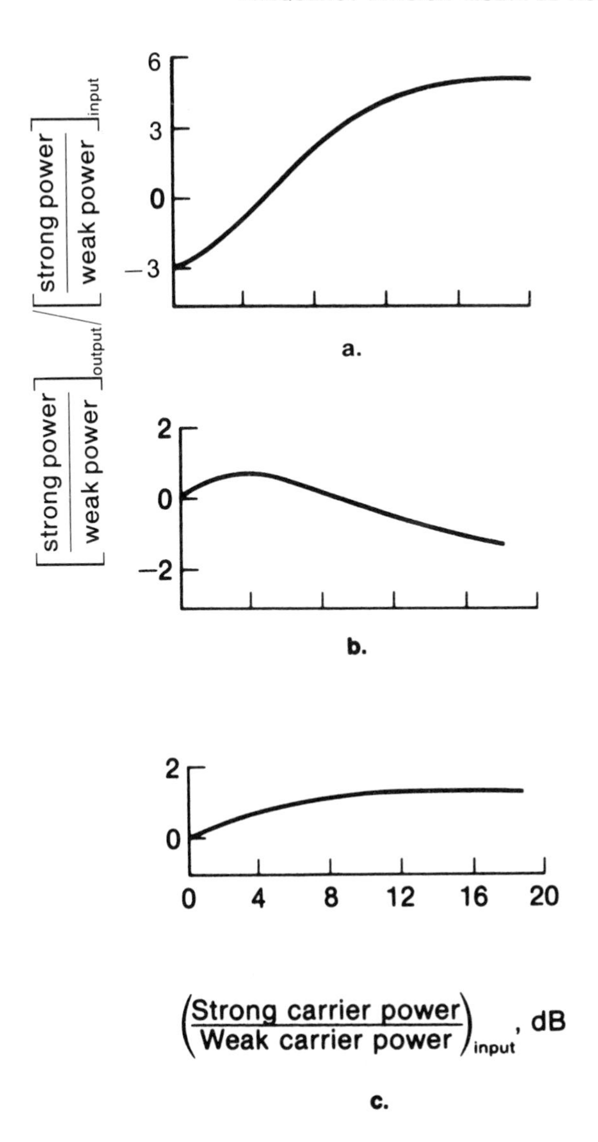

FIGURE 5.7 Four-carrier power suppression with hard-limiting nonlinearities [from Shaft (1965)]: (a) one strong, three weak; (b) two strong, two weak; (c) one weak, three strong.

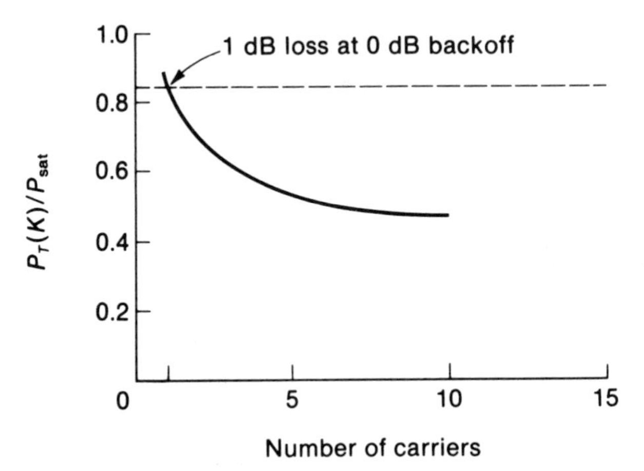

FIGURE 5.8 Decrease in total available output carrier power versus number of carriers; hard-limiting nonlinear amplifier.

amplified. Note that the largest loss occurs during an increase from one to two carriers, and the output power tends to remain relatively constant with K for values beyond about 5. We emphasize that this is the total amplifier power available to all carriers, and therefore each amplified carrier will obtain $1/K$ of this power level.

The remaining terms of Eq. (5.2.4), other than the carrier terms, constitute the intermodulation terms of the amplified output. The important terms, however, are only those whose frequencies fall within the satellite input bandwidth, since the remaining ones will be filtered in the downlink RF bandpass filtering. To determine these interfering terms, it is necessary to compute the actual spectral distribution of the output $y(t)$, noting that Eq. (5.2.3) involves modulated sine waves distributed at all the beat frequencies of the input. The interfering intermodulation terms are those components producing frequencies in the amplifier bandwidth (i.e., the bandwidth occupied by the K input carriers). A particular index vector $(m_1, m_2, \ldots, m_K)$ will generate a frequency

$$\sum_{q=1}^{K} m_q \omega_q \tag{5.2.10}$$

with a total power of

$$h^2(|m_1|, |m_2|, \ldots, |m_k|) = P_{\text{sat}}\left[\int_0^\infty \prod_{q=1}^{K} J_{m_q}(Au)\,\frac{du}{u}\right]^2 \tag{5.2.11}$$

assuming equal amplitude carriers. The sum of terms of Eq. (5.2.3) for all index vectors such that the spectra located at Eq. (5.2.10) falls within the amplifier bandwidth will generate the total interfering intermodulation spectrum. The value of such a sum will depend on the location of the carrier frequencies ω_q in Eq. (5.2.10), and its computation theoretically requires a search over all possible index vectors. However, since Bessel functions decrease quite rapidly with their index, it would be expected that only interfering intermodulation terms with the smaller $|m_q|$ values will contribute most significantly to the in-band interference.

The *order* of a particular intermodulation term is defined as

$$\text{Order} = \sum_{q=1}^{K} |m_q| \tag{5.2.12}$$

With the condition of Eq. (5.2.10), we see that the order of the interfering intermodulation is always odd, with the lowest order being 3. The most significant interference would be derived from terms corresponding to the lower orders, if such intermodulation terms fall in-band. If the carrier frequencies are equally spaced over the amplifier bandwidth, as shown in Figure 5.9, then it is evident that third-order intermodulation terms of the form $(\omega_i + \omega_q - \omega_K)$ and $(2\omega_i - \omega_q)$ can always be found in-band. For equal amplitude carriers, each intermodulation term of the former type will have power level

$$h^2(1, 1, 1, 0, \ldots, 0) = P_{\text{sat}}\left[\int_0^\infty J_1^3(Au)J_0^{K-3}(Au)\,\frac{du}{u}\right]^2 \tag{5.2.13}$$

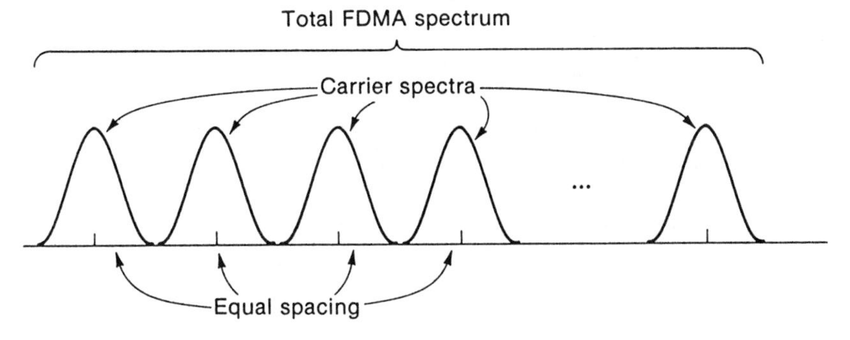

FIGURE 5.9 FDMA spectra.

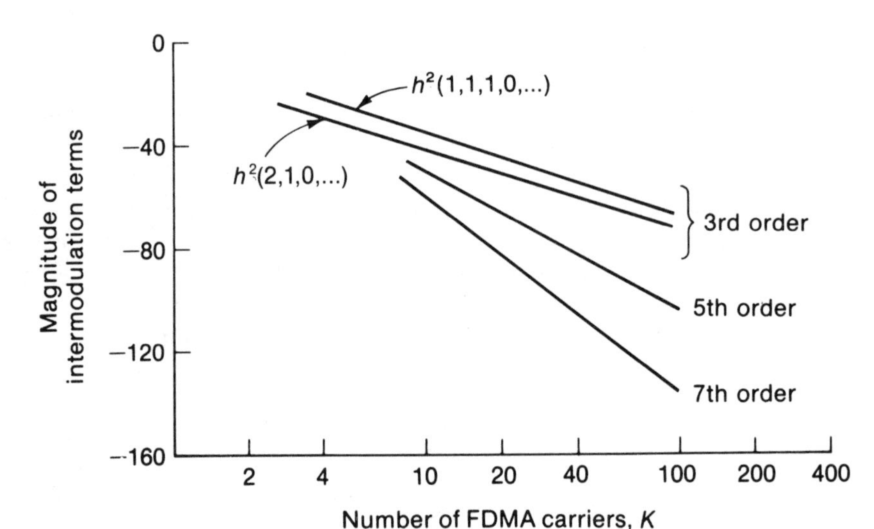

FIGURE 5.10 Magnitude of intermodulation terms versus number of FDMA carriers; hard-limiting nonlinearity.

whereas those of the latter type have power of

$$h^2(2, 1, 0, \ldots, 0) = P_{\text{sat}}\left[\int_0^{\infty} J_2(Au)J_1(Au)J_0^{K-3}(Au)\,\frac{du}{u}\right]^2 \qquad (5.2.14)$$

Equations (5.2.13) and (5.2.14) are plotted in Figure 5.10 as a function of K. Terms corresponding to fifth and seventh orders are included for comparison. Note that the intermodulation power of a particular term decreases approximately as $1/\alpha$, where α is the order. The total intermodulation power of a particular order requires an accumulation of all such terms of each type that produce in-band interference. In general, for equally spaced carriers and large K, approximately α combinations of order α will produce in-band interference (Sevy, 1966). This means that the total intermodulation interference of a particular order tends to remain fairly constant with K, for large numbers of carriers.

Computation of the total intermodulation power in the RF bandwidth neglects the fact that each carrier is immersed in only a portion of the total interference. To determine the interference per carrier bandwidth, it is necessary to determine the actual intermodulation frequency distribution. This requires knowledge of the exact spectral shape of each modulated sine wave in Eq. (5.2.3) in order to determine its frequency contribution. A typical result is shown in Figure 5.11 showing the summed spectral distribution for $K = 20$ Gaussian-shaped carrier spectra in Figure 5.9, with the spectral values normalized to the power of a single carrier. Note that the intermodulation tends to be concentrated in the center of the RF bandwidth, so that center carriers receive the most intermodulation interference. Note also the predominance of the third- and fifth-order intermodulation interference. The peak value of the intermodulation spectrum is often defined as the *intermodulation spectral level.*

Studies have been performed to determine if intermodulation levels can be reduced by separating the individual FDMA carrier spectra so as to lessen the possibility and degree of spectral overlap. The studies have shown that if uneven spacing of the carrier spectra is used and if the overall RF bandwidth is extended, spectral locations can be found such that all possible intermodulation terms of a given order and type will not overlap any carrier spectrum (Babcock, 1953; Westcott, 1967). However, these methods are extremely wasteful of RF bandwidth, and the total required bandwidth is much larger than that occupied by the sum of the carrier bandwidths, because wide spacings must be inserted between the spectra in order to achieve significant intermodulation reduction.

When the input power is not strong enough to force operation into saturation, the actual soft-limiting characteristics of the amplifier must be taken into account. If the amplifier nonlinearity is modeled as in Eq. (4.7.1), then our previous equation [Eq. (5.2.9)] must be examined with $b \neq 0$ in order to determine power loss and intermodulation interference.

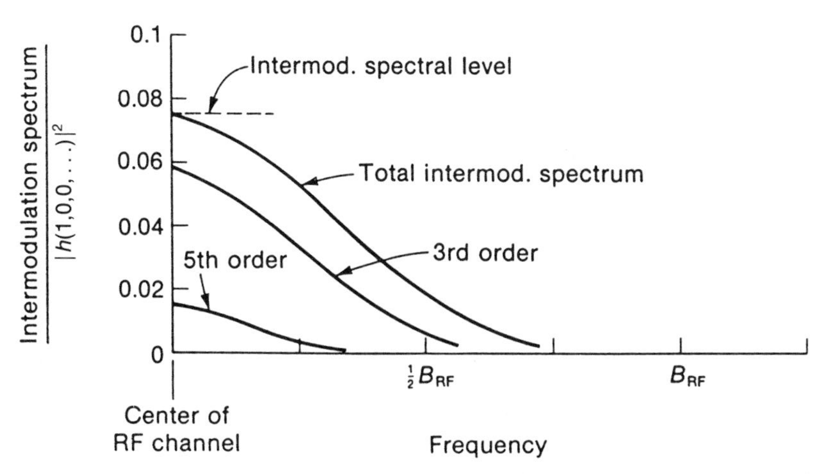

FIGURE 5.11 Intermodulation spectral distribution over FDMA bandwidth; hard-limiting nonlinearity, Gaussian-shaped carrier spectra.

Recall from Eq. (4.7.3) that the parameter b defines the input saturation power and the input backoff as

$$\text{Input } P_{\text{sat}} = \tfrac{1}{2}b^2 \tag{5.2.15}$$

$$\beta_i = \frac{\frac{1}{2}b^2}{P_{\text{in}}} \tag{5.2.16}$$

where P_{in} is the total input power. The amplifier output backoff, β_o, is defined as the ratio of the maximum achievable output power with the given number of carriers to that actually obtained. Hence,

$$\beta_0 = \frac{\max P_T(K)}{P_T(K)} \tag{5.2.17}$$

The maximal power occurs when the amplifier is driven into saturation and is given by the hard-limiting result in Eq. (5.2.6). The actual soft-

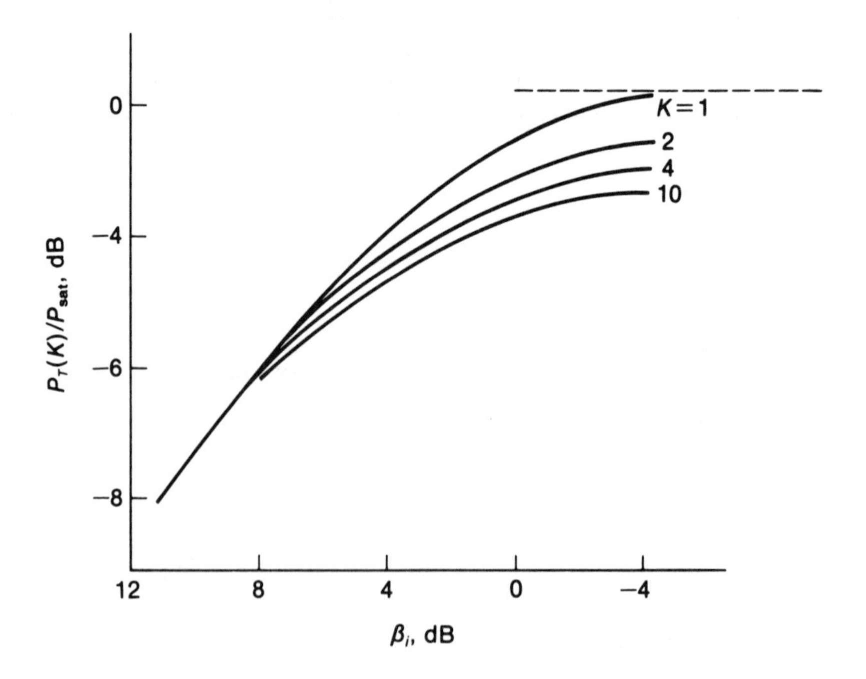

FIGURE 5.12

limited output power obtained with K equal amplitude carriers is

$$P_T(K) = KP_{\text{sat}}\left[\int_0^\infty J_1(Au)J_0^{K-1}(Au)e^{-b^2u^2/2}\, du/u\right]^2 \tag{5.2.18}$$

Using Eq. (5.2.16), and the fact that $P_{\text{in}} = (KA^2/2)$, this can be rewritten directly in terms of the input backoff as

$$P_T(K) = KP_{\text{sat}}\left[\int_0^\infty J_1(Au)J_0^{K-1}(Au)e^{-K\beta_i^2u^2/2}\, du/u\right]^2 \tag{5.2.19}$$

Equation (5.2.19) is plotted in Figure 5.12 as a function of input backoff for several values of K. The asymptotic upper bound is the hard-limited result of Eq. (5.2.9) and is reached only if the amplifier is driven well into saturation. We see, therefore, that the output backoff parameter in Eq. (5.2.17) is a function of both the input backoff and the number of carriers and depends explicitly on the amplifier characteristic. It is obvious that output backoff increases with input backoff for all values of K, and that the two can be related through the power characteristic for a given value of K.

Besides increasing output backoff, increasing the input backoff forces the amplifier to operate more in its linear region, reducing the effective nonlinearity and associated intermodulation. A particular intermodulation term will now have a power contribution of

$$\begin{aligned} &P_{\text{sat}}\left[\int_0^\infty \prod_{q=1}^{K} J_{|m_q|}(Au)e^{-b^2u^2/2}\, du/u\right]^2 \\ &\qquad = P_{\text{sat}}\left[\int_0^\infty \prod_{q=1}^{K} J_{|m_q|}(v)e^{-K\beta_i v^2/2}\, dv/v\right]^2 \end{aligned} \tag{5.2.20}$$

Using the same Gaussian carrier spectra of Figure 5.9, the resultant intermodulation spectrum, similar to that in Figure 5.11, can be computed for a given value of K and various degrees of input backoff. Figure 5.13 shows a plot of the carrier power-to-intermodulation power ratio of the center channel as a function of the input backoff for several values of K. The curve shows the decrease in the intermodulation interference as backoff is increased and the amplifier is operated more as a linear amplifier. Thus, a natural trade-off exists in satellite systems between decreasing downlink carrier power levels and reducing the accompanying intermodulation interference. Analysis of the various design alternatives

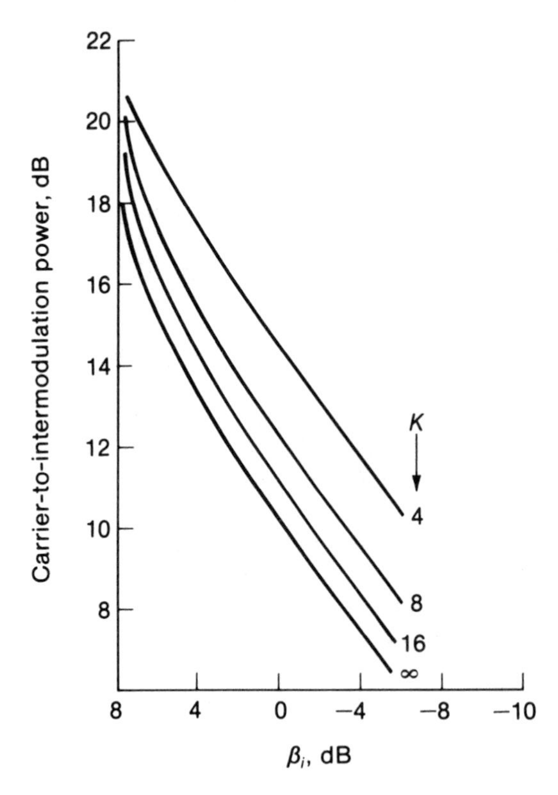

FIGURE 5.13 Carrier-to-intermodulation power ratio versus input backoff; center FDMA channel assumed (peak intermodulation).

must be made to assess these trade-off directions. This we do in the next section.

5.3 FDMA NONLINEAR ANALYSIS

In Section 5.1 we computed carrier CNR_d for an FDMA system operating with a linear amplifying transponder. In this section we extend these results to include the nonlinear amplifier. We again assume K equal bandwidth uplink carriers, and a total uplink power as in Eq. (5.1.5). Assume the driver power control for the satellite amplifier is provided by a bandpass limiter and possibly a low-level linear amplifier. By adjusting the output limiting level (including amplifier gain), the input power to the power amplifier can be controlled. This allows adjustment of the power amplifier input backoff defined in Eq. (5.2.16). We again let max $P_T(K)$

be the maximum available satellite output power with K simultaneous carriers, and let

$$P_T = \frac{\max P_T(K)}{\beta_0} \tag{5.3.1}$$

be the usable power when an output backoff of β_0 occurs. The ith carrier downlink receiver power, as apportioned by the driver stage, is then

$$P_{di} = P_T\left(\frac{P_{ui}}{P_u}\right)\alpha_s^2 L \tag{5.3.2}$$

where α_s^2 accounts for the additional carrier suppression imposed on a carrier by the nonlinear amplification of the combined carriers plus noise (as given in Figure 5.7), and L is again the downlink loss factor. Downlink receiver noise and interference in the carrier bandwidth B is due to (1) downlink receiver noise, (2) uplink retransmitted noise, (3) intermodulation interference caused by the nonlinear power amplifier, and (4) crosstalk spectral overlap. The total receiver interference is then

$$\begin{aligned} \text{Receiver noise power} &= N_{0d}B \leftarrow \text{Downlink} \\ &+ P_T\left(\frac{P_{un}}{P_u}\right)\alpha_n^2 L \leftarrow \text{Uplink} \\ &+ N_{0I}BL \leftarrow \text{Intermodulation} \\ &+ c_i P_{di} \leftarrow \text{Crosstalk} \end{aligned} \tag{5.3.3}$$

where

α_n^2 = Noise suppression factor of the nonlinear amplifier,
N_{0d} = receiver noise level,
P_{un} = uplink noise power in bandwidth B,
N_{0I} = intermodulation noise spectral level at the satellite amplifier output,
c_i = fraction of carrier power falling into adjacent carrier bandwidths.

The resulting downlink receiver carrier-power-to-total-interference ratio, CNR_d, for the ith carrier is then the ratio of Eq. (5.3.2) to Eq. (5.3.3):

$$\text{CNR}_d = \frac{P_T(P_{ui}/P_u)\alpha_s^2 L}{N_{0d}B + P_T(P_{un}/P_u)\alpha_n^2 L + N_{0I}LB + c_i P_T(P_{ui}/P_u)\alpha_s^2 L} \tag{5.3.4}$$

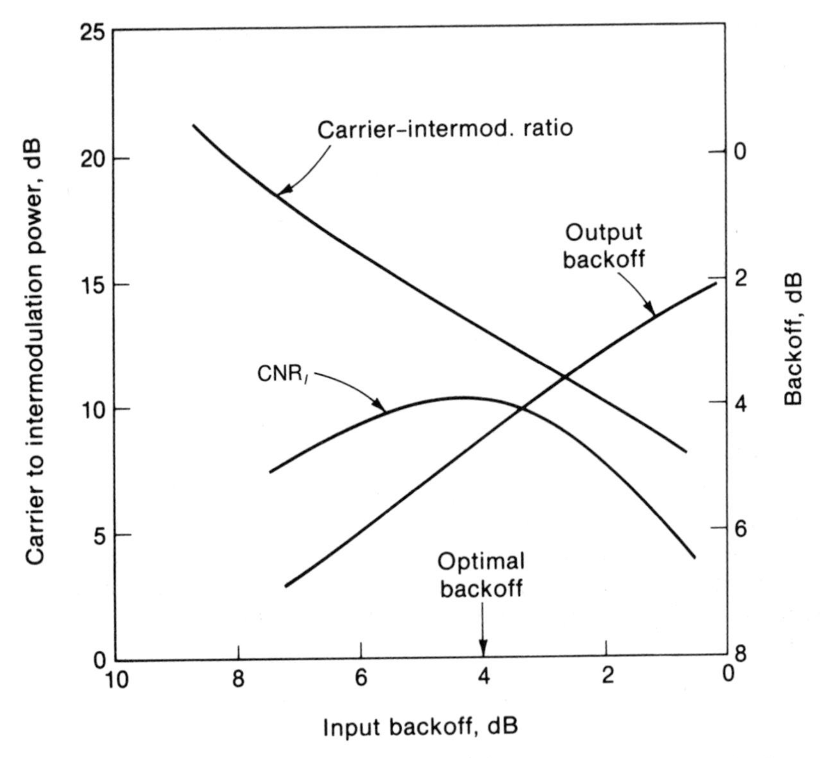

FIGURE 5.14 CNR_I versus input backoff ($K = 10$ FDMA carriers).

This represents the downlink receiver CNR_d of a *particular* FDMA carrier. As such it represents the extension of the single carrier result in Eq. (4.6.10) to the general multiple carrier, nonlinear transponder case. To emphasize this, we divide through by the numerator to obtain

$$(\text{CNR}_d)^{-1} = (\Gamma \text{CNR}_u)^{-1} + (\text{CNR}_r)^{-1} + (\text{CNR}_I)^{-1} + (\text{CNR}_c)^{-1} \quad (5.3.5)$$

where

$\text{CNR}_u = P_{ui}/P_{ni}$ = uplink carrier CNR at the satellite limiter input,
$\text{CNR}_r = P_{di}/N_{0d}B$ = downlink carrier CNR due to available satellite power,
$\text{CNR}_I = [P_T(K)/K]/N_{0I}B$ = carrier-to-intermodulation ratio,
$\text{CNR}_c = 1/c_i$ = carrier-to-crosstalk ratio,
$\Gamma = \alpha_s^2/\alpha_n^2$ = nonlinear suppression of the satellite limiter.

Equation (5.3.5) extends Eq. (4.6.10) by inserting the backoff suppression [via $P_T(K)$], intermodulation (via N_{0I}) and crosstalk (via c_i) effects for multiple carriers. The first two effects are both dependent on the input backoff (operating drive power) of the nonlinear power amplifier. When uplink and downlink power levels are sufficient, and crosstalk is negligible, downlink CNR is determined primarily by the intermodulation, and we can approximate:

$$\mathrm{CNR}_d \approx \mathrm{CNR}_I = \frac{(\max P_T)/K\beta_o}{N_{0I}B} \tag{5.3.6}$$

The numerator term of CNR_I decreases with input backoff (see Figure 5.12), while the intermodulation level N_{0I} is likewise reduced with backoff (Figure 5.13). When these results are superimposed, the ratio of the two behaves as shown in Figure 5.14. As backoff increases, the intermodulation

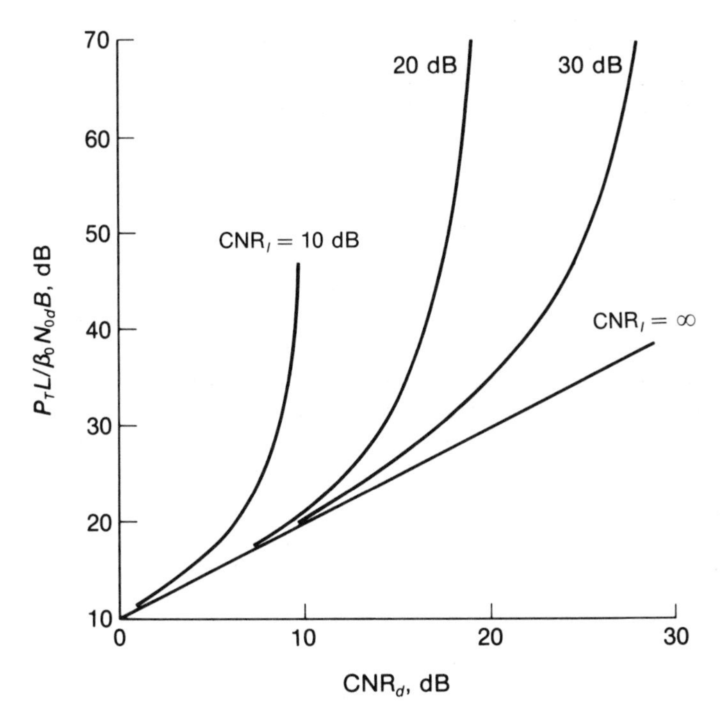

FIGURE 5.15 Required satellite power to achieve CNR_d in an FDMA channel, with specified carrier to intermodulation ratio ($K = 10$ carriers, $\mathrm{CNR}_u \gg 1$).

decreases, increasing CNR_I; eventually, however, the available satellite power will decrease, producing a backoff point at which CNR_I is maximized. Plots of this type are most convenient for locating optimal backoff operating points in FDMA systems. We emphasize that the desired operating backoff is strongly dependent on the shape of the backoff curves in Figures 5.12 and 5.13, and therefore will vary with the particular power amplifier involved.

For digital carriers, decoding performance can be determined by computing E_b/N_0 for that carrier channel. This can be obtained by replacing bandwidth B by bit time T_b^{-1} in Eq. (5.3.4). That is,

$$\left(\frac{E_b}{N_0}\right) = \mathrm{CNR}_d \Big|_{B=1/T_b} \tag{5.3.7}$$

Note that the resulting error-probability performance now depends on the four CNR in Eq. (5.3.5), and each must be evaluated to determine the combined CNR_d in order to plot PE performance curves.

Again, as we did in Section 5.1, we can determine the amount of satellite power required to support an FDMA system with K carriers. Letting $P_T = P_T(K)$ and solving for P_T in Eq. (5.3.4) yields

$$P_T = \frac{\beta_o}{L}\left(\frac{N_{0d}B(P_u/P_{ui})\mathrm{CNR}_d}{\alpha_s^2 - \mathrm{CNR}_d[(P_{un}/P_{ui}) + (N_{0I}B/P_{ui}) + c_i]}\right) \tag{5.3.8}$$

Note again that P_T depends on the amplifier output backoff and the intermodulation level N_{0I}, both of which depend on the input drive level, that is, β_i. Equation (5.3.8) yields the required amplifier power, P_T, at the satellite in order for the ith carrier downlink to operate with the given CNR_d. This must be evaluated separately for each carrier of the system. To guarantee that every carrier achieves its desired CNR_d during operation, it is necessary that P_T correspond to the maximum value of all P_T computed from Eq. (5.3.8).

As an example, consider the case where $P_{ui} = P$ (equal uplink carrier power); $P/P_{un} \gg 1$ (high uplink carrier CNR_u); and there is negligible crosstalk ($c_i = 0$). Under these conditions, Eq. (5.3.8) simplifies to

$$P_T = \frac{\beta_0 \, \mathrm{CNR}_d(N_{0d}B)K/L}{1 - \mathrm{CNR}_d(\mathrm{CNR}_I)^{-1}} \tag{5.3.9}$$

where α_s^2 is taken as 1. Equation (5.3.9) is plotted in Figure 5.15 as a function of desired CNR_d for several values of the parameter CNR_I. We

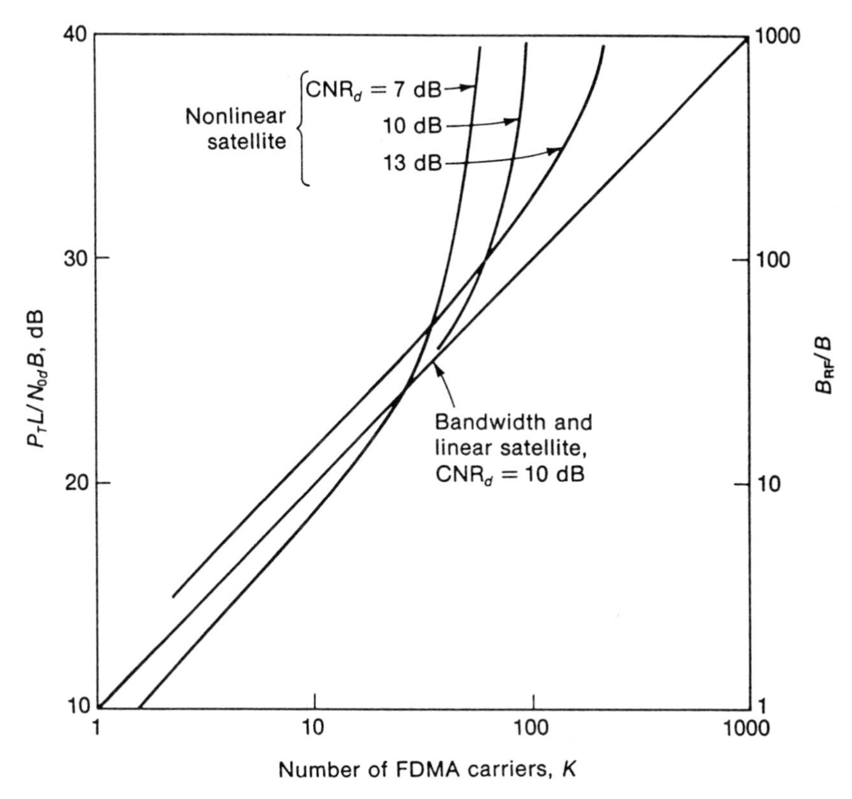

FIGURE 5.16 Required satellite power and bandwidth to support K FDMA carriers; linear and nonlinear satellites.

see that the required P_T increases linearly with CNR_d as long as $\mathrm{CNR}_d/\mathrm{CNR}_I \ll 1$. However, the required power increases more rapidly as $\mathrm{CNR}_d \to \mathrm{CNR}_I$, and it becomes infinite at the value $\mathrm{CNR}_d = \mathrm{CNR}_I$. This simply means the desired CNR_d becomes more difficult to maintain as it approaches the intermodulation ratio CNR_I. The entire allowed interference is being provided by the intermodulation alone, and additional noise interference cannot be tolerated. The inclusion of crosstalk into Eq. (5.3.9) would decrease further the denominator and cause the asymptotic increase to occur at a lower value of CNR_d.

In Figure 5.15 the number K of uplink carriers was considered a fixed parameter. Another aspect of design is to determine the number of carriers that can be supported in a nonlinear FDMA satellite system. As we saw in Section 5.1, this number is determined by either the available satellite bandwidth or the available satellite power P_T. The number of FDMA carriers permitted by the nonlinear satellite power can be obtained from Eq. (5.3.8) [or Eq. (5.3.9) if the stated conditions are satisfied]. However, we see that K directly affects the numerator in Eq. (5.3.9) and indirectly affects the denominator through the parameter N_{0I}. (Recall that the degree of intermodulation in a given bandwidth depends on the number of carriers appearing in the satellite bandwidth.) As K increases, CNR_I decreases, and a larger value of P_T is required over that predicted by the numerator in Eq. (5.3.9). The result produces a variation of P_T, with K similar to that sketched in Figure 5.16 for the parameters stated. Also included are the bandwidth and linear power curves from Figure 5.4. We see that the satellite nonlinearity eventually causes a more rapid increase in P_T with the number of carriers. Conversely, when operating with a fixed satellite power, fewer FDMA carriers can be supported than when a linear satellite amplifier is used. Again, the actual number of carriers permitted will be the smaller of the number allowed by the power P_T and the number allowed by the satellite bandwidth.

5.4 FDMA CHANNELIZATION

We have found that when dealing with an FDMA system using nonlinear satellite amplifiers, the available satellite power in the downlink must be divided among all carriers. Furthermore, strong carriers tend to suppress weak carriers in the downlink. This means that, when a mixture of both strong and weak carriers are to use the satellite simultaneously, we must ensure that the weaker carriers can maintain a communication link, especially if the mixture is to be transmitted to a relatively small (small $g/T°$) receiving station. One way in which weak carrier suppression can be reduced in FDMA formats is by the use of satellite *channelization*. In channelization, the strong and weak carriers are assigned frequencies so that they can be received in the satellite in separate RF bandwidths. That is, the total available satellite RF bandwidth (B_{RF}) is divided into smaller bandwidths, and the uplink carriers are assigned frequencies so as to be grouped in a bandwidth with other carriers of the (approximate) same satellite power level. These individual RF bandwidths are called satellite *channels*, and they can be used in two basic ways. One is to permit each channel to have a separate RF filter and amplifier, but to use only a single power amplifier (Figure 5.17a). The outputs of all channel amplifiers are summed prior to limiting and power amplification. The advantage of the channelization is that the amplifier gains in each channel can be individually adjusted so that all carriers will have roughly the same power levels when they appear at the amplifier input. This prevents suppression effects

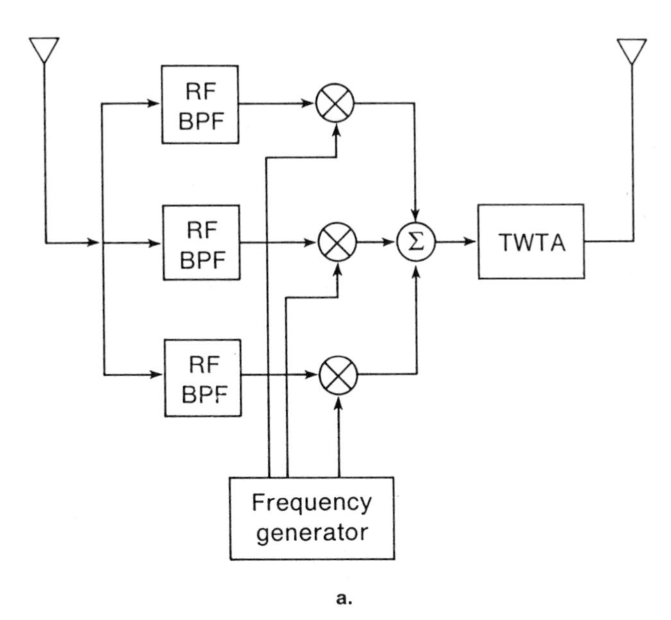

a.

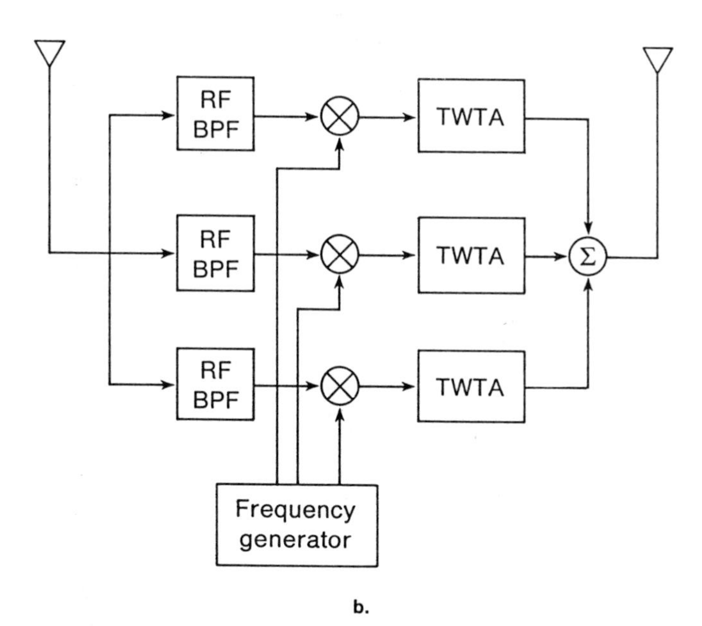

b.

FIGURE 5.17 Channelized satellite: (a) single TWTA; (b) multiple transponders.

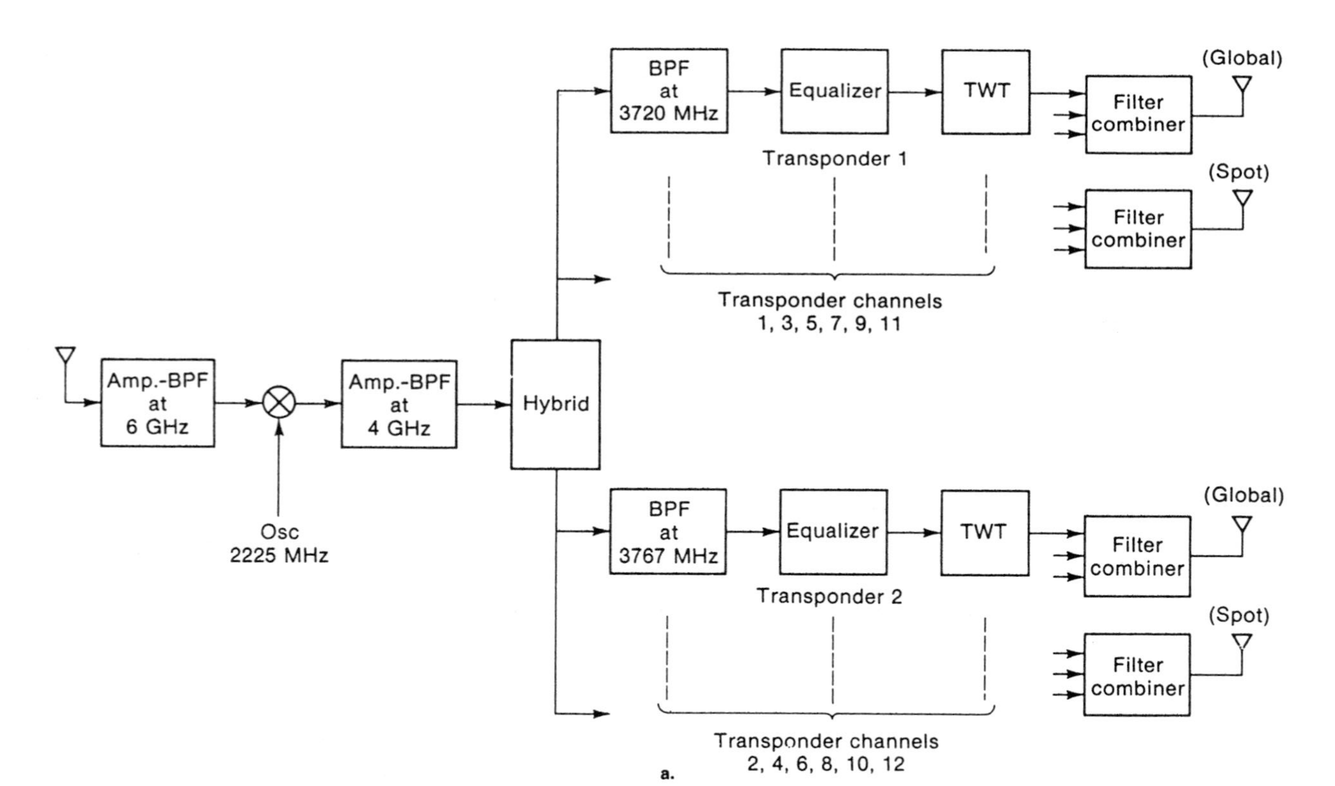

FIGURE 5.18 Intelsat channelized satellite: (a) block diagram; (b) frequency plan.

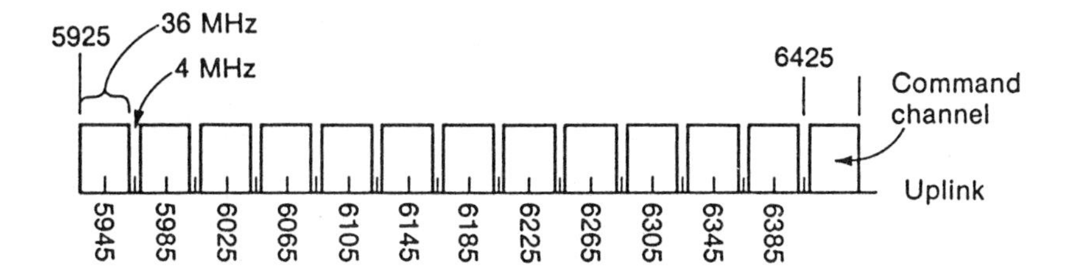

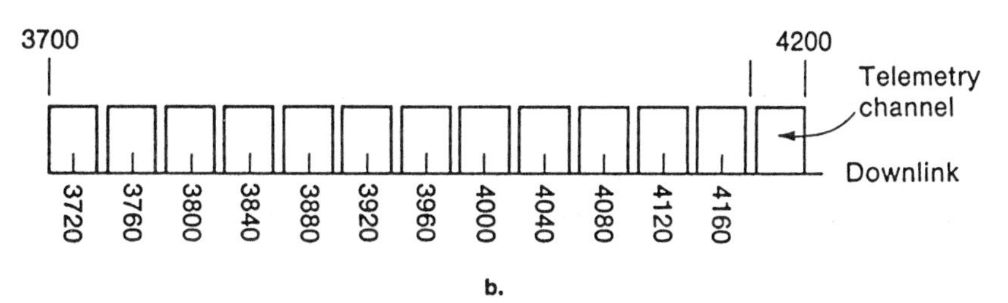

b.

FIGURE 5.18 Intelsat channelized satellite: (a) block diagram; (b) frequency plan.

due to strong uplink carriers, although the total number of carriers and the total amount of noise remains the same. In essence, uplink power control is obtained at the satellite instead of at the earth stations.

The second channelization method is to use separate power amplifiers for each channel (Figure 5.17b). Each satellite channel then becomes an independent transponder. Only carriers of the same power are used in the same channel. The power of each amplifier is therefore divided only among the carriers in its own bandwidth. The uplink noise per channel is reduced because of the smaller bandwidths, thus leading to improved CNR for the downlink. In addition, the intermodulation and power suppression effects are reduced since there are fewer carriers in each transponder. The limit, of course, is when each uplink carrier is assigned its own transponder channel, which is the so-called SCPC (*single channel per carrier*) format and all nonlinear effects are removed. The advantages of channelization are achieved, of course, at the expense of a more complex satellite, since the weight of not only the additional power amplifiers and filters must be included but also that of the supporting auxiliary primary power. The advantages in performance of the increased number of independent transponders must be carefully weighed against the additional satellite cost.

The use of increasing numbers of satellite transponders is an obvious trend in modern satellite design (see Table 1.1). Figure 5.18a shows the processing block diagram for the 12-transponder Intelsat satellite. The uplink and downlink RF bandwidth is divided, as shown in Figure 5.18b. Each individual transponder has a 36-MHz bandwidth, with each channel center frequency separated by 40 MHz. The 12 transponders therefore utilize the entire 500-MHz RF bandwidth. Satellites may employ additional channels by making use of antenna beam separation or antenna polarization separation in the uplink and the downlink. Recall that this allows frequency reuse, in which the RF bandwidth can be used simultaneously by two separate carriers at the same uplink and downlink frequencies.

5.5 AM/PM CONVERSION WITH FDMA

In addition to the intermodulation interference produced by FDMA carriers, nonlinear power amplifiers introduce AM/PM conversion. To examine this effect analytically, consider the input to the amplifier to be

$$x(t) = A \sum_{q=1}^{K} a_q \cos[\omega_q t + \theta_q(t)] \tag{5.5.1}$$

corresponding to a set of K FDMA carriers of various forms. We write the total envelope variation of $x(t)$ as in Eq. (4.7.7),

$$\alpha(t) = A[1 + e(t)] \tag{5.5.2}$$

Using a linear AM/PM conversion model, the conversion causes the amplifier output to be

$$y(t) = g[\alpha(t)] \sum_{q=1}^{K} a_q \cos[\omega_q t + \theta_q(t) + \eta A e(t)] \tag{5.5.3}$$

where η is the AM/PM conversion coefficient in Figure 4.24b. Note that the amplitude modulation is coupled into the phase of *each* carrier, with the value of η dependent on the degree of amplifier nonlinearity (backoff). Since this is itself phase modulation, typically having frequency components directly in the bandwidth of all other modulation, the converted phase modulation appears as an additive carrier crosstalk, rather than as noise. This crosstalk can then be demodulated along with the desired carrier modulation, which we referred to as *intelligible crosstalk* (Bryson, 1971; Chapman and Millard, 1964; Stette, 1975). With voice-modulated carriers, for example, this intelligible crosstalk corresponds to direct voice interference of one voice circuit onto another.

As an example of the AM/PM conversion effect, consider the simplified case of two equal amplitude carriers at two separate RF frequencies, one of which has baseband amplitude modulation. The waveform in Eq. (5.5.1) simplifies to

$$x(t) = A\cos[\omega_1 t + \theta_1(t)] + A[1 + m(t)]\cos(\omega_2 t) \tag{5.5.4}$$

where $\omega_2 = \omega_1 + \omega_d$, and ω_d is the frequency separation. After trigonometrically expanding, Eq. (5.5.4) can be rewritten as

$$\begin{aligned} x(t) = A\{1 + [1 + m(t)]\cos[\omega_d t - \theta_1(t)]\}\cos[\omega_1 t + \theta_1(t)] \\ + A\{[1 + m(t)]\sin[\omega_d t - \theta_1(t)]\}\sin[\omega_1 t + \theta_1(t)] \end{aligned} \tag{5.5.5}$$

The envelope of $x(t)$ expands out as

$$\alpha(t) = \sqrt{2}A[1 + f(t)]^{1/2} \tag{5.5.6}$$

where

$$f(t) = [1 + m(t)]\cos[\omega_d t - \theta_1(t) + \tfrac{1}{2}m^2(t) + m(t)] \tag{5.5.7}$$

The amplifier output in Eq. (5.5.3) is then

$$\begin{aligned} y(t) = g[\alpha(t)]\cos[\omega_1 t + \theta_1(t) + 2A\eta e(t)] \\ + g[\alpha(t)]\cos[\omega_2 t + 2A\eta e(t)] \end{aligned} \tag{5.5.8}$$

Since $e(t)$ has a component due to the modulation $m(t)$, $e(t)$ in Eq. (5.5.8) introduces intelligible crosstalk onto the phase of the angle-modulated carrier, the strength of which depends on the coefficient η. It is for precisely this reason that amplitude-modulated carriers are usually not used in multiple-carrier FDMA satellite systems having TWT amplification.

Often the undesired amplitude modulation $m(t)$ in Eq. (5.5.4) appears unintentionally. In FDMA formats with channelization, FM carriers of the uplink are filtered and combined for downlink amplification. During the filtering, if the RF filters are somewhat narrow, the filter gain characteristics cause the FM to be converted to undesired AM. For example, if $|H_c(\omega)|$ has a constant slope of V_ω V/rps (called the filter *gain slope*) in the vicinity of carrier center frequency, then an FM carrier with frequency variation $\Delta_\omega m(t)$ produces an amplitude variation on this carrier of

$$e(t) = V_\omega \Delta_\omega m(t) \tag{5.5.9}$$

This envelope variation is then coupled into the phase of all other channelized carriers using the same transponder. Hence, the frequency modulation of one carrier is transferred to all other carriers through the filter gain slope and amplifier AM/PM effect. The strength of this FM crosstalk depends on the filter gain slope and the AM/PM coefficient. The former is reduced by better control of the filter functions, while the latter is reduced by backing off the amplifier.

5.6 SATELLITE-SWITCHED FDMA

A channelized FDMA format is particularly suited for operation with multiple spot beams. Consider the multiple-beam model in Figure 5.19. Each spot beam illuminates a particular set of earth stations, and within each beam FDMA is used. The frequency separation of the carriers in each uplink beam can be used to channelize each beam, with a separate carrier filter at the satellite for each carrier within a beam. An uplink carrier designated for a downlink earth station is then routed to the

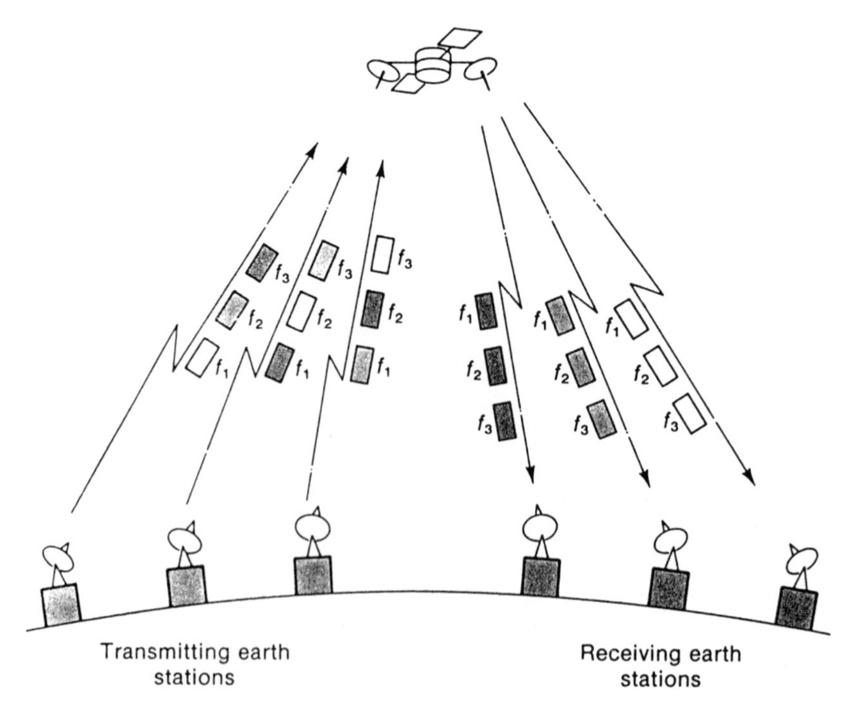

FIGURE 5.19 Satellite-switched FDMA model.

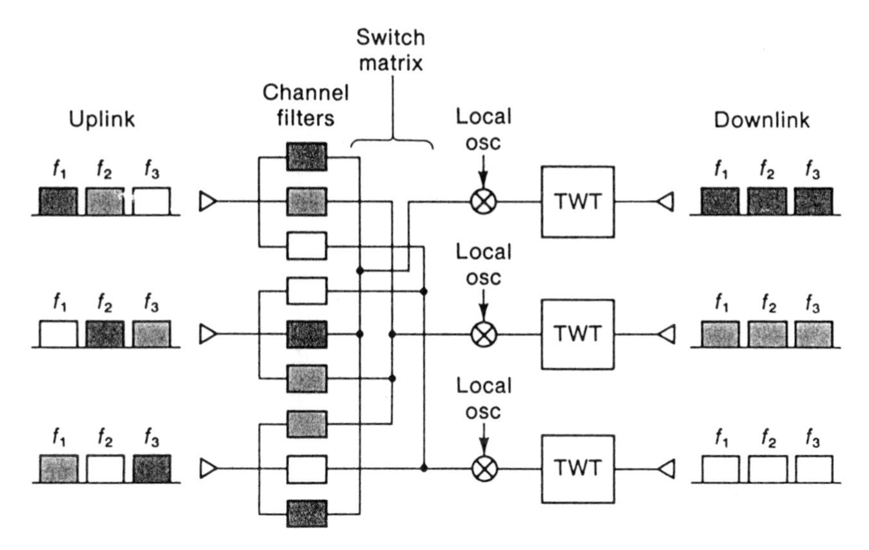

FIGURE 5.20 SS-FDMA satellite block diagram.

particular spot beam covering that station. Hence, each uplink channel filter output must be switched to the proper downlink beam channel. Since all uplinks must be switched simultaneously, an onboard switching matrix is needed to provide all possible routing directions. This switching can be suitably provided by a microwave diode gate matrix that allows signal flow in only specified directions. An FDMA satellite system of this type is referred to as *satellite-switched FDMA* (SS-FDMA).

Figure 5.20 shows a simplified channelized diode switch matrix implementation for a three-beam SS-FDMA satellite. Each beam uses the same frequency bands, and an uplink carrier to be routed to a particular downlink spot beam is assigned a specific frequency band in each uplink beam. This band is different for every uplink beam. The uplink beams are then channelized, and each band filtered off. The diode switch matrix connects each band to a different downlink. In other uplink beams, the same bands are switched to different downlinks. In this way, the same frequency bands from different beams are never superimposed on the same downlink. An uplink earth station need only select the appropriate band for the desired downlink earth-station beam. Thus, at any one time each uplink in any beam has connectivity to any downlink earth station in any beam.

Note that the system utilizes complete frequency reuse, with the same bands being used in each uplink beam. In addition, channelization of the uplinks in SS-FDMA allows the filters to include gain adjustment as well, providing power control over the downlink carriers in the same beam.

This reduces suppression effects on weaker carriers being routed to the same beam as stronger carriers.

A basic disadvantage with SS-FDMA is that the routing must be incorporated *a priori*; that is, the switch matrix is "hardwired in." Thus, the frequency distribution pattern is then fixed for that matrix, and traffic patterns cannot be altered. (We shall find this is not the case with TDMA systems.) In addition, care must be used to prevent the same frequency band in two different beams from appearing in the same downlink and causing crosstalk. Preventing this crosstalk interference requires a significant isolation from the channel filters, and negligible leakage in the diode matrix. Isolation as high as 60 dB is possible with microwave switches. Another disadvantage is that the number of filters increases directly with the product of the number of one-way beams and frequency bands. Hence, if 3 bands are used with 4 uplink and downlink beams, a total of 12 carrier filters is required in the satellite.

REFERENCES

Babcock, W. (1953), "Intermodulation Interference in Radio Systems," *BSTJ* (January), 63–69.

Bryson, J. (1971), "Intelligible Crosstalk in FM Systems with AM-PM Conversion," *IEEE Trans. Comm. Tech.*, **Com-19** (June), 366–368.

Chapman, R., and Millard, J. (1964), "Intelligible Crosstalk with FM Carriers Through AM-PM Conversion," *IEEE Trans. Comm. Syst.*, **CS-12** (June), 160–166.

Sevy, J. (1966), "The Effect of FM Signals Through a Hard Limiting TWT," *IEEE Trans. Comm. Tech.*, **COM-14**, (October), 568–578.

Shaft, P. (1965), "Limiting of Signals and Its Effect on Communication," *IEEE Trans. Comm. Tech.*, **COM-13** (December), 504–512.

Spoor, T. (1967), "Intermodulation Noise in FDMA Communications Through a Hard Limiter," *IEEE Trans. Comm. Tech.*, **Com-15** (August), 557–565.

Stette, G. (1975), "Intelligible Crosstalk in Nonlinear Amplifiers: Calculation of AM-PM Transfer," *IEEE Trans. Comm.*, **52** (February), 256–268.

Van Trees, H., ed. (1979), "Demand Assignment," in *Satellite Communications*, IEEE Press, New York, Section 3.6.5.

Westcott, R. (1967), "Investigation of Multiple FDMA Carriers Through a Satellite TWT Operating Near Saturation," *Proc. IEEE*, **114** (6) (June), 726–740.

PROBLEMS

5.1. Given two adjacent BPSK carriers, each with power P, data rate R bps, and carrier frequency separation Δf Hz. Derive an expression for the crosstalk interference of one carrier spectrum onto the other's main-hump bandwidth in terms of the $\mathrm{Si}(x)$ function:

$$\mathrm{Si}(x) = \int_0^x \left(\frac{\sin u}{u}\right)^2 du$$

5.2. An FDMA system transmits carriers with a 50-MHz bandwidth, and uses a linear satellite with a 500-MHz RF bandwidth. Each uplink carrier has a $\mathrm{CNR}_M = 20$ dB. Let the satellite power be 5 W, the net downlink loss be 140 dB, and the downlink receiver have $N_0 = -200$ dB W/Hz. (a) Determine if this FDMA system is power- or bandwidth-limited. (b) Repeat if the carrier uplink CNR_u is reduced to 11 dB. For both parts use $\mathrm{CNR}_d = 10$ dB.

5.3. An FDMA-linear satellite system is designed to accommodate a total of K carriers with a satellite power of P_r. (a) Show that if only Q of the K carriers are active at one time (and the satellite knows it), the satellite power can be reduced without lowering performance (i.e., all active downlinks will still have the required CNR_d). (b) If each carrier has a 60% activity time, determine the reduction in average satellite power P_r if Q can be continuously monitored.

5.4. Given K equally spaced carrier spectral lines at frequency f_i, $i = 1, 2, \ldots, K$. (a) Determine the number of intermodulation lines of the form $f_i + f_j - f_m$ that will fall on top of the rth line. (b) Show that for the middle line ($r = K/2$) and high K, this number is approximately $3K^2/8$.

5.5. A TWTA saturates with an input power of 8 mW and has the intermodulation performance of Figure 5.13. If a 16-carrier FDMA system is operated through this amplifier, what will be the carrier-to-intermodulation ratio if the total input power is maintained at 2 mW?

5.6. Consider the FDMA system in Problem 5.2 with the same bandwidths, downlink losses, and N_0, except that a nonlinear TWTA is used. The required CNR_d is 13 dB. Using Figure 5.16, what is the required value of satellite power P_T to operate with the maximum number of carriers?

5.7. An FDMA repeater is to be channelized as in Figure 5.17a. Consider the following simple five-carrier system with a linear satellite amplifier:

Carrier	P_{uc} (mW)	P_{sd} (W)
1	10	10
2	10	5
3	20	10
4	20	2
5	20	4

Here P_{uc} is the uplink power, and P_{sd} is the required satellite downlink power for each carrier. (a) What is the minimum value of P_T that can satisfy all downlinks? (b) What P_T is needed for a single transponder (no channelization)? (c) Repeat (b) for a two- and three-channel transponder. (d) How many channels are needed to achieve the minimum P_T in (a)?

5.8. Consider a filter function $|H(\omega)| = \omega^n$, operating with an FM carrier at its input. The carrier has frequency ω_0 rps and frequency modulation $\Delta_\omega m(t)$ rps. The filter is followed by an amplifier with AM/PM coefficient η rad/V. By expanding around ω_0, show that the AM/PM power varies at $(\eta n)^2$.

5.9. In nonlinear FDMA, optimal backoff points for maximizing CNR_d depend on the parameter Q, the number of carrier signals passing through the satellite at any one time, as in Problem 5.3. Devise a satellite block diagram (onboard equipment) that will optimally adapt an FDMA satellite for optimal operation, as the number of active carriers change in time.

CHAPTER 6
Time-Division Multiple Access

In multiple accessing through a satellite, uplink carriers can be separated in time rather than in frequency. Instead of assigning a frequency band to each uplink, we assign a specific time interval, and a given station transmits only during its allotted interval. This type of operation is referred to as *time-division multiple access* (TDMA). As we shall see, TDMA theoretically avoids the problem of many carriers trying to pass through the satellite at the same time, thereby avoiding the intermodulation problem of FDMA. However, while FDMA involves relatively simple frequency tuning for accessing, and providing essentially independent channel on–off operation, TDMA requires communication concepts that are relatively new. To accommodate many users, TDMA time intervals must necessarily be short, requiring burst-type transmissions, and the time intervals of all users must be properly and accurately synchronized, requiring several levels of timing control. The required high-speed hardware for these operations is relatively new and, in many cases, is still under development.

Experimental TDMA systems were first developed in 1965 (Sekimoto and Puento, 1968). These early systems proved that time-interleaved, short-interval communications were in fact technically feasible. Later systems (IEEE, 1979; Kwan, 1973; Maillet, 1972; Thompson, 1983; Watt, 1986) established that advanced operational TDMA concepts, such as high-accuracy, fast-acquisition synchronization, and high-capacity data formats, were both possible and advantageous. In this chapter we examine TDMA satellite communications.

6.1 THE TDMA SYSTEM

In time-division multiple-accessing systems, each uplink earth station is assigned a prescribed time interval in which to relay through the satellite. During its interval, a particular station has exclusive use of the satellite, and its uplink transmission alone is processed by the satellite for the downlink. This means each carrier can use the same carrier frequency and make use of the entire satellite bandwidth during its interval. Since no other carrier uses the satellite during this time interval, no intermodulation or carrier suppression occurs, and the satellite amplifier can be operated in saturation so as to achieve maximal output power. Thus, TDMA downlinks always operate at full saturation power of the satellite. However, the entire TDMA system must have all earth stations properly synchronized in time so that each can transmit through the satellite only during its prescribed interval, without interfering with the intervals of other stations. This time synchronization between satellite and all earth stations is called *network synchronization.* A downlink earth station, wishing to receive the transmissions from a particular uplink, must gate in to the satellite signal during the proper time interval. This means that all earth stations, whether transmitting or receiving, must be part of the synchronized network.

Since there may be many users of the TDMA satellite, each wishing to establish a communication link in approximately real time, the total transmission time must be shared by all users. Thus, the time intervals of each station must be relatively short, and repeated at regular epochs. This type of short-burst, periodic operation is most conducive to digital operation, where each station transmits bursts of data bits during its intervals. However, TDMA systems using digital transmissions require that all receiving stations must obtain decoder synchronization in each interval, in addition to the required network synchronization for slot timing. For phase-coherent decoding, decoder synchronization requires establishing both a coherent phase reference and a coherent bit timing clock before any bits can be decoded within a slot. Also, word sync may be needed to separate the digital words occurring during a slot. This hierarchy of decoding synchronization must be established at the very beginning of each slot if the subsequent slot bits are to be decoded. Furthermore, since each slot contains data from a different source, synchronization must be separately established for each slot being received. In fact, even when receiving from the same station, synchronization must generally be reestablished from one periodic burst to the next. Hence, digital communications with TDMA has an inherent requirement for rapid synchronization in order to perform successfully. The technology for short-burst communications is rather new and, of course, will be closely linked to the development of high-speed digital processing hardware.

A TDMA satellite system is shown in Figure 6.1a. Each uplink station is assigned one of a contiguous set of time slots, and the group of all such time slots forms a transponder time frame (Figure 6.1b). We assume a

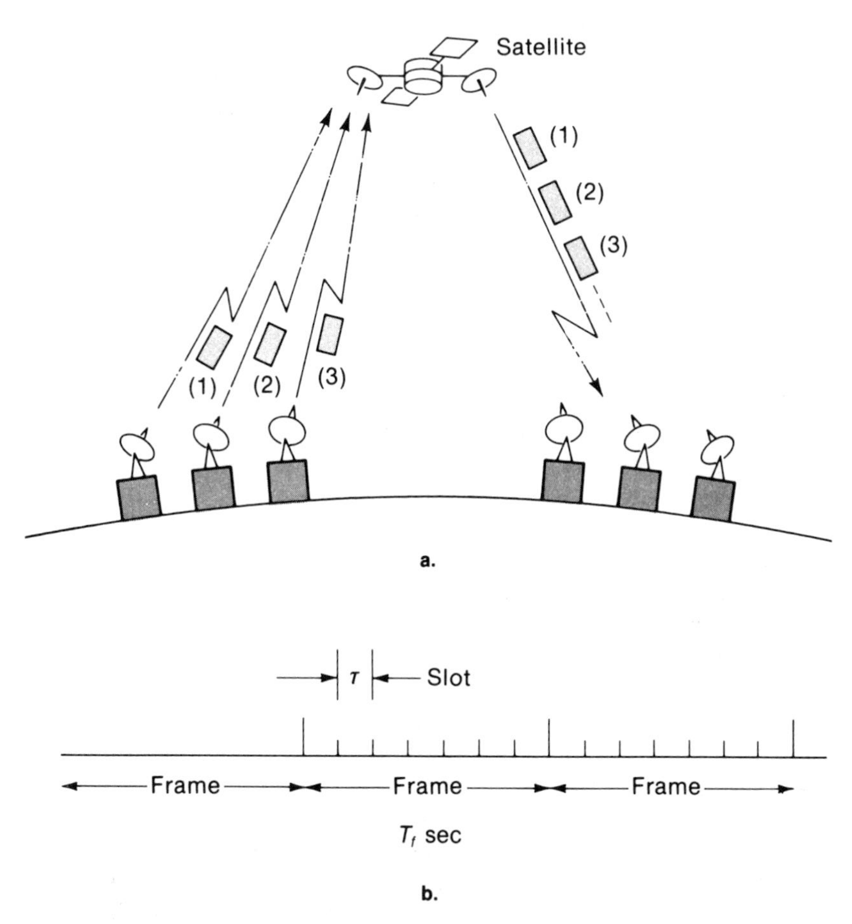

FIGURE 6.1 TDMA satellite system: (a) system model; (b) time-frame structure.

single transponder operating at the satellite. During each successive frame, each transmitting station has its own specific slot. Hence, if slots are τ s, a given station uses the satellite τ s during each T_f s frame. The station transmits bursts of data bits during its τ-s transmission time, while using the remaining frame time to generate the next set of data bits. If the frame time corresponds exactly to the sampling time of the earth-station message, then the bit transmissions during the τ-s intervals would correspond to the real-time transmission of a quantization word. For this to be true for all stations of the network simultaneously, all stations must use the same sampling times—that is, they must transmit message waveforms with approximately the same bandwidth. Hence, TDMA operation is most applicable to digital encoding, with stations operating at approximately the same bit rate. For example, if all earth-stations were sending digital voice through the satellite, all would require voice sampling at a rate of approximately 2×4 kHz $= 8 \times 10^3$ samples/s. If the TDMA frame time is $\frac{1}{8} \times 10^{-3} = 0.16$ ms, then ideally each station in a frame can use the satellite once each sampling time, transmitting the A–D voice samples as they are generated in real time. To accommodate stations with widely varied transmission bit rates in a common frame requires station buffering and storage. It may therefore be more advantageous to provide several separate TDMA transponder channels with stations of approximately the same rates grouped in a common frame.

A transponder TDMA frame is typically formatted as in Figure 6.2. The frame is divided into slots, each assigned to an uplink station. Each slot interval is then divided into a *preamble* time and a data-transmission time. The preamble time is used to send a synchronization waveform so that a receiving station gated to the slot can lock up its receiver decoder. The preamble generally contains guard time (to allow for some errors in slot timing), a phase referencing and bit-timing interval (to allow a phase-coherent decoder to establish carrier and bit synchronization), and a unique code word (to establish word sync). Observation of this unique word can be used to set the word markers for the data transmission during the remainder of that slot interval. Each frame slot can be formatted in this way. Any station can be received by gating in at the proper slot time, referred to the earth-station time axis. The latter must be separately referenced by each earth-station to the satellite time axis through network synchronization. All transmitting stations use the same carrier frequency, so that every receiving station has (approximately) frequency synchronization throughout a frame. Thus, the preamble time is needed primarily to adjust to the phase and bit timing of a given station, which can possibly drift from one frame to the next.

In general, preamble time should be long enough to establish reliable synchronization, but should be short compared to the data transmission

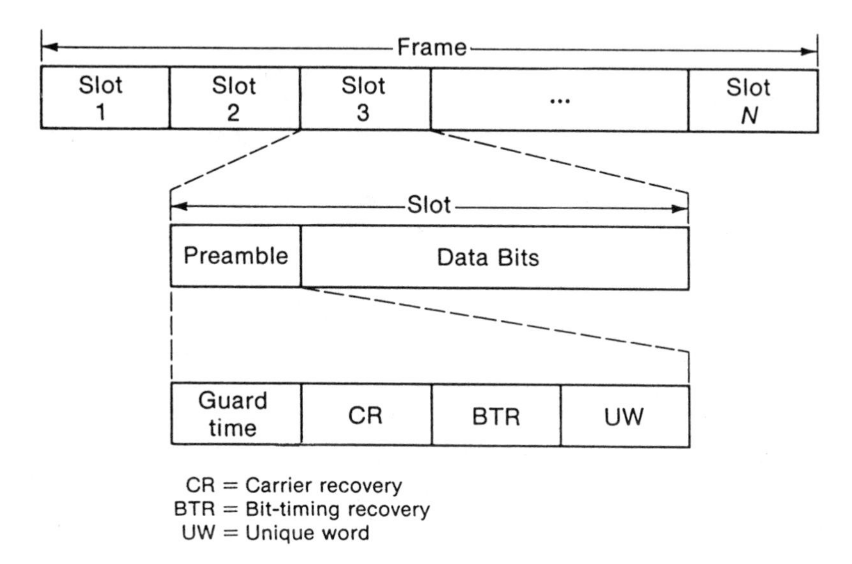

FIGURE 6.2 Frame formatting in TDMA.

time. The ratio of preamble time to total slot time is sometimes called the *preamble efficiency*, or *overhead.* We often measure these times in numbers of bits or symbols and write this efficiency as

$$\eta_p = \frac{\text{Number of preamble symbols}}{\text{Total number of symbols per slot}} \tag{6.1.1}$$

Typically, TDMA systems are designed with overhead efficiencies of about 10% or less.

Assume that a preamble requires a total of P transmitted bits for sync, including guard-time, carrier referencing, bit timing, and unique word transmission. For a preamble efficiency of η_p in Eq. (6.1.1), a total of D data bits must be sent in each slot, where

$$D = \left(\frac{1 - \eta_p}{\eta_p}\right) P \tag{6.1.2}$$

Let the data to be transmitted during a slot correspond to transmissions from digital sources producing b bits/slot and operating at a rate of R_c bits/s. This requires that the frame time be

$$T_f = \frac{b}{R_c} \text{ s} \tag{6.1.3}$$

For example, a digital voice source operating at 64 kbps, sending four bits in each burst, would require a frame time of $T_f = 4/64 \times 10^3 =$ 0.063 ms, while eight-bit bursts would require 0.125 ms and 16-bit bursts, 0.25 ms. Thus, we see that frame times in TDMA voice formats are usually relatively short, which of course complicates the slot synchronization operation. Increasing the number of bits per burst lengthens the frame time, but as the frame is lengthened, more time elapses between bursts from a given station, allowing more drift time to produce larger station oscillator phase shifts. Hence, shorter frames tend to produce better burst-to-burst sync coherency. Note that b refers to the number of bits per burst from a single source (we refer to this as a single channel), whereas D in Eq. (6.1.2) refers to the total number of data during the entire slot.

Assume the satellite channel has an RF capability of sending bits at a rate R_{RF} bps, based on the satellite bandwidth and power levels. This means the slot time τ must be long enough to allow $D + P$ bits. Hence, $R_{\mathrm{RF}}\tau = D + P$, or

$$\tau = \frac{D + P}{R_{\mathrm{RF}}} \tag{6.1.4}$$

The number of slots in a frame is then

$$Q = \frac{T_f}{\tau} = \frac{bR_{\mathrm{RF}}}{R_c(D + P)} \tag{6.1.5}$$

The total number of data channels (i.e., separate sources operating at b bits/burst and rate R_c bps) that can be accommodated in the TDMA frame is then

$$K = \frac{QD}{b} = \frac{D}{R_c\tau} = \frac{R_{\mathrm{RF}} - (P/\tau)}{R_c} \tag{6.1.6}$$

Note that a TDMA frame can generally support many more channels than the number of slots, since D is generally much larger than b. Equation (6.1.6) is plotted in Figure 6.3, showing the number of TDMA channels that can be supported by particular satellite bit rates, for different preamble efficiencies.

As an example, suppose a QPSK TDMA system requires a total preamble of 40 QPSK symbols (in the next section we consider how to estimate the required preamble length) and is to operate with a 9% preamble efficiency. Sampled voice transmissions at a rate of 64 kbps and 8 bits/burst are to be transmitted over a TDMA satellite with an RF bit

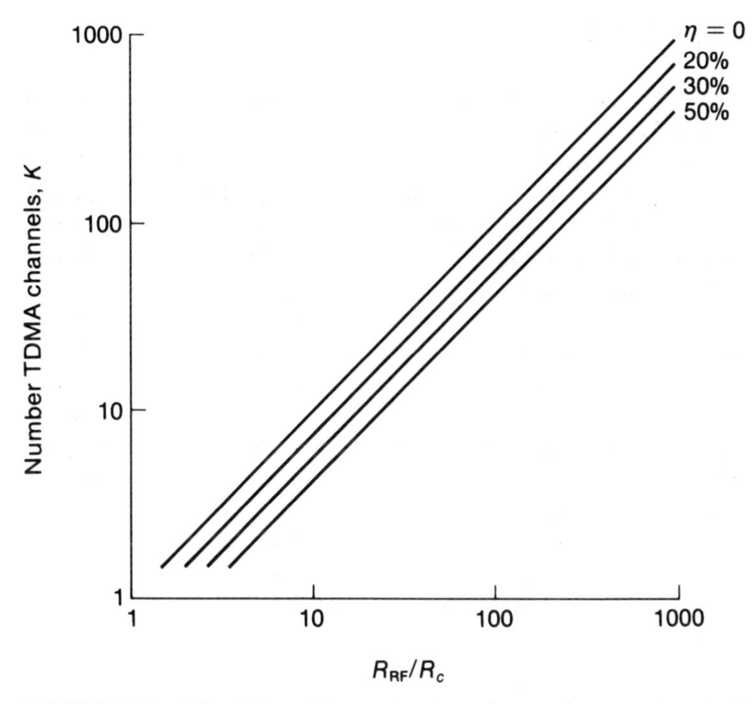

FIGURE 6.3 Number of TDMA channels versus RF bandwidth and frame efficiency.

rate of 60 Mbps. The total number of QPSK data bits per slot should be $40/0.1 = 400$ symbols $= 800$ bits. The required frame length must then be $8/64 \times 10^3 = 0.12$ ms. The slots are $(800 + 80)/60 \times 10^6 = 14.6$ μs long. This means each frame has $Q = 0.12 \times 10^{-3}/14.6 \times 10^{-6} \approx 9$ slots per frame, and a total of

$$K = \frac{(800)9}{(64 \times 10^3)(0.12 \times 10^{-3})} \approx 937 \tag{6.1.7}$$

voice channels can be sent each frame. Each slot can therefore be divided into $937/9 \approx 117$ separate voice channels that are serially multiplexed and modulated onto the station carrier for that slot. With the same frame format, a single earth station can alternatively send data at a combined rate of $800/0.12$ ms $= 6.7$ Mbps through the TDMA link.

6.2 PREAMBLE DESIGN

We have seen from Eq. (6.1.6) that the number of channels that can be supported in a TDMA system increases as the preamble time (number of preamble bits) decreases. The smaller the number of sync bits for a given slot size, the larger the number of data bits. Hence, we strive to operate TDMA systems with the least possible preamble length, which implies that shortest possible sync time. This makes the general topic of short-burst synchronization vital to overall TDMA capacity.

Preamble synchronization is achieved by a receiver subsystem operating in parallel with the data-recovery channel of the slot burst, as shown in Figure 6.4. The network sync subsystem establishes the gating times for an earth station to tune into a particular transmitting station. Assuming accurate network sync, the receiver will gate in at the beginning of preamble reception. A carrier reference system locks to the received carrier to establish a phase-coherent reference, which immediately begins demodulating the superimposed bits. Bit timing is then established on the preamble bits (for example, alternating 1s and -1s) and the unique word is detected when it arrives to indicate commencement of data transmission.

Preamble time is the time necessary to achieve an acceptable level of receiver synchronization for decoding data bits. As stated earlier, the total synchronization time can be separated into guard time, carrier reference and bit sync times, and unique word recovery time. We investigate each of these separately.

The required guard time is set by the accuracy of the network sync clocks (to be discussed in Section 6.4). These clocks set the overall slot timing of the earth stations, and errors in this timing will tend to offset

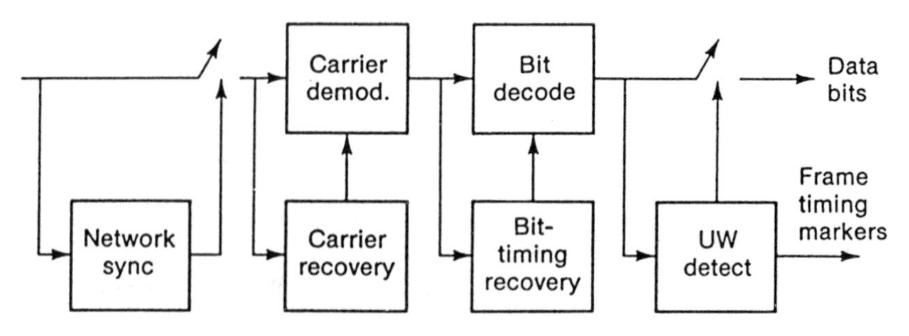

FIGURE 6.4 Preamble synchronization subsystem.

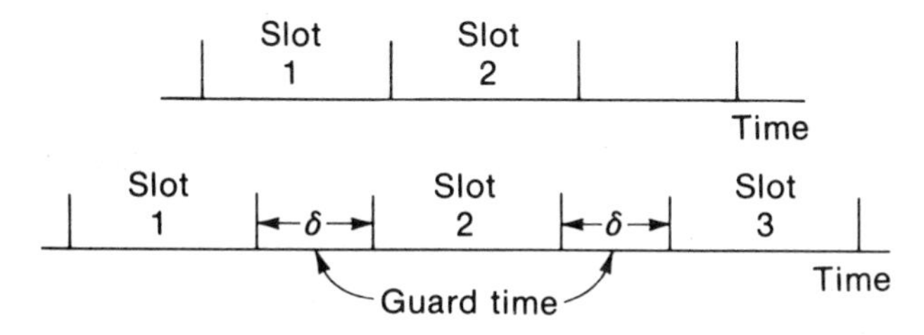

FIGURE 6.5 Slot timing and guard spacing.

the slot gate time (Figure 6.5). By allowing some guard time between slots, these errors will not significantly degrade TDMA operation. If there is an inherent rms network timing error e_τ, then guard time should be about ± 10 times e_τ in order to fully compensate. Hence, guard time is generally taken as

$$\delta = 20e_\tau \tag{6.2.1}$$

Expressed in data bit times,

$$\frac{\delta}{T_b} = 20\left(\frac{e_\tau}{T_b}\right) \tag{6.2.2}$$

Since e_τ can generally be maintained within a fraction of a bit, guard times of only one or two bits are usually all that is needed.

At the beginning of a slot, the carrier burst from the satellite is received at the earth station with a known frequency but an arbitrary phase offset.

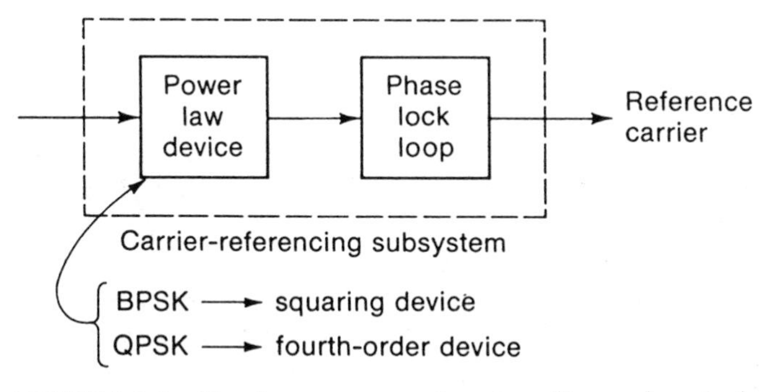

FIGURE 6.6 Carrier-recovery subsystem. Power law device depends on data modulation format.

Phase reference is achieved by phase locking a local reference to the received reference burst during the preamble. The referencing time is therefore the time needed to achieve phase pull-in. The theory of referencing loops operating with typical digital formats is discussed in Appendix B. For BPSK or QPSK carriers, reference systems are usually power-law devices (squaring or fourth-order loops) followed by tracking loops, as shown in Figure 6.6. The power-law device removes the modulation, and the loop locks to the resultant carrier phase. Phase lock is essentially achieved when the phase error between the unmodulated carrier and the local carrier is reduced to about 5–10°. The time to reduce an initial phase error of ϕ_0 to these values depends on the loop transient time, which is related to the loop *natural frequency* ω_n and *damping factor* ξ. These parameters are related to the *loop noise bandwidth* B_L by

$$B_L = \frac{\omega_n}{8\zeta}(1 + 4\zeta^2) \tag{6.2.3}$$

The *loop response time* (time to reduce an offset to about 10% of its initial value) is about $0.5/\omega_n$. Hence,

$$\text{Loop response time} \approx \frac{1 + 4\zeta^2}{16\zeta B_L} \tag{6.2.4}$$

For loop damping of $\xi = 0.707$ (a common value for trading off steady-state and noise-bandwidth values), $B_L = 0.53\omega_n$, and when converted to bits, Eq. (6.2.4) becomes

$$\begin{bmatrix} \text{Loop response time} \\ \text{in numbers of} \\ \text{bits} \end{bmatrix} \approx \frac{0.26}{B_L T_b} \tag{6.2.5}$$

Equation (6.2.5) is plotted in Figure 6.7, showing the number of bits that must generally be allowed for carrier referencing. For example, if $B_L T_b = 0.01$, then the response time will be about 26 bit times. However, while increasing B_L reduces response time, it also increases the loop-noise bandwidth, which makes it more difficult to maintain loop lock-up. To examine this lock-up effect, it is necessary to determine the loop carrier-to-noise ratio CNR_L produced from the retransmission through the satellite. The latter can be computed via the analyses in Chapter 4, since the satellite acts as a single-carrier, hard-limiting transponder during each

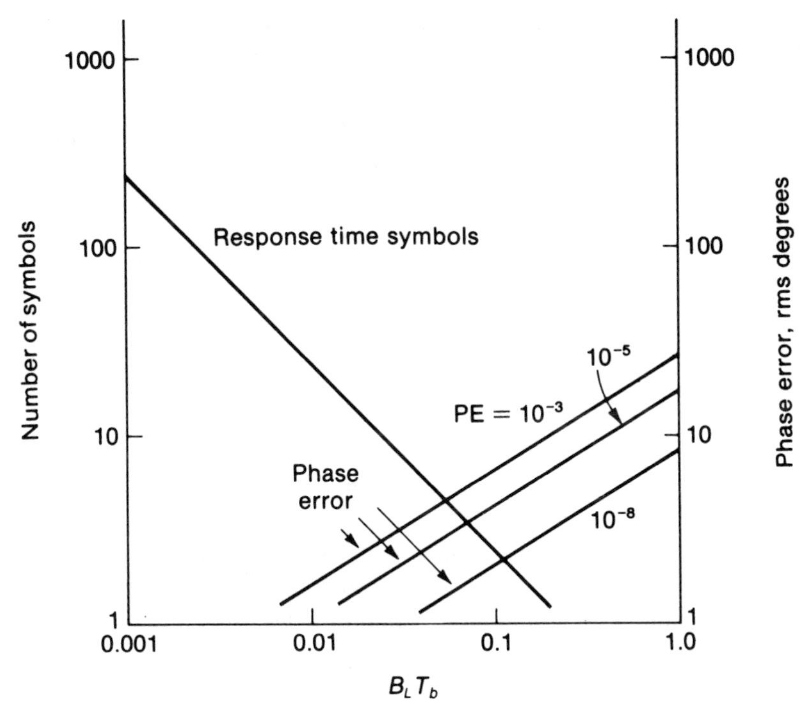

FIGURE 6.7 Carrier-referencing performance: number of symbols to pull in and resulting rms phase error.

slot time. After uplink–downlink retransmission, we can use Eq. (4.6.10) to compute the downlink CNR in the RF bandwidth as

$$\mathrm{CNR}_{\mathrm{RF}} = \frac{P_T \alpha_s^2 L}{\alpha_n^2 P_T L + N_{0d} B_{\mathrm{RF}}} \tag{6.2.6}$$

where P_T is the saturation power, and α_s^2 and α_n^2 are the signal and noise suppression factors of a hard-limiting TWT [from (4.6.15)];

$$\alpha_s^2 = \frac{\Gamma(\mathrm{CNR}_u)}{1 + \Gamma(\mathrm{CNR}_u)} \tag{6.2.7a}$$

$$\alpha_n^2 = \frac{1}{1 + \Gamma(\mathrm{CNR}_u)} \tag{6.2.7b}$$

where CNR_u is the uplink CNR in the satellite RF bandwidth. By inserting Eq. (6.2.7) we rewrite Eq. (6.2.6) as

$$\text{CNR}_{\text{RF}} = \frac{E_b/N_0}{B_{\text{RF}} T_b} \mathscr{S}_T \tag{6.2.8}$$

where

$$\frac{E_b}{N_0} = \frac{P_T L T_b}{N_{0d}} \tag{6.2.9}$$

and

$$\mathscr{S}_T = \frac{\Gamma(\text{CNR}_u)}{1 + \Gamma(\text{CNR}_u) + (P_T L / N_{0d} B_{\text{RF}})} \tag{6.2.10}$$

Here E_b/N_0 represents the downlink bit energy to downlink noise level that can be provided by the saturated satellite, and $\mathscr{S}_T$ represents the total degradation of the satellite transponding. The operation of modulation removal produces a tracking loop input CNR in the loop bandwidth B_L of

$$\text{CNR}_L = \text{CNR}_{\text{RF}} \left(\frac{B_{\text{RF}}}{B_L} \right) \mathscr{S}_q \tag{6.2.11}$$

where $\mathscr{S}_q$ is the squaring loss (see Appendix B),

$$\mathscr{S}_q = \left[1 + \frac{\frac{1}{2}}{\text{CNR}_{\text{RF}}} \right]^{-1} \quad \text{for BPSK} \tag{6.2.12a}$$

$$= \left[1 + \frac{4.5}{\text{CNR}_{\text{RF}}} + \frac{6}{(\text{CNR}_{\text{RF}})^2} + \frac{1.5}{(\text{CNR}_{\text{RF}})^3} \right]^{-1} \quad \text{for QPSK} \tag{6.2.12b}$$

Rewriting,

$$\text{CNR}_L = \frac{E_b/N_0}{B_L T_b} \mathscr{S}_T \mathscr{S}_q \tag{6.2.13}$$

The rms loop-phase error, which should be maintained at no larger than about 5° for minimal decoding degradation, is then given by

$$\sigma_\phi = \left(\frac{1}{\mathrm{CNR}_L}\right)^{1/2}$$

$$= \left[\frac{B_L T_b}{(E_b/N_0)\mathscr{S}_T \mathscr{S}_q}\right]^{1/2} \tag{6.2.14}$$

Thus, increasing loop bandwidth B_L shortens acquisition time, but it also increases loop noise, which can degrade phase lock-in if the phase error caused by the noise becomes too excessive. Figure 6.7 also plots Eq. (6.2.14) for various values of BPSK bit-error probability [which sets the value of the denominator in Eq. (6.2.14)]. Plots of this type allow a proper balance to be obtained for both acquisition time and loop-phase error.

A serious problem that arises with short-burst phase referencing with phase-lock loops is loop *hang-up*. If the initial phase offset that is to be pulled in during preamble sync is more than 90°, the nonlinear nature of the loop causes the pull-in time to be significantly longer than that predicted by Eq. (6.2.5), as shown in Figure 6.8. This can be attributed to the fact that an unstable null exists in the phase plane trajectories of phase-tracking loops at 180° offsets (Gardner, 1977). Since the null is unstable, the loop will eventually slide away to a stable lock point, but the restoring force is relatively weak in the vicinity of the 180° null, and the pull-in time may be relatively long. This slow pull-in will be extremely detrimental to the short-burst phase referencing needed in TDMA operation.

Several alternative schemes have been suggested for avoiding hang-up. One is to modify the standard phase-lock loop to aid pull-in when the initial phase offset is large. Gardner (1974, 1977) suggests inserting an additional restoring voltage of the proper polarity when the phase offset is greater than 90° (Figure 6.9a). This requires a logic circuit to determine polarity and time of voltage insertion, and therefore needs high CNR to ensure correct decisioning. A variation of this concept is the *limit-switched loop* in Figure 6.9b proposed by Taylor et al. (1982). Instead of applying an external voltage, the loop VCO is internally shifted by 180° when a large offset is detected. This converts unstable phase offsets near 180° to stable offsets near 0°, and eliminates hang-up. The loop is basically a decision-directed loop, which again uses phase decisions to modify error signals. Figure 6.9c shows the reported results of hang-up reduction by limit-switched operation under several operating conditions.

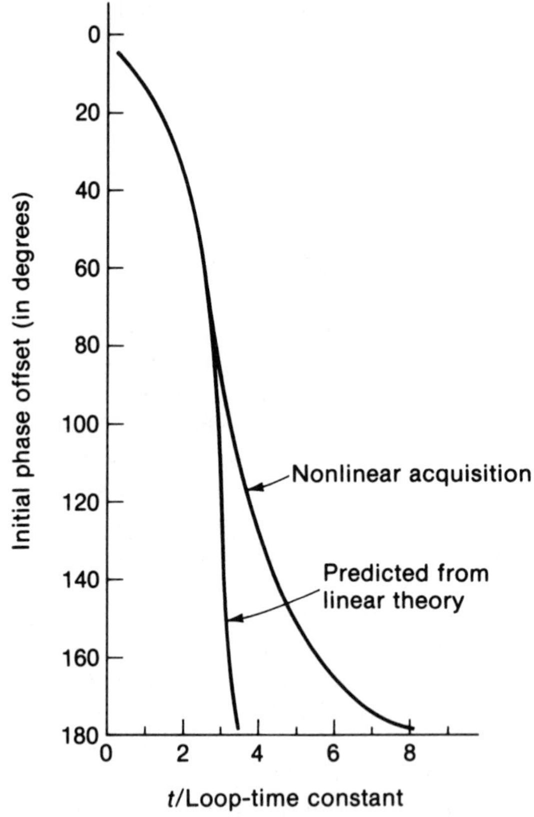

FIGURE 6.8 Required acquisition time versus initial phase offset, as predicted from linear and nonlinear theories.

Another technique to avoid hang-up is to not use a loop at all, but instead simply use a narrow-band tuned filter following the power-law device in Figure 6.6. Since the power-law-device output is generally at the desired frequency, and if the CNR is sufficiently high, a tuned filter at the correct frequency will produce a reasonably accurate carrier reference. This avoids the nonlinear-loop operation at large offset angles, but generates a noisy carrier reference instead of the relatively clean reference of a phase-locked VCO.

After carrier referencing is achieved, the phase-modulated carrier can be demodulated to baseband, but the data bit cannot be decoded until bit timing is achieved. Bit timing is obtained by locking a decoder timing block to the transition points in the baseband waveform. A transition occurs when the baseband waveform changes sign at the end of each data

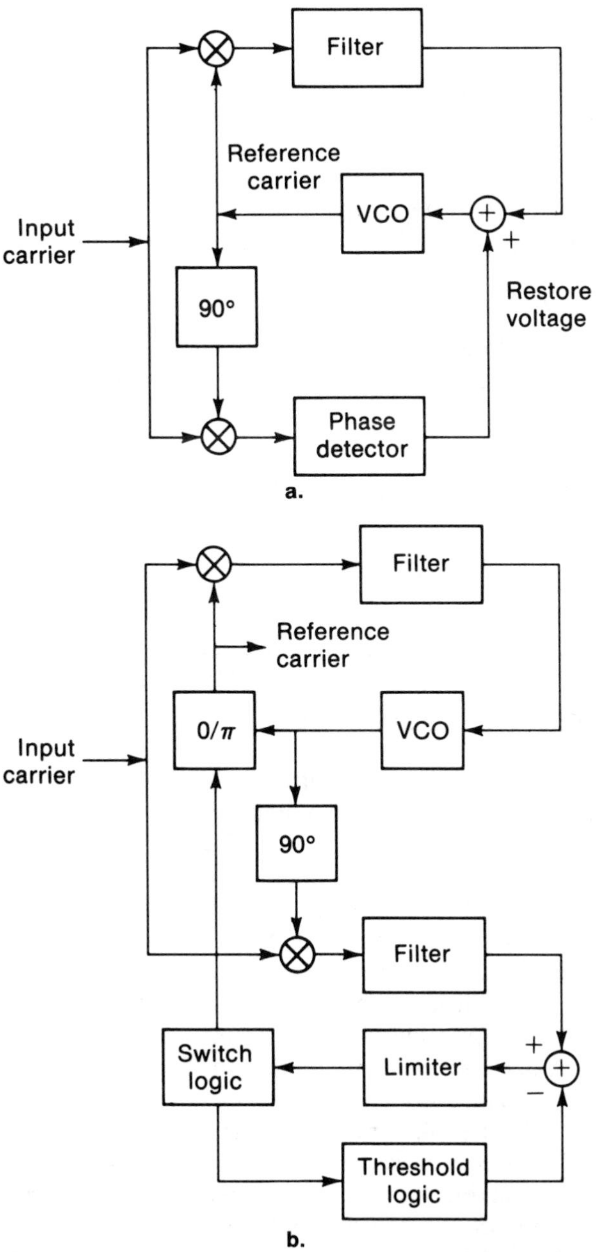

FIGURE 6.9 Loop hang-up control: (a) phase offset correction; (b) limit-switched loop; (c) acquisition improvement with limit-switched loops. [from Taylor et al. (1982)] (initial offset = 180°).

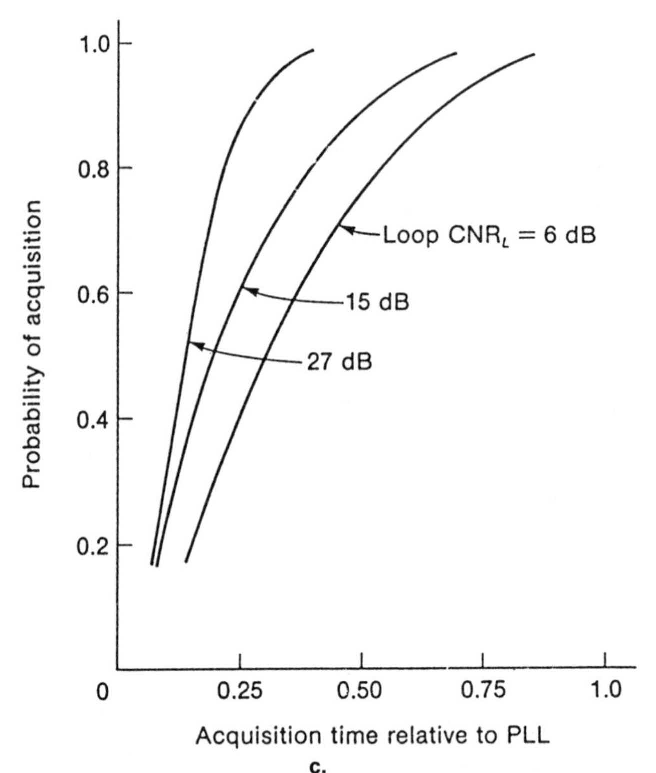

FIGURE 6.9 (continued) Loop hang-up control: (a) phase offset correction; (b) limit-switched loop; (c) acquisition improvement with limit-switched loops. [from Taylor et al. (1982)] (initial offset = 180°).

symbol. Bit timing is achieved by a symbol synchronizer subsystem that can take on various forms (see Appendix B, Section B.5). The most common methods are the transition tracking loop and the filter-square synchronizer (Figure 6.10). Transition trackers measure the timing error between local clock markers and the measured transition times of the demodulated baseband, and use the error to pull the local clock into synchronization. The local clock markers then time the subsequent bit decoder, as was shown in Figure 6.4. Variations in the form of the symbol synchronizer loops differ primarily in the way in which the transition errors are measured, and in the way the bit modulation is removed. Filter-square synchronizers use a baseband nonlinear operation to generate an unmodulated tone at the bit-rate frequency. A phase-lock loop can then be locked to that tone, producing timing markers in phase with the transitions. The design of the prefilter and the square (usually accomplished by a delay and multiply circuit) are the key aspect of bit synchronizer implementation (McCallister and Simon, 1981).

The number of bit times needed to achieve adequate bit timing depends on the number of symbol transitions that must be observed for accurate local clock control. Since each symbol transition is measured in noise, increasing the number of symbols accumulated prior to timing adjustment reduces the effective timing-error variance. The rms timing error per symbol time of a symbol-synchronizing loop is given by

$$e_t = \frac{1}{[m(E_b/N_0)\mathscr{S}_T\mathscr{S}_q]^{1/2}} \tag{6.2.15}$$

where m is the number of symbols accumulated, $\mathscr{S}_T$ is again the transponder loss, and $\mathscr{S}_q$ is the squaring loss, depending on whether a transition-tracking loop or filter-squarer is used [see Eq. (B.5.8) and (B.5.18) of

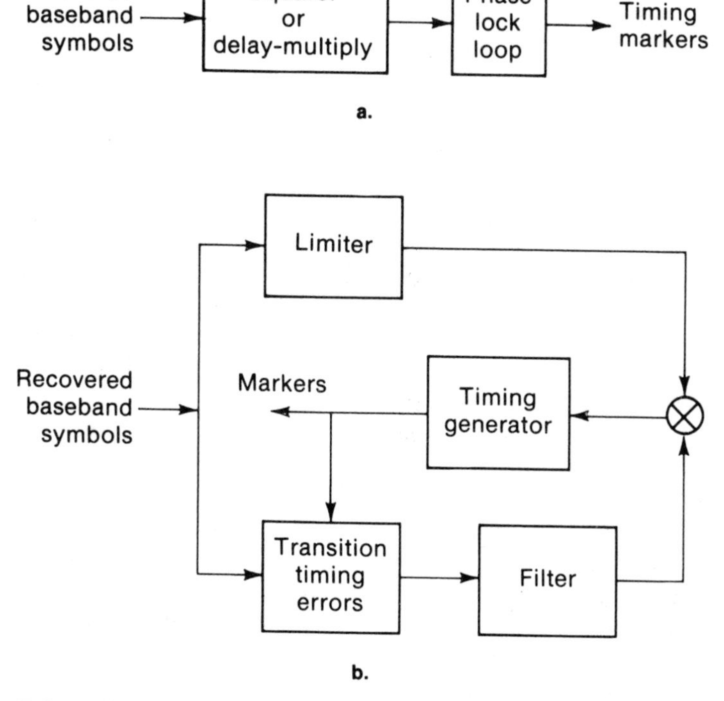

FIGURE 6.10 Bit-timing subsystems: (a) squaring system; (b) transition tracking.

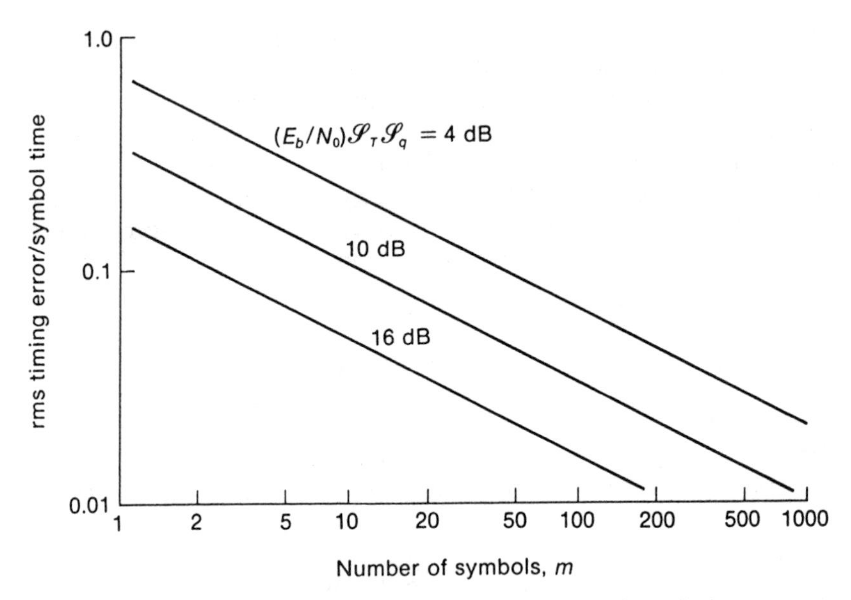

FIGURE 6.11 Bit-timing rms error versus number of symbols accumulated (E_b = baseband symbol energy).

Appendix B]. Figure 6.11 shows how bit-timing rms error is reduced as the number of accumulations is increased, for several values of desired bit-error probability. Bit-timing errors should be no larger than about 10% of the bit period in order to avoid significant decoding degradation. Figure 6.11 can be used to determine the necessary number of symbols m needed to achieve this value. It should be remembered that for BPSK carriers, a symbol corresponds to a bit, and a symbol transition can occur every bit time. In QPSK a symbol time corresponds to two bit times, and the number of symbols is half the number of bits.

A problem that occurs with symbol synchronization via transition tracking is that an abnormally long sequence of like bits will have no transition, and the loop will essentially "float" with no error updates during this time. During preamble lock-up this can be avoided by intentionally using bit patterns with many transitions (e.g., alternating 1s and -1s). During a TDMA data burst, however, transitions cannot be guaranteed unless alternating sync symbols are purposely inserted periodically in place of data bits. The problem is entirely avoided by use of Manchester baseband signals (Section 2.3), which have guaranteed transitions at the middle of each symbol for any bit sequence.

After obtaining both phase referencing and bit timing, subsequent carrier bits can now be decoded. To mark the data words in the frame, and to signal the beginning of data transmission, a *unique word* (UW) is sent immediately following the sync symbols. The decoder contains a digital word correlator that stores the UW (Figure 6.12). As the decoded bits are shifted through the register bit by bit, a word correlation is made with the stored word. When the transmitted UW is received, its correlation will produce a large signal output that can be noted as a threshold crossing. This crossing can be used to generate a time marker that marks all subsequent data words of the frame. During the next station burst in the next frame, the preamble is resynchronized, the UW is detected, and the word timing is again established for the new frame.

The primary concern in UW detection is that a false correlation producing a threshold crossing will occur when random bits fill the correlator (false alarm), or that the true UW is not recognized (miss). The latter will occur if the UW bits are decoded incorrectly, which will decrease the correlator value during UW arrival. By lowering the threshold, however, the miss probability can be reduced. For example, if only M of the N UW bits are needed for detection, and if PE is the decoded bit-error probability, the miss probability is then

$$\text{Prob[UW miss]} = \sum_{i=N-M+1}^{N} (\text{PE})^i (1 - \text{PE})^{N-i} \binom{N}{i} \qquad (6.2.16)$$

On the other hand, decreasing M to lower the miss probability will increase the probability of a false alarm. The probability that a random

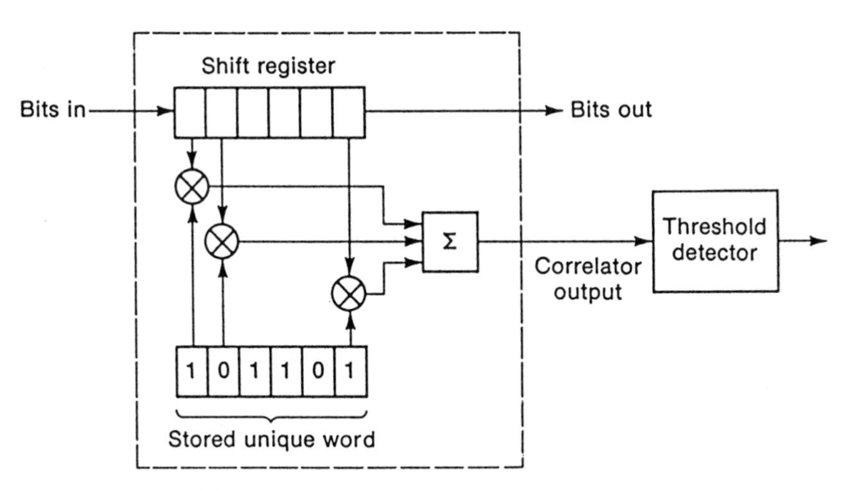

FIGURE 6.12 Unique word detector.

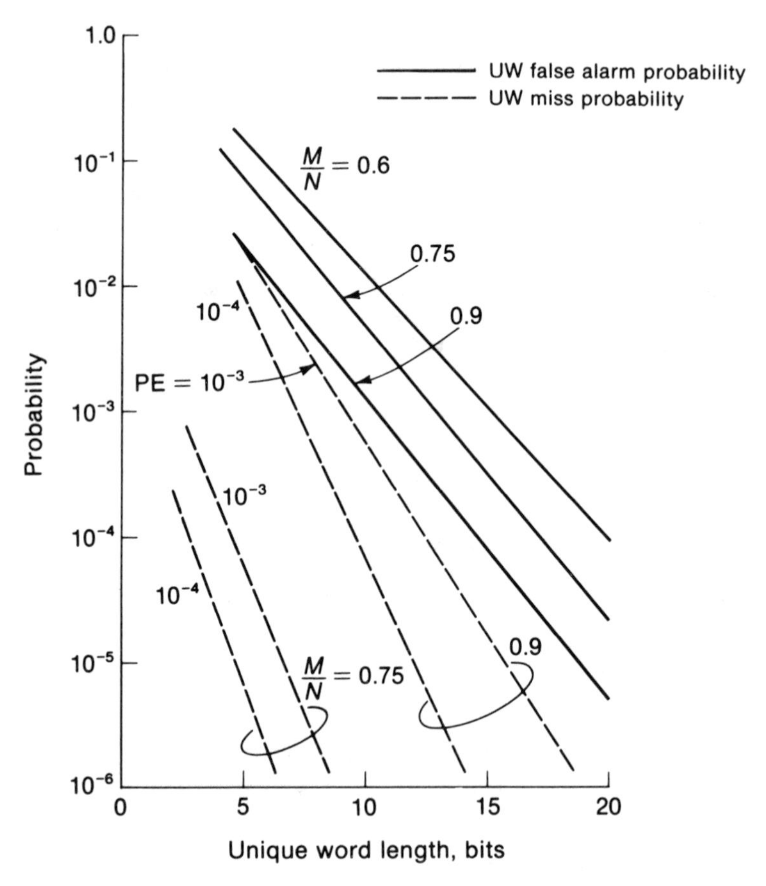

FIGURE 6.13 Miss and false alarm probabilities for unique word (UW) detection versus word length.

set of bits in the correlator will match M bits or more of the N-bit UW is

$$\text{Prob[UW false alarm]} = \frac{1}{2^N} \sum_{i=M}^{N} \binom{N}{i} \tag{6.2.17}$$

If we assume M is selected as a fixed fraction of the word length N, then increasing N decreases both the false alarm and miss probabilities of UW detection. Figure 6.13 plots Eqs. (6.2.16) and (6.2.17) as a function of word length N. Of course, increasing UW length adds to the preamble length.

False alarm probabilities also can be reduced by the use of *forward windows,* in which the correlation peak of the unique word is examined

TABLE 6.1 Preamble length computation

Sync Operation	*Constraints*	*Number Bits*
Carrier referencing	$B_L T_b = 0.01$ PE $= 10^{-3}$ Phase error $= 2°$ (Figure 6.7)	20
Bit timing	10% accuracy $E_b/N_0 = 10$ dB (Figure 6.11)	10 symbols =10 bits BPSK =20 bits QPSK
Unique word detection	75% threshold PE $= 10^{-3}$ Miss probability $= 10^{-4}$ False alarm probability $= 10^{-4}$ (Figure 6.13)	17

only in a specific time interval around its expected occurrence time. Only false alarms during this time window are allowed to occur. The window can be obtained by estimating forward in time from the last UW detection by an amount corresponding to the next UW arrival period. Further discussion of this windowing technique can be found in Campanella and Shaefer (1983).

The results of this section can now be used to estimate total preamble length, based on desired constraints on synchronization accuracy. Table 6.1 summarizes a typical computation, and shows that preambles of several tens of bits are usually adequate in TDMA operation.

6.3 SATELLITE EFFECTS ON TDMA PERFORMANCE

In Figure 6.3 it was shown that the number of individual data channels that can be supported in a TDMA frame increases with the bit rate that the satellite can support. The latter depends on the available satellite bandwidth and the CNR that the satellite can deliver to the ground decoder. A satellite RF bandwidth B_{RF} will allow a bit rate

$$R_{RF} = \eta_T B_{RF} \tag{6.3.1}$$

where η_T is the satellite channel throughput, which depends on the encoding format of the TDMA carriers, as was discussed in Section 2.8. The bit rate permitted by the power level of the satellite depends on the receiver decoder E_b/N_0 and on the desired bit-error probability PE.

The receiver downlink CNR, after transponding through the satellite, is again obtained as in Eq. (6.2.8). Substituting with Eq. (6.2.9), the decoder E_b/N_0 is then

$$\frac{E_b}{N_0} = \left(\frac{P_T L T_b}{N_{0d}}\right)\mathscr{S}_T \tag{6.3.2}$$

with $\mathscr{S}_T$ given in Eq. (6.2.10). The allowable RF bit rate produced by the satellite is then

$$R_{\mathrm{RF}} = \frac{P_T L \mathscr{S}_T}{\gamma N_{0d}} \tag{6.3.3}$$

where γ is the value of E_b/N_0 required to achieve the desired PE. The bit rate that can be supported by the satellite is the smaller of Eqs. (6.3.1)

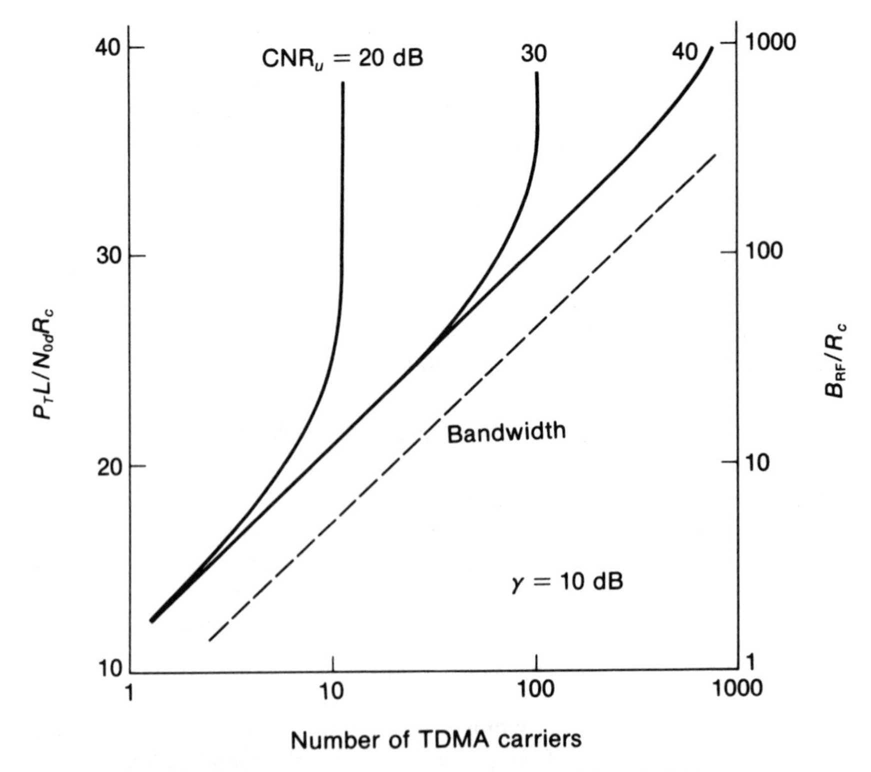

FIGURE 6.14 Required satellite power and RF bandwidth to support K TDMA channels.

and (6.3.3). This can then be converted to allowable number of channels using Figure 6.3.

Suppose we wish to send K QPSK digital carriers, each with bit rates of R_c bps through the TDMA system. Neglecting overhead bits, the required RF bandwidth is

$$B_{\mathrm{RF}} = KR_c \tag{6.3.4}$$

If the desired PE is 10^{-5}, we require a decoder $E_b/N_0 = \gamma \approx 10$ dB. We wish to determine the required satellite power P_T that will support the K channels. Assume a given CNR_u, and assume that limiter suppression effects in the satellite are negligible ($\Gamma = 1$). Solving Eq. (6.3.3) yields

$$P_T \mathscr{S}_T = KR_c N_{0d} \gamma / L \tag{6.3.5}$$

Equations (6.3.4) and (6.3.5) are plotted in Figure 6.14 for several values of PE and CNR_u. As the satellite power is increased, the number K of channels operating at R_c bps and PE increases. Eventually a point is reached when the number of channels is limited by the uplink CNR_u rather than the satellite power. At the same time, the satellite bandwidth only permits a fixed number of channels. Thus, the overall system capability is determined by the smaller number of allowable channels determined by each of these constraints.

6.4 NETWORK SYNCHRONIZATION

Network synchronization is achieved by clocking together all the transmitting and receiving stations of the TDMA system. Theoretically, if each transmitting station knew its range to the satellite precisely, network synchronization could be achieved by a single ground master clock used to time all earth stations. Accurate timing could be obtained from the master clock to each station using terrestrial links, and each station could be assigned a satellite time slot beginning at a fixed-time epoch relative to the master clock. Each station then need only adjust its uplink transmission so as to arrive at the satellite at exactly the correct time interval. The earth station would merely compensate in the uplink for the time delay due to its range to the satellite. This is referred to as *open-loop timing*. In practice, however, range values to a satellite cannot be determined precisely due to inherent uncertainty in satellite location. Recall from Eq. (1.4.11) that a typical 40 km uncertainty in satellite slant range will cause a timing uncertainty of hundreds of microseconds in transmission time. Unfortunately, timing accuracy for network sync must be maintained to within a small fraction of a slot time, which, from the previous examples, may be on the order of several microseconds. In addition, maintaining common clocks with microsecond accuracy over remote earth stations for long periods of time may be difficult. It is therefore necessary to combine a common, simultaneous range measurement for each station with a timing marker transmission from the satellite in order to achieve the desired timing. The synchronization markers are sent in the downlink, and network synchronization is initiated directly from the satellite.

The timing diagram in Figure 6.15 illustrates how this can be accomplished easily. The satellite transmits continually a periodic sequence of timing markers (in the form of some convenient waveform). These timing markers must be sent over a separate satellite bandwidth, and cannot interfere with the TDMA channels. The markers can be self-generated on-board by a stable satellite clock or can be relayed through the satellite from an earth-station clock. Each transmitting station is assigned a time slot at the satellite with respect to the marker points; that is, a station is assigned a time length t_s after each marker initiation, indicating where its time slot begins at the satellite. A station wishing to transmit first sends up its own ranging markers, which are retransmitted by the satellite and received back at the transmitting station, along with the satellite markers. By measuring the total two-way transmit time, t_r, of its range signal, a transmitting station can then adjust its own uplink transmission so it arrives exactly in its own time slot. From Figure 6.15 we can easily

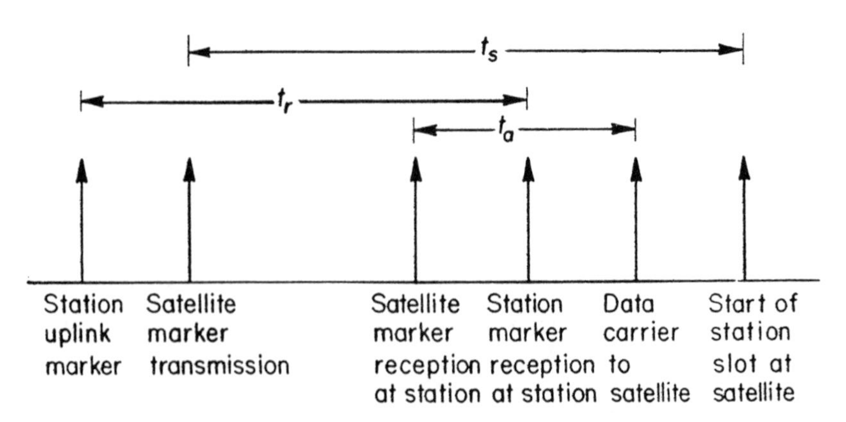

FIGURE 6.15 Timing diagram for TDMA network synchronization.

establish that, after receiving the satellite markers, a station must wait t_a sec before transmitting, where

$$t_a = \begin{cases} t_s - t_r & \text{if } t_s > t_r \\ t_s - t_r + T_f & \text{if } t_s < t_r \end{cases} \tag{6.4.1}$$

This now allows the earth station to acquire the correct slot, and transmission of the slot burst can begin. By monitoring the return time of the unique word of its own burst preamble (relative to the received timing marker of the satellite), closed-loop control of the slot timing can be maintained. In effect, the earth station replaces the ranging signal by the preamble word, and adjusts uplink burst transmission so that the returned sync word falls exactly in the proper slot position t_s s after marker reception. Since the satellite may drift slowly, the return preamble measurement must be updated continually to provide station-keeping and maintain the transmitting station synchronized within the network. This discussion points out a basic disadvantage of this type of TDMA operation. The necessity for continual network synchronization requires that each transmitting station have the capability of receiving its own transmissions. This can occur only with global downlink antennas, and requires all transmitting stations to have reception capability also. Note, however, that a receiving earth station does not require a ranging operation, but only the ability to receive the satellite markers. If it wishes to recover the transmission from a station using a time slot t_s s after marker generation, it need only adjust to the time slot beginning t_s s after marker reception.

The role of the ranging signal is only to aid in the initial acquisition of the slot [i.e., in the initial estimate of the parameter t_a in Eq. (6.4.1)]. While the preamble unique word can be used to maintain slot timing, it cannot be used for the initial acquisition. This is due to the fact that without some *a priori* indication of slot location, transmission of the preamble burst from an earth station would directly interfere with other slots. For this reason, the ranging operation is generally performed in an adjacent satellite band so as not to interfere with the TDMA channels (e.g., ranging through a TT&C subsystem). An alternative that avoids use of other bands is use of low-level range codes that are spread over the entire satellite bandwidth. If the power level of the range code is low enough, it will not interfere with any TDMA slot, and its length can be correlated to provide accurate range markers. However, each station would then have to have its own recognizable range code.

An alternative to in-band range codes is the use of a single-frequency tone burst to achieve the initial acquisition. The tone frequency is selected within the satellite bandwidth, and the tone burst is transmitted at a low

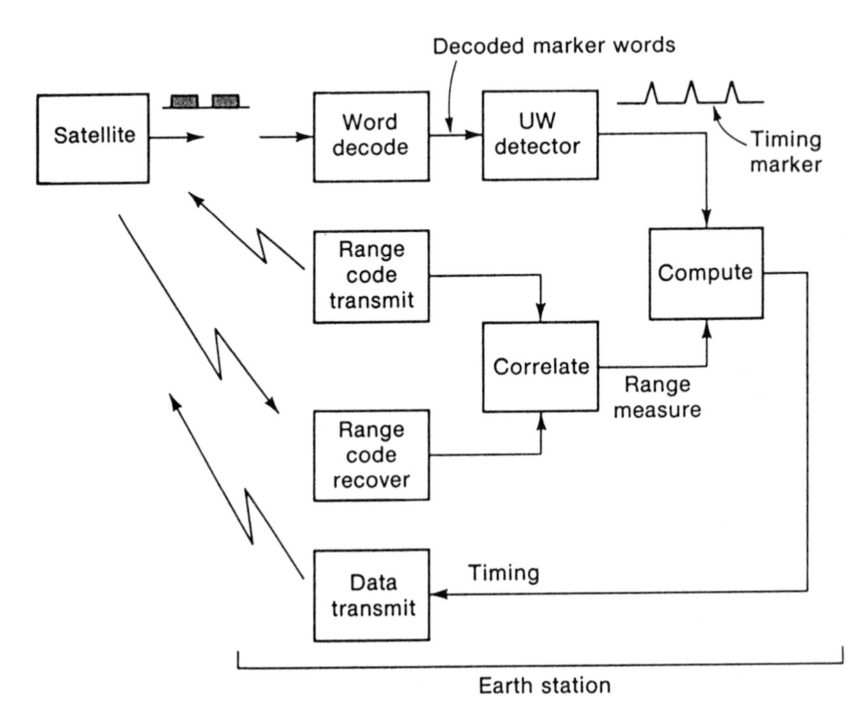

FIGURE 6.16 Earth-station network synchronization subsystem.

power level from the earth station through the satellite. Such tones will not interfere with the existing TDMA links because of its low power value. The earth station listens for its return by observing continually the output of a narrow filter tuned to the frequency. The tone burst therefore represents the station range markers in Figure 6.15. Although the tone burst is of low level, the narrow-band-tuned filter provides the noise and carrier interference reduction that allows detection of the burst arrival.

The accuracy to which the network timing must be maintained is directly determined by the selection of the size of the station slot time τ in Figure 6.1. Inaccuracies in network timing will cause station transmissions to fall into adjacent time slots, causing station crosstalk. To prevent this, time guard bands at the end of each time slot must be provided, as shown in Figure 6.5.

The network synchronization operations require a synchronization subsystem at the earth station similar to that shown in Figure 6.16. The satellite generates timing markers in the form of digital words that are broadcast continuously from the satellite to all earth stations. These marker words, for example, can be transmitted as BPSK modulation on a separate downlink carrier, usually located in a satellite band outside the TDMA band. These markers are produced at the TDMA frame rate, and correspond to a fixed word size. They can be produced from a satellite clock, or can be transponded from an earth-control station. At the earth station the marker bits are decoded and correlated with stored marker words, as in unique word detection. When the decoded marker bits fill the receiver correlator, a timing marker is produced, corresponding to the time points t_s in Figure 6.15. Since the satellite marker channel must operate at low bit-error probability, the primary effect of the marker detection is bit-timing errors, which cause timing offsets in the satellite marker location.

Simultaneous with the marker reception is the range measurement via round-trip transmissions through the satellite and back, using either a range code (see Appendix C) or burst preamble. The timing markers and range markers are then used to adjust the subsequent uplink burst transmissions for the TDMA data link. The time differential measurement is usually made by starting a digital clock counter with the satellite markers, and stopping the count with the returned range marker. The number of counted clock ticks is then a numerical indication of the required uplink adjustment. Since the TDMA slot bursts are generally timed by a digital clock as well, the numerical clock count can be easily converted to fractional TDMA slot delays or advances by deleting or inserting clock cycles. However, when timing differentials are measured in clock counts, an inherent quantization error appears in the clock count. If Δ is the clock period, and the quantization error is assumed uniform, the rms count error will be $\Delta/\sqrt{12}$. This quantization error adds directly to the timing errors of the satellite and range markers. Hence, the total slot timing mean-squared error that will occur in the network sync operation is

$$\overline{e_\tau^2} = \overline{e_m^2} + \overline{e_r^2} + \Delta^2/12 \text{ s}^2 \tag{6.4.2}$$

where $\overline{e_m^2}$ and $\overline{e_r^2}$ are the mean-squared timing error due to satellite marker reception and range delay measurement. For the marker word detection subsystem using a bit-synchronizing loop, the timing error relative to slot time is obtained as in Eq. (6.2.15):

$$\frac{\overline{e_m^2}}{\tau^2} = \frac{0.5(T_b/\tau)^2}{N_m(E_b/N_0)_s} \tag{6.4.3}$$

where T_b is the marker bit time, N_m is the marker word length, τ is the slot time, and $(E_b/N_0)_s$ is associated with the satellite downlink. It is

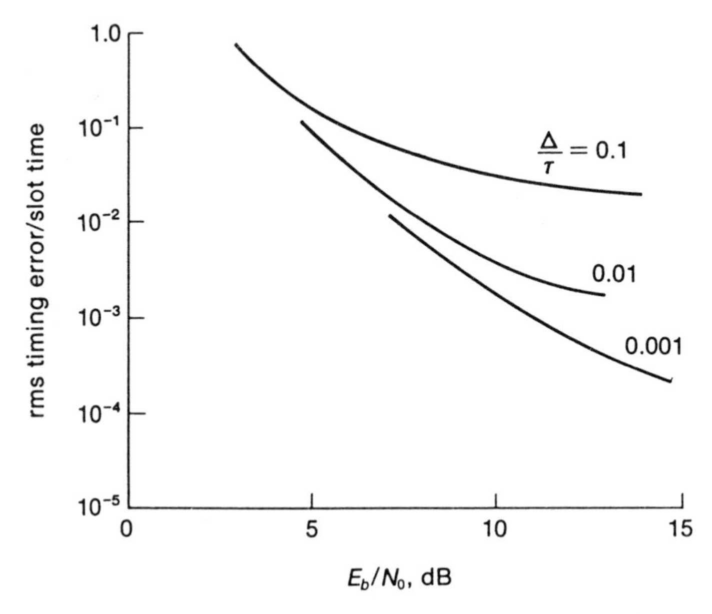

FIGURE 6.17 Timing error (rms) versus downlink E_b/N_0 and clocking rates. Assumes $e_m = e_r$, $T_b/\tau\sqrt{N_M} = 0.001$.

evident that the marker channel should use bit times much smaller than a slot time (if the bandwidth is available) and long marker words.

The range error will be that due to a turnaround satellite link, using range codes or preamble words. Hence, following the analysis of Eq. (6.2.14), we have

$$\frac{\overline{e_r^2}}{\tau^2} = \left(\frac{T_c}{\tau}\right)^2 \left(\frac{1}{N_c(E_c/N_0)\mathscr{S}_T}\right) \quad \text{Range code} \qquad (6.4.4a)$$

$$= \left(\frac{T_b}{\tau}\right)^2 \left(\frac{1}{N_m(E_b/N_0)\mathscr{S}_T}\right) \quad \text{Preamble word} \qquad (6.4.4b)$$

where E_b and E_c are the bit and code chip energies, N_c and N_m are the number of code and marker symbols, and T_c and T_b are the code chip or bit times. Figure 6.17 plots $\overline{e_\tau^2}$ in Eq. (6.4.2), using Eqs. (6.4.3) and (6.4.4), as a function of downlink E_b/N_0 and several clock rates. Note that operation with network sync errors well below a fraction of a slot time is quite feasible. Further discussions of TDMA synchronization methods can be found in Nuspl et al. (1977).

6.5 SS-TDMA

The advantage of spot beams in satellite links was pointed out in Section 3.6. It is therefore natural to extend the spot beam advantage to TDMA operation. The system will appear as in Figure 6.18, with selected spot beams for uplink and downlink transmission during each burst. The beams must be generated by moving a single beam to various locations (*beam switching*), or by selecting sequentially a different beam from a multiple-beam array. An immediate consequence of spot beaming is the loss of interconnectivity among all earth stations that was available during global-beam operation. While a single spot beam is in operation, earth stations not in that beam will not have access to that satellite burst. In addition, a given uplink station in one beam may not be able to receive its own downlink transmissions (if they are in another beam) and therefore cannot acquire and maintain network synchronization by the technique previously described. To circumvent these problems, and still obtain the TDMA spot beam advantages, it is necessary to apply spot-beam switching at the satellite. This concept is referred to as *satellite-switched TDMA* (SS-TDMA).

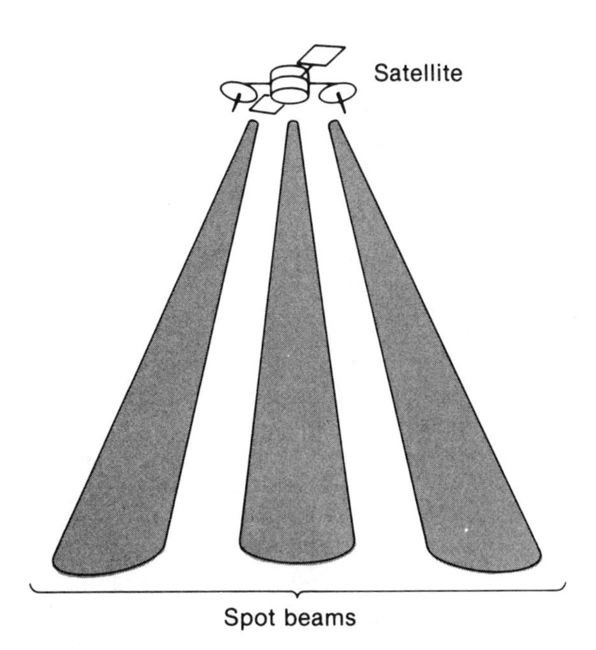

FIGURE 6.18 TDMA spot-beam model.

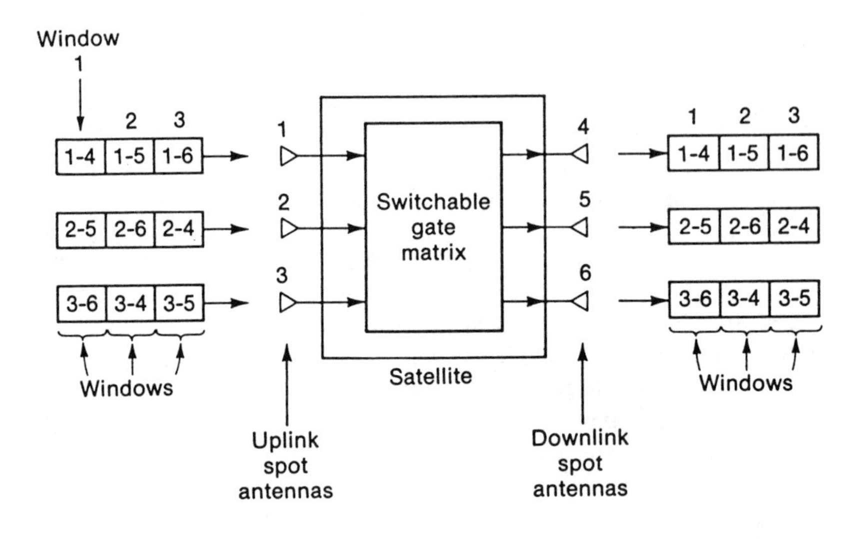

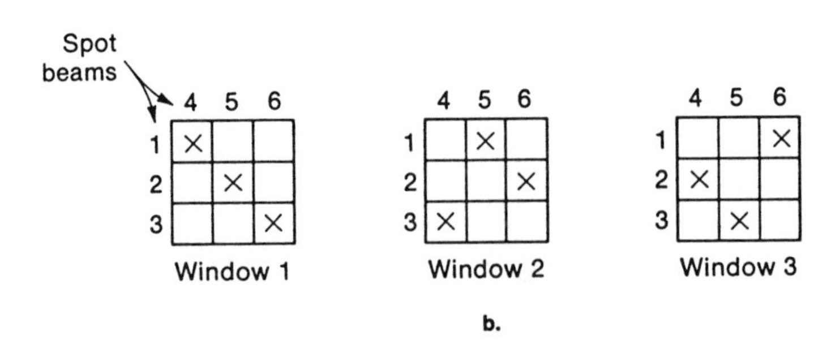

FIGURE 6.19 SS-TDMA system: (a) block diagram; (b) switching matrix.

Satellite switching restores the network connectivity lost by spot beaming. A microwave switch is implemented at the satellite to interconnect sequentially specific uplink to specific downlink beams. Consider the diagram in Figure 6.19a. A diode switch matrix (as in SS-FDMA) establishes the necessary connections of uplink spots to downlink spots. During the first switching "window" of the frame, uplink spot 1 is interconnected to downlink spot 4, uplink spot 2 is interconnected to downlink spot 5, and so on, as indicated by the matrix in Figure 6.19b. Frame slots of earth stations in the uplink spots (divided among the number of stations in that spot) are therefore available during that window to receiving stations in the interconnected downlink spot. During the second window the matrix switch is reconnected so that uplink spot 1 is interconnected to downlink spot 5, uplink 2 to downlink 6, and so on. Thus, during the second window a different set of earth stations has connectivity to a given uplink station. If the matrix cycles through all windows during a frame, an uplink station cycles through all possible receiving stations during a frame. In essence, the SS-TDMA system operates as a set of parallel TDMA links, with the parallel interconnections switching each window. Note that a transmitting earth station must transmit during each window of the frame, rather than just one slot per frame, as in standard TDMA.

The SS-TDMA system requires multiple-spot beams, and the switching matrix must be programmed to switch at fixed window times. If a single uplink earth station operates in each uplink beam, then only one slot per window is needed, and in fact the windows can correspond to slots. In this case, the switching is done at slot rates, and a given earth station transmits in every slot, but to a different receiver in each slot. Since a switch occurs once each slot, the programmable switch matrix must operate at microsecond rates. Clearly, the development of high-speed switching technology will greatly influence the capabilities of this type of operation. Note that an inherent advantage of SS-TDMA is that a transmitting station can adjust its transmissions in each slot for the type of receiver that will occur for that slot—transmitting at slower rates and with higher power for weak receivers, and at faster rates for stronger receivers. Since both transmitters and receivers are synchronized to the switching program, the SS-TDMA system can therefore be made instantaneously adaptive with respect to matching stations.

SS-TDMA complicates the network sync operation, since a transmitting earth station no longer can arbitrarily receive its own transmissions. A proposed solution (Carter, 1980; Camponella and Shaefer, 1983; Nupsl et al., 1977) is to allow a sync window in the matrix switching in which each uplink beam is returned on the same downlink beam for each slot of that window. This return transmission is referred to as *loop-back*, and it allows a transmitting station to receive itself during that window. This means that all network synchronization must be accomplished during this loop-back. The acquisition and tracking of the network timing must therefore be modified from that discussed in Section 6.4 when global antennas were available (a transmitter could receive any of its transmissions at any time). With loop-back, a transmitting station has only one slot per frame to receive itself. When first entering the network it must determine where this slot is. Since the satellite provides no markers, this can only be accomplished by randomly transmitting a slot burst at a fixed

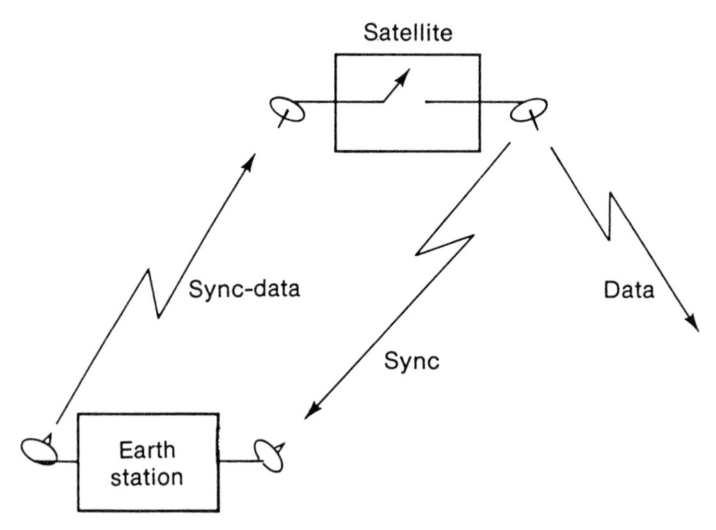

FIGURE 6.20 Satellite gating model for slot synchronization with loop-back.

point per frame, and listening for its return. The satellite matrix switch acts as a gate, which returns the uplink only for the loop-back slot. During all other slots, the gate is open and nothing is received (see Figure 6.20).

If the transmitter receives no return from its initial transmissions, it must move to another slot position and repeat. The station therefore searches throughout a frame until loop-back provides a return. This

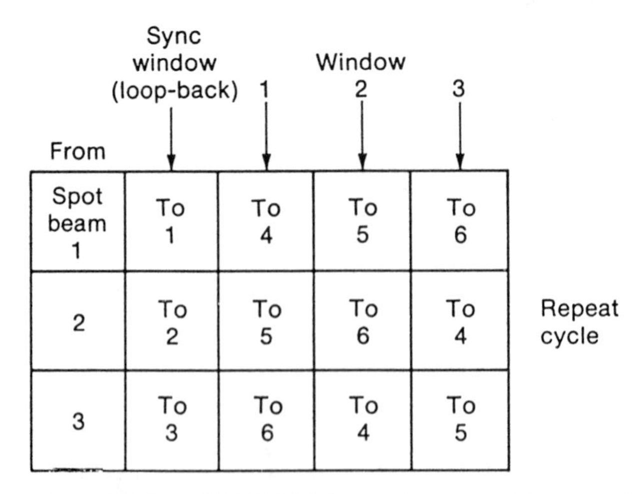

FIGURE 6.21 SS-TDMA frame formats with sync window.

means the uplink is arriving at the satellite when the switch-gate is closed. This lost transmission point in the frame identifies the loop-back slot of the sync window. By counting down to the next window, the station has acquired network synchronization, and its data-burst transmission can begin. To maintain slot sync, the station continues to transmit in both the sync loop-back slot and in the data slots, with the returned sync used to update network slot sync. Thus, an SS-TDMA transmitting station acquires and tracks with the frame formats shown in Figure 6.21. Note that prior to acquisition of the satellite gate, the uplink transmission may fall into the slots of other stations. Hence, acquisition signals during loop-back search must be of lower power level (about 20 dB below). The

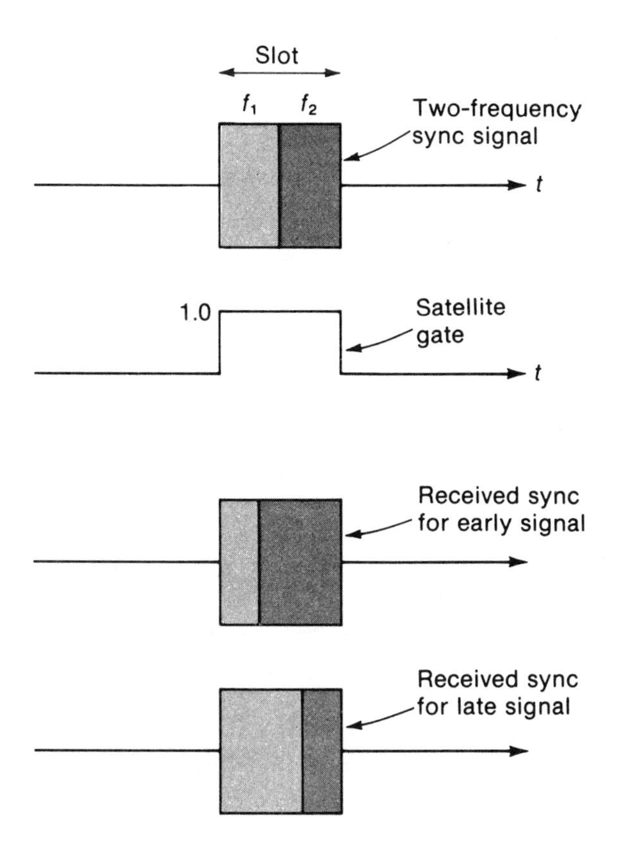

FIGURE 6.22 SS-TDMA slot sync using two-frequency transmission; frequency time length depends on sync error.

data bursts, and acquisition search requires a dwell time of several frames at each slot position in order to integrate up for acquisition detection when the proper slot is selected. For this reason SS-TDMA may require relatively long (several seconds) of network acquisition at station turn-on.

The initial acquisition signal need only be a carrier slot burst at a known frequency whose return can be identified. However, the loop-back signal for maintaining slot sync is usually designed as shown in Figure 6.22. The transmitted signal for the sync slot is divided into equal bursts of two different frequencies. As the sync signal passes through the gate in Figure 6.20, an early or late arrival will have either the beginning or end of the burst truncated during loop-back. By measuring the time difference of the two frequencies at the ground receiver, a measure of correction is obtained for subsequent slot transmission. An alternative scheme could use positive and negative BPSK bits to replace the two frequencies. Note the difference in these methods from the marker system described in Section 6.4. The latter uses global beams to return all uplink transmissions, and sends satellite markers to allow all receivers to compute slot timing errors. In SS-TDMA no markers are used, but the gating action of satellite switch effectively measures the timing error at the satellite, and loops back the result for timing correction.

Spot-beam operation of TDMA can be operated with a single switchable satellite beam instead of with multiple beams, but several of its advantages are lost. The satellite would relocate the single spot beam each slot, thereby interconnecting all uplink stations (each using a different slot) to a set of downlink stations. In the next frame, the spot beam is switched to relocate the beam in a different order, so that a given uplink slot is redirected to a different downlink beam location at each frame. This is often referred to as a beam-switched TDMA. This operation attains the spot beam advantage and maintains the interconnectivity, but an earth station is connected to a particular receiver only once in many frames, instead of once each frame as in SS-TDMA. This slows the data rate between station pairs, and may require buffer and storage hardware to maintain continuity. This topic will be considered again in Chapter 9.

An SS-TDMA format with multiple beams was used successfully on the WESTAR satellite, and is an integral part of the INTELSAT VI satellite. Figure 6.23a shows a schematic of the communication payload of the WESTAR. Separate spots were used for geographic portions of the United States, with a 4×4 TDMA switch used to obtain the necessary connectivity. The INTELSAT payload (Figure 6.23b) uses a 10×6 dynamic matrix switch to interconnect both C-band and K-band channels of the hemisphere, zonal, and spot beams of its multiple antenna assembly (recall Figure 3.31).

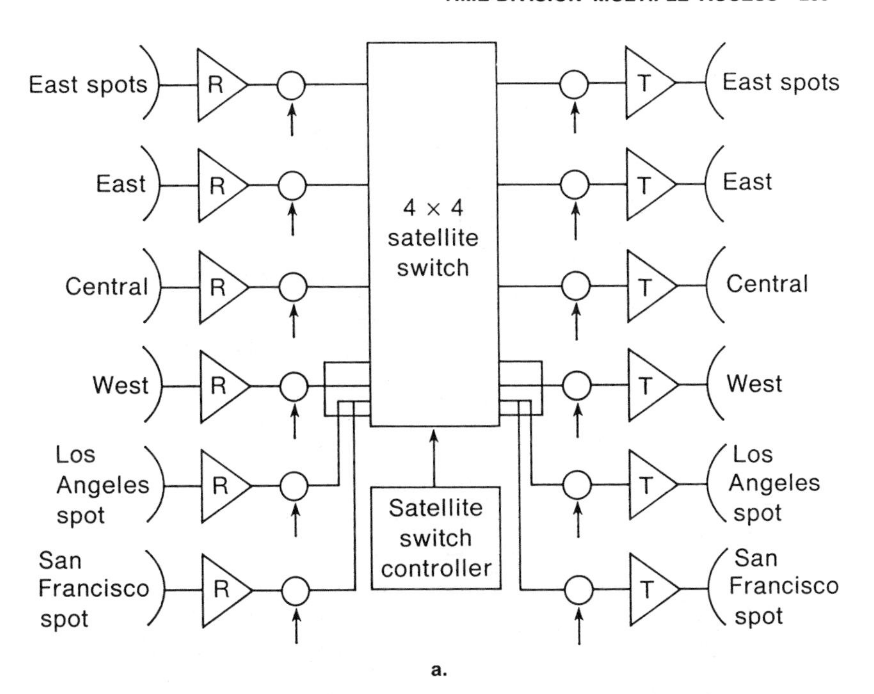

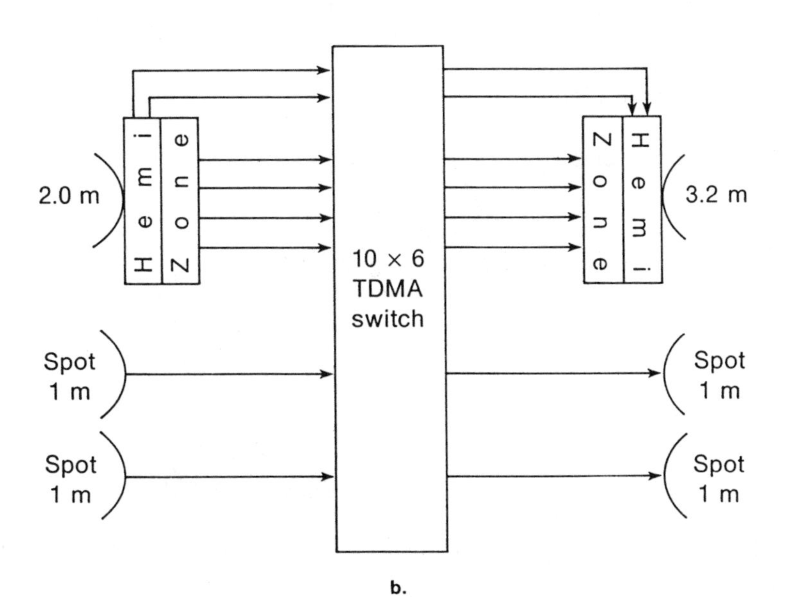

FIGURE 6.23 SS-TDMA payload examples: (a) advanced WESTAR; (b) INTELSAL/VI.

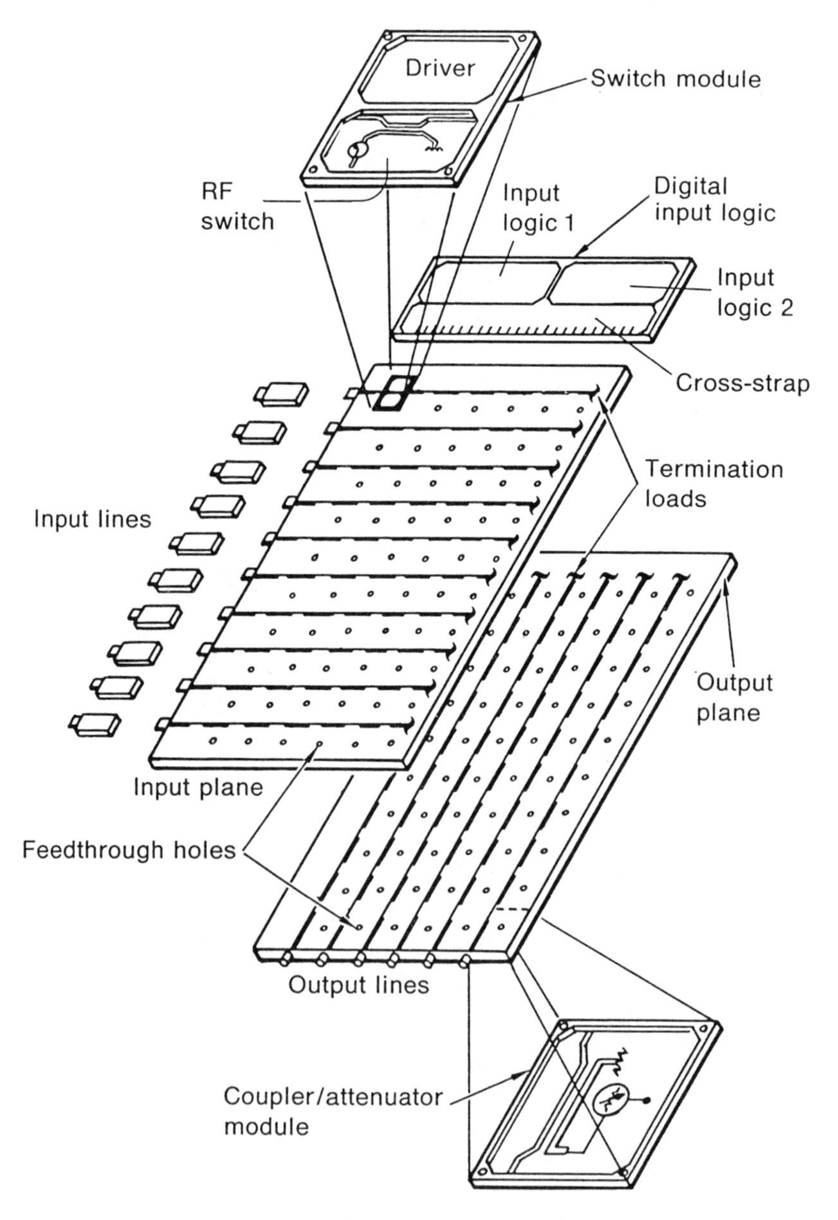

FIGURE 6.24 INTELSAT SS-TDMA microwave switch. (Courtesy of Intelsat Corp.)

Figure 6.24 shows the design of the microwave TDMA switch used in the INTELSAT VI. The input and output plates are sandwiched together with aligned feed holes interconnecting each. A series of drive circuits are used to control individual RF switches that cause the input field to either pass over or pass through the feed holes to the output plates. By properly driving the individual switches, an input RF line can be interconnected to any one (or many) output lines. These are then fed into the particular antenna electronics, completing the uplink–downlink SS-TDMA operation. The INTELSAT TDMA format uses 2 ms frame times with capability of switching at a 1-μs rate.

REFERENCES

Carter, C. (1980), "Survey of Synchronization Techniques for a SS-TDMA System," *IEEE Trans. Comm.*, **COM-28** (August), 1291–1302.

Camponella, J., and Shaefer, D. (1983), "TDMA Synchronization," in *Digital Communications—Satellite Earth Station Engineering*, edited by K. Feher, Prentice-Hall, Englewood Cliffs, NJ, Chapter 8.

Gardner, F. (1974), "Clock Recovery for QPSK-TDMA Receivers," *Proc. of the ICC*, Paper 43B, Minneapolis, MN (June).

Gardner, F. (1977), "Hang up in Phase Lock Loops," *IEEE Trans. Comm.* (October), 1210–1213.

IEEE (1979), "Special Issue on Satellite Communications," *IEEE Trans. Comm.*, **COM-27** (October), 1381–1423.

Kwan, R. K. "A TDMA Application in the Telesat Satellite Systems," *Proceedings of NTC*, 1973.

McCallister, R. D., and Simon, M., (1981), "Cross Spectrum Symbol Synchronization," *Proceedings of the NTC*, Houston, TX.

Maillet, W. (1972), "INTELSAT's 50 Mbits/s TDMA-2 system," *Proceedings of the 1972 International Conference on Digital Satellite Communications*, Paris, France, 1972.

Nuspl, P., Brown, K., Steenaart, W., and Ghicopoulis, B. (1977), "Synchronization Methods for TDMA," *Proceedings of the IEEE*, **65** (March), 631–642.

Sekimoto, T., and Puento, J. G. (1968), "A Satellite Time-Division Multiple Access Experiment," *IEEE Trans. Comm. Tech.*, **COM-16** (August), 581–588.

Taylor, D., Tang, S., and Marivz, S. (1982), "The Limit Switched Loop—A PLL For Burst Mode Operation," *IEEE Trans. Comm.*, **COM-30** (February), 396–407.

Thompson, P. and Johnston, E. (1983). "Intelsat VI—A New Generation for 1986." *Intern. J. Satellite Comm.*, **1** (January), 3–14.

Watt, N. (1986), "Multi-beam SS-TDMA Design Considerations Related to Olympus Payload," *IEE Proc.* **133**, Part F (February), 210–222.

PROBLEMS

6.1. A BPSK TDMA system is to transmit 1000 digital voice channels, each with four bits per sample, at a 64 kbps rate. The system must accommodate 1000 data bits/slot, at a frame efficiency of 10%. (a) What satellite bandwidth is needed? (b) How long is a TDMA frame? (c) How many slots in a frame? (d) How many preamble bits can be used?

6.2. (a) Derive Eqs. (6.2.16) and (6.2.17) for unique word detection. (b) Recompute the false alarm probability, assuming a window is used that observes the expected correlation peak to within $\pm q$ bits around its true arrival time.

6.3. A TDMA system requires timing to occur within 5% of a slot time. Plot a curve showing the available slot length versus satellite location uncertainty. [*Hint*: Recall Eq. (1.4.11).]

6.4. A TDMA sync system uses 1-ms slots, and the satellite transmits 10-bit marker words of bit length 5×10^{-4} s, with $(E_b/N_0)_s = 10$ dB. Assume no range timing error. What timing clock frequency is needed to ensure an rms timing error of no more than 5% of a slot time?

6.5. A TDMA slot acquisition system uses a frequency burst of 20 μs to determine its range. The satellite has a 500-MHz bandwidth and can produce a carrier downlink $\text{CNR}_{\text{RF}} = 10$ dB. How much lower in power can the frequency burst be, relative to a TDMA carrier, to produce the same CNR in its filter bandwidth?

CHAPTER 7
Code-Division Multiple Access

In *code-division multiple access* (CDMA) satellite systems, uplink stations are identified by uniquely separable address codes embedded within the carrier waveform. Each uplink station uses the entire satellite bandwidth and transmits through the satellite whenever desired, with all active stations superimposing their waveforms on the downlink. Thus, no frequency or time separation is required. Carrier separation is achieved at an earth station by identifying the carrier with the proper address. These addresses are usually in the form of periodic binary sequences that either modulate the carrier directly or change the frequency state of the carrier. Address identification is accomplished by carrier correlation operations. CDMA carrier crosstalk occurs only in the inability to correlate out the undesired addresses while properly synchronizing to the correct address for decoding. As in TDMA, CDMA carriers have the use of the entire satellite bandwidth for their total activity period, and CDMA has the advantage that no controlled uplink transmission time is required, and no uniformity over station bit rates is imposed. However, system performance depends quite heavily on the ability to recognize addresses, which often becomes difficult if the number of stations in the system is large.

Digital addresses are obtained from code generators that produce periodic sequences of binary symbols. A station's address generator continually cycles through its address sequence, which is superimposed on the carrier along with the data. If the address is modulated directly on the carrier, the format is referred to as *direct-sequence CDMA* (*DS-CDMA*). If the digital address is used continually to change the frequency of the carrier, the system is referred to as *frequency-hopped CDMA* (*FH-CDMA*). Superimposing addresses on modulated uplink carriers generally produces a larger carrier bandwidth than that which will be generated by the modulation alone. This spreading of the carrier spectrum has caused CDMA systems to be referred to also as *spread-spectrum multiple access* (*SSMA*) *systems*. Spreading of the carrier spectrum has an important application in military satellite systems, since it produces inherent antijam advantages (see Section 7.5). For this reason the designation SSMA is generally used in conjunction with military systems, whereas CDMA is usually reserved for commercial usage.

In this chapter we consider direct-sequence-CDMA systems, and discuss frequency hopping in Chapter 8.

7.1 DIRECT-SEQUENCE CDMA SYSTEMS

A digital version of a direct sequence CDMA link is shown in Figure 7.1a. The ith uplink station has assigned to it a digital address code, $q_i(t)$, the latter is a periodic binary sequence with binary symbols (chips) of width w s, and code length k (i.e., k chips per period). Each earth station has its own such address code. Digital information bits are transmitted by superimposing the bits onto the address code. If the ith station is to transmit the binary data waveform $d_i(t)$ $[d_i(t) = \pm 1]$, it forms the binary sequence

$$m_i(t) = d_i(t)q_i(t) \tag{7.1.1}$$

If T_b is the bit time of $d_i(t)$, then T_b may correspond to either a full period or $q_i(t)$, or to a fraction of a period. If each T_b is less than one address code period, then the data bits are modulating the polarity of a portion of a code period. Figure 7.1b shows an example of these situations for a particular address code and data sequence. In either case the address code $q_i(t)$ is serving as a subcarrier for the source data. The binary sequence in Eq. (7.1.1) is then PSK directly onto the station RF carrier located at the center frequency of the satellite uplink RF bandwidth. Since each station can use the entire satellite bandwidth and since Eq. (7.7.1) has a code chip rate of $1/w$ chips per second, each BPSK carrier will utilize a RF bandwidth of approximately

$$B_{\text{RF}} = 2/w \text{ Hz} \tag{7.1.2}$$

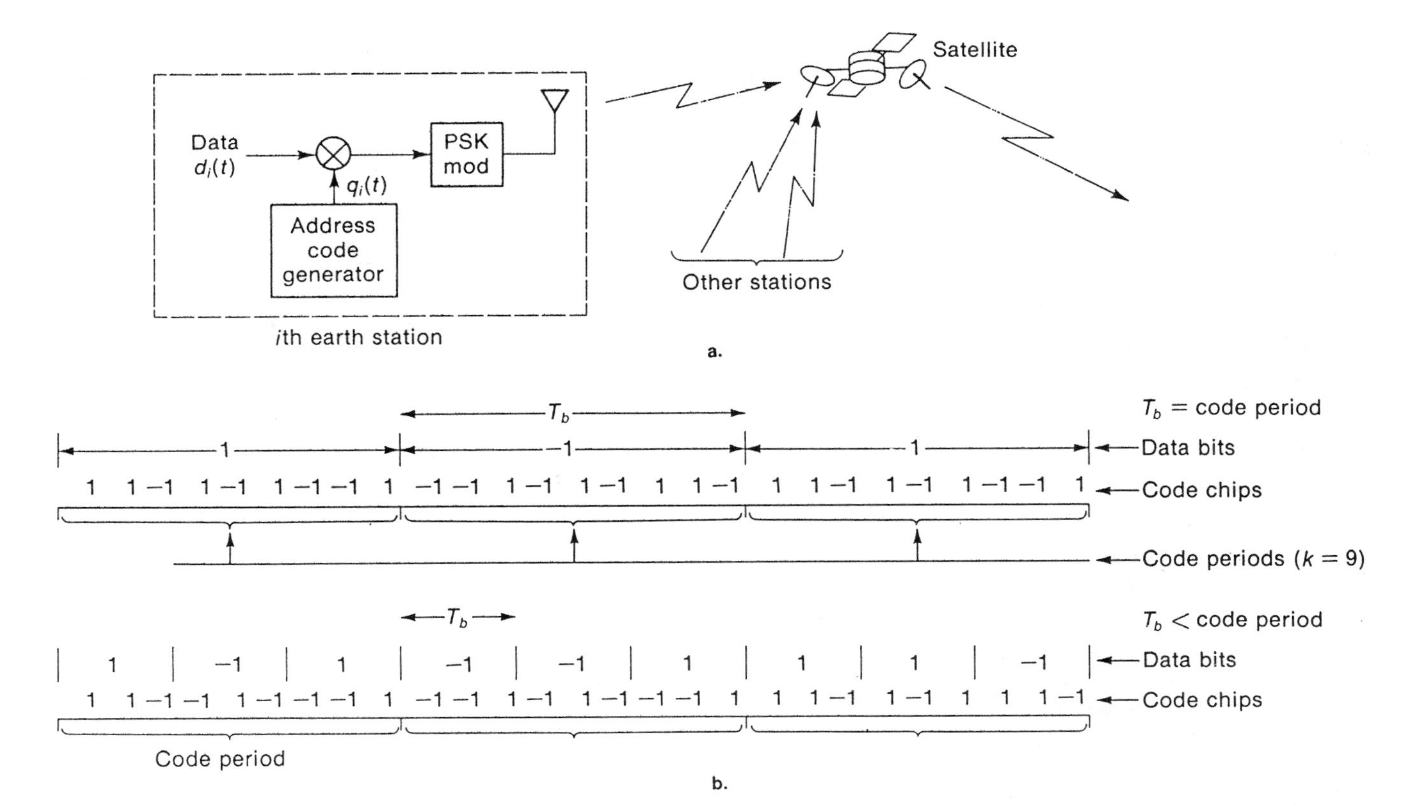

FIGURE 7.1 CDMA system. (a) Block diagram. (b) Bit and code period alignments.

Conversely, the available satellite RF bandwidth determines the minimum chip width, while the code period determines its relation to the bit times. That is, the number of code chips per bit is given by

$$k = \frac{T_b}{w} = \frac{B_{\text{RF}}}{2R_b} \tag{7.1.3}$$

The satellite RF bandwidth therefore determines the code length per bit. The ratio $B_{RF}/2R_b$ is often referred to as the CDMA *bandwidth expansion factor*, or simply, the *spreading ratio* of the code modulation. It indicates how much the RF bandwidth must be spread relative to the bit rate in order to accommodate a given address code length. Each station forms its PSK carrier in exactly the same way, each using the same RF carrier frequency and RF bandwidth, but each with its own address code $q_i(t)$. At the satellite, the frequency spectra of all active carriers are superimposed in the RF bandwidth.

The satellite repeater retransmits the entire uplink RF spectrum in the downlink, using straightforward RF–RF or RF–IF–RF conversion. Since all active carriers pass through the satellite simultaneously, limiting driver stages for the satellite amplifier will produce power robbing due to uplink noise, and weak carrier suppression by strong carriers, just as in FDMA. Hence, uplink power control is usually required with CDMA. In addition, if nonlinear amplification is used, intermodulation interference will be produced in the downlink, and the available satellite power must be shared by all stations. The amount of intermodulation can be controlled by adjustment of the satellite amplifier backoff.

In CDMA no attempt is made to synchronize or align the bit intervals, and the various uplink carriers operate independently with no overall network timing. Each active station simply transmits its modulated addressed carrier through the satellite into the downlink. A ground receiver must again obtain phase, bit, and code coherency with the desired uplink transmission in order to detect coherently, in the presence of the undesired carriers, the bits of the desired addressed carrier.

Temporarily ignoring all noise and intermodulation, the satellite combined BPSK carrier signals in the downlink is

$$\begin{aligned} x(t) &= \sum_{i=1}^{K} A_i \sin[\omega_c t + \tfrac{1}{2}\pi m_i(t - \tau_i) + \psi_i] \\ &= \sum_{i=1}^{K} A_i m_i(t - \tau_i) \cos(\omega_c t + \psi_i) \end{aligned} \tag{7.1.4}$$

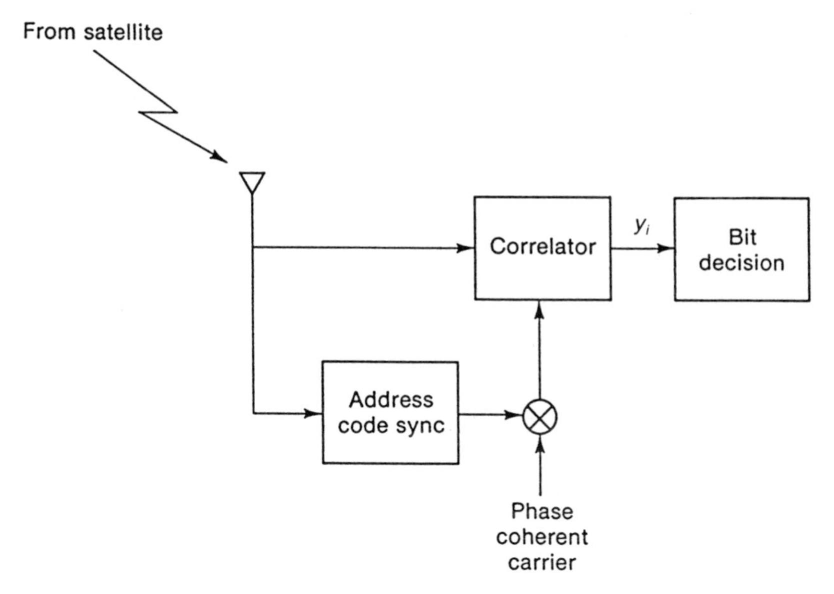

FIGURE 7.2 The CDMA receiver.

where the sum is over all active carriers and where $\{\tau_i\}$ and $\{\psi_i\}$ account for the different time shifts and phase shifts in passing through the satellite. At a particular earth receiving station, a corresponding decoder is used to recover the message bits of a particular uplink station, as shown in Figure 7.2. A receiver decoding the jth uplink therefore would generate the coherent coded reference

$$r_j(t) = 2q_j(t - \tau_j)\cos(\omega_c t + \psi_j) \tag{7.1.5}$$

Note that the generation of this receiver reference signal requires a time referenced code and a phase-coherent RF carrier. Time and phase coherency of this type requires a combined phase-locking–code-locking loop, as will be discussed in Section 7.7. In order to maintain the required address code synchronism, the code-locking subsystem must first acquire, then accurately track, the received address code. It does this by operating its own address code generator (identical to that of the transmitting station to be decoded) in time synchronization with the arriving address.

Assuming perfect code acquisition and lock-up, a decoder for the jth uplink correlates over a bit period the received waveform in Eq. (7.1.4) with the coded reference in Eq. (7.1.5), generating

$$\begin{aligned} y_j &= \frac{1}{T_b}\int_{\tau_j}^{\tau_j+T_b} r_j(t)x(t)\,dt \\ &= d_jA_j + C_j \end{aligned} \tag{7.1.6}$$

where $d_j = \pm 1$ is the desired data bit being received during the jth decoding bit time and

$$C_j = \sum_{\substack{i=1 \\ i\neq j}}^{K} A_i \cos(\psi_i - \psi_j)c_{ij} \tag{7.1.7}$$

$$\begin{aligned} c_{ij} &= \frac{1}{T_b}\int_0^{T_b} m_i(t-\hat{\tau}_i)q_j(t)\,dt \\ \hat{\tau}_i &= \tau_i - \tau_j \end{aligned} \tag{7.1.8}$$

The parameter C_j is the carrier crosstalk term arising in the jth decoder during the bit interval $(0, T_b)$ and is composed of a crosstalk contribution due to each active carrier in the downlink. The effect of each depends on the modulated address code cross-correlation c_{ij} in Eq. (7.1.8). The crosstalk C_j appears as an effective data-dependent interference term that adds to, or subtracts from, the desired correlation value d_jA_j in Eq. (7.1.6). Note that the value of C_j depends on (1) the data bits $\{d_i(t)\}$ being sent in the various uplinks, (2) the properties of the addressing sequences during the bit correlation, and (3) the amplitudes of the interfering stations. The latter is important, since an interfering earth station may arrive with more power than the desired station, resulting in a significant cross-correlation contribution even though the code correlation c_{ij} is minimal. This is the so-called *near–far problem* in CDMA, which arises when a near transmitter (shorter propagation distance) produces more receiver power than a desired transmitter located further away, even though both stations transmit with the same power. In satellite relay systems the near–far effect may not be that significant, since overall propagation distances through satellites tend to be comparable. The primary effect of an uplink power imbalance is the power robbing produced in the satellite amplification.

In typical DS-CDMA system operation, each active station simply transmits its modulated addressed carrier through the satellite into the downlink. A ground receiver obtains phase, bit, and code coherency with the desired uplink transmission in order to detect coherently, in the presence of the undesired carriers, the bits of the desired addressed carrier. The crosstalk during decoding is given by Eq. (7.1.8), where $\{\hat{\tau}_i\}$ must now be considered a set of random, independent time shifts occurring between the various uplink carriers as they arrive at the satellite. This means that the crosstalk parameter C_j evolves as a random disturbance. We proceed by assuming all carriers use the same bit time T_b. In this case, during a decoding interval of the jth carrier, the ith carrier arrives with relative delay $\hat{\tau}_i$. This means we can write

$$d_i(t - \hat{\tau}_i) = \begin{cases} d_{i_1}, & 0 \le t \le \hat{\tau}_i \\ d_{i_0}, & \hat{\tau}_i \le t \le T_b \end{cases} \tag{7.1.9}$$

where d_{i_1} and d_{i_0} are the previous and present bits, respectively, of the ith carrier during $(0, T_b)$. Because of the equal bit rate assumption, only two adjacent bits can cause crosstalk interference. This means that the crosstalk parameter C_j, conditioned on the data bits and the set of random shifts $\hat{\tau}_i$, is again given by Eq. (7.1.8) only now with

$$c_{ij} = d_{i_1}\zeta_{ij}(\hat{\tau}_i) + d_{i_0}\zeta'_{ij}(\hat{\tau}_i) \tag{7.1.10}$$

where

$$\zeta_{ij}(\hat{\tau}_i) = \frac{1}{T_b}\int_0^{\hat{\tau}_i} q_i(t - \hat{\tau}_i)q_j(t)\,dt \tag{7.1.11a}$$

$$\zeta'_{ij}(\hat{\tau}_i) = \frac{1}{T_b}\int_{\hat{\tau}_i}^{T_b} q_i(t - \hat{\tau}_i)q_j(t)\,dt \tag{7.1.11b}$$

The terms ζ_{ij} and ζ'_{ij} are partial cross-correlations of the address codes, each involving integrations over only a portion of the T_b s interval. Averaging over equal likely data bits and uniform phase angles shows that the mean of $C_j = 0$ for all $\{\hat{\tau}_i\}$, whereas the mean squared value follows as

$$\overline{C_j^2} = \sum_{\substack{i=1 \\ i \ne j}}^{K} \frac{A_i^2}{2}\mathscr{E}[c_{ij}^2] \tag{7.1.12}$$

Substituting from Eq. (7.1.10) and averaging over the data bits yields

$$\mathscr{E}[c_{ij}^2] = \int_{-\infty}^{\infty} [\zeta_{ij}^2(\tau) + \zeta_{ij}'^2(\tau)]p(\tau)\,d\tau \tag{7.1.13}$$

where $p(\tau)$ is the probability density of the delay τ_i. If each delay is uniformly distributed over $(0, T_b)$, implying that the relative carrier delays are independent and completely random, we then have

$$\overline{C_j^2} = \sum_{\substack{i=1 \\ i \neq j}}^{K} \frac{A_i^2}{2} \gamma_{ij}^2 \tag{7.1.14}$$

where now

$$\gamma_{ij}^2 = \frac{1}{T_b} \int_0^{T_b} [\zeta_{ij}^2(\tau) + \zeta_{ij}'^2(\tau)]\, d\tau \tag{7.1.15}$$

We see that γ_{ij}^2 plays the role of a code cross-correlation parameter and the mean squared value of the address code interference $\overline{C_j^2}$ increases directly with the set of address code cross-correlations $\{\gamma_{ij}^2\}$. Note that performance depends on the squared integrated cross-correlation functions in Eq. (7.1.15). It is this complexity that makes the CDMA case difficult to analyze. Investigation of maximum correlation values alone can be extremely misleading here, since it is the integrated squared value that is of prime importance. The interested reader is referred to IEEE (1982), Pursley (1977), Pursley and Roefs (1979), Sarwate and Pursley (1980) for further discussion of the properties of γ_{ij}^2, including some useful bounds for estimating Eq. (7.1.15).

An interesting result is that of Welch (1974), who showed that when cross-correlations involve k code symbols and K separate codes, the maximum value of γ_{ij} in Eq. (7.1.15) must always be at least

$$\max \gamma_{ij} \geq \left[\frac{K-1}{kK-1}\right]^{1/2} \tag{7.1.16}$$

For large K, this implies that the worst case cross-correlation for any code set behaves as $1/\sqrt{k}$, where k is the number of code chips per bit. Hence, maximum cross-correlation can only be reduced by increasing the code length per bit. From Eq. (7.1.3), this can be restated as

$$\max \gamma_{ij}^2 \geq \frac{2R_b}{B_{\text{RF}}} \tag{7.1.17}$$

and shows that code cross-correlation can only be reduced by increasing the spreading ratio of the code modulation.

7.2 CODE GENERATION FOR DS-CDMA SYSTEMS

The digital address codes used in the DS-CDMA must have other favorable properties in addition to the low cross-correlation. Each address code must be periodic, easily generated in a repetitive manner, and easily time-shifted and phase controlled for synchronization. In addition, all codes used in the same DS-CDMA system should be efficient in that they have cross-correlation values reasonably close to that satisfying the bound in Eq. (7.1.16). Later in Section 7.7 it will also be shown that each address code must individually have good autocorrelation properties in order to aid the initial acquisition of the code at the receiver. To establish a large (many user) CDMA system, it is necessary therefore to find large families of such codes having these specific properties.

A class of address code sets that meet many of these requirements are derived from *feedback shift registers*. Such a code generator is shown in Figure 7.3a. Digital sequences are shifted through the register one bit at a time at a preselected shifting rate. The register output bits form the address code. At any shifting time, the contents of the register are binary combined and fed back to form the next input bit to the register. Once started with an initial bit sequence stored in the register, the device continually regenerates its own inputs while shifting bits through the register to form the output code. The feedback logic and the initial bit sequence determine the structure of this output code. With properly designed logic, the output codes can be made to be periodic, and the register acts as a free-running code oscillator. Codes generated in this manner are called *shift register codes*. The class of periodic shift register codes having the binary waveform correlation of Figure 7.3b are called *pseudorandom-noise* (*PRN*) codes. PRN codes are known to have the longest period associated with a given register size and are therefore also called *maximal-length* shift register codes. If the register length is l stages, the PRN code will have a period of $n_l = 2^l - 1$ chips.

The autocorrelation function in Figure 7.3b is convenient for synchronizing two versions of the same code, since a high correlation occurs only when the codes are aligned, and a low (approximately zero) correlation occurs at all offsets. The underlying mathematical structure of shift register codes, and the technique for determining proper feedback logic, are well documented in the literature (Golomb, 1967; Golomb et al., 1964; Lee and Smith, 1974) and will not be considered here.

A property of shift register codes well suited to acquisition is that time shifting of the code can be easily implemented by externally changing the bits in the register at any time shift. Changing these bits causes the output to jump to a new position in the code period, which then continues

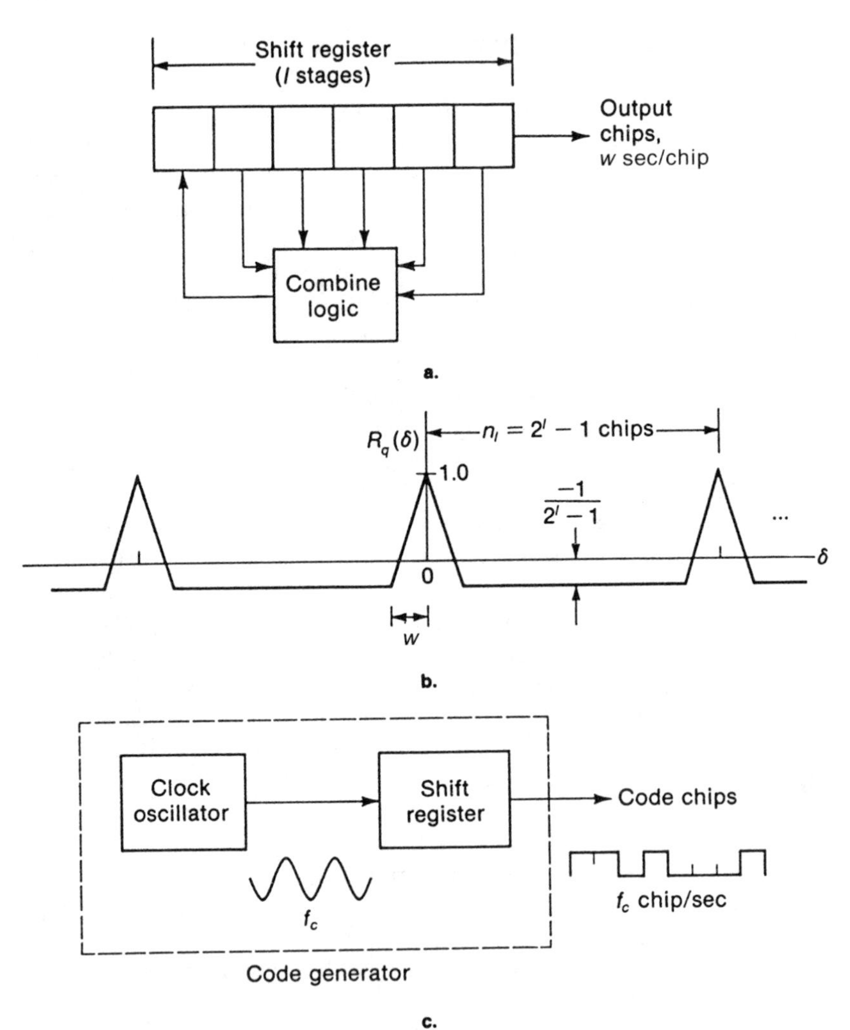

FIGURE 7.3 Shift register code generator. (a) Block diagram. (b) Code correlation function. (c) Clock-drive shift register for code generation.

periodic generation from that point on. Thus, within one shift time the output code can be forced to skip ahead in its cycle. Practical code generators are obtained by driving the shift register with a clock oscillator, Figure 7.3c. Each cycle of the clock provides the register shifting rate, so that an oscillator frequency of f_c would produce a code rate of f_c chips/s, and a chip width of $w = 1/f_c$ s. Hence, code rates are determined by the maximal rate at which a shift register can be driven.

To operate in a DS-CDMA environment a large family of shift register codes must be available, one to be assigned to each transmitter. However, there is a limit to the number of distinct maximal-length codes available from different connections of a register of a given length. Table 7.1 lists the number of such sequences produced from a register of the stated length. It is evident that the size of the code set grows rather slowly with register length, and one would have to resort to long registers to achieve large code sets. (Later it shall be shown that long registers may have abnormally long acquisition times.) More seriously, however, is that distinct maximal-length codes generated from different tap connections of registers of the same length will not necessarily have low pairwise cross-correlations needed in CDMA. Table 7.1 also lists the maximum normalized cross-correlations found among members of the corresponding set, and the result is considerably larger than that predicted by the Welch bound in Eq. (7.1.16). If only a subset of the register code set is used (only those exhibiting the lowest pairwise correlations), the available code set size will be significantly reduced. Hence, families of PRN codes of the same length, although easily generated from shift registers and each having desirable autocorrelation properties, are in general not suited for CDMA addressing codes, owing to either the restricted code set size or the intolerably high cross-correlations.

TABLE 7.1 Maximal-length PRN code sequences

Register Length (l)	*Code Length* (n_l)	*Number of PRN sequences*	*Maximum Pairwise Cross-correlation among Sequences (normalized to* n_l*)*
3	7	2	0.71
4	15	2	0.60
5	31	6	0.35
6	63	6	0.36
7	127	18	0.32
8	255	16	0.37
9	511	48	0.22
10	1023	60	0.37
11	2047	176	0.14
12	4095	144	0.34

As a result studies have been pursued to find modified register sequences that have improved correlation properties. Classes of such code sets were developed by Gold (1968) and Kasami and Lin (1976). These code sets are now universally suggested for modern CDMA usage. Gold developed a periodic address set by combining pairwise a specific PRN code with each shifted versions of another PRN code to generate the set members. If the register of the original PRN code had length l, so that the PRN code had length $n_l = 2^l - 1$, the new code set will have $n_l + 2$ distinct members. Furthermore, Gold showed that if the original two PRN sequences were properly selected, the pairwise cross-correlation among all members of the new set will be no larger than

$$\begin{aligned} \max \gamma_{ij} &\leq \sqrt{2^{l+1}}, \qquad l \text{ odd} \\ &\leq \sqrt{2^{l+2}}, \qquad l \text{ even} \end{aligned} \tag{7.2.1}$$

The family of codes generated in this way is called *Gold codes.* If the entire code length is used for a data bit, the normalized cross-correlation in Eq. (7.2.1) is

$$\begin{aligned} \frac{\gamma_{ij}}{n_l} &\leq \frac{\max \gamma_{ij}}{n_l} \\ &\approx \sqrt{\frac{2}{n_l}} \qquad \text{for } l \text{ odd} \\ &\approx \frac{2}{\sqrt{n_l}} \qquad \text{for } l \text{ even} \end{aligned} \tag{7.2.2}$$

Hence, Gold codes have a maximum pairwise cross-correlation that is within a factor of $\sqrt{2}$ (or 2) of the lowest achievable, as indicated in Eq. (7.1.16).

Gold codes can be generated by combining properly the outputs of two l-stage registers (Figure 7.4a) or from modified shift registers of length $2l$. Gold codes therefore retain the advantages of shift register generation, while having the near-optimal cross-correlation properties. However, each Gold code is not itself a maximal-length code, and does not have the idealized autocorrelation function in Figure 7.3b. Instead, Gold codes can have offset autocorrelation values (sidelobes) that may be as large as that in Eq. (7.2.1), as shown for example in Figure 7.4b. These sidelobes may lead to false lock-on points during the code acquisition phase of the receiver synchronization operation.

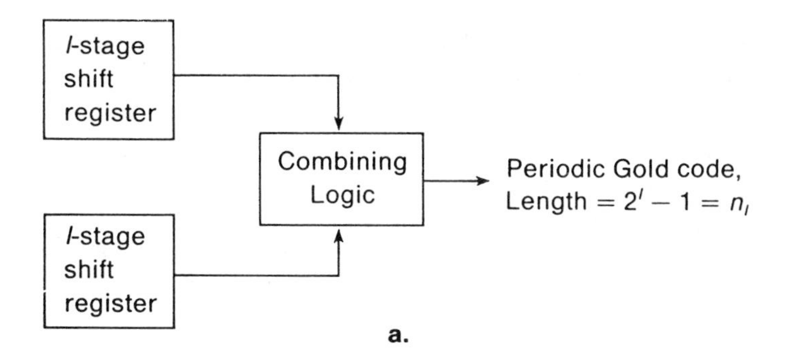

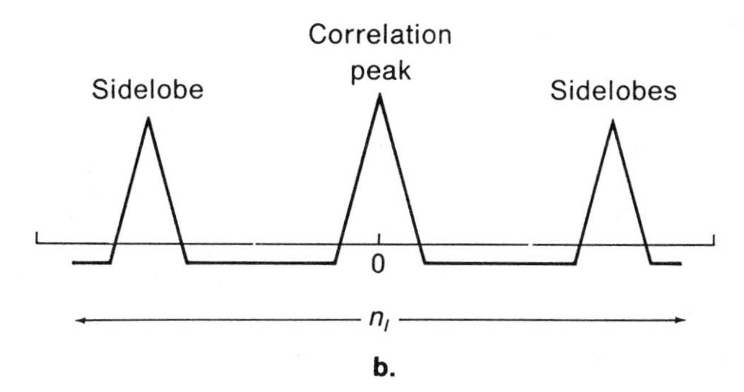

FIGURE 7.4 Gold code generator. (a) Block diagram; (b) code autocorrelation function.

Another set of periodic shift register sequences with improved pairwise cross-correlation are referred to as *Kasami codes*. These are formed by combining two maximal-length PRN codes from registers of length l in a slightly different manner. The Kasami code set is formed by adding to one PRN code of length $n_l = 2^l - 1$ the bits of another sequence formed by taking every other $\sqrt{2^l}$ bit of another PRN code, and using all shifted versions of the latter sequence. This produces a code set of only $\sqrt{n_l}$ members, as opposed to the $n_l + 2$ of a Gold code of the same length. However, the Kasami set has a maximum pairwise cross-correlation that never exceeds $\sqrt{2^l}$, so that the normalized maximum value in Eq. (7.2.2) is always identically $1/\sqrt{n_l}$. Hence, Kasami code sets are smaller in size, but achieve a cross-correlation that exactly meets the Welch bound in Eq. (7.1.16). For this latter reason, Kasami code sets are considered to be "optimal" addressing sequences. Like Gold codes, Kasami codes also are generated from shift register sequences.

Some recent study work (Boztas and Kumar, 1990) has shown that code addresses using QPSK modulation, instead of BPSK, can provide improved cross-correlation performance over the binary Gold codes (the maximum cross-correlation can be reduced by $\sqrt{2}$ for the same size code set). The QPSK modulation permits the address code sequences to be selected with symbols from a four phase alphabet, instead of the two phases (± 1) symbols of a BPSK code. The data bits would then directly multiply a sequence of QPSK waveforms, instead of BPSK waveforms, to represent the transmitted coded signals.

7.3 PERFORMANCE OF DS-CDMA SATELLITE SYSTEMS

When the addressing code set is selected for a DS-CDMA system, the cross-correlation interference can be estimated, and the performance of any individual CDMA receiver can be determined. This can be obtained by applying the correlation results in Section 7.1 to the satellite link analyses presented in Chapter 3.

The effect of code cross-correlation during CDMA decoding is to add an interference term to the receiver correlator output during bit decoding. The bit correlator output used for the receiver decisioning in Figure 7.2 is then

$$y_j = d_j A_j + n_j + C_j \tag{7.3.1}$$

where n_j is a zero mean Gaussian noise variable with variance $N_{0T}/2T_b$. Here N_{0T} is the total one-sided noise spectral level due to both receiver input noise (downlink and retransmitted uplink noise) and any intermodulation noise produced during the satellite transponding. This spectral level can therefore be written as

$$\begin{aligned} N_{0T} &= N_{0d} && \text{Downlink receiver noise} \\ &\quad + P_T L\left(\frac{N_{0u}}{P_u}\right)\alpha_n^2 && \text{Retransmitter uplink noise} \\ &\quad + N_{0I} L && \text{Intermodulation noise} \end{aligned} \tag{7.3.2}$$

where all parameters are defined in Eq. (5.3.3). The downlink carrier amplitudes $\{A_i\}$ are evaluated from the inherent power division at the satellite, and therefore

$$P_{di} = \frac{A_i^2}{2} = P_T L\left(\frac{P_i}{P_u}\right)\alpha_s^2 \tag{7.3.3}$$

If there were no cross-correlation interference (orthogonal system), the decoder would operate with

$$\left(\frac{E_b}{N_0}\right)_o = \frac{P_{dj} T_b}{N_{0T}}$$

$$= \frac{P_T L T_b (P_j/P_u)\alpha_s^2}{N_{0d} + P_T L(N_{0u}/P_u)\alpha_n^2 + N_{0I} L} \quad (7.3.4)$$

After substituting Eqs. (7.3.2) and (7.3.3) into Eq. (7.3.4), it is convenient to rewrite more compactly as

$$\left(\frac{E_b}{N_0}\right)_o = [\Gamma \mathrm{CNR}_u^{-1} + \mathrm{CNR}_r^{-1} + \mathrm{CNR}_I^{-1}]^{-1} \quad (7.3.5)$$

where the CNR terms are defined in Section 5.3. Equation (7.3.5) is identical in form to the FDMA result in Eq. (5.3.5) when no spectral crosstalk occurs, and when $B = 1/T_b$. Hence, orthogonal CDMA performs similarly to FDMA, and Figures 5.14 and 5.15 can be used directly to determine satellite power, with $(E_b/N_0)_o$ interpreted as CNR_d. The resulting carrier bit-error probability can then be obtained from $(E_b/N_0)_o$ using standard PE curves. It should be noted, however, that the intermodulation in CDMA is due to the nonlinear amplification of overlapping carrier spectra. This is in contrast to the case in FDMA where the carrier spectra are contiguous. Nevertheless, intermodulation spectra for the two cases have been observed to be quite similar, and intermodulation effects for the FDMA case are often used in CDMA analysis as well. Note also that the number of users (number of active stations) enters into Eq. (7.3.5) only through its effect on the intermodulation level N_{0I}. Furthermore, the near–far problem does not appear in the cross-correlation effect in an orthogonal system, and the only concern of unequal station powers is its effect on power robbing at the satellite.

It should be pointed out that the previous analysis was based on the assumption that the predominate intermodulation effect is to produce a noise-type interference. From the discussion in Chapter 5, this assumption is essentially valid if there are many approximately equal power carriers passing through the satellite, or if the uplink noise (which mixes with the carriers) is relatively significant. However, these are situations that could cause a few dominant direct sequence carriers to mix in the satellite nonlinearity to produce intermodulation terms that are not noiselike but retain the structure of the addressed carrier. This effect may, in fact, be more severe than the code cross-correlation of other carriers. Consider two equal powered, addressed carriers:

$$c_1(t) = Ad_1(t)q_1(t)\cos(\omega_c t + \psi_1)$$
$$c_2(t) = Ad_2(t)q_2(t)\cos(\omega_c t + \psi_2) \qquad (7.3.6)$$

In passing through a hard-limiting nonlinear amplifier, a third-order intermodulation will be generated having the form

$$\begin{aligned}\text{[third-order term]} &= h_3 d_1(t)q_1(t)[d_2(t)q_2(t)]^2 \cos(\omega_c t + \psi_3)\\ &= h_3 d_1(t)q_1(t)\cos(\omega_c t + \psi_3) \qquad (7.3.7)\end{aligned}$$

where h_3 is the third-order harmonic amplitude. This intermodulation term is therefore identical in form (but out of phase) with the carrier that will be decoded in the $c_1(t)$ channel. Subsequent address correlation with $q_1(t)$ will, however, correlate with the intermodulation in Eq. (7.3.7) as well, and produce an interference that will not correlate out as the other codes. Furthermore, there may be several of these terms, as well as possible fifth- and seventh-order intermodulation terms with the same effect. An interesting study of these effects was reported in Anderson and Wintz (1969), Baer (1982), and Kochevar (1977).

To determine performance with C_j present in Eq. (7.3.1) we must proceed formally by treating receiver correlations as intersymbol interference. The bit error probability for the jth receiver must then be obtained by averaging over all possible bit sequences and code delays of all interfering stations. Although the resulting PE_j will then represent the exact bit-error probability for a particular receiver, its computation may become quite long if the number of interfering carriers is large. This is due to the fact that the number of possible data vectors that must be averaged grows exponentially with the number of carriers K. This computation can be circumvented, however, when K is large, since we can model C_j as an additive Gaussian crosstalk variable that simply adds to the Gaussian interference of the downlink carrier. Justification and conditions for the validity of this Gaussian assumption were studied in depth in IEEE (1977/1982), Weber et al. (1981), and Yao (1977). Using Eq. (7.1.14), we can establish that C_j has a zero mean and a variance given by

$$\overline{C_j^2} = \sum_{\substack{i=1\\ i\neq j}}^{K} P_{di}\gamma_{ij}^2 \qquad (7.3.8)$$

We can interpret $\overline{C_j^2}$ as an interference power, spread over the data bandwidth. Hence, the decoder E_b/N_0 is now modified from Eq. (7.3.4) to

$$\frac{E_b}{N_0} = \frac{P_{dj} T_b}{N_{0T} + (\overline{C_j^2}/R_b)} \tag{7.3.9}$$

It is evident from the denominator that the code cross-correlations act to increase the noise level of the receiver. Eq. (7.3.9) can be rewritten in the form

$$\left(\frac{E_b}{N_0}\right) = \frac{(E_b/N_0)_o}{1 + (E_b/N_0)_o[\overline{C_j^2}/P_{dj}]} \tag{7.3.10}$$

This relates the (E_b/N_0) of a CDMA system to that that would occur if the system were truly orthogonal (i.e., no station cross-correlation). The latter system depends only on the intermodulation and power division that occurs in passing through the satellite with the addressed carriers. Note that the operating E_b/N_0 of a CDMA receiver is always less than that with the same number of orthogonal carriers, because of the denominator in Eq. (7.3.10). Thus, a CDMA system must be designed with higher power levels than an orthogonal system if it is to have the same performance. It should be noted that Eq. (7.3.10) can be written in the form of Eq. (7.3.5) as

$$\left(\frac{E_b}{N_0}\right) = [\Gamma \text{CNR}_u^{-1} + \text{CNR}_r^{-1} + \text{CNR}_I^{-1} + \text{CNR}_c^{-1}]^{-1} \tag{7.3.11}$$

where the cross-correlation CNR_c is now defined as

$$\text{CNR}_c = \frac{P_{dj}}{\overline{C_j^2}} \tag{7.3.12}$$

In this interpretation, address cross-correlations simply add an auxiliary CNR_c term to the orthogonal satellite E_b/N_0 result.

To further examine the cross-correlation effect, consider the simplified case where

$$P_i = P \qquad \text{(Equal carrier power}$$

$$\gamma_{ij} = \gamma \qquad \text{(Equal address correlation)} \tag{7.3.13}$$

$$\left.\begin{aligned} \frac{N_{0u}B_{\mathrm{RF}}}{P_u} &= 0 \\ \alpha_s^2 &= 1 \end{aligned}\right\} \quad \text{(High uplink CNR)}$$

Under these conditions,

$$\frac{P_i}{P_u} = \frac{1}{K}$$

$$\overline{C_j^2} = (K-1)\gamma^2 P \tag{7.3.14}$$

Eq. (7.3.9) becomes

$$\boxed{\left(\frac{E_b}{N_0}\right) = \frac{PT_b}{N_{0T} + [(K-1)P\gamma^2/R_b]}} \tag{7.3.15}$$

Substituting with the lower bound in Eq. (7.1.17) produces

$$\boxed{\left(\frac{E_b}{N_0}\right) = \frac{PT_b}{N_{0T} + \dfrac{(K-1)P}{B_{\mathrm{RF}}/2}}} \tag{7.3.16}$$

Equation (7.3.16) shows that the cross-correlation noise level is equal to the total interference power divided by half* the RF bandwidth. This noise level therefore increases with the number of interferers, but is directly reduced by the spread bandwidth. Equation (7.3.15) can be rewritten as in Eq. (7.3.10) as

$$\left(\frac{E_b}{N_0}\right) = \frac{(E_b/N_0)_o}{1 + (E_b/N_0)_o[(K-1)\gamma^2]} \tag{7.3.17}$$

where now $(E_b/N_0)_o = P_T L T_b / K N_{0T}$. This represents the operating E_b/N_0 (from which PE can be determined) for a CDMA system of K simultaneous carriers and fixed cross-correlation γ^2. Equation (7.3.17) can be

---

* The factor of a half occurs due to the assumption of BPSK modulation and the definition of RF bandwidth in Eq. (7.1.2). This factor may change slightly for other types of carrier modulation and bandwidth definitions.

solved for the required satellite power P_T needed to achieve a desired (E_b/N_0) of Υ. Hence,

$$P_T = \frac{1}{L}\left[\frac{\Upsilon(N_{0T}/T_b)K}{1 - \Upsilon[(K-1)\gamma^2]}\right] \tag{7.3.18}$$

Equation (7.3.18) is plotted in Figure 7.5 showing a normalized P_T as a function of the number of transmitting stations K for several values of cross-correlation γ^2, with $\Upsilon = 10$ dB. If a perfectly orthogonal system ($\gamma = 0$) can be maintained, P_T is linearly related to K. However, for K binary sequences to be exactly orthogonal, each must have length $k \geq K$, that is, there must be at least K chips per bit. From Eq. (7.1.3), this limits the number of orthogonal CDMA carriers to

$$K \leq \frac{B_{\text{RF}}}{2R_b} \tag{7.3.19}$$

Thus the maximum number of carriers is limited by the spreading ratio of the modulation. This limit to K is also included in Figure 7.5. Therefore, we see that a perfectly orthogonal CDMA system can either be power-limited or bandwidth-limited.

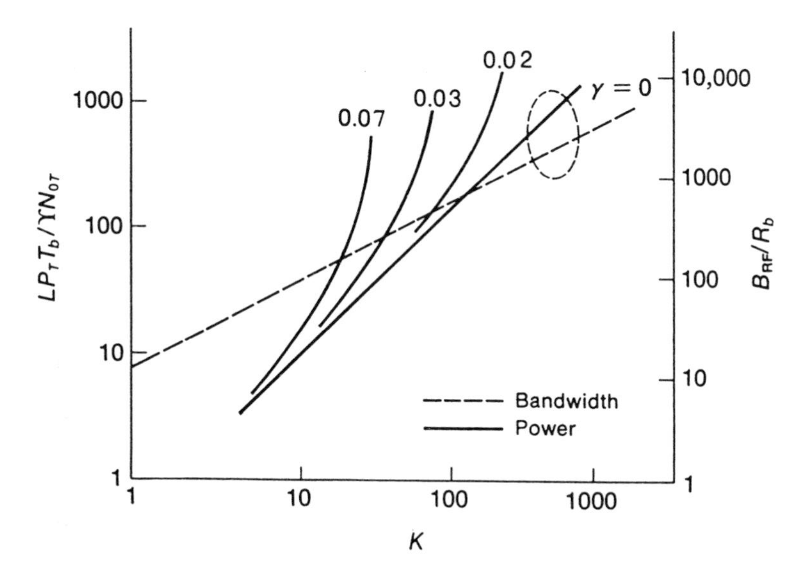

FIGURE 7.5 Required satellite power (normalized) and bandwidth to support K CDMA carriers.

A nonsynchronized CDMA cannot guarantee that the code sequences will remain orthogonal for all arrival time shifts and, as stated earlier, the code correlation must be at least $\gamma \geq 1/\sqrt{k}$. For the case of $\gamma \neq 0$, the buildup of interference with γ will further limit the number of CDMA carriers, as shown in Figure 7.5. In fact, an infinite satellite power is required in Eq. (7.3.18) if K is larger than

$$K_{\max} = \frac{1}{\Upsilon \gamma^2} + 1 \tag{7.3.20}$$

If we insert the lower bound for γ, this is approximately

$$\boxed{K_{\max} \approx \frac{B_{\mathrm{RF}}}{\Upsilon(2R_b)}} \tag{7.3.21}$$

With $\Upsilon = 10$ dB, the maximum number of CDMA carriers is about one-tenth the number of potential orthogonal carriers with the same spreading ratio. Hence, a CDMA system, although having the advantages of simplicity, nonsynchronism, and independent operation, appears less efficient in bandwidth usage. However, it should be emphasized that Eq. (7.3.21) refers to the actual number of CDMA carriers passing through the satellite at one time, which may be considerably fewer than the total CDMA carriers assigned in the system, if each operates only a small percentage of time. To see this, suppose each transmitter operates only a fraction v of the time. Then $\overline{C_j^2}$ in Eq. (7.3.8) and (7.3.14) must be multiplied by the averaging parameter v, since the cross-correlation power is only present v of the time. This means the value for K in Eq. (7.3.21) now becomes

$$\boxed{K_{\max} = \left(\frac{1}{v}\right)\left[\frac{B_{\mathrm{RF}}}{\Upsilon(2R_b)}\right]} \tag{7.3.22}$$

Thus, the maximum number of CDMA transmitters permitted in the entire system can be increased by the factor $1/v$. For example, if each CDMA transmitter operates only ten percent of the time (a few hours per day) the number of potential users in Eq. (7.3.22) theoretically can increase by a factor of 10 over that permitted by the bandwidth limitation. That is, the CDMA system can increase its total number of users by taking account of the fact that only a fraction will be operating at any one time. An FDMA system, with its dedicated frequency bands, generally, cannot compensate for individual transmitter off-times, unless some form of demand accessing is utilized. This ability to adjust can be a significant advantage for CDMA in applications involving low duty cycle operation, which tends to be characteristic of VSAT and mobile telephone systems. This point will be further discussed in Chapter 11.

7.4 COMBINED ERROR CORRECTION AND CODE ADDRESSING IN DS-CDMA

The fact that CDMA operation inherently increases the noise level suggests the possible use of error correction coding in conjunction with the code addressing to improve individual link performance. As shown in Section 2.6, the use of error correction via channel coding permits a specified PE at lower E_b/N_0 values, and thereby can handle higher noise levels, which in turn should allow more CDMA users. Since channel coding either reduces data rate or increases link bandwidth, a question arises concerning the additional effect of the code addressing on the overall bandwidth spreading.

In the CDMA transmitter in Figure 7.1, the data bits were directly superimposed on the address code. With error correction coding, the transmitter would be designed as shown in Figure 7.6. The data bit sequence is first channel coded into the code symbol sequence, using either block or convolutional coding techniques, as described in Section 2.6. The code symbols are then superimposed (modulated) on to the address code for BPSK carrier transmission. At the receiver the coherent address code is used to despread the received signal and recover the code symbols, after

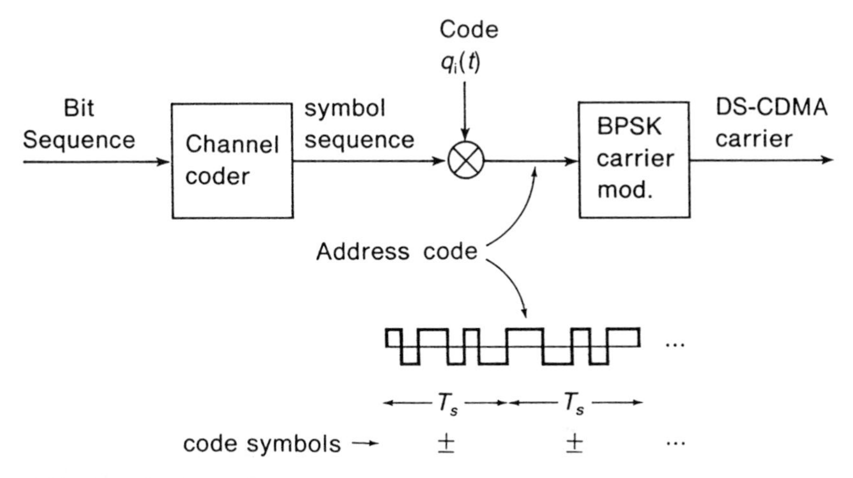

FIGURE 7.6 CDMA transmitter with channel coding.

which channel decoding algorithms are applied to recover the data bit sequence. Assume the channel coding is at rate r_c bits per symbol, and there are k_s address chips per code symbol. Then for the transmitted CDMA carrier,

$$\begin{aligned}\frac{B_{\mathrm{RF}}/2}{R_b} &= \frac{\text{bit time}}{\text{chip time}} \\ &= \left(\frac{\text{code symbols}}{\text{bit}}\right)\left(\frac{\text{address chips}}{\text{symbol}}\right) \\ &= \frac{k_s}{r_c} \end{aligned} \tag{7.4.1}$$

Thus, the spreading ratio is determined solely by the code ratio k_s/r_c, and can be achieved with either low rate channel codes or long address codes, or both. Note that if T_s is the symbol time in Figure 7.6, then

$$\begin{aligned} k_s &= T_s B_{\mathrm{RF}}/2 \\ r_c &= 1/T_s R_b \end{aligned} \tag{7.4.2}$$

Following the development leading to Eq. (7.3.15), the code symbol energy-to-noise level at the receiver becomes

$$\begin{aligned}\left(\frac{E_s}{N_0}\right) &= \frac{PT_s}{N_{0T} + [PT_s(K-1)/k_s]} \\ &= \frac{PT_s}{N_{0T} + [2(K-1)P/B_{\mathrm{RF}}]} \end{aligned} \tag{7.4.3}$$

The equivalent coded bit energy to noise level is then

$$\begin{aligned}\left(\frac{E_b}{N_0}\right) &= \left(\frac{1}{r_c}\right)\left(\frac{E_s}{N_0}\right) \\ &= \frac{PT_b}{N_{0T} + \dfrac{(K-1)P}{B_{\mathrm{RF}}/2}} \end{aligned} \tag{7.4.4}$$

Equation (7.4.4) is identical in form to Eq. (7.3.16), and again reduces cross-correlation interference by the total bandwidth spreading. That is, the CDMA interference is reduced by the combined spreading effects of the channel coding and the code addressing via Eq. (7.4.1). However, the required value of E_b/N_0 is that needed to produce the specified PE at the coding rate r_c inserted. This is lower than that for the uncoded case by the coding gain achieved. If the E_b/N_0 is set equal to Υ as in Eq. (7.3.18), and solved for the K with coding, Eq. (7.3.21) is again obtained, except Υ is reduced from the uncoded case. Thus, the maximum number of CDMA carriers with coding inserted is directly increased over the uncoded case by the coding gain of the error correction. Since coding gain increases with lower code rates, this suggests a basic CDMA design direction using as low a code rate as feasible, in conjunction with long addressing codes (high k_s) to achieve the maximum spreading ratio. Design studies of this type have been reported (Boudreau et al., 1990; Viterbi, 1990).

7.5 ANTIJAM ADVANTAGES OF DIRECT CODE ADDRESSING

The use of direct-sequence modulation allows station addressing and reduced interference in a CDMA format. However, the resulting spectral spreading associated with these methods also affords some defensive advantages against intentional jamming of the system by an external source. For this reason spread-spectrum communication play an important role in data links operating in a military environment.

To see the inherent advantages of spectral spreading in a jamming situation, consider the system in Figure 7.7a. We assume a communication link between two stations, with the presence of an external source intentionally transmitting noise into the receiver to attempt to destroy the link. We initially neglect both receiver noise and the interference effect of other stations, and omit the presence of a satellite relay. Hence, the only form of interference at the receiver is due to the jamming signal alone. Refer to the spectral diagram in Figure 7.7b. The transmitter uses a DS-CDMA modulated carrier in which a data rate R_b is spread to occupy a carrier bandwidth of B_{RF} Hz. The jammer produces a noise spectrum of width B_j Hz and power P_j at the receiver input. If $B_j \geq B_{RF}$, we say the jammer is *broadband.* When $B_j < B_{RF}$, we refer to the jammer as being *partial band.*

Under these assumptions the received waveform is

$$x(t) = Ad(t)q(t)\cos(\omega_c t + \psi) + j(t) \tag{7.5.1}$$

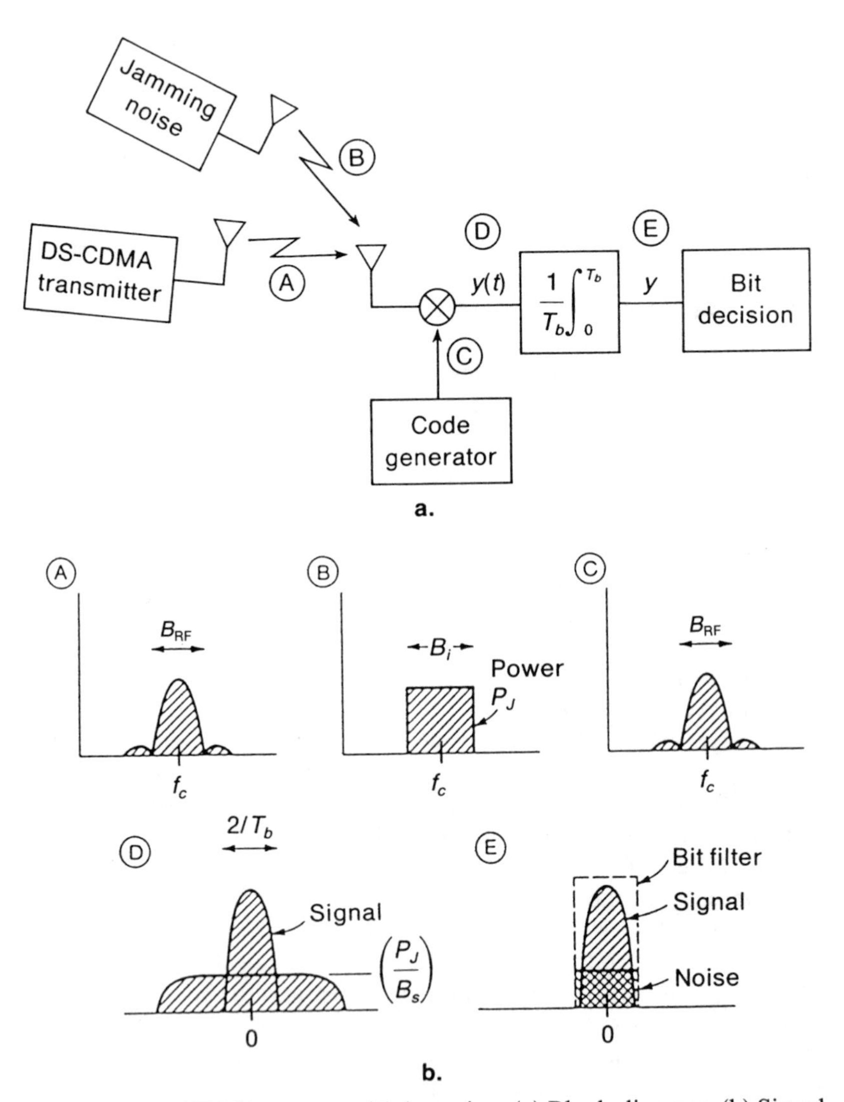

FIGURE 7.7 CDMA system with jamming. (a) Block diagram. (b) Signal spectra.

where $j(t)$ is the jamming noise. The received carrier therefore has power $P_c = A^2/2$, and the receiver-input-signal-to-jammer ratio is then

$$\mathrm{SJR_{in}} = \frac{P_c}{P_J} \tag{7.5.2}$$

The receiver uses a coherent addressed carrier $r(t) = 2q(t)\cos(\omega_c t + \psi)$ to recover the data $d(t)$, as described in Section 7.1. The decoder multiplier output in Figure 7.7 produces

$$y(t) = Ad(t) + j(t)r(t) \tag{7.5.3}$$

Bit integration over $(0, T_b)$ now generates the decoding variable

$$\begin{aligned} y &= \frac{1}{T_b}\int_0^{T_b} y(t)\,dt \\ &= \pm A + \frac{1}{T_b}\int_0^{T_b} j(t)r(t)\,dt \end{aligned} \tag{7.5.4}$$

where the $\pm$ sign depends on the data bit $d(t)$ over $(0, T_b)$. Thus, y is identical to Eq. (7.3.1), with jamming noise $j(t)$ replacing the receiver interference. The second term in Eq. (7.5.4) can be interpreted as the filtering (with bandwidth of $1/T_b$) of the random process $j(t)r(t)$. This latter process is known to have a spectrum given by the convolution of $S_j(\omega)$, the spectrum of the jammer, with $S_r(\omega)$, the spectrum of the addressed carrier, $r(t)$. Since $r(t)$ has a spread spectrum, then convolution will always produce a spectral spreading over the baseband bandwidth $(B_{RF} + B_j)/2$, with a spectral level $2P_J/B_{RF}$ W/Hz. Hence, Eqs. (7.5.3) and (7.5.4) have the spectral interpretation shown in Figure 7.7b.

The receiver correlation operation using the coherent addressed carrier has reduced (despread) the bandwidth of the modulated carrier, but has spread out the bandwidth of the jammer. The bit correlation now filters to the bit rate bandwidth $1/T_b$, producing a decoding SNR of

$$\begin{aligned} \frac{E_b}{N_0} &= \frac{P_c}{(2P_J/B_{RF})(1/T_b)} \\ &= \left(\frac{P_c}{P_J}\right)\left(\frac{B_{RF}T_b}{2}\right) \\ &= (\text{SJR})_{\text{in}}\left(\frac{B_{RF}}{2R_b}\right) \end{aligned} \tag{7.5.5}$$

Thus, the spreading and despreading with the addressed carriers has produced an effective gain in SNR by the factor $B_{RF}/2R_b$. This gain is often referred to as the *processing gain* of the spread-spectrum system,

$$\mathrm{PG} \triangleq \frac{B_{\mathrm{RF}}/2}{R_b} \tag{7.5.6}$$

The processing gain is therefore identical to the spreading ratio defined previously. Hence, the more the carrier is spectrally spread, the larger is the effective processing gain, and the lower the input SJR can be in achieving a desired E_b/N_0. Figure 7.8a shows the required value of P_J/P_c needed to achieve the stated bit-error probabilities for different processing gains. Figure 7.8b shows the required BPSK bandwidth B_{RF} needed to achieve a specified PG at different bit rates. Processing gain, or spreading

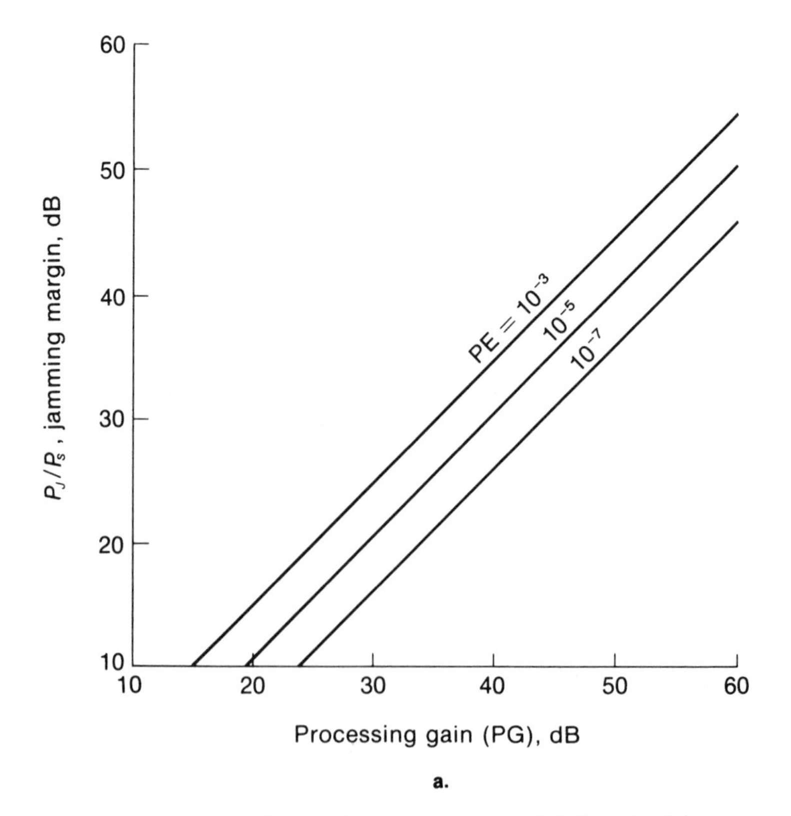

FIGURE 7.8 Jamming performance curves. (a) Required jammer power needed to overcome. (b) Required satellite bandwidth to achieve given processing gain and bit rate.

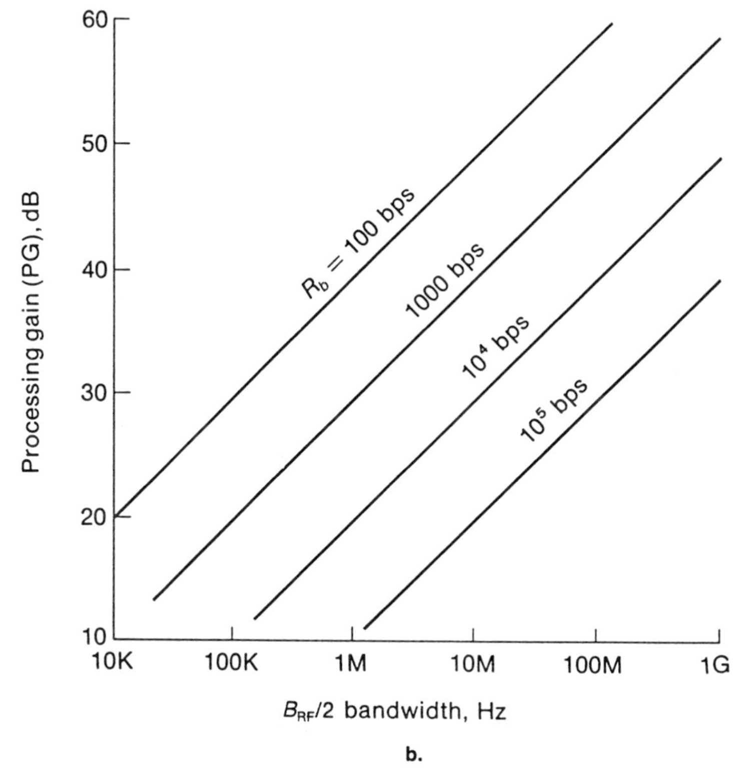

FIGURE 7.8 (continued) Jamming performance curves. (a) Required jammer power needed to overcome. (b) Required satellite bandwidth to achieve given processing gain and bit rate.

ratio, therefore acts to overcome a strong power advantage that a jammer may have over a transmitter. In a satellite link, the bandwidth is of course restricted to the available satellite RF bandwidth. These facts tend continually to influence military spread-spectrum systems toward the higher satellite frequency bands.

We emphasize that the concept of processing gain simply follows from the fact that a synchronized receiver, having an addressed replica of the carrier, can correlate the data carrier while spreading out the jammer. The result is completely independent of the jammer bandwidth—that is, of whether the jammer is broadband or partial band, as long as the jammer has finite power.

It is common to define P_J/P_c in Figure 7.8a as the *jamming margin*, which in decibels is

$$(\mathrm{JM})_{\mathrm{dB}} = (\mathrm{PG})_{\mathrm{dB}} - \left(\frac{E_b}{N_0}\right)_{\mathrm{dB}} \tag{7.5.7}$$

This margin indicates how much stronger in power the jammer can be relative to the transmitter, referred to the receiver input while achieving a desired (E_b/N_0). The latter is dependent on the desired bit-error probability and the modulation format. Jamming margin can be increased by either increasing processing gain (spreading the carrier bandwidth or reducing bit rate) or by reducing the required E_b/N_0 via coding. However, E_b/N_0 reduction through coding can only produce several decibels of margin improvement (recall Figure 2.27), whereas processing gain increases continually with available RF bandwidth.

7.6 SATELLITE JAMMING WITH DS-CDMA

In a satellite system, the jammer may have the opportunity to jam the satellite instead of the receiving earth station (Figure 7.9). By transmitting strong, broadband noise up to the satellite, the jammer can control the limiting power suppression that occurs in the satellite processing. In effect, the jammer power P_J replaces the uplink noise power in our earlier

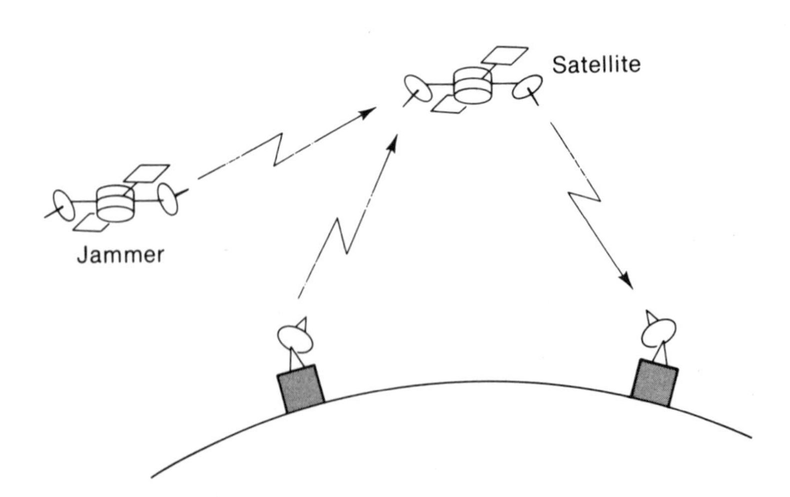

FIGURE 7.9 Jamming of a satellite.

analysis. If we assume limiter suppression models as in Eq. (4.6.15), the downlink SNR in a bandwidth B_{RF} is, from Eq. (3.6.6),

$$\mathrm{SNR}_d = \frac{[P_c/(P_c + P_J)]P_T L}{P_T L[P_J/(P_c + P_J)] + N_{0d}B_{\mathrm{RF}}} \tag{7.6.1}$$

If $P_J/P_c \gg 1$, this is

$$\mathrm{SNR}_d \simeq \left(\frac{P_c}{P_J}\right)\left(\frac{\mathrm{CNR}_d}{\mathrm{CNR}_d + 1}\right) \tag{7.6.2}$$

where $\mathrm{CNR}_d = P_T L/N_{0d}B_{\mathrm{RF}}$. If we solve for the required jamming margin to achieve a specified despread SNR of Υ, we have

$$\frac{P_J}{P_c} = \left(\frac{\mathrm{PG}}{\Upsilon}\right)\left(\frac{\mathrm{CNR}_d}{\mathrm{CNR}_d + 1}\right) \tag{7.6.3}$$

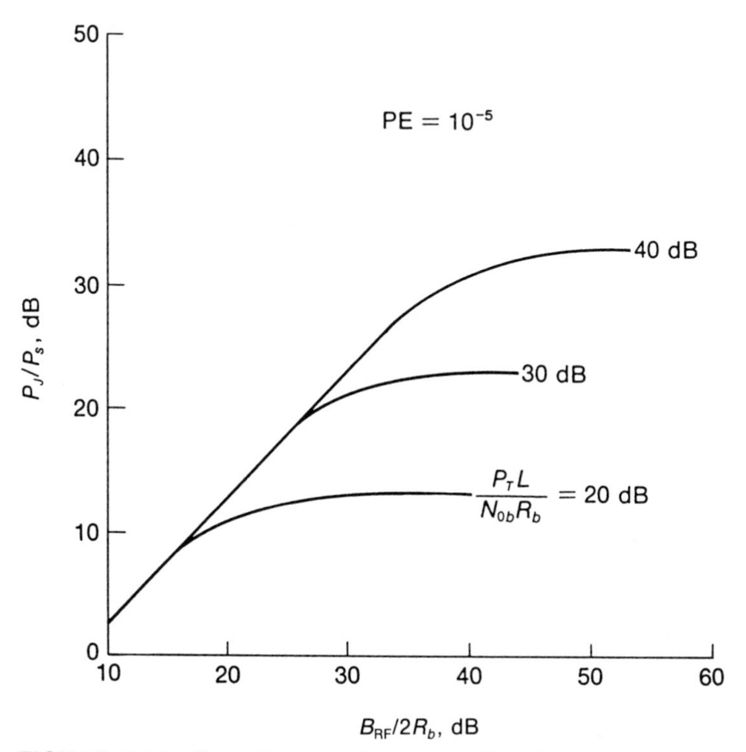

FIGURE 7.10 Jamming margin vs. satellite bandwidth and downlink CNR.

That is the maximum uplink P_J/P_c ratio that can be tolerated in achieving the Υ performance. If the downlink is strong ($\text{CNR}_d \gg 1$), Eq. (7.6.3) simply states that the uplink determines downlink performance, and the entire link has the full advantage of the receiver processing gain. However, as the bandwidth is increased, the downlink is eventually weakened, and the bracketed term begins to reduce the effective receiver PG. In fact, when $\text{CNR}_d \ll 1$,

$$\frac{P_J}{P_c} = \frac{B_{\text{RF}}/2R_b}{\Upsilon}\left(\frac{P_T L}{N_{0d} B_{\text{RF}}}\right)$$
$$= \frac{1}{\Upsilon}\left(\frac{P_T L}{N_{0d} 2R_b}\right) \tag{7.6.4}$$

and the maximum acceptable P_J/P_c no longer depends on B_{RF}. Continually increasing satellite bandwidth beyond this point provides no processing advantages. Figure 7.10 shows this behavior, plotting jamming margin versus available satellite bandwidth B_{RF} for different downlink CNR values. In effect the limit of antijam performance is determined by the downlink parameters, and the advantage of having a strong downlink (high satellite power and large receiver g/T° values) is apparent.

7.7 DS-CDMA CODE ACQUISITION AND TRACKING

We have shown that multiple accessing and spectral spreading can be achieved by using digital codes to either encode directly or generate frequency-hopping patterns. To decode in either format, however, it is necessary to have at the receiver a time-coherent replica of the same code sequence in order to despread the address or dehop the frequency pattern. This code synchronization is obtained by a code-locking subsystem operating in parallel with the data decoding channel, as we showed in Figure 7.2. A local version of the code is generated at the receiver, using the identical code generator. Initially, however, the local code is out of alignment with the received code from the transmitter, and it is necessary to bring the two codes into synchronization before data can be decoded. This synchronization is accomplished by an acquisition operation that basically applies a search-and-test aligning procedure. When an indication appears that the codes are nearly aligned, a tracking subsystem is activated to maintain the code synchronization throughout the subsequent data transmissions. The specific implementation of the acquisition and tracking subsystems depends on the addressing format used.

The code acquisition subsystem appears as in Figure 7.11. The output code sequence from the receiver code generator is multiplied with the received modulated carrier. The mixer output is then bandpass filtered, enveloped detected, and integrated for a fixed interval T_{in}. The voltage value of the integrator output serves as an indication of whether the two codes are in alignment. If it is concluded that the codes are not aligned, the local code sequence can be delayed or advanced, and another correlation measurement is performed. This testing can be repeated until code alignment is indicated.

The fact that the integrated envelope voltages serve as an indicator of code alignment can be shown as follows. Assume the received addressed carrier $c(t)$ arrives with a δ s code offset relative to the local code $q(t)$. Thus, neglecting interfering carriers,

$$c(t) = Ad(t + \delta)q(t + \delta)\cos(\omega_c t + \psi) \tag{7.7.1}$$

where $d(t)$ is the data waveform. The multiplier in Figure 7.11 multiplies $c(t)$ by the local code. The bandpass filter is centered at ω_c and is wide enough to pass the data $d(t)$, but not the code $q(t)$. Hence, the filter output is

$$f(t) = \overline{Aq(t)q(t + \delta)}d(t + \delta)\cos(\omega_c t + \psi) \tag{7.7.2}$$

where the overbar denotes the averaging effect of the BPF on the code.

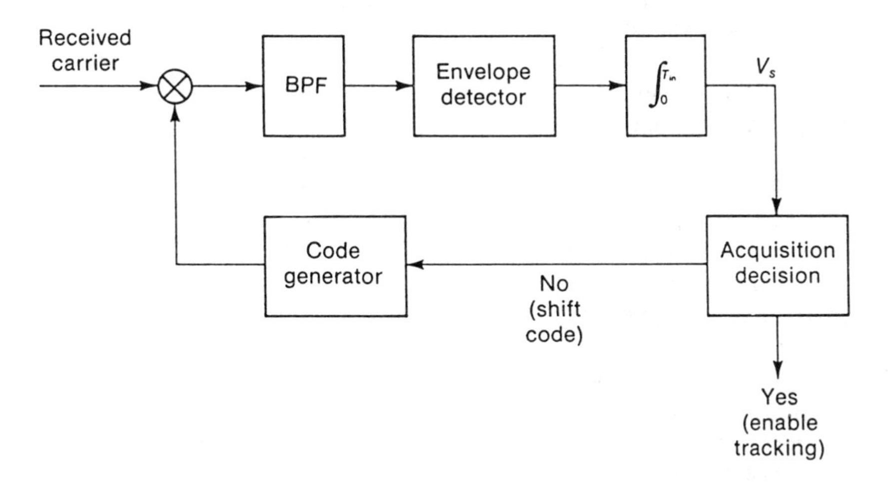

FIGURE 7.11 CDMA noncoherent code acquisition subsystem.

The output of the envelope detector is then

$$\begin{aligned}\text{Envelope detector output} &= |f(t)| \\ &= |\overline{Aq(t)q(t+\delta)}|\end{aligned} \tag{7.7.3}$$

where we have used the fact that the envelope of the PSK carrier is unity. The envelope detector output is then integrated to produce the voltage v_s. If we take the integration time to be an integer multiple of a code period T_c, we have $T_{in} = \eta T_c$, and

$$v_s = \left(\frac{AT_c}{2}\right)\eta R_q(\delta) \tag{7.7.4}$$

where we have written

$$\begin{aligned}R_q(\delta) &= \overline{q(t)q(t+\delta)} \\ &= \frac{1}{T_c}\int_0^{T_c} q(t)q(t+\delta)\,dt\end{aligned} \tag{7.7.5}$$

The function $R_q(\delta)$ is the periodic autocorrelation function of the code. Thus, the integrator voltage reading is directly related to the code correlation function evaluated at the offset δ. It is desirable to have this voltage v_s low when the codes are out of alignment, and high when the codes are perfectly aligned. This requires that the code autocorrelation function be (ideally) zero for $\delta \neq 0$ and maximum when $\delta = 0$. Such a function would allow observations of the voltages v_s to indicate directly when the codes are aligned. Note that this autocorrelation condition places an additional requirement on the codes selected for addressing in the CDMA format. We have previously found that address codes should be long (many chips per data bit) and that sets of codes should exhibit low pairwise cross-correlation, or partial cross-correlation, with other address codes. We now see that the acquisition operation further constrains the autocorrelation properties of each address code as well. Note that the correlation values v_s are generated without knowledge of the carrier phase ψ and independent of the specific data $d(t)$ being sent. This is referred to as *noncoherent* code acquisition.

The desired code autocorrelation for $R_q(\delta)$ matches approximately that produced by maximal-length shift registers sequences, as shown in Figure 7.3b. The fact that these shift register codes have these desirable autocorrelation functions is a fundamental reason why they are advantageous in code address systems. Gold codes, formed from combinations of maximal-length register codes, are themselves not maximal length, and often have autocorrelation sidelobes, as was shown in Figure 7.4b. When the sidelobes are relatively low, relative to the maximum peak, the overall acquisition operation can generally be analyzed as an ideal maximal-length sequence.

Code Acquisition

The code correlation voltages in Eq. (7.7.2) are observed in the presence of additive noise arriving with the addressed carrier. The noise is due to the satellite downlink, and therefore will have the spectral level given in Eq. (7.3.2), neglecting station interference. The correlation integral will actually produce an output dependent on the envelope of a combined carrier plus noise waveform, instead of simply the signal term in Eq. (7.7.2) while generating these noisy envelope values. The code acquisition system must decide when true acquisition has been achieved. There are two basic procedures for deciding the correct code position from these noisy voltage observations. One method is called *maximum-likelihood* (ML) acquisition

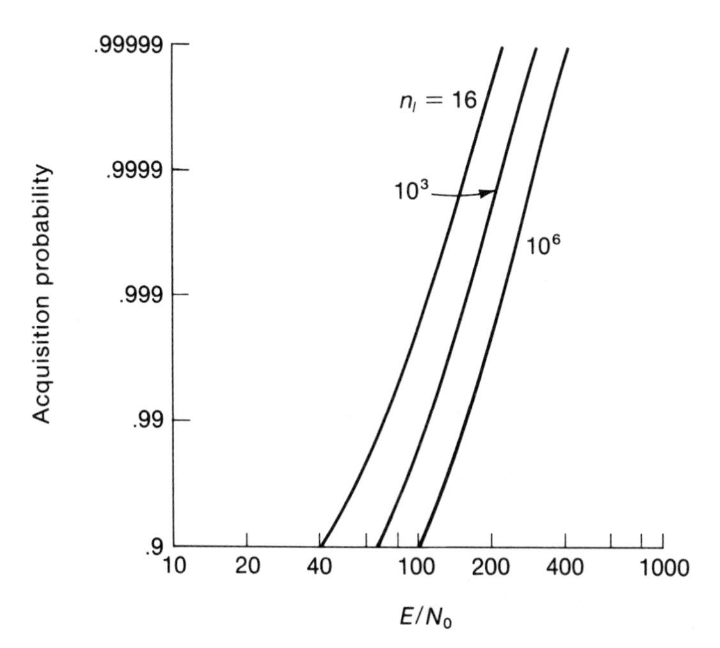

FIGURE 7.12 Maximum likelihood code acquisition probability vs. integrated code energy (n_l = code length, $E = A^2T_{in}/2$).

and involves an observation of the correlation of all code positions in a period, selecting the position with the maximum voltage for acquisition. The second method, called *threshold acquisition*, selects the first position whose voltage exceeds a fixed threshold value as the acquisition position. Note that in maximum-likelihood acquisition every code position is examined, and a decision is always made after completing the entire examination. In threshold acquisition, code positions are continually examined until a threshold crossing occurs.

In ML acquisition, incorrect acquisition occurs if an incorrect position voltage exceeds the correct one. As such, the effect is similar to that occurring in an M-ary noncoherent orthogonal decoder. In this case, word decisions are also based on maximal envelope samples, where all incorrect words have a zero signal voltage. (For shift register and Gold codes, it is assumed that the codes are long enough that the offset code correlation values in Figure 7.3b can be considered zero.) The probability of incorrect ML acquisition is therefore given by the word error probability of a corresponding noncoherent orthogonal M-ary system, where M is equal to the number of possible code positions; that is, the code length n_l. Such curves were shown in Figure 2.33 as a function of the bit energy and noise spectral level. For the acquisition case, the energy corresponds to the integrated carrier envelope energy collected in the T_{in} s integrator,

$$\begin{aligned} \frac{E}{N_0} &= \frac{(A^2/2)T_{\text{in}}}{N_{0T}} \\ &= \eta\left(\frac{E_c}{N_{0T}}\right) \end{aligned} \tag{7.7.6}$$

where η is the number of code periods in T_{in} and E_c is the carrier energy in a code period. Making the change of variable allows us to replot noncoherent word-error curves as equivalent acquisition probability curves, as shown in Figure 7.12. The curves relate the ML acquisition probability, the integrated acquisition energy, and the code length. With fixed acquisition carrier power, Figure 7.12 serves as a basis for determining the required envelope integration time.

If E_c is not satisfactory for the desired acquisition probability, we must integrate over more periods in the correlation detector in order to integrate up the acquisition power to the desired energy value. This means we spend more time examining a code shift position. Since each position must be examined, the total acquisition time to perform the ML test is then

$$T_{\text{acq}} = n_l T_{\text{in}} = n_l(\eta T_c) \tag{7.7.7}$$

Hence, acquisition time depends directly on the code length. We see the obvious trade-off of long-length address codes (many chips per bit) versus the disadvantage of increased acquisition time.

For threshold acquisition with a threshold value of Υ V, a false threshold crossing will occur with envelope samples with probability

$$\mathrm{PFC} = Q(0, \Upsilon) \tag{7.7.8}$$

where $Q(a, b)$ is defined in Eq. (A.8.5). A correct threshold crossing occurs with probability

$$\mathrm{PCC} = Q\left(\frac{E}{N_0}, \Upsilon\right) \tag{7.7.9}$$

with E/N_0 given in Eq. (7.7.6). If the jth code position is correct, and the acquisition code search begins with the first, correct acquisition will occur with probability

$$\mathrm{PAC}_j = \mathrm{PCC}(1 - \mathrm{PFC})^{j-1} \tag{7.7.10}$$

The average acquisition probability is then

$$\begin{aligned} \mathrm{PAC} &= \frac{1}{n_l} \sum_{j=1}^{n_l} \mathrm{PAC}_j \\ &= \frac{\mathrm{PCC}}{n_l} \left(\frac{1 - (1 - \mathrm{PFC})^{n_l}}{\mathrm{PFC}} \right) \end{aligned} \tag{7.7.11}$$

If $\mathrm{PFC} \ll 1/n_l$, then $\mathrm{PAC} \approx \mathrm{PCC}$, and Eq. (7.7.9) can be used to adjust threshold and integration time T_{in} to the desired acquisition probability.

The length of code acquisition time for the threshold test depends on the number of code periods that are searched before acquiring. The probability that the test will go through a complete period without reporting an acquisition is

$$\mathrm{PNC} = (1 - \mathrm{PCC})(1 - \mathrm{PFC})^{n_l - 1} \tag{7.7.12}$$

The probability that it will take exactly i periods to successfully acquire the jth position is then $\mathrm{PAC}_j(\mathrm{PNC})^{i-1}$. The average acquisition time is then

$$T_{\text{acq}} = \frac{1}{n_l} \sum_{j=1}^{n_l} \sum_{i=1}^{\infty} T_{\text{in}}[(i-1)n_l + j]\mathrm{PAC}_j(\mathrm{PNC})^{i-1} \tag{7.7.13}$$

When $\text{PFC} \ll 1/n_l$, $\text{PAC} \approx \text{PCC} \approx 1 - \text{PNC}$, and

$$T_{\text{acq}} \approx T_{\text{in}}\left(\frac{n_l(1 - \text{PAC})}{\text{PAC}} + \frac{n_l}{2}\right)$$

$$\approx T_{\text{in}} \frac{n_l}{2} \tag{7.7.14}$$

Thus, on the average, approximately one-half the code chips will be searched before acquisition. We see, therefore, that threshold testing also has an acquisition time dependent on the code length. However, it must be remembered that T_{acq} is an average time, and individual acquisition operations may run considerably longer.

Since coded acquisition with shift register codes may have an abnormally long acquisition time, there has been interest in finding modifications that can reduce that time. One method is to use *preacquisition detection* of the code chips. Recall that shift register codes are generated at the register output by a binary sequence, passing through the register. If the register had l stages, then each l chip subsequence of the address code must have been a binary sequence in the register at one time. Furthermore, if we observed any l consecutive chips of the code and loaded them in order into an identical register, the feedback chip generated at the next register shift time must be the $(1 + l)$st chip of the code. This immediately suggests an acquisition aid in which we first attempt to determine l consecutive chips of the incoming address code. If these l chips are determined correctly, and loaded into the receiver register, the sequence of chips generated from the feedback logic from then on will exactly match the incoming code. Hence, no position search is required and acquisition is immediately achieved.

The method, however, requires the correct detection of l consecutive input address chips. That is, we must treat the received addressed carrier as a modulated PSK carrier, and perform binary detection. Since this detection requires carrier phase referencing, we must first construct a squaring or *Costas loop* to remove the address modulation. The phase-referenced carrier can then be used for address code detection, and the decisions used to load the receiver shift register. When loaded, the register is then switched into the code acquisition loop of Figure 7.11. If the l address chips were all correctly decoded, the loop is immediately in time synchronism. If an error was made, the two codes are not aligned (which is so indicated by the correlation detection). The detection of l new chips must then be repeated. The task of correctly decoding l successive code chips, however, may not be trivial. This is because the energy used for the code chip decisioning is that of only a single chip in the entire code. If the code has length n_l, then only $1/n_l$ of the available code energy is used for this decoding. That is, if E_{cc} is the code chip energy, $E_{cc} = E_c/n_l$, which produces an (E_{cc}/N_{0T}) that is n_l times smaller than E_c/N_0. Unless the acquisition energy is large, this reduction with long codes will produce such a poor chip detection probability that the l chip detection may have to be repeated many times. In such cases the usefulness of the predetection operation is suspect, and must be carefully evaluated.

Code Tracking

Once code acquisition has been accomplished (local and received codes brought into alignment), the received code must be continually tracked to maintain the synchronization. A code-tracking loop for discrete sequences is shown in Figure 7.13. It contains two parallel branches of a multiplier, bandpass filter, and envelope detector. (One branch can be that used in the acquisition operation, so that initiation of tracking requires only the insertion of the second branch.) The two-branch system in Figure 7.13a is referred to as a *delay-locked loop* (Spilker, 1977). A delay-locked tracking loop uses the local code to generate error voltages $e(t)$ that are fed back to correct any timing offsets of the local code generator as they arise. Timing-error voltages are obtained by using the local code to generate two offset sequences advanced and delayed by one-half chip (see Figure 7.13b). With shift register codes, these are obtained from different taps within the same register. Thus, if $q(t)$ is the local code, the sequences $q(t + w/2)$ and $q(t - w/2)$ are simultaneously generated from the register. These offset codes are each separately multiplied with the input in the two channels of the delay-locked loop. The multiplied outputs are then bandpass filtered and envelope detected to produce output voltages that are subtracted to form the correction signal for code-timing adjustment. Again, each BPF passes the data bits but averages the code product, which is then envelope detected. Following the discussion in Eq. (7.7.3), the advanced code channel output is

$$V_a = A\left|R_q\left(\delta - \frac{w}{2}\right)\right| \tag{7.7.15}$$

where $R_q(\delta)$ is defined in Eq. (7.7.5). Similarly, the delayed code channel produces

$$V_d = A\left|R_q\left(\delta + \frac{w}{2}\right)\right| \tag{7.7.16}$$

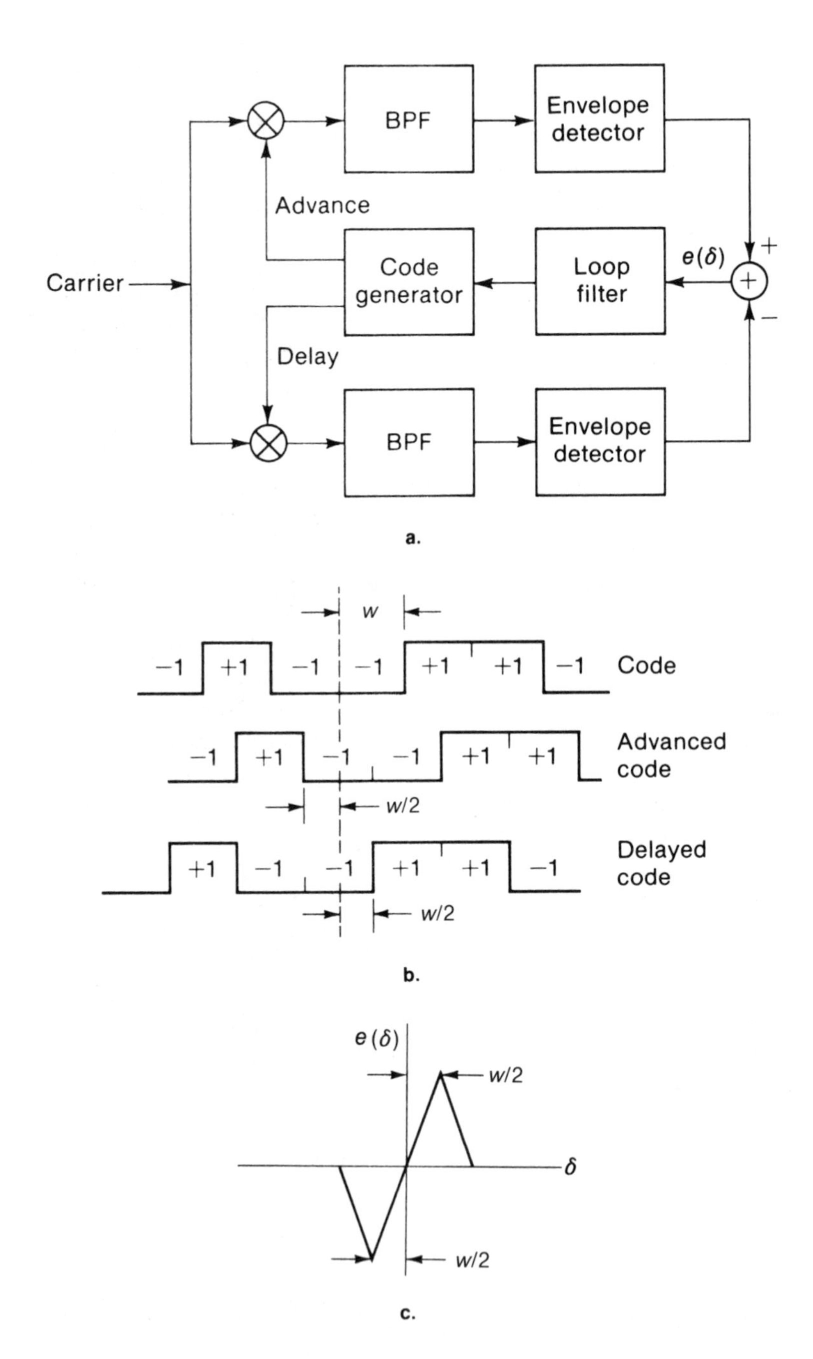

FIGURE 7.13 Delay-locked code-tracking subsystem. (a) Block diagram. (b) Advanced and delayed code alignment. (c) Code-error voltage vs. timing offset.

The output of the subtractor is therefore the correction voltage

$$\begin{aligned} e(\delta) &= V_d - V_a \\ &= A\left|R_q\left(\delta + \frac{w}{2}\right)\right| - A\left|R_q\left(\delta - \frac{w}{2}\right)\right| \end{aligned} \qquad (7.7.17)$$

for any offset δ.

This function is plotted in Figure 7.13c for the code correlation functions depicted in Figure 7.3b. We see that regardless of whether δ is positive or negative, a voltage proportional to δ is produced having the proper sign to adjust correctly the local code timing to reduce δ. That is accomplished by using $e(\delta)$ to adjust the code clock that drives the code register (slow it down or speed it up). Note that a proportional correction voltage is generated only if $\delta \leq w/2$, that is, only if we are within a half chip time of code lock. This is why an initial acquisition procedure is necessary to first align the codes to within this accuracy. The delay-locked loop then operates continually to correct for subsequent timing errors. Note that, as a result of the envelope detection, the delay-locked loop achieves code tracking noncoherently, that is, without knowledge of the carrier phase or frequency (as long as the multiplier outputs pass through the BPF).

The acquisition system in Figure 7.11 and the delay-locked loop in Figure 7.12a can be combined to form the total DS-CDMA acquisition and tracking subsystem shown in Figure 7.14. When the tracking error is zero, the local code generator (without the offset) is exactly in phase with the received code and can therefore be used to despread the received carrier. Thus, in the data channel in Figure 7.14, we use the fact that $q(t) \cdot q(t) = q^2(t) = 1$ to form the product

$$[d(t)q(t)\cos(\omega_c t + \psi)]q(t) = d(t)\cos(\omega_c t + \psi) \qquad (7.7.18)$$

The despread PSK carrier can now be processed in a standard PSK decoder for RF carrier phase referencing, bit timing, and data recovery.

An alternative to the two-channel delay-locked loop is to use a *tau-dither loop* (Hartmann, 1974). This loop uses only a single channel to perform serially the advanced and delayed multiplication, as shown in Figure 7.15. Although the same code offsets are generated, only one is used at a time with a single multiplier channel. This is accomplished by gating in each offset code over separate periodic intervals. In effect we generate the gated local code

$$q'(t) = g(t)q\left(t - \frac{w}{2}\right) + [1 - g(t)]q\left(t + \frac{w}{2}\right) \qquad (7.7.19)$$

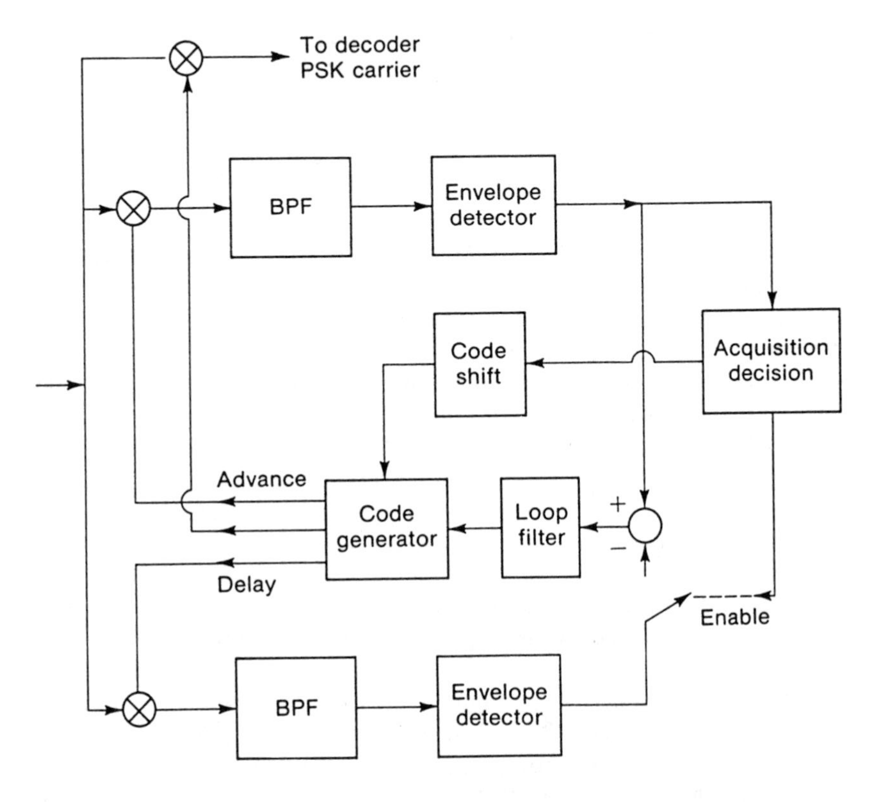

FIGURE 7.14 Combined code acquisition and tracking subsystem.

where $g(t)$ is the gate function (see Figure 7.15b). Multiplying this by the input carrier, the envelope detector output is

$$\begin{pmatrix}\text{Envelope detector}\\ \text{output}\end{pmatrix} = A|\overline{q'(t)q(t+\delta)}|$$

$$= A\left[g(t)\left|R_q\left(\delta + \frac{w}{2}\right)\right| + [1 - g(t)]\left|R_q\left(\delta - \frac{w}{2}\right)\right|\right] \tag{7.7.20}$$

By multiplying in the gating waveform $[2g(t) - 1]$ and averaging (in the loop filter) we generate the control signal

$$e(\delta) = \frac{A}{2}\left[\left|R_q\left(\delta + \frac{w}{2}\right)\right| - \left|R_q\left(\delta - \frac{w}{2}\right)\right|\right] \tag{7.7.21}$$

This is one-half of the delay-lock loop control signal in Eq. (7.7.17). Hence, the tau-dither loop, using only one code multiplier, generates the same

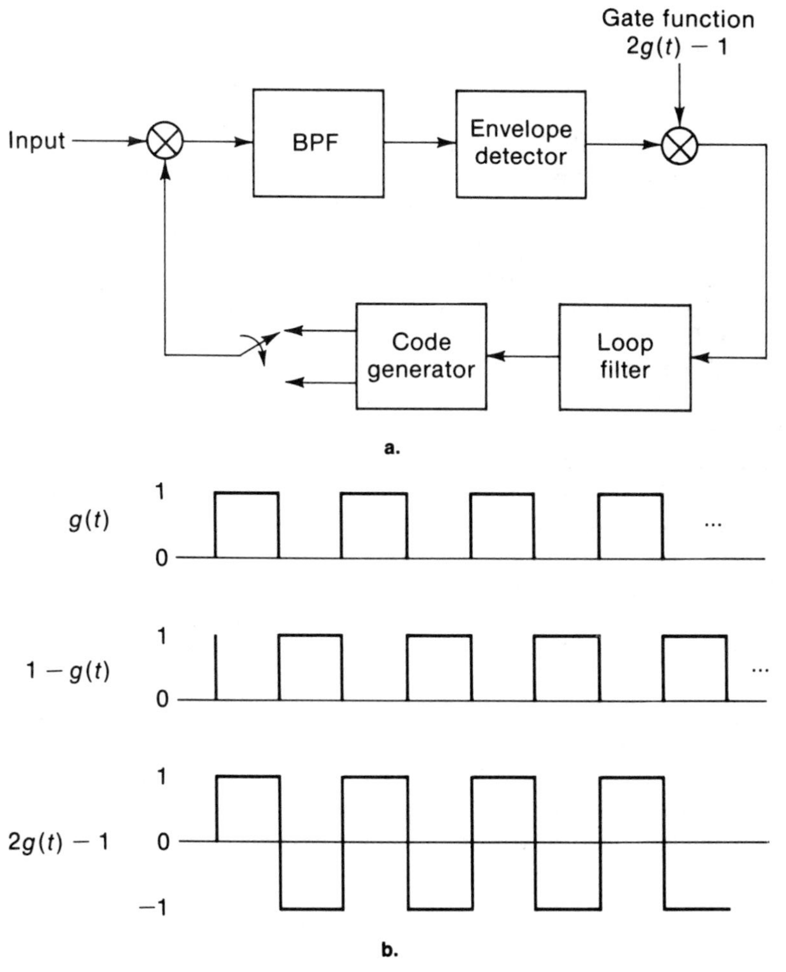

FIGURE 7.15 Tau-dither code-tracking loop. (a) Block diagram. (b) Gating functions.

error voltage as the two channel, delay-locked loop. If the one-half amplitude reduction is tolerable, this hardware simplification may be advantageous.

While the tracking subsystems for either discrete-sequence or frequency-hopped CDMA operate to keep any offset δ small, it should be pointed out that even a relatively small offset can seriously degrade data decoding. For example, consider again Eq. (7.7.18) when the codes used for despreading or dehopping are slightly offset. The signal portion of the data decoder waveform is now given by

$$A[\overline{q(t)q(t+\delta)}]d(t)\cos(\omega_c t+\psi)=AR_q(\delta)d(t)\cos(\omega_c t+\psi) \quad (7.7.22)$$

Hence, the effective PSK amplitude is reduced by the value $R_q(\delta)$ of the code correlation function. If the latter function falls off rapidly, a relatively small value of δ can produce a significant amplitude degradation. For example, with the correlation function of Figure 7.13b, an offset of $\delta = 0.2w$ produces $R_q(\delta) = 0.8$, and the decoding power is reduced by 0.64. Hence, only a 20% of a chip offset in timing can cause a 1.93 dB loss in decoding E_b/N_0.

This also shows the potential difficulties that can occur with Gold codes if the tracking system false locks to a sidelobe peak in Figure 7.4b. This will again translate to a reduced value of $R_q(\delta)$ in Eq. (7.7.22) and a corresponding degradation in decoding E_b/N_0. In fact, the E_b/N_0 will be reduced by the ratio of the square of the false lock peak to the maximum in-phase peak. This is why preventive measures are often inserted to prevent sidelobe false locking.

Tracking accuracy requires a careful analysis of the noise effect on the tracking operation. This is obtained by a timing-system model, similar to the phase model for phase lock loops. Figure 7.16a shows such a model. The input to the system is the time location of the input code, and the feedback variable is the corresponding location of the receiver code. The difference generates the timing error δ in sequence over the code period. This timing error is converted to voltage via the gain function in Figure 7.13c. Envelope noise is added to the error voltage, which is then filtered to drive the local generator. Envelope noise has the effective spectral level [see Eq. (B.2.4)]

$$N_0 = 2(N_{0T})^2 B_{RF} + 2N_{0T}A^2/2 \quad (7.7.23)$$

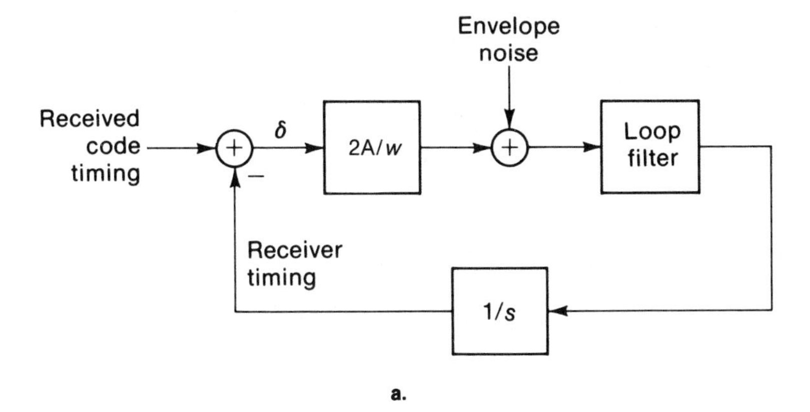

a.

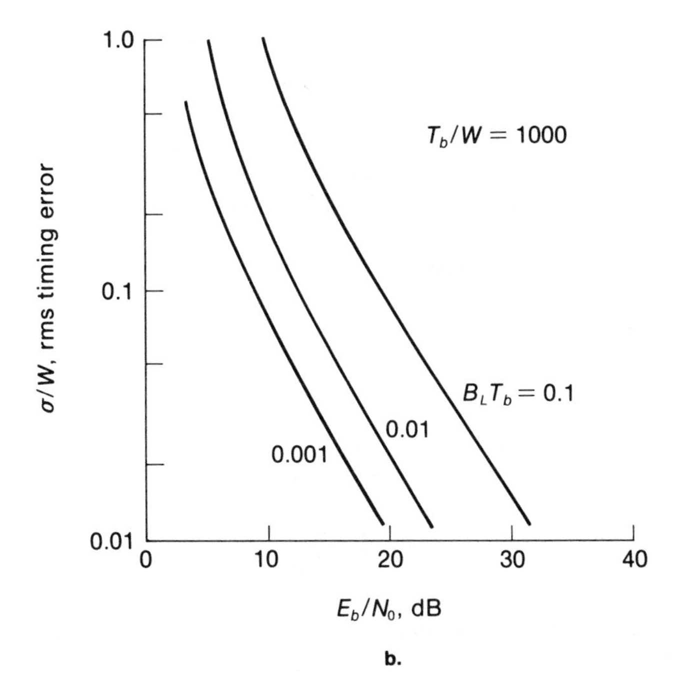

b.

FIGURE 7.16 Code-tracking loop analysis. (a) Timing block diagram. (b) rms loop-timing error due to noise.

When this noise is inserted into the delay-lock loop model in Figure 7.16, the variance of the timing error δ due to the noise is

$$\sigma_t^2 = \frac{w^2 N_0 2B_L}{4(A_2/2)^2} \tag{7.7.24}$$

where B_L is the tracking loop noise bandwidth. Substituting from Eq. (7.7.23) we can rewrite this variance as

$$\begin{aligned}\sigma_t^2 &= \frac{4[N_{0T}A^2/2 + N_{0T}^2 B_{\mathrm{RF}}]w^2 B_L}{4(A^2/2)^2} \\ &= \frac{w^2 N_{0T} B_L}{A^2/2}\left[1 + \frac{N_{0T} B_{\mathrm{RF}}}{A^2/2}\right]\end{aligned} \tag{7.7.25}$$

This variance can be normalized to the chip time and written as

$$\left(\frac{\sigma_t}{w}\right)^2 = \frac{B_L/R_b}{E_b/N_{0T}}\left[1 + \frac{N_{0T}}{2E_{\mathrm{cc}}}\right] \tag{7.7.26}$$

where E_b is the bit energy and E_{cc} is the code chip energy. Figure 7.16b plots the fractional rms timing error in Eq. (7.7.26) as a function of the operating (E_b/N_0) (which determines the data-error probability), for several values of B_L/R_b. For typical design conditions ($E_b/N_{0T} \approx 10$ dB), the tracking error is negligible as long as the loop bandwidth is a small fraction of the bit rate.

REFERENCES

Anderson, D. and Wintz, P., (1969), "Analysis of Spread Spectrum System with a Hard Limiter," *IEEE Trans. Comm. Tech.*, 285–290.

Baer, H. (1982), "Interference Effects of Hard Limiters in PN Spread Spectrum Systems," *IEEE Trans. Comm.*, **COM-30.**, 1010–1018.

Boudreau, G., Falconer, D., and Mahmoud, S. (1990), "A Comparison of Trellis Coded Versus Convolutionally Coded Spread Spectrum Systems." *IEEE Trans. on Sel. Areas Comm.*, **8**, No. 4 (May), 628–641.

Boztas, S. and Kumar, P. (1990), "Near-optimal Four Phase Sequences For CDMA," *CSI Report* 90-03-01, Dept. of Electrical Engineering, Univ. of Southern Calif., Los Angeles CA, 90089-0272 (March).

Gold, R. (1968), "Maximal Recursive Sequence with Multi-valued Cross Correlation," *IEEE Trans. Inf. Theory*, **IT-14**, 154–156.

Golomb, S. (1967), *Shift Register Sequences* Holden-Day, San Francisco.
Golomb, S., et al. (1964), *Digital Communications with Space Applications* Prentice-Hall, Englewood Cliffs, NJ.
Hartmann, H. (1974), "Analysis of a Dithering Loop for Code Tracking," *IEEE Trans. Aerospace Electronic Systems*, **AES-10**, 24–33.
IEEE (1977/1982) "Special Issue on Spread Spectrum Communications," *IEEE Trans. Comm.*, **COM-25** and **COM-30**.
IEEE (1982), see section on "Coded Division Multiple Access" in *IEEE Trans. Comm.*, **COM-30**.
Kasami, T., and Lin, L., (1976), "Coding for a Multiple Access Channel," *IEEE Trans. Inf. Theory*, **IT-22**, 123–132.
Kochevar, H., (1977), "Spread Spectrum Multiple Access Experiment Through a Satellite," *IEEE Trans. Comm.*, 853–856.
Lee, J., and Smith, D., (1974), "Families of Shift Register Sequences with Impulse Correlation Registers," *IEEE Trans. Inf. Theory*, **IT-20**, 321–330.
Pursley, M. (1977), "Performance Evaluation for Phase-Coded Spread Spectrum Multiple-Access—Part I: System Analysis," *IEEE Trans. Comm.*, **COM-25**, 795–799.
Pursley, M. B., and Roefs, H. F. A. (1978), "Numerical Evaluation of Correlation Parameters for Optimal Phases of Binary Shift-Register Sequences," *IEEE Trans. Comm.*, **COM-27**, 1597–1604.
Sarwate, D. V., and Pursley, M. B. (1980), "Crosscorrelation Properties of Pseudorandom and Related Sequences," *Proc. IEEE*, **68**, 593–619.
Spilker, J., (1977), *Digital Communication by Satellite*, Prentice-Hall, Englewood Cliffs, NJ.
Viterbi, A. (1990), Very Low Rate Convolutional Codes For Maximum Performance in Spread Spectrum, *IEEE Trans. Sel. Areas Comm.* **8**, No. 4 (May), 641–650.
Weber, C., Huth, G., and Batson, B. (1981), "Performance Considerations of CDMA Systems," *IEEE Trans. Vehicle Tech.*, **VT-30**, 183–192.
Welch, L., (1974), "Lower Bounds on the Maximum-Cross Correlation of Signals," *IEEE Trans. Inf. Theory*, **IT-20**, 397–399.
Yao, K. (1977), "Error Probability of Asynchronous Spread Spectrum Multiple-Access Communication Systems," *IEEE Trans. Comm.*, **COM-25**, 803–809.

PROBLEMS

7.1. Given two periodic, binary-coded (± 1) NRZ waveforms $m_1(t)$ and $m_2(t)$, with chip period τ s and n chips per period. Show that the cross-correlation of these waveforms becomes

$$\gamma = \frac{1}{n\tau}\int_0^{n\tau} m_1(t)m_2(t)\,dt$$

$$= \frac{1}{n}\,[(\text{number of matching symbols}) - (\text{number of mismatched symbols})]$$

where the matching refers to corresponding symbols of the same chip period.

7.2. Given the two periodic code sequences:

Code 1: 1 −1 1 1 −1 −1 1 1 (one period)

Code 2: −1 1 1 −1 −1 −1 −1 1

(a) Compute the *periodic* cross-correlation of the sequences at each chip shift time. [*Hint*: Use the results of Problem 7.1.] (b) Without shifting, compute the *partial* cross-correlations of length three.

7.3. Define a binary, square 2 × 2 matrix H_1 as

$$H_1 = \begin{bmatrix} 1 & -1 \\ 1 & 1 \end{bmatrix}$$

Define its nth-order extension as

$$H_n = \begin{bmatrix} H_{n-1} & \hat{H}_{n-1} \\ H_{n-1} & H_{n-1} \end{bmatrix}$$

where $\hat{H}_{n-1}$ is obtained by changing the signs of H_{n-1}. Show that the rows of H_n, for any n, represent a set of n orthogonal binary sequences.

7.4. A DS-CDMA system is to operate with PE $= 10^{-5}$ and transmit a carrier data rate of 10^4 bps through a satellite with a 10-MHz bandwidth. A jammer can deliver a power of 10^{-6} W to the receiver. How much power must the carrier provide at the receiver to operate the system, assuming the full satellite bandwidth is used?

7.5. Consider the shift register in Figure P7.5. Start with all zeros in the register, and shift in one bit from the left. Compute the output sequence from the register as the bits are clocked through.

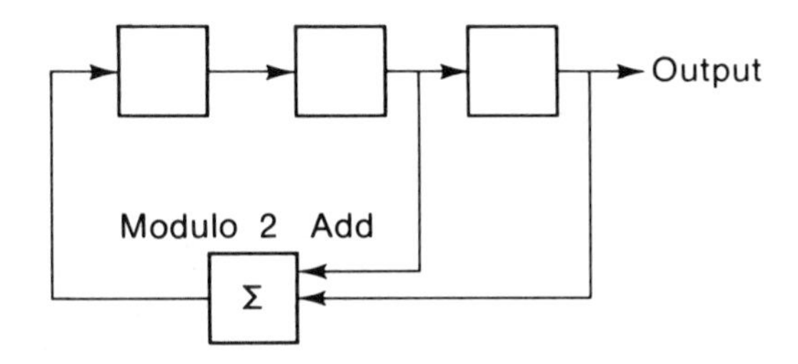

FIGURE P7.5

7.6. Show that the power spectrum of a PRN binary waveform of period n bits and pulse width w is given by

$$S(\omega) = \left(\frac{n+1}{n^2}\right)\left[\frac{\sin(\omega w/2)}{(\omega w/2)}\right]^2 \sum_{\substack{i=-\infty \\ i \neq 0}}^{\infty} \delta\left(\omega - \frac{2\pi i}{nw}\right) + \frac{2\pi}{n^2}\,\delta(\omega)$$

7.7. In *pulse-addressed multiple accessing* (PAMA), a station pair uses a τ-s burst of the PSK carrier modulation placed in a specific location in a T-s data frame. The burst locations are selected randomly by each station pair, and any station transmits when desired, without knowledge of other station pairs. (a) Show that the mean square cross-correlation for any station pair is $(\tau/3T)$. (b) For equal amplitude carriers, show that the maximum number of users is $K \leq 3T/\tau$.

7.8. In *slotted PAMA*, only specific τ-s slots can be used. Assume the T-s frame is divided into T/τ disjoints slots and each station pair selects its slot randomly. Show that the probability that r of k users will lie in the same slot is

$$P(r) = \binom{k}{r}\left(\frac{1}{\mu}\right)^r\left(\frac{\mu-1}{\mu}\right)^{k-r}$$

where $\mu = T/\tau$.

7.9. A channel coded CDMA system uses a rate 1/2, 7-stage convolutional encoder before addressing at each transmitter. (a) By what factor can the total number of users in the system be increased over the uncoded system? [*Hint*: see Section 2.6.] (b) Repeat if a rate 1/3 encoder is used. (c) What is the maximum factor achievable, assuming a rate $1/n$ encoder with $n \to \infty$?

CHAPTER 8
Frequency-Hopped Communications

Spread-spectrum communications may also be achieved by using frequency hopping of the RF carrier. Instead of direct-phase shift modulation of a digital sequence onto the RF carrier, as in DS-CDMA, the sequence is used to frequency hop the carrier. In this chapter we examine the basic elements of this type of communication link.

8.1 THE FREQUENCY-HOPPED SYSTEM

In frequency-hopped communications, the available satellite bandwidth is partitioned into frequency bands, and the transmission time is partitioned into time slots, as shown in Figure 8.1. A hopping pattern in this frequency–time matrix is defined as a sequence of specific frequency bands, one for each time slot, as shown. A transmitter assigned a particular hopping pattern jumps from one band to the next according to the pattern, readjusting its carrier frequency from one time slot to the next. Such a frequency-hopping transmitter is shown in Figure 8.2a. During each band transmission, the transmitter sends some form of modulated carrier that occupies only the designated band. Thus, a station with a hopping pattern appears to utilize the entire satellite bandwidth when observed over a long time although transmitting only within a specified band at any one time. In a military environment, it is often further required that the

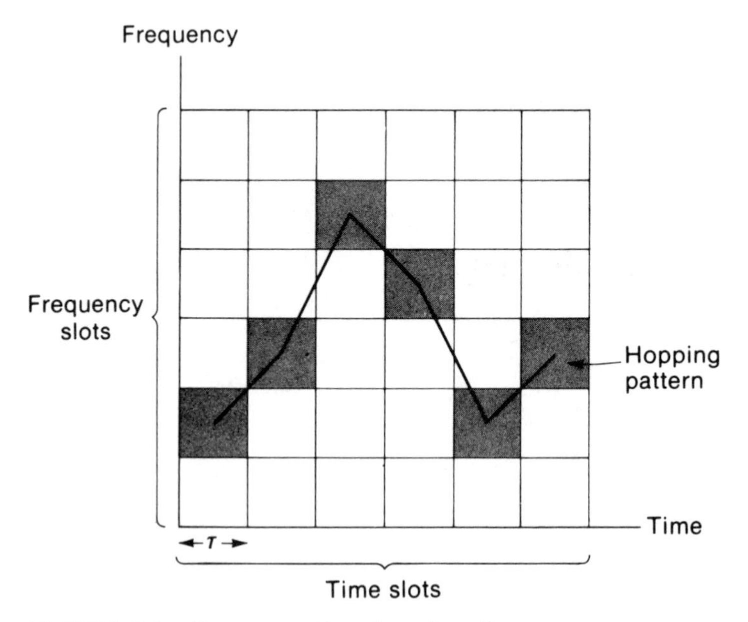

FIGURE 8.1 Frequency-time hopping diagram.

hopping patterns appear random, so that future values of the pattern cannot be predicted.

Hopping patterns can be obtained from periodic binary sequences similar to the address codes. If such a sequence is partitioned into blocks, each block can designate a particular frequency band in the frequency–time matrix (Figure 8.2b). Thus, a particular binary sequence specifies a specific hopping pattern, and as the code sequence periodically repeats, the hopping pattern will likewise repeat. If there are n frequency bands in the satellite bandwidth B_s, then a code sequence of length k chips generated at the rate R_c chips/s, will produce a hop every

$$\tau = \frac{\log_2 n}{R_c} \text{ s} \tag{8.1.1}$$

with a total of $k/\log_2 n$ hops per code period. A code generator of this sequence would then drive the frequency hopping (carrier frequency shifting) indicated in Figure 8.2a. Since the hopping rate R_H is $1/\tau$, Eq. (8.1.1) relates the chip rate to the hopping rate as

$$R_c = (\log_2 n) R_H \tag{8.1.2}$$

To increase the hopping rate with a given number of frequency bands, it is therefore necessary to generate the code at a faster rate. Hence, hopping rate is directly related to code rate.

Frequency-hopped systems commonly use some form of noncoherent modulation, since a coherent system would require phase referencing following each hop. The noncoherent modulation is generally frequency shift keying (FSK) or differential phase shift keying (DPSK). In M-ary FSK, one of M frequencies within each band is used, as shown in Figure 8.3a. In DPSK, a single-phase shifted carrier in the center of the band is used, as in Figure 8.3b. When transmitting in a given band, the transmitter sends the modulated carrier (either FSK or DPSK) for that band during a slot time τ. This corresponds to sending a data symbol as a carrier of fixed frequency for τ s, producing a transmitter carrier of bandwidth of

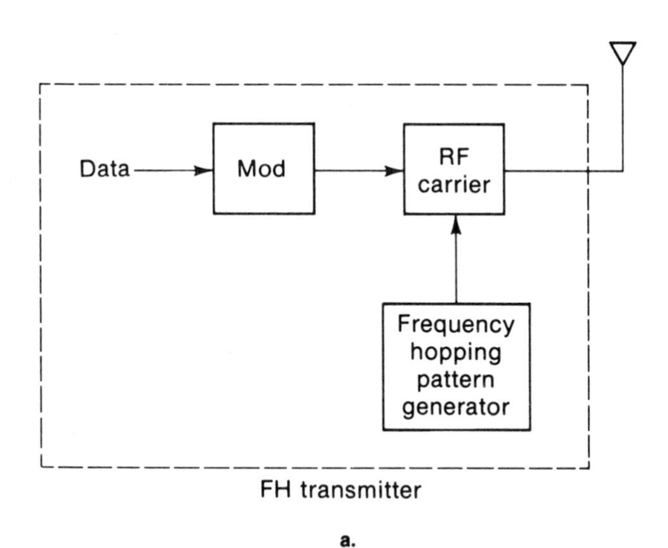

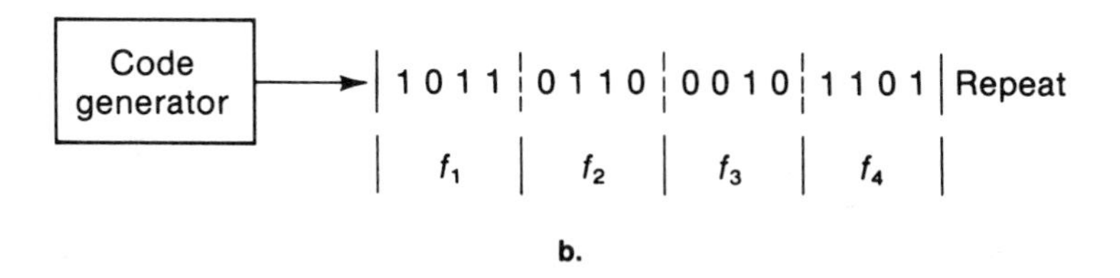

FIGURE 8.2 Frequency-hopping system. (a) Transmitter. (b) Code generator sequence ($\{f_i\}$ = frequencies).

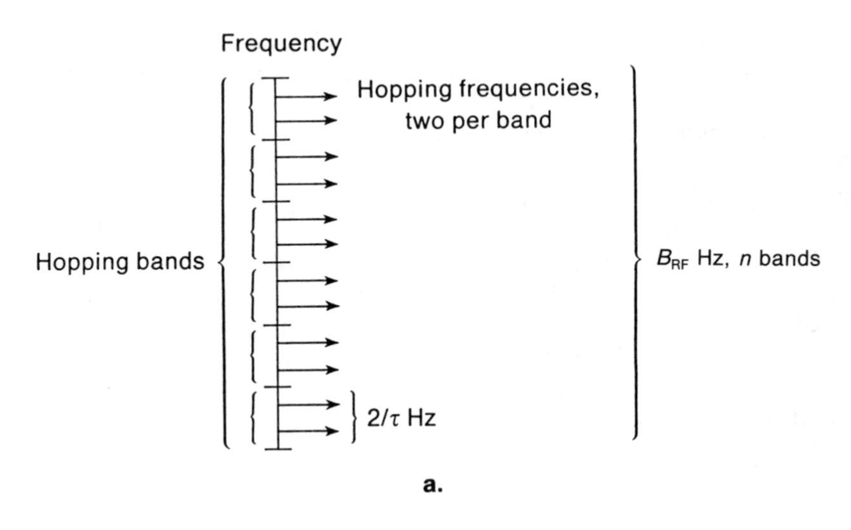

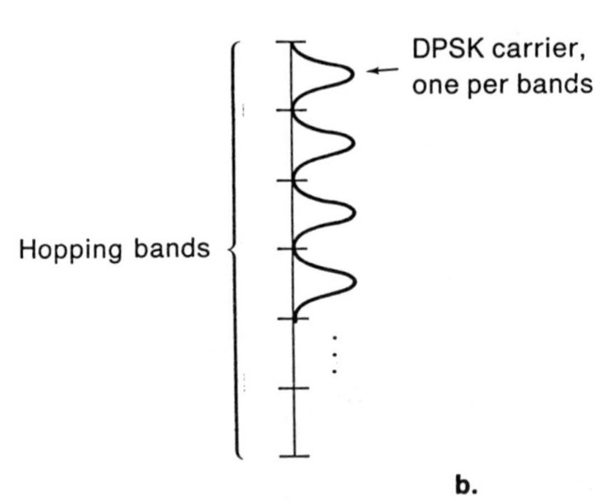

FIGURE 8.3 Frequency diagram for binary data with frequency hopping. (a) FSK encoding. (b) DPSK encoding.

approximately $2/\tau$ Hz around each frequency. Hence, the FSK or DPSK frequencies must be separated by $2/\tau$ Hz to guarantee sufficient separation during decoding. With n satellite hopping bands, the required RF bandwidth must be

$$B_{\mathrm{RF}} = nM(2/\tau) \tag{8.1.3}$$

where $M = 1$ for DPSK. If the hopping time τ corresponds to a single-symbol time, then the hopping system is producing one symbol per hop, or equivalently, operating at one hop per symbol. This is usually referred to as *slow* frequency hopping. If a symbol time is several hop times (i.e., a sequence of hops represents one data symbol), then the transmitter is

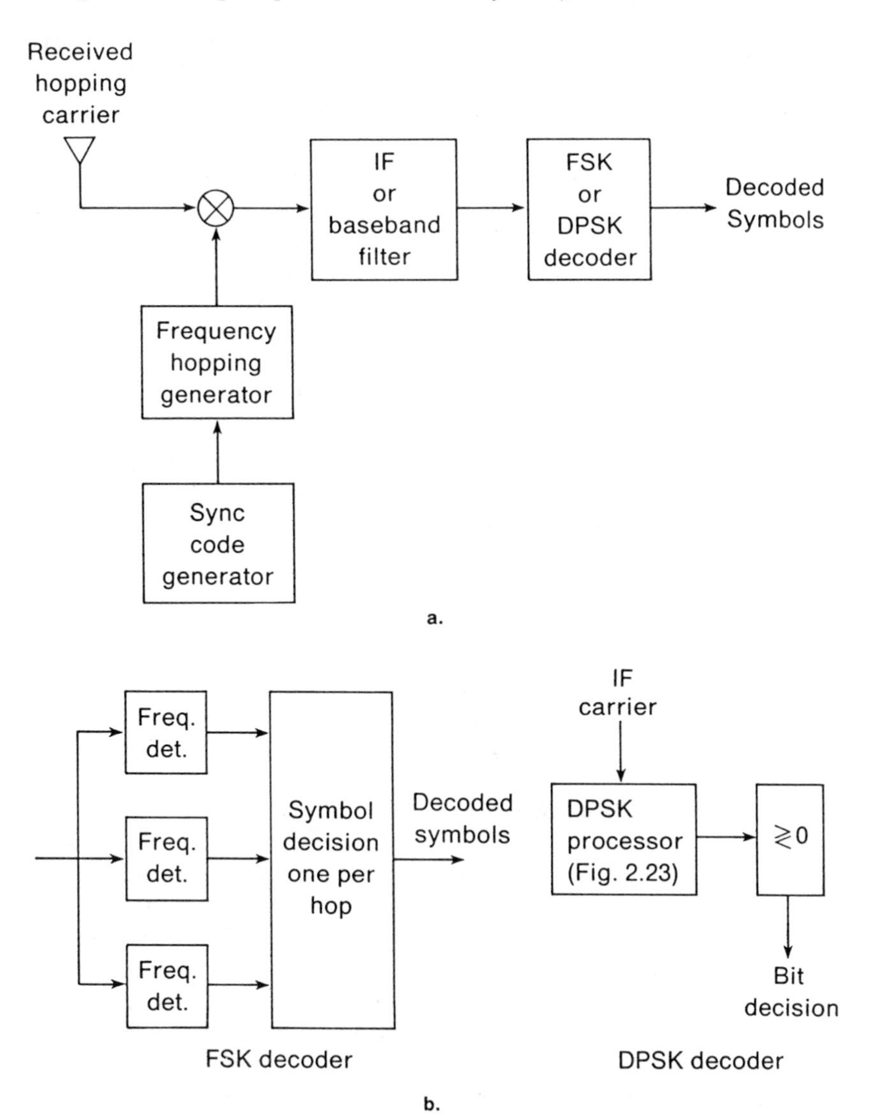

FIGURE 8.4 Receiver for frequency hopped carrier. (a) Block diagram. (b) Decoders.

hopping several times per symbol and is called *fast* frequency hopping. If there are L hops per symbol, the transmitted data rate is then

$$\begin{aligned} R_b &= [(\log_2 M)/L]R_H, \qquad && \text{MFSK} \\ &= R_H/L, && \text{DPSK} \end{aligned} \tag{8.1.4}$$

Hence, Eq. (8.1.4) relates the bit rate and hopping rate through the FSK modulation parameter M (the number of FSK frequencies per channel), whereas Eq. (8.1.3) relates bandwidth, hopping rate, and the number of hopping frequencies.

A receiver for a frequency-hopped FSK carrier is shown in Figure 8.4. In order to decode, the receiver must have a synchronized version of the hopping pattern. This is achieved by running an identical code sequence generator in time synchronism with the code sequence producing the hopping pattern of the received carrier. It hops its local carrier frequency in synchronism with the received carrier, so that only the modulation frequency (FSK shift in each band or DPSK IF carrier frequency) appears as a difference frequency. The modulation can then be decoded directly, as shown in Figure 8.4b. In FSK, a bank of filters tuned to each possible FSK frequency noncoherently decodes the symbol modulation in each time slot. In DPSK the IF carrier is decoded using the processor in Figure 2.23. The advantage of noncoherent FSK or DPSK is that only slot timing is needed for decoding, precluding the necessity of achieving phase coherency. Note that with synchronized patterns, each transmitted frequency band is individually mixed to baseband, and the same noncoherent decoder can be used successively during each slot. In effect, the synchronized local pattern removes the hopping from the received carrier, and we say the received carrier has been *dehopped*.

8.2 FREQUENCY-HOPPING SYNTHESIZERS

In frequency-hopped systems a key element is the frequency generators that produce the hopping and dehopping RF frequencies at both the transmitter and receiver in Figures 8.2 and 8.4. Since the frequencies are commanded to hop over the entire RF bandwidth, the frequency generators must have the capability of changing frequency both accurately and rapidly over an extremely wide frequency band. The accuracy is necessary because any unresolved frequency offset of the dehopped carrier will lead directly to decoding errors, since the data frequency will be misaligned with the decoding filters by the same offset. Frequency accuracy is directly related to the frequency resolution capability of the

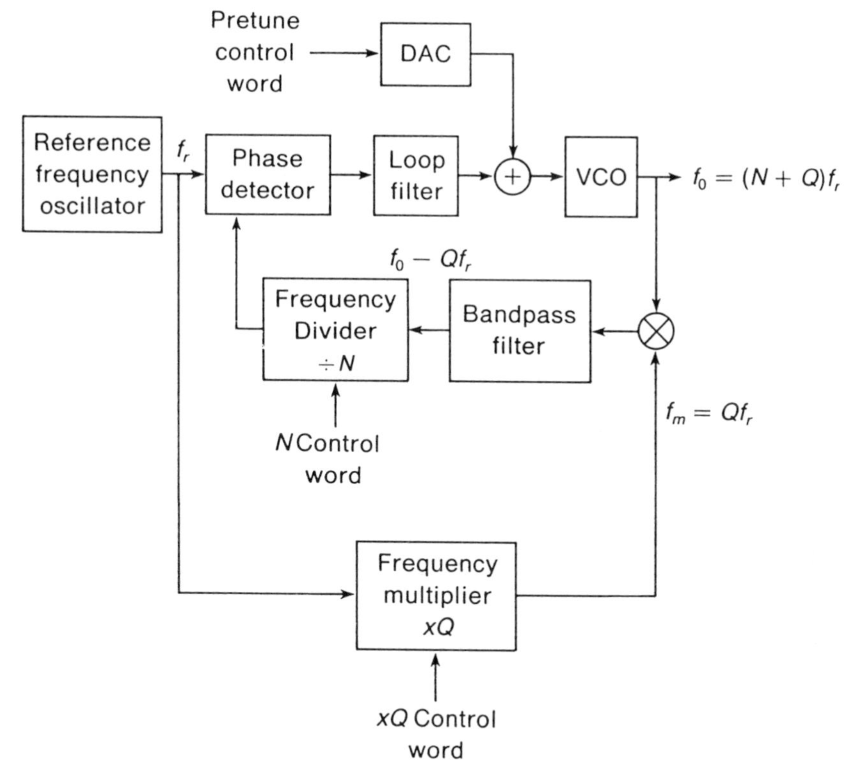

FIGURE 8.5 Frequency hopping synthesizer.

frequency generators; that is, how close its frequency can be set to a prescribed value.

A typical frequency hopping synthesizer is shown in Figure 8.5. Such devices are usually used when wideband frequency hopping with many frequency states are desired (Alexovich and Gagliardi, 1989; Egan, 1981a; Manassewitsh, 1976; Rhode, 1983). An RF voltage controlled oscillator (VCO) has its output mixed down to a lower frequency, then frequency divided to phase lock to a referencing frequency via a standard phase lock loop, similar to the operation of the frequency generators in Figure 4.18a. The mixing frequency, f_m, is provided by a multiple of the same reference. If f_o is the VCO output frequency and f_m is Q times the reference frequency, f_r, then the mixing produces sidebands at $f_o \pm Qf_r$. The loop bandpass filter extracts the lower sideband, which is then frequency divided by N and phase locked to the reference. The loop therefore locks at the frequency where $f_r - (f_o - Qf_r)/N = 0$ or at

$$f_o = (N + Q)f_r \tag{8.2.1}$$

The loop divider is generally a programmable integer count-down divider. The external multiplier Q is generally designed as $Q = Q_I + Q_F$, where Q_I is fixed or variable integer multiplier and Q_F is a variable fractional multiplier. Since Q directly multiplies the reference frequency f_r in Figure 8.5, it is generally implemented by upconverting the output of a digital frequency synthesizer (DFS) clocked by the reference. Recall from Section 4.5 that a DFS can produce divided-down subharmonics of its input frequency, thereby producing the fractional multiples of the reference frequency. The result is that the VCO output is locked to a frequency that is an integer multiple $(N + Q_I)$ of the reference, plus a fraction, Q_F, of the reference. Thus, the integer part of Q adds to the loop divide factor N, whereas the fractional part provides improved resolution. More important, the integer part, Q_I, increases the separation of the feedback mixer sidebands, easing the bandpass filtering of the lower sideband.

Frequency hopping is achieved by controlling the values of N and Q with the code-hopping generator. The divide-down range of N as well as the range of Q_I and Q_F determine the hopping band $[(Q_I + Q_F + N)_{\min} f_r$ to $(Q_I + Q_F + N)_{\max} f_r]$. In addition, by proper design of Q_F, the output frequency can be set across the band to within a frequency resolution equal to the DDS resolution. For example, a DDS with a 23-bit phase accumulator driven by a 5-MHz clock can produce frequency resolution to less than 1 Hz.

Figure 8.5 also shows an external *digital-to-analog converter* (DAC). Its purpose is to provide a course pretune voltage to the VCO each hop time. This pretune voltage reduces the frequency step that must be acquired by the PLL when stepping the output frequency. As such it becomes an important aid to the frequency-hopping operation.

Hopping synthesizers are sometimes designed with the ideal phase detectors in Figure 8.5 implemented by digital detectors that measure zero crossing time differentials, rather than phase differences (Egan, 1981; Rohde, 1983). The phase error between sine waves is converted to pulse width modulation on a sequence of reference clock pulses. Digital detectors are almost always used in synthesizers that incorporate a feedback divider, owing to the fact that the divide circuit that feeds the detector is itself implemented with digital circuitry, and a digital detector simplifies the interface. In addition, digital phase detectors have an extended linear range and are less sensitive to input level variations.

The use of the digital detector generates the equivalent synthesizer

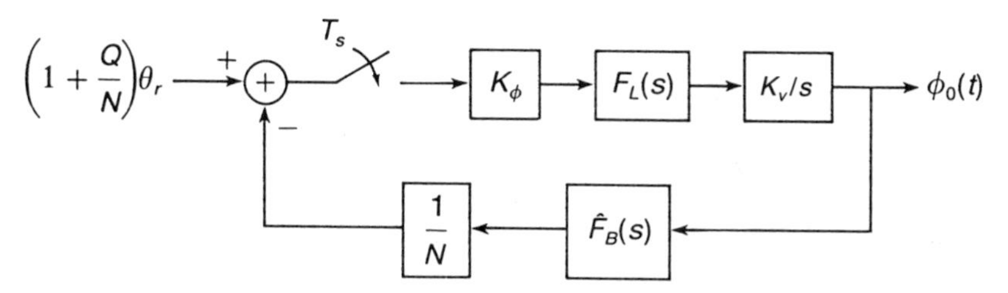

FIGURE 8.6 Equivalent synthesizer block diagram.

block diagram shown in Figure 8.6. The output of the system is the instantaneous phase of the VCO, $\phi_0(t)$. Assuming an ideal mixing frequency, and neglecting the bandpass filter, the output phase is divided down by N and subtracted from the equivalent loop reference phase, $(1 + Q/N)\theta_r$, to generate the loop phase error. The phase error is then sampled at the sampling rate of the detector and filtered by the loop filter $F_L(s)$ to drive the VCO frequency, thereby completing the loop.

When a step change is made in the divider to produce a VCO frequency hop, the synthesizer loop undergoes a step change in frequency and phase. This causes a transient effect during its hop to the next frequency. This transient will be exhibited in the phase output process of the VCO, which in turn affects the dehopping operation at the receiver.

To step to a new frequency, the DAC in Figure 8.5 outputs a voltage step (pretune voltage) that causes the VCO to step to a new frequency and phase. Simultaneously, the divide by N factor (or Q or both) are changed such that the synthesizer desired output is the frequency f_o in Eq. (8.2.1). The frequency difference between the stepped VCO frequency, determined by the DAC pretune voltage, and the desired output frequency, f_o, is the output frequency error, Δf_o. This frequency error is divided by N and appears as an input to the phase detector. Therefore, a frequency step of $\Delta f_o/N$ is applied effectively at the phase detector input at each hop time.

A frequency step appears as an input to the equivalent synthesizer loop in Figure 8.6. If the loop filter is narrow relative to the sampling rate, the effective detector sampling can be neglected, and the loop modeled by a continuous system with the sampler removed. In addition, the feedback mixer and bandpass filter can be neglected, as long as the bandpass filter is symmetrical and wideband relative to the loop filter, which is generally true. We now apply linear analysis to solve for the output transient effect due to a frequency step at the phase detector input. The VCO output phase is given by N times the loop phase error plus a constant. Since the noncoherent decoder is not affected by a constant-phase offset, the

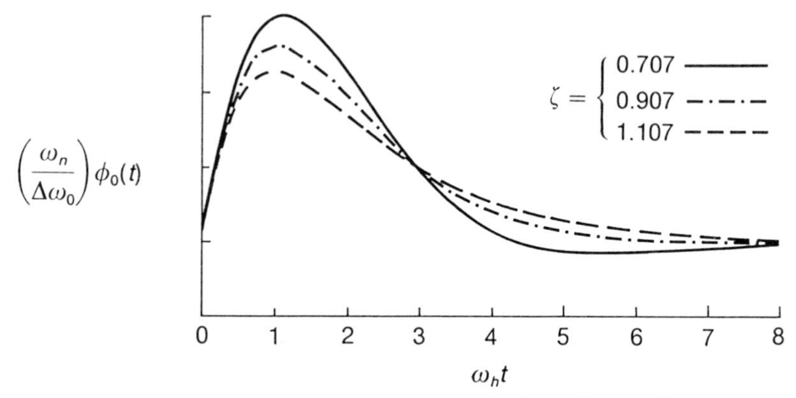

FIGURE 8.7 Synthesizer output phase transient. $\Delta\omega_0$ = frequency step size, ω_n = loop natural frequency, ζ = loop damping.

constant-phase portion of $\phi_0(t)$ is neglected. The phase error transient may be evaluated by inverse transforming

$$\Phi_e(s) = \frac{1}{1 + K_\phi K_v F_L(s)/N_s} \times \frac{\Delta\omega_o}{Ns^2} \tag{8.2.2}$$

where the K values are the component gains in Figure 8.6, $F_L(s)$ is the loop filter function, and $\Delta\omega_o$ the frequency step in radians per second after pretuning. For a second-order loop with a perfect integrating loop filter, the phase error time transient immediately following a hop is

$$\phi_e(t) = \frac{\Delta\omega_o}{N\omega_n} e^{-\zeta\omega_n t} \frac{1}{\sqrt{1-\zeta^2}} \sin(\omega_n\sqrt{1-\zeta^2}t) \tag{8.2.3}$$

where ω_n and ζ are the loop natural frequency and damping factor. The synthesizer output phase then varies as

$$\phi_o(t) = N\phi_e(t) \tag{8.2.4}$$

The phase transient $\phi_o(t)$ is plotted in Figure 8.7 with the time axis normalized. This phase error function $\phi_o(t)$ occurs each time a frequency hop occurs. It should be pointed out that the latter assumes the synthesizer loop always responds linearly to the input frequency step. This requires that the step size $\Delta\omega_o/N$ is less than the "lock-in" frequency of the loop (Gardner, 1979). (The lock-in frequency is the largest frequency step permitted at the phase-detector input, such that no phase-detector cycles are slipped.) Note the phase transient has an inherent settling time, during which the phase (and therefore the frequency) is changing rapidly, before settling to a steady state. During this settling time, the received carrier is not being properly dehopped, and the available symbol energy for decoding is lost. Hence, FSK or DPSK performance will be degraded by the decoding time lost during this hopping transient. From Figure 8.7 we see that transient time in sec can only be reduced by increasing loop natural frequency ω_n, but this directly increases the loop bandwidth and therefore impacts its noise contributions.

Transient effects can also be reduced or eliminated by ping-ponging multiple synthesizers, but at the expense of redundant hardware. In a dual-synthesizer design, two separate synthesizers are connected by a control switch that permits either to be inserted, as shown in Figure 8.8. While one synthesizer is providing dehop, the other is commanded by the synchronized code to the next hopping frequency. As long as the hop period exceeds the transient time of each synthesizer, the second synthesizer will have settled to its proper frequency by the time it is switched in at the next hop. It then serves as the dehop, while the first synthesizer is being reset to the next hop frequency, and operation "ping-pongs"

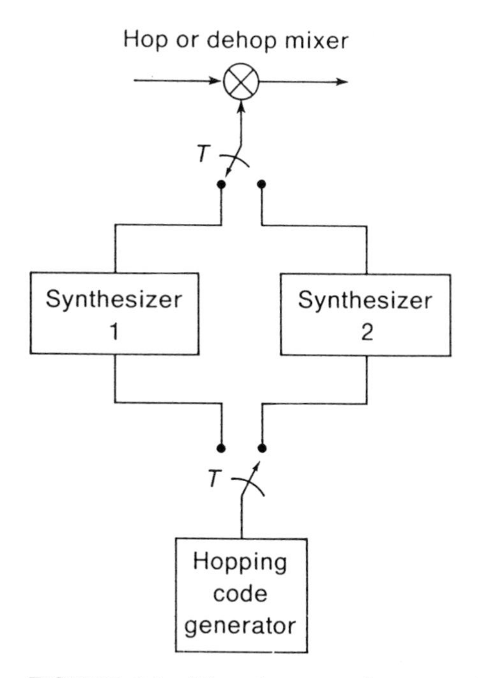

FIGURE 8.8 The ping-pong frequency hopping synthesizer.

between the two synthesizers. If the hop time is reduced (faster hopping rate) so as to approach the synthesizer transient time, some transient degradation may still exist in spite of the ping-ponging. By ping-ponging more than two synthesizers, a longer transient time can be tolerated relative to the hop time.

In addition to the transients, a synthesizer loop is susceptible to spurious tone (spur) generation. A spur entering the loop may phase modulate the VCO output, thereby producing unwanted spectral lines during dehopping. These spectral lines may appear in the modulation band to degrade decoding, or may appear as a mixing frequency that can possibly beat with out-of-band jamming tones to degrade performance.

The primary sources of these spurs are the frequency multiplier, the feedback mixer, and the phase detector. When the multiplier is implemented by upconverting the output of a DFS, spurs about the desired output frequency are produced at multiples of the DFS clock frequency. These spurs are the result of the nonideal digital-to-analog conversion and sampling effects within the DDS. In addition, any product or mixer circuit will produce an output consisting of the desired signal product, plus beat frequency components at the sum and difference of all multiples of its input frequencies (Egan, 1981b). As a result, the feedback mixer in Figure 8.5 produces at its output all products of its input signals (VCO output and the multiplier output). These product terms contain not only the desired mixing terms but all the cross-product terms that effectively appear as added spurs. These spur signals enter the loop in the feedback path and can possibly circulate back to the VCO input.

Spur lines can also be produced by the phase detector. A digital phase detector modulates the pulse width of a pulse train with the instantaneous phase error between the reference and the divider output. The output of the detector therefore contains all the harmonic lines of the pulse train itself. These detector spurs enter the forward path of the synthesizer loop.

The preceding discussion extends the equivalent loop phase model in Figure 8.6 to that in Figure 8.9, where spurs enter the loop as shown. The feedback spurs, owing to the multiplier and mixer, enter at the mixer point and are filtered by the lowpass equivalent [$\hat{F}_B(s)$] of the bandpass filter. The filter output is divided-down to form the phase detector input. The phase detector samples the instantaneous error signal, $\phi_e(t)$ (at a sampling rate, equal to the reference frequency) and adds in the phase detector spurs to form the loop filter input.

The transform of the synthesizer output phase process, from sampled loop analysis, is given by

$$\Phi_o(s) = [\Phi_e^*(s) + \Phi_{\mathrm{PD}}(s)]K_\Phi F_L(s)\frac{K_v}{s} \tag{8.2.5}$$

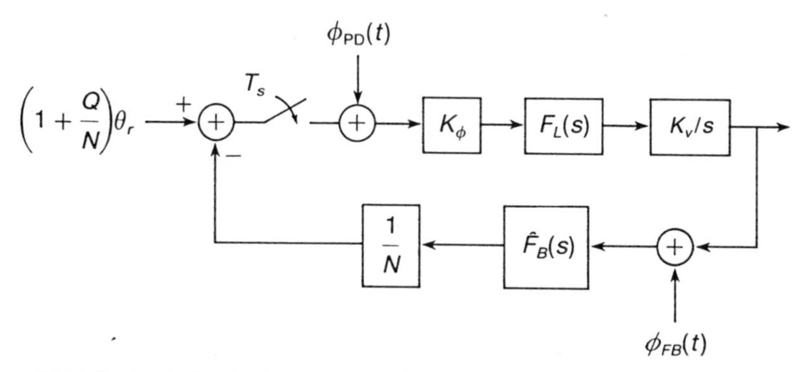

FIGURE 8.9 Equivalent block diagram for synthesizer spur analysis. $\phi_{PD}(t)$ = phase detector spurs, $\phi_{FB}(t)$ = feedback loop spurs.

where $[\cdot]^*$ denotes a time-sampled function, and $\Phi_{PD}(s)$ is the transform of the phase-detector spurs. Since the bandwidth of

$$G(s) \triangleq K_\Phi F_L(s) \frac{K_v}{s} \hat{F}_B(s) \frac{1}{N} \tag{8.2.6}$$

is generally much lower than the sampling frequency, f_s, the effect of sampling the phase-detector spurs may be neglected. The transform of the synthesizer output phase process (neglecting any constant-phase terms) can therefore be approximated as

$$\Phi_o(s) \approx [\Phi_{FB}(s)\hat{F}_B(s)]^* H(s) + \Phi_{PD}(s) K_\Phi F_L(s) K_v/s \tag{8.2.7}$$

where $\Phi_{FB}(s)$ is the transform of the spurs input at the feedback mixer and

$$H(s) \triangleq \frac{G(s)/\hat{F}_B(s)}{1 + G(s)} \tag{8.2.8}$$

From Eq. (8.2.7) we see the phase-detector spurs are not affected by the sampler and appear at the synthesizer output filtered by the feed-forward transfer function of the loop. The feedback spurs, however, are filtered [by the lowpass equivalent filter $\hat{F}_B(s)$], sampled and then loop filtered by $H(s)$ before adding to the synthesizer output phase. The sampling produces replicas of the feedback spur spectrum, shifted by all multiples of the sampling rate, at the phase-detector output. This means any spurs in the vicinity of the sampling harmonics are always aliased into the loop filter and therefore will appear at the output. This somewhat subtle point is extremely important, since it means that feedback spurs located well outside the PLL bandwidth, can still affect the synthesizer performance if they fall in the vicinity of a sampling harmonic. Effectively the loop collects spur lines by filtering the original spur distribution in the vicinity of each sampling harmonic with filter function, $H(s)$. The filtered spurs in the vicinity of any harmonic then appear at the synthesizer output at its difference frequency from the harmonic. It is therefore important to remove as many input spurs as possible with the filter $F_B(s)$ prior to phase-detector sampling.

Spur lines entering the loop that flow around to the VCO input can accumulate to degrade the subsequent decoding after dehopping. Any spur signal at the VCO input will frequency modulate the VCO, producing both unwanted harmonic lines at the spur frequency and a suppression effect (modulation loss) on the desired hopping carrier. The harmonic lines can mix with the receiver RF carrier at any hop, and produce interference terms that may fall in the dehopped decoding bandwidth. Since the spurs are generally of low amplitude (30–50 dB below the hopping carrier), and should be uniformly distributed over the decoding band, their presence should have only minimal effect on decoding. More serious, however, is the suppression effect of the spur modulation on the desired hopping carrier. If there are N random spurs at the VCO inputs with amplitudes $\{a_i\}$, the dehopped carrier at the detector input will be suppressed by

$$D = \prod_{i=1}^{N} J_0^2(a_i) \tag{8.2.9}$$

where $J_o(a)$ is the zero-order Bessel function. Note that the carrier suppression term is a function of *all* the synthesizer output spur modula-tion indices (not just those falling in the detector bandwidth) and is independent of the actual spur frequency and phase.

The carrier suppression term in Eq. (8.2.9) is plotted in Figure 8.10 versus the total number of spurs with the modulation index of all spurs varied as a parameter. The carrier suppression increases rapidly with increasing number of spurs, with the rate of increase determined by the value of $J_o(a)$. It is interesting to note that a 10-dB reduction in spur level (for example, 40 dBc to 50 dBc) results in a factor of about 10 reduction in the carrier suppression. Since most of the significant spurs are limited to the bandwidth, B_L [the bandwidth of $H(s)$] and the minimum spur spacing is f_Δ, the total number of spurs is generally less than B_L/f_Δ. Therefore, decreasing f_Δ (improving resolution) or increasing B_L (faster loop) results in the potential increase in the total number of spurs, and therefore a possible increase in the suppression factor.

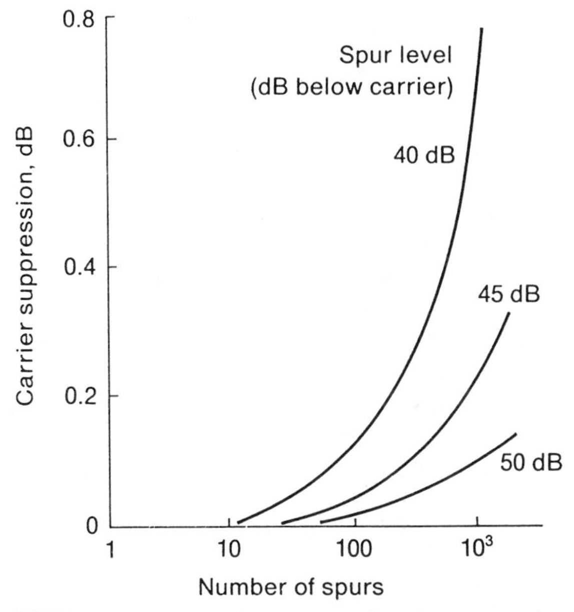

FIGURE 8.10 Carrier suppression due to synthesizer spurs.

8.3 PERFORMANCE OF FREQUENCY-HOPPED SYSTEMS

If the receiver hopping pattern exactly matches the received pattern, and the synthesizer is ideal, the RF carrier will be perfectly dehopped. The resulting decoding operation will be transparent to the hopping, and error probability performance can be obtained from the curves in Chapter 2. The DPSK PE was given in Table 2.4 as

$$\mathrm{PE} = \tfrac{1}{2}e^{-(E_b/N_0)} \tag{8.3.1}$$

while the MFSK symbol error probability in Table 2.6 was

$$\mathrm{PSE} = 1 - e^{-\rho_s^2/2}\sum_{q=0}^{M-1}(-1)^q\binom{M-1}{q}\frac{\exp\,[(\rho_s^2/2)/(1+q)]}{1+q} \tag{8.3.2}$$

where $\rho_s^2 = 2E_s/N_0$ and E_s is the symbol energy. The resulting PSE can be approximated and bounded by the simpler expression

$$\mathrm{PSE} \approx \frac{M-1}{2}\,e^{-E_s/2N_0} \tag{8.3.3}$$

The MFSK bit-error probability PE is then obtained from PSE by

$$\mathrm{PE} = \left(\frac{M/2}{M-1}\right)\mathrm{PSE} \tag{8.3.4}$$

In fast hopping with FSK modulation, the individual FSK symbols are repeated over many hops, and decoding is achieved by combining the energy variable over all hops. Each hop is referred to as a "chip" of the FSK symbol, and the operation of processing the hop sequence is referred to as *chip combining*. Symbol decoding is obtained by selecting the maximum of the combined chip variable after all hops have been accumulated. Thus, if there are L hops per symbol, and if x_{ij}^2 is the energy variable collected at the ith FSK time detector in Figure 8.4b during the jth hop, the chip combiner produces the set of FSK variables

$$V_i = \sum_{j=1}^{L} x_{ij}^2 \tag{8.3.5}$$

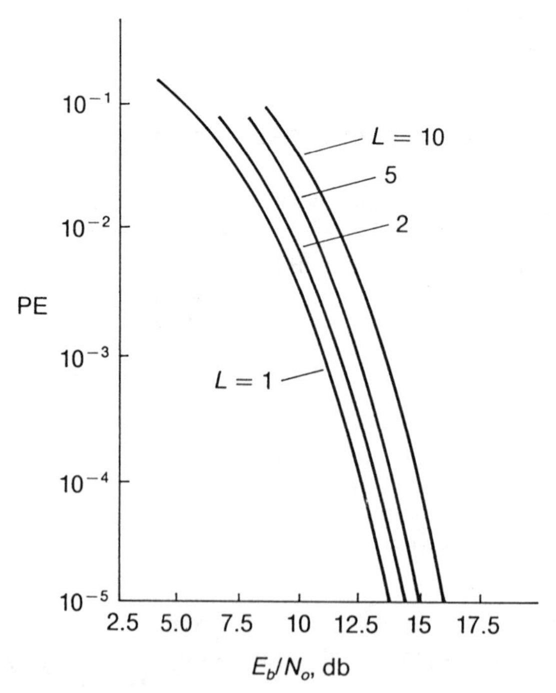

FIGURE 8.11 Bit error probability versus E_b/N_o for binary FSK. L = number chips per bit.

The FSK symbol is decoded from the set $\{V_i\}$ by selecting the maximum. The probability of symbol error after chip combining is obtained by determining the probability that the correct V_i is not the largest. In the presence of Gaussian noise, each V_i in Eq. (8.3.5) is a noncentral chi-squared random variable with $2L$ degrees of freedom (Abromovitz and Stegun, 1965; Miller, 1964), with the noncentrality parameter depending on the signal and noise energies collected in each channel over all L hops. The probability of symbol error among these variables can be written as (Proakis, 1989)

$$\mathrm{PSE} = 1 - \int_0^{\infty} \left[1 - e^{-\beta_s V} \sum_{j=0}^{L-1} \frac{V^j}{j!}\right]^{M-1} \beta_s V^{L-1} e^{-\beta_s(1+V)} I_{L-1}(2\beta_s \sqrt{V})\, dV \tag{8.3.6}$$

where β_s is the symbol energy to noise level ratio over L chips, and $I_i(x)$ is the Bessel function of order i. For the binary ($M = 2$) case, Eq. (8.3.6) simplifies to

$$\mathrm{PE} = \frac{1}{2^{2L-1}} e^{-\beta_b/2} \sum_{j=0}^{L-1} K_j \left(\frac{\beta_b}{2}\right)^j$$

$$K_j = \frac{1}{j!} \sum_{r=0}^{L-1-j} \binom{2L-1}{r}$$

$$\beta_b = E_b/N_0 \tag{8.3.7}$$

A plot of Eq. (8.3.7) is shown in Figure 8.11. The $L = 1$ curve corresponds to slow hopping (one hop per symbol). The curves demonstrate that higher (poorer) PE occurs with increasing L with the same total bit energy E_b. Thus, a performance degradation occurs when the bit energy of a noncoherent FSK system is spread over many hops, and combined for decoding. This is referred to as the *noncoherent combining loss*, and reflects the fact that dividing the available bit energy E_b into L segments always results in poorer performance than if the energy was concentrated in a single hop. This combining loss can be plotted directly as a function of diversity L, for specific values of E_b/N_0, as shown in Figure 8.12. The curves in Figure 8.11 would directly apply to DPSK with L chip combining, if the bit energy E_b is divided by 2. The MFSK results for $M > 2$ can be estimated by multiplying the corresponding PE in Figure 8.11 by $M/2$.

Frequency-hopping performance may be further degraded if the frequency synthesizer is not ideal. If the synthesizer has a hopping transient,

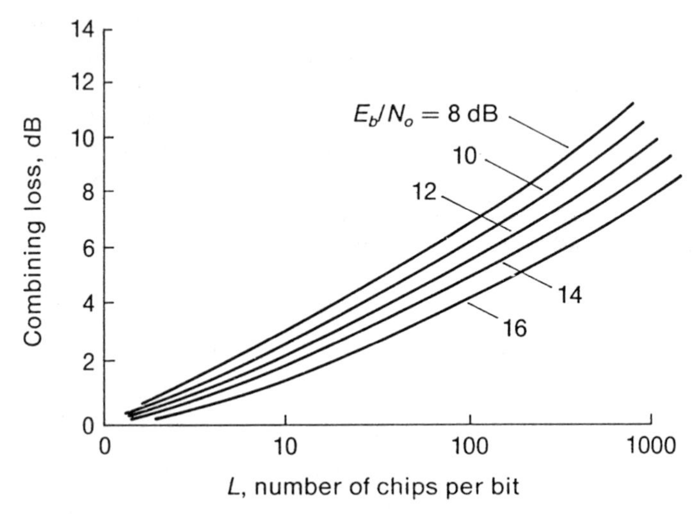

FIGURE 8.12 Combining loss versus diversity L for binary FSK with fast hopping.

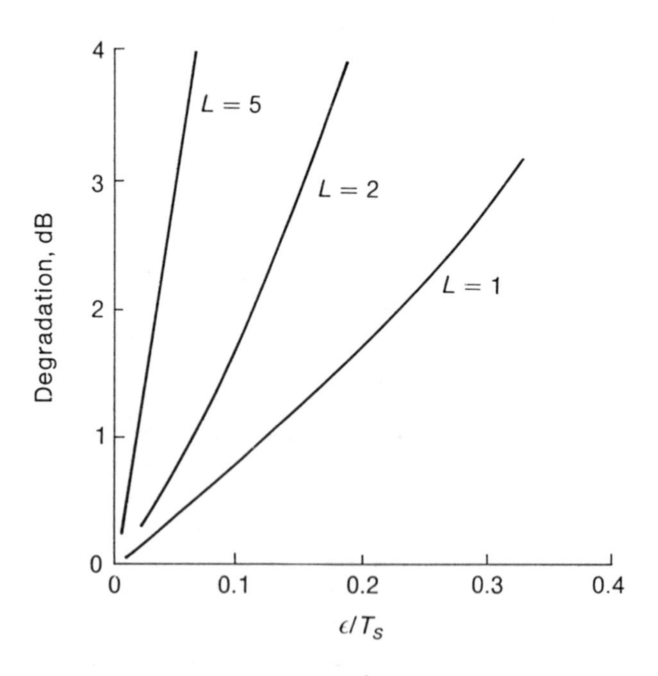

FIGURE 8.13 Degradation $\hat{E}_s/E_s$ in symbol energy versus normalized transient time per hop. $L =$ number chips per symbol.

as in Figure 8.7, the decoding energy is effectively lost during the transient time. If the transient lasts for ϵ s after each hop, and if the receiver uses L-ary chip combining (i.e., L hops per FSK symbol), the decoding symbol energy E_s in Eq. (8.3.6) is reduced to

$$\hat{E}_s = E_s\left[1 - \left(\frac{L\epsilon}{T_s}\right)\right]^2 \tag{8.3.8}$$

where T_s is the total symbol time. Figure 8.13 plots this degradation in dB for E_s as a function of the fractional hopping transient time for various L's. This is a direct loss to decoding symbol energy which must be added (in dB) to the combining losses from Figure 8.11. In addition, if the synthesizer has excessive spur generation, the energy degradation must also include the spur suppression effect obtained via Eq. (8.2.9). The latter requires an accurate spur assessment to be performed.

8.4 FREQUENCY-HOPPED CDMA SYSTEMS

Frequency hopping can also be used to derive a multiple access system, similar to the DS-CDMA operation in Chapter 7. Rather than using the low cross-correlation of the addressing codes, the frequency-hopped system would use the disjointness of separate hopping patterns to achieve the access separability. Different transmitters, using the same satellite bandwidth but with different hopping patterns, can theoretically communicate at the same time with negligible crosstalk if the patterns have little or no frequency overlaps.

A complete frequency-hopped CDMA satellite system would be obtained by assigning each transmitting station a different hopping pattern. Each active station operates by generating its own pattern, encoding the data symbols at each hop, and accessing through the satellite whenever desired. A station receives a transmission by hopping in sync with the desired station, and decoding each symbol after dehopping. In effect, the hopping pattern is playing the role of an address, with all interfering stations having different patterns. The latter will be rejected if they hop into other bands at each slot. If no other station hopped into the same band at the same slot, the system would be truly orthogonal, and no station interference would exist. If there are n frequency bands per slot, there can be at most $K \leq n$ distinct patterns that will never overlap. Hence, the number of orthogonal, frequency-hopped CDMA users is, from

Eqs. (8.1.2) and (8.1.3),

$$K \leq n = \frac{B_{\text{RF}}\tau}{2M}$$
$$= 2^{R_c/R_H} \tag{8.4.1}$$

The number of orthogonal users therefore increases linearly with the satellite bandwidth, or equivalently, increases *exponentially* with the code rate generating the hopping. Recall that the number of orthogonal users in DS-CDMA increased only *linearly* with code rate [see Eq. (7.1.3)]. Hence, when code rate is the limiting factor, frequency hopping has a decided capacity advantage over direct sequence modulation.

If more than n user stations are involved, they cannot be assigned patterns without some overlap occurring. If the patterns are selected without regard to the patterns of other stations, then it can be assumed that during any slot time locations of all stations are randomly distributed in frequency. In an FSK system, receiver performance can be estimated by taking into account the chance of a pattern overlap. If no other active station overlaps a given station operating in a particular band, the error probability for that time slot is that of a standard noncoherent link, and

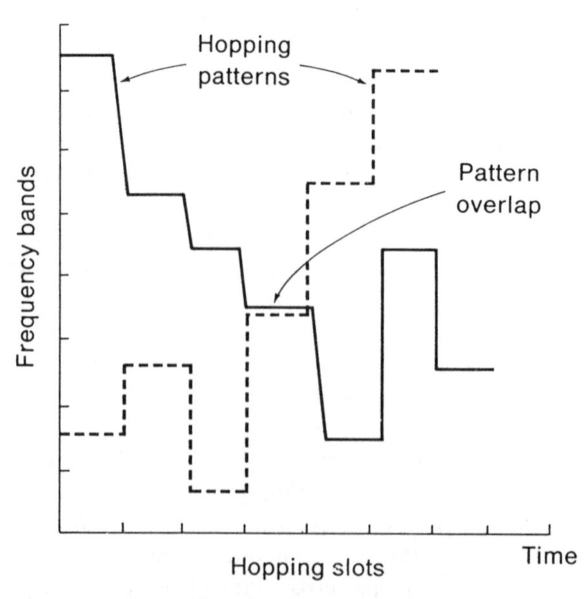

FIGURE 8.14 Hopping patterns with band overlap.

the clear (no interference) symbol error probabilities are those in Eqs. (8.3.1)–(8.3.4). If another carrier overlaps during that band (Figure 8.14), its modulation may either aid or hinder the FSK decoding in that band. Assuming a hindrance will always cause a bit error, the frequency-hopped symbol error probability is then

$$\text{PSE} = \left(\frac{M-1}{M}\right)P_H + (1 - P_H)\text{PSE}_0 \tag{8.4.2}$$

where P_H is the probability that another carrier hops into the same band and PSE_0 is the symbol error probability with no interference. For K active carriers and N bands,

$$P_H = 1 - \left(1 - \frac{1}{n}\right)^{K-1} \tag{8.4.3}$$

As the number of users increases, $P_H \to 1$ and PSE increases from PSE_0 to $(M-1)/M$, essentially destroying the data link. As a rule of thumb, $\text{PSE} \leq 2(\text{PSE}_0)$ when $P_H \leq \text{PSE}_0$, which requires

$$K \lesssim n\text{PSE}_0 \tag{8.4.4}$$

Hence, the number of users employing randomly selected hopping patterns is considerably less than the number of possible orthogonal pattern users.

Inserting fast hopping with chip combining helps to alleviate the effects of the CDMA interference. Since a data bit is extended over many hops, any CDMA interference during any one particular hop will have a lesser effect after the chip combining. The more the diversity, the less the probability that random CDMA interference will overlap enough times to degrade decoding significantly. If we again assume CDMA interference will dominate over the thermal noise during any hop that overlap occurs, the probability of symbol error, after L chip combining, can be upper bounded and approximated by (Proakis, 1989)

$$\text{PSE} \approx \left[\left(\frac{M-1}{M}\right)P_H + (1 - P_H)\left(\frac{M-1}{2}\right)e^{-(E_s/2LN_0)}\right]^L \tag{8.4.5}$$

As long as P_H is less than, say, 10^{-1}, then as L increases the second bracketed term will eventually be larger than the first, and PSE approaches the result in Eq. (8.3.6). Hence, the use of fast hopping at increasing diversity levels overcomes the degradation effect of any CDMA interference. The chip combining has therefore improved the orthogonality of the CDMA system by reducing the effect of CDMA interference.

The price to be paid for this accessing improvement, besides the L-ary combining loss, is immediately obvious from Eq. (8.1.4). An increased diversity L at a fixed hopping rate will produce a reduction in data rate available for each access. Conversely, the hopping rate will have to be increased in order to maintain a fixed data rate when increasing diversity (chip combining) is inserted.

8.5 JAMMING IN FREQUENCY-HOPPED SYSTEMS

A frequency-hopped system is also susceptible to external jamming of the receiver. However, the effect is slightly different than in DS-CDMA, where the receiver achieves the inherent advantage of diluting the jamming interference via the spectral spreading, as was discussed in Section 7.4. With frequency hopping it may in fact be advantageous for the jammer to concentrate his power rather than operate broadband. Suppose, for example, the transmitter uses a frequency-hopped, binary FSK system, and assume the jammer concentrates his available power P_J over q FSK bandwidths, each $1/\tau$ Hz wide (Figure 8.15). (The jammer may actually change the q bands from one time slot to the next, so the jammed bands

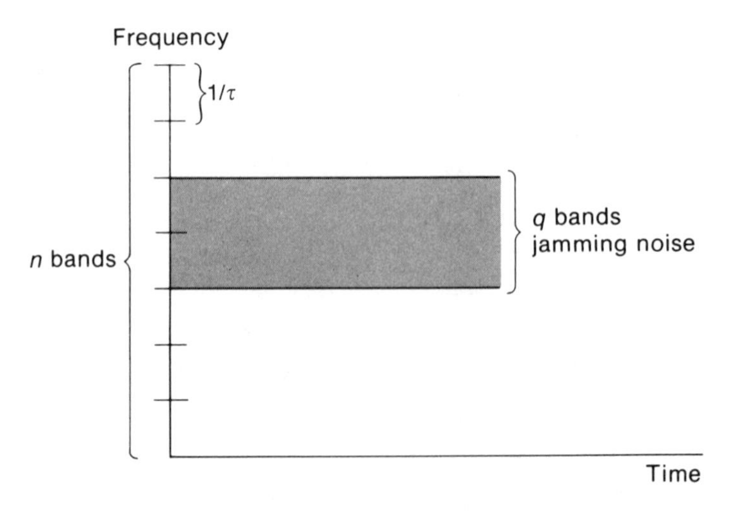

FIGURE 8.15

cannot be known in advance.) The jammer, therefore, introduces an effective noise level

$$J_0 = P_J(\tau/q) \text{ W/Hz} \tag{8.5.1}$$

If the frequency-hopped carrier is not in these bands, decoding is achieved with the PE associated with the receiver thermal noise. For the binary FSK case we write this as

$$\text{PE}_0 = \tfrac{1}{2}e^{-E_b/2N_0} \tag{8.5.2}$$

If the jammer overlaps the carrier band, it adds its noise level to the thermal noise level, and now

$$\begin{aligned} \text{PE}_J &= \tfrac{1}{2}e^{E_b/2(N_0+J_0)} \\ &= \tfrac{1}{2}\exp\left\{\frac{1}{2}\left[\frac{N_0}{E_b} + \frac{P_J}{qP_c}\right]^{-1}\right\} \end{aligned} \tag{8.5.3}$$

The average error probability is then

$$\text{PE} = P_H(\text{PE}_J) + (1 - P_H)\text{PE}_0 \tag{8.5.4}$$

where now P_H is the probability that the carrier band is jammed. Assuming the jammer randomly selects his q bands,

$$P_H = q/n \tag{8.5.5}$$

By increasing q (jamming over a wider bandwidth), the probability of overlapping is increased, but the jammer dilutes his power over the jammed bands. To see this trade-off, let the thermal noise be neglected so that Eq. (8.5.4) is approximately

$$\text{PE} \approx \left(\frac{q}{n}\right)e^{-q(P_c/P_J)/2} \tag{8.5.6}$$

A simple calculation shows that the largest PE (best jamming strategy) occurs for the choice

$$\frac{q}{n} = \frac{2}{n(P_c/P_J)} \qquad \text{if } \frac{nP_c}{P_J} \geq 2 \tag{8.5.7a}$$

$$= 1 \quad \text{if } \frac{nP_c}{P_J} < 2 \tag{8.5.7b}$$

If the jammer has excessive power, broadband jamming should be used as indicated in Eq. (8.5.7b). Otherwise, the jammer power should be confined to the fractional band in Eq. (8.5.7a), for which

$$\mathrm{PE} = \frac{e^{-1}}{nP_c/P_J} \tag{8.5.8}$$

Thus, in a binary FSK frequency-hopping system with n hopping frequencies, optimal jammer strategy would use partial band jamming, with the fraction of total bandwidth to be jammed given by Eq. (8.5.7). This produces an error probability that at best varies inversely with transmitter power (instead of exponentially, as when combating thermal noise alone). Note that the denominator in Eq. (8.5.8), after substituting from Eq. (8.1.3), is directly proportional to the same processing gain defined in Eq. (7.5.6) for the DS-CDMA system. We see that in frequency hopping, as well, the ability to increase the link bandwidth-bit time product leads directly to improved performance.

The error probability in Eq. (8.5.8) can be written in terms of the bit energy by noting that

$$\frac{nP_c}{P_J} = \frac{B_{\mathrm{RF}} P_c}{R_b P_J} = \frac{E_b}{P_J/B_{\mathrm{RF}}} = \frac{E_b}{J_0} \tag{8.5.9}$$

where J_0 is the equivalent broadband noise level [Eq. (8.5.1)] inserted by the jammer with power P_J. Thus, the jammed bit-error probability in Eq. (8.5.8) is equivalent to

$$\mathrm{PE} = \frac{e^{-1}}{E_b/J_0} \tag{8.5.10}$$

The use of diversity with chip combining can aid the frequency-hopped jamming link, just as in the case of CDMA interference in Section 8.4. By increasing the number of hops per symbol, the probability of a partial band jammer overlapping all hops of a given symbol is significantly reduced. Since the jammers will often have a power advantage in a given hopping band, the chip combining in Eq. (8.3.5) should be modified to incorporate a weighted chip energy summation. That is, each individual x_{ij} in Eq. (8.3.5) should be multiplied by a weighting factor before summing, with the weighting factor ideally selected to be inverse to the noise power of that sample. This has the effect of "dividing down" any large sample value contributed by the jammer. Weighting of this type can

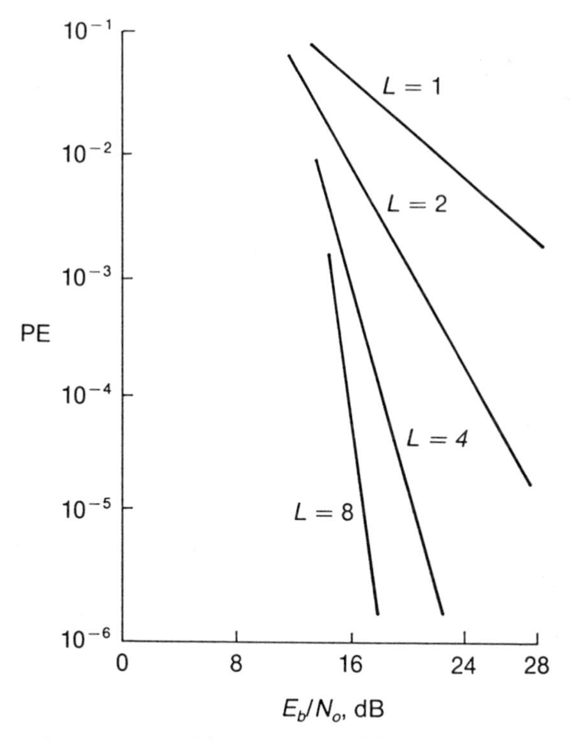

FIGURE 8.16 Bit error probability in binary FSK with optimal partial band jamming and L-ary diversity.

be approximated by the use of automatic gain control (AGC) or limiting circuitry inserted prior to the FSK decoder.

If the available bit energy is divided into L hops, and if a bounding procedure similar to Eq. (8.4.5) is applied to the weighted chip summations, the bit-error probability of the binary jammed FSK system is Eq. (8.5.9), when L-diversity is inserted, now becomes (Proakis, 1989; Simon et al., 1985)

$$\begin{aligned} \mathrm{PE} &\lesssim \left[\frac{4/e}{E_b/LJ_0}\right]^L \\ &= \left[\frac{1.47L}{E_b/J_0}\right]^L \end{aligned} \qquad (8.5.11)$$

This PE is plotted in Figure 8.16 for several values of L, along with the broadband result in Eq. (8.3.6). The $L = 1$ case refers to the no-diversity result in Eq. (8.5.8). We see that the insertion of fast hopping improves the link PE, and begins to approach the broadband performance. Hence, hopping diversity tends to overcome the jamming advantage obtained by partial banding in a frequency-hopped system.

It can also be noted from Eq. (8.5.11) that the insertion of too much diversity can eventually degrade the link. (When the bracket exceeds unity, the PE bound is no longer reduced with increasing L.) The value of L that in fact minimizes PE in Eq. (8.5.11) can be found as

$$L_{\text{opt}} \cong \frac{E_b/J_0}{3} \tag{8.5.12}$$

Hence, the optimal diversity is directly related to the expected energy ratios according to Eq. (8.5.12).

It should be pointed out here that the jammer has other options as well, such as sweep frequency jamming, repeat-back jamming, and so on, to further combat these alternatives. A complete discussion of jamming and antijamming strategies is outside the scope of our discussion here, and the reader is referred to discussions on these subjects in Houston (1975), IEEE (1977/1982), Kullstan (1977), and Pettit (1982).

8.6 CODE ACQUISITION AND TRACKING IN FH SYSTEMS

Frequency dehopping requires a local code generator, which generates the synthesizer hopping pattern, to be locked to the code pattern generating the received hopping pattern. This code pattern must be acquired and continually tracked, just as in the DS-CDMA system in Chapter 7. However, the frequency hopped system must utilize an acquisition operation completely different from that described in Section 7.7.

If the local and received codes generating the hopping patterns are not in perfect alignment, the partitioned words specifying the frequency band during a given time slot will be different. As a result, the mixing of the local and received carrier will no longer mix to baseband, and no energy will be detected in the noncoherent decoder. Hence, any misalignment of frequency-hopping codes will always produce a zero signal voltage in any of the FSK envelope detectors. Only when the codes are perfectly aligned, and the carrier frequencies are hopping in synchronism, will signal energy be observed continually in the decoder. Acquisition is therefore achieved by mixing the local and received carriers as they hop, and observing FSK decoder energy for a fixed integration time. The local code can then be

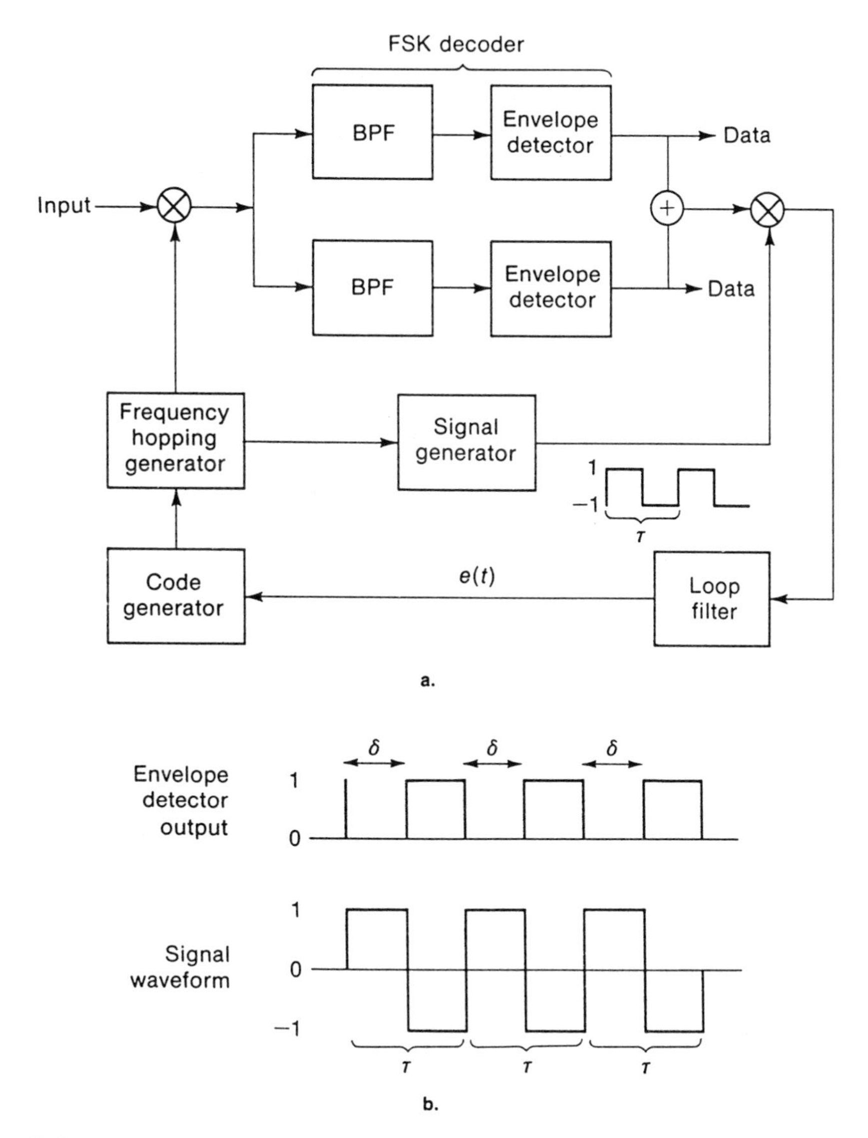

FIGURE 8.17 Frequency hopping acquisition and tracking subsystem. (a) Block diagram. (b) Waveforms.

shifted one chip at a time (shifting the local hopping pattern) until decoding energy is observed at one position.

Frequency-hopped acquisition performance is governed by FSK decoder variables, which also correspond to envelope samples, as in noncoherent DS-CDMA acquisition. Hence, the decoding E_c/N_0 is again given by Eq. (7.7.6), except that E_c is now the carrier energy per hop time, and the integer η here indicates the number of hops over which the decoder energy is integrated. Acquisition probabilities can be computed using either the maximum-likelihood or threshold procedures discussed previously.

Tracking in frequency-hopped systems is obtained via the system diagram in Figure 8.17a. Again, acquisition is used to bring the hopping patterns in near alignment, and the tracking subsystem is used to remove any residual offsets and hold the patterns aligned. As in direct-sequence tracking, the frequency-hopped tracking operates by generating a correction voltage to adjust the timing of the local code generator that drives the local hopping pattern. This correction voltage is obtained using the fact that if the hopping patterns are offset by δ s ($|\delta|$ less than a chip time) there will be a δ-s interval in each slot time when the frequencies of the hopping patterns are not aligned (see Figure 8.17b). During this δ-s interval no signal energy appears in the FSK decoder. This δ-s interval will be either at the front end or rear end of the slot time, depending on whether the local pattern is early or late. By multiplying the energy detector output in each slot time by the periodic square wave in Figure 8.17a and integrating, we generate a correction voltage with the proper sign and magnitude for timing control. If $\delta = 0$ (codes perfectly aligned), the square wave clock causes the product to contain equal positive and negative areas, and the integrator output sums to zero. If a δ-s interval of zero voltage arises (codes not perfectly aligned), a portion of the product area will cancel, reducing the integrator value by an amount proportional to δ. Whether the zero area subtracts from the positive or negative area depends on whether the δ-s interval is at the beginning or ending of the slot. Hence, a filtered correction voltage $e(t)$ in Figure 8.17a will be proportional to δ with the proper sign for continually correcting the local code. The frequency-hopped code tracking system can thus be converted to the equivalent tracking loop in Figure 7.16, and analyzed as in Section 7.7.

REFERENCES

Abromowitz, A., and Stegun, I. (1965), *Handbook of Mathematical Functions*, National Bureau of Standards, Washington, DC, Chapter 26.

Alexovich, J., and Gagliardi, R. (1989), "Effect of PLL Frequency Synthesizer in FSK Frequency Hopped Communications," *IEEE Trans. Commun.*, **COM-37** (3) (March) 268–276.

Egan, W. F. (1981a), *Frequency Synthesis by Phaselock*, Wiley, New York.
Egan, W. F. (1981b), "The Effects of Small Contaminating Signals in Nonlinear Elements Used in Frequency Synthesis and Conversion," *IEEE Proc.*, **69** (7) (July) 797–811.
Gardner, F. M. (1979), *Phaselock Techniques*, Wiley, New York.
Houston, S. (1975), "Tone and Noise Jamming of Spread Spectrum FSK and DPSK Systems," *Proc. of the IEEE Aerospace and Electronic Conference*, Dayton, OH.
IEEE (1977/1982), "Special Issue on Spread Spectrum Communications," *IEEE Trans. Comm.*, **COM-25** and **COM-30**.
Kullstam, P. (1977), "Spread Spectrum Performance in Arbitrary Interference," *IEEE Trans. Comm.*, **COM-25** (August) 848–853.
Manassewitsh, V. (1976), *Frequency Synthesizers: Theory and Design*, Wiley, New York.
Miller, K. (1964), *Multidimensional Gaussian Distributions*, Wiley, New York.
Pettit, R. (1982), *ECM and ECCM Techniques For Digital Communications*, Lifetime Learning, Belmont, CA.
Proakis, J. (1989), *Digital Communications*, 2nd ed., McGraw-Hill, New York, Chapter 8.
Rohde, U. (1983), *Digital Frequency Synthesizers*, Prentice-Hall, NJ.
Simon, M., Omura, J., Scholtz, R. and Levitt, B., (1985), *Spread Spectrum Communications*, Computer Science Press, Rockville, MD., Vols. II and III.

PROBLEMS

8.1. An FH FSK system uses a frequency bandwidth of 160 MHz, and transmits three-bit words at 1000 hops per second. The decoding FSK $E_b/N_0 = 7$ dB. (a) How many separate carriers can be supported? (b) What will be the symbol-error probability, PSE?

8.2. Assume the reference source in Figure 8.5 has the phase noise spectrum $S_\psi(\omega)$. Determine the spectrum of the resulting phase noise that will appear on the hopping synthesizer output carrier. Use the model in Figure 8.6 and assume the phase detector sampling is extremely fast compared to the phase noise bandwidth.

8.3. A FH system hops over a 50-kHz bandwidth. The frequency generator is simply an oscillator that is swept at a rate of 50 MHz/s. What is the worst case decoding energy degradation that can be expected in a system that hops at 150 hops per second?

8.4. An acquisition waveform in FH systems is composed of a specific sequence of frequency bursts, $\mathbf{F} = \{f_i\}$, where each f_i is one of a set of n frequencies. Acquisition is achieved by frequency correlating the input waveform with the expected frequency sequence **F**. Good acquisition requires high correlation only when the true **F** completely arrives, and zero correlation at all other times.

Assume during each burst two identical frequencies will correlate to one, while two different frequencies correlate to zero. Show that the desired correlation property is satisfied if and only if no two f_i of **F** are identical. What does this mean about the length of **F**?

8.5. An *array* is a matrix of cells, with dots in some cells. A *Costas array* is a square array with only one dot in any row or column, and having the property that any shift of the array horizontally and/or vertically will produce no more than one dot overlap with the original array.

Explain how the solution of this mathematical problem has direct application to the acquisition problem of a frequency hopping system. [*Hint*: Observe Figure 8.1.]

8.6. An 8-ary FSK frequency hopped system operates with slow hopping ($L = 1$), $E_b/N_0 = 10$ dB, and with a probability of hop overlap of $P_H = 10^{-3}$. (a) Estimate the PSE using Eq. (8.4.5). (b) Repeat, assuming fast hopping is inserted, with $L = 10$.

8.7. Show that Eq. (8.5.11) is minimized with the selection of L in Eq. (8.5.12).

CHAPTER 9
On-Board Processing

The previous chapters concentrated on relatively straightforward satellite carrier processing, in which uplink carriers were either directly turned around on the downlink or switched between satellite antennas. The current trend, however, is toward more sophisticated processing at the satellite prior to retransmission, an area referred to as *on-board processing*. By designing satellite payloads to accomplish more internal processing, a more efficient overall satellite link can be realized, which invariably leads to a simplification of the required earth stations. In this chapter we examine some of the present directions in on-board processing, and the expected emphasis in the next generation of communication satellites.

9.1 ON-BOARD PROCESSING SUBSYSTEMS

The on-board processors planned for satellites can be roughly classified into two basic types, as shown in Figure 9.1. The first, in Figure 9.1a, is referred to as a *carrier*, or *microwave*, *processor*, in which uplink modulated RF carriers are directly distributed to downlink antennas via microwave connecting circuitry and gate switches. Besides the standard RF-RF transponders, this includes the class of SS-FDMA and SS-TDMA system described in Section 5.6 and 6.5. The basic concept is to orient the uplink-to-downlink connectivity at the satellite using fixed or reconfigure-

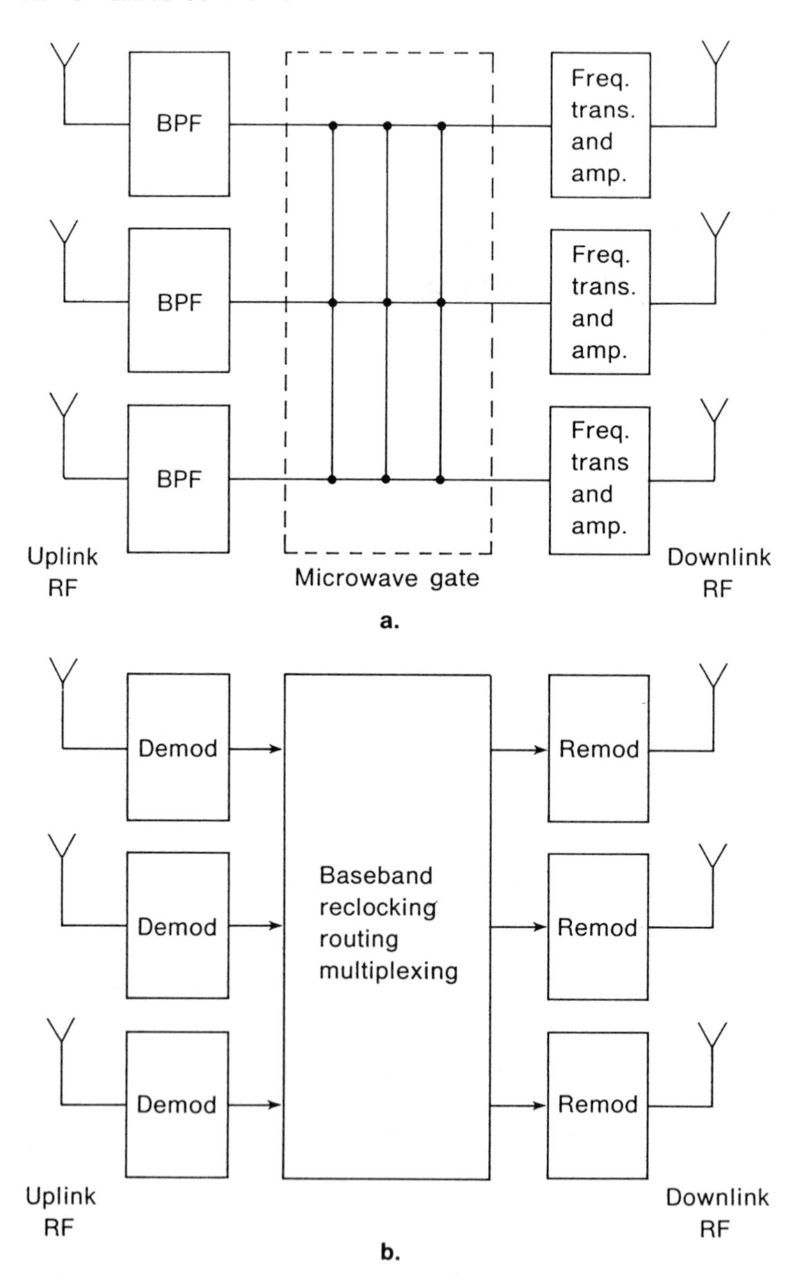

FIGURE 9.1 Satellite on-board processors. (a) Carrier processor. (b) Baseband processor.

able waveguide gates. The interconnection is accomplished at either RF or IF frequencies using microwave circuitry, and no attempt is made to recover the baseband data during the processing. This means there can be no changes in modulation formats, and all the uplink carriers are directly connected to the downlink. The key parameters of such processors are the gate isolation (which determines carrier crosstalk) and the interconnect isolation losses (which determines power losses during processing). In reconfigureable gates the interconnections between uplink and downlink carriers are restructured periodically, and gate switching speeds and settling times become important in determining connection and revisit time between specific earth stations.

The second type of on-board processor (Figure 9.1b) is referred to as a *baseband processor*. Here the uplink RF carriers are first demodulated to baseband data bits, which are then demultiplexed, routed, and remodulated on the downlink carriers. During the routing, different groups of data bits (packets) from a single carrier can be addressed and separately redistributed to different downlinks, producing more complete data distribution capability. Also, decoding directly at the satellite permits data reformatting and improved routing, and allows a complete alteration of the downlink modulation scheme from that of the uplink for improved performance. Baseband processing also employs the demod-remod concept discussed previously in Section 4.4, and as a digital link, benefits from the improved performance achieved by avoiding the retransmission of uplink noise.

Baseband processing has benefited from the rapid advances in solid-state digital decoding and routing hardware, which provide significant advantages in satellite complexity and processing speed. The insertion of digital hardware reduces the number of analog components, whose internal capacitances tend to limit processing bandwidths and switching times, and are more susceptible to component aging, temperature variations, and element failures over time. Digital circuitry can be made to perform faster decoding operations with more accuracy and, when designed with solid-state circuitry, yield advantages in overall satellite weight and power. In the following sections we consider some of this digital hardware.

9.2 BASEBAND DIGITAL DECODING

On-board digital decoding on satellites is based on performing accurate analog-to-digital (A-D) conversion. Here satellite waveforms are converted to digital sequences, with all subsequent processing accomplished

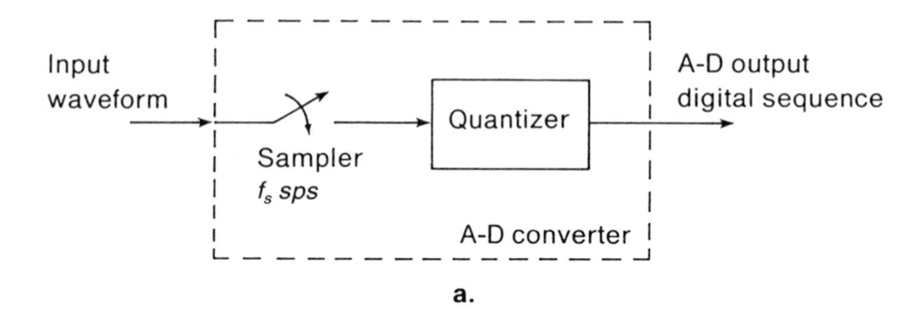

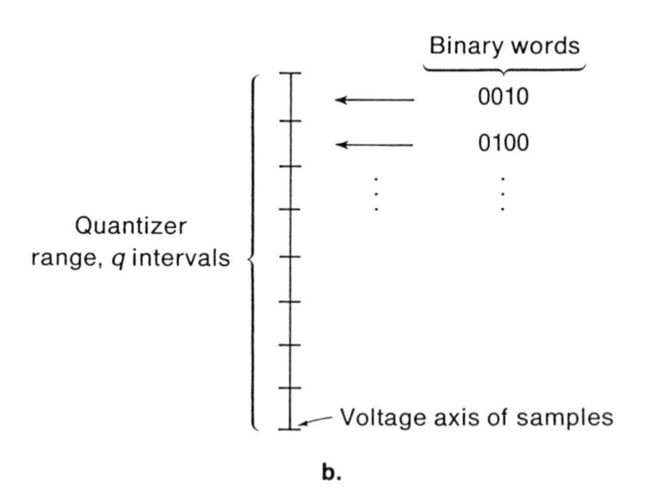

FIGURE 9.2 A-D conversion. (a) Block diagram. (b) Quantizer.

entirely in software. The basic A-D converter is shown in Figure 9.2a, and consists of a waveform sampler followed by a sample quantizer. By sampling the input waveform periodically, we obtain a continued sequence of samples, each of which is mapped, or quantized, into a digital word. The sequence of these digital words then represents the A-D output. Quantization is achieved by dividing the signal sample range into preselected intervals, or slices (Figure 9.2b), with each such interval designated by a binary word. All sample values occurring in a specific interval are quantized to that word. The input waveform is therefore "digitized" into the binary word sequence for signal processing.

The selection of the binary word for each quantizer interval is arbitrary, but clearly a distinct word must be available for each interval. If there are q intervals, then $\log_2 q$ (or the next largest integer) bits are needed to represent all intervals uniquely. Thus, $\log_2 q$ bits are produced at each sample time, and if the sampler operates at f_s samples per second, the A-D subsystem generates bits* at the rate

$$R_{\text{AD}} = f_s \log_2 q \quad \text{bits/s} \tag{9.2.1}$$

The basic type of quantizer is the *uniform* quantizer in which all intervals are preselected to have equal width over a prescribed range. If the sample value exceeds the quantizer range, the quantizer is said to be *saturated.*

All the analog decoding operations discussed in Chapter 2 can be carried out with digital signal processing using A-D conversion. Figure 9.3 shows the block diagram of several analog decoders, and their equivalent A-D implementations. While most of the high-frequency mixing is done with standard oscillators and mixers, the significant portion of the baseband processing can be done in software. In each case the baseband signals are first A-D converted, then combined in the accumula-or or digital filter to produce the data bit variable used for decoding, just as in analog decoding.

The design of the digital processor must be related to both the data rate and the data waveforms being decoded. In particular:

1. The A-D sampling rate f_s must be at least equal to the data bit rate in order to ensure at least one sample per data bit.
2. The quantizer range (Figure 9.2b) must cover the maximum range of the input signal sample to prevent limiting the baseband data by the A-D conversion. If the decoding carrier power is P_r, then the baseband signals will have amplitude $\pm\sqrt{2P_r}$, and the quantizer range should be at least this wide. When the input carrier has a large dynamic range, the quantizer range must be aligned with the maximum expected value of the carrier power P_r.
3. The quantizer intervals must be smaller than the input-noise standard deviation. The noise variance is given by the input noise power P_n in the decoder input filter bandwidth. If the quantizer has q intervals, the interval widths are $2\sqrt{2P_r}/q$. Therefore we require

$$\frac{2\sqrt{2P_r}}{q} \leq \sqrt{P_n} \tag{9.2.2}$$

* The reader must carefully distinguish here between the A-D output bit rate and the data bit rate of the input waveform being processed.

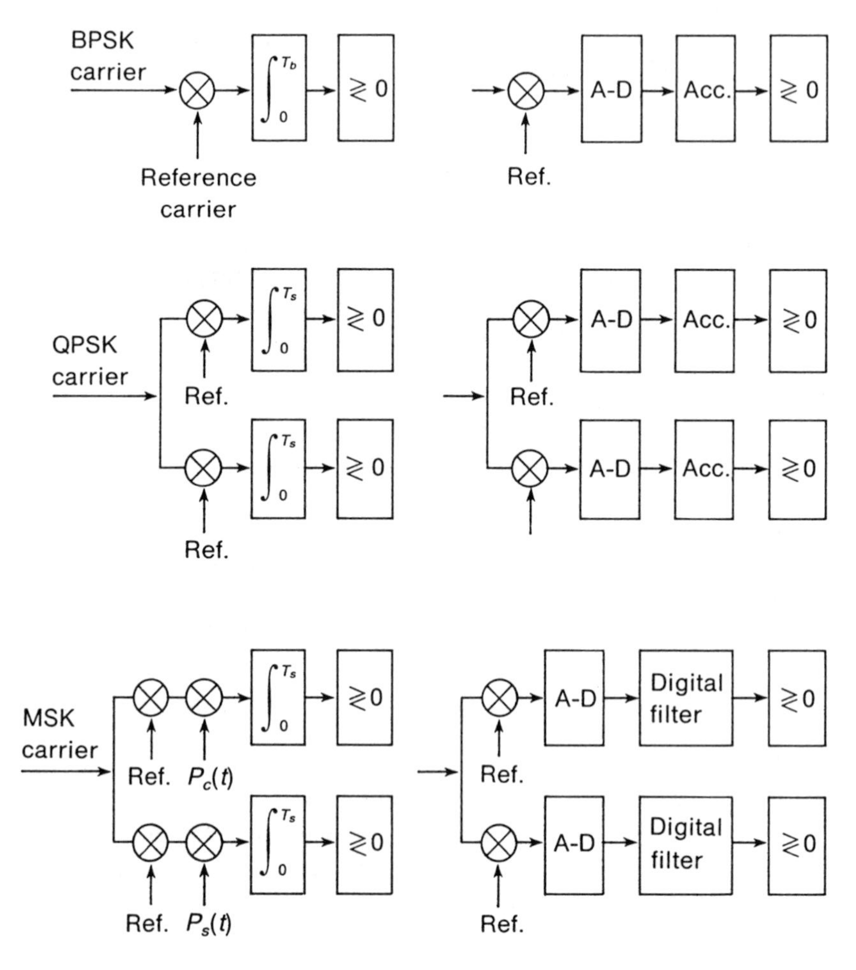

FIGURE 9.3 Analog decoders and equivalent digital decoders.

or

$$q^2 \geq 8\left(\frac{P_r}{P_n}\right) \tag{9.2.3}$$

Hence, the quantizer design must be related to the maximum decoder input CNR. We immediately see a direct trade-off occurring between the A-D parameters, controlled by the input data rate and CNR, and the processing A-D rate in Eq. (9.2.1). Since software algorithms are limited in the A-D output rate that can be processed in real time, digital decoders will likewise be limited in the data rate and dynamic range that can be handled.

Baseband decoding can be improved by using more samples per data bit, and inserting digital accumulation (integration) prior to data bit decisioning. If the decoder input noise bandwidth is B_c, then time samples of this noise process spaced $1/2B_c$ apart will be uncorrelated. This means there will be

$$m = \frac{T_b}{1/2B_c} = 2B_c T_b \tag{9.2.4}$$

independent time samples per data bit. If the A-D filter is properly designed, the decoding squared mean to noise ratio after accumulation will be

$$\begin{aligned} \Upsilon &= m\left(\frac{P_r}{P_n}\right) \\ &= 2B_c T_b\left(\frac{P_r}{N_0 B_c}\right) \\ &= 2\left(\frac{P_r}{N_0 R_b}\right) \end{aligned} \tag{9.2.5}$$

provided the A-D does not limit the baseband data waveform. The resulting BPSK bit-error probability is then

$$\text{PE} = Q(\sqrt{\Upsilon}) \tag{9.2.6}$$

which corresponds to the analog result in Chapter 2. In order to achieve this PE, the digital filter must process at the A-D rate

$$R_{\text{AD}} = mR_b \log_2 q \tag{9.2.7}$$

with q satisfying Eq. (9.2.3). The A-D rate can therefore limit the data bit rate that can be decoded, and the latter must be traded against the Υ value via the parameter m.

If the A-D converter limits the baseband data signal (i.e., operates in saturation) the resulting Υ in Eq. (9.2.5) will be reduced. This is because the saturation prevents the noise averaging during the m sample accumulation, and produces a noise bias that accumulates with m. The resulting

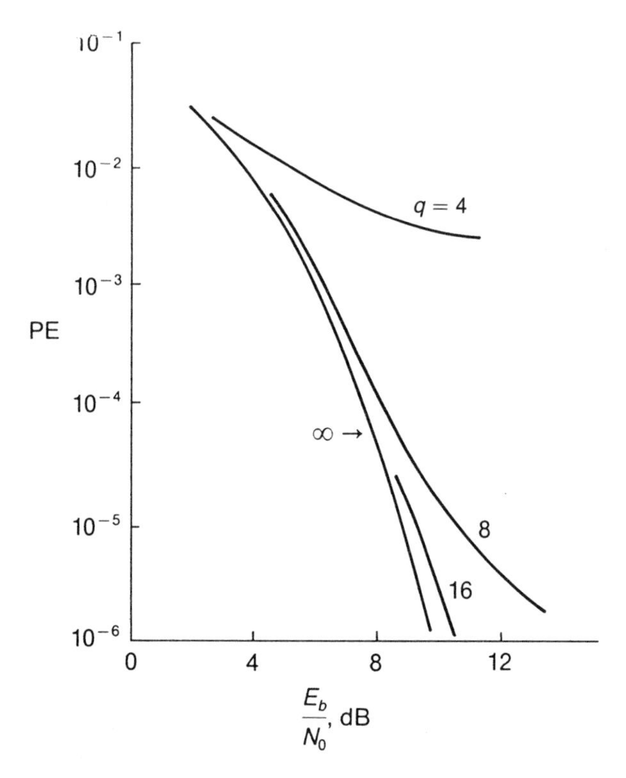

FIGURE 9.4 PE degradation due to A-D saturation in BPSK decoding. (q = number of quantization intervals.).

Υ approaches the limit $\Upsilon \rightarrow q^2/2$ at high P_s/P_n. Figure 9.4 shows this saturation effect in a BPSK system when A-D is used during decoding. The degradation in performance is due to the insufficient range of the quantizer as the input signal power is increased for each value of q selected. This converts directly to a decoder implementation loss (power loss) that must be accounted for in any link budgeting.

Bit timing for A-D processing can be obtained directly from the quantized time samples used for the bit decoding. The timing is accomplished by a digital version of the bit timing subsystems discussed in Appendix B.5. Such devices are called *digital transition tracking loops* (DTTL) (Lindsey and Anderson, 1968; Simon, 1970). The block diagram of a DTTL is shown in Figure 9.5a, and its operation can be explained by the time line in 9.5b. The baseband data bit waveform is A-D converted, with samples at the midbit point and at the bit edge used for the timing. These samples are processed in the DTTL software to generate a timing-error voltage e_i. If successive midbit samples are pairwise subtracted, they will indicate whether a bit transition has occurred, as well as the direction of the transition. The edge samples (even samples in Figure 9.5b) will be zero if exactly at a bit edge crossover, but will generate positive or negative voltages if it is offset in one direction or the other, or if no transition has occurred. This latter information is provided by the midbit samples, and the difference value is used to multiply the edge sample value to achieve the correct voltage sign.

The multiplication is done in software, usually using three-state logic $(+1, -1, 0)$, so the error sample is generated as

$$e_i = \text{sign}[(a_i - a_{i+2})]\text{sign}[a_{i+1}], \qquad i = 1, 3, 5, \ldots \qquad (9.2.8)$$

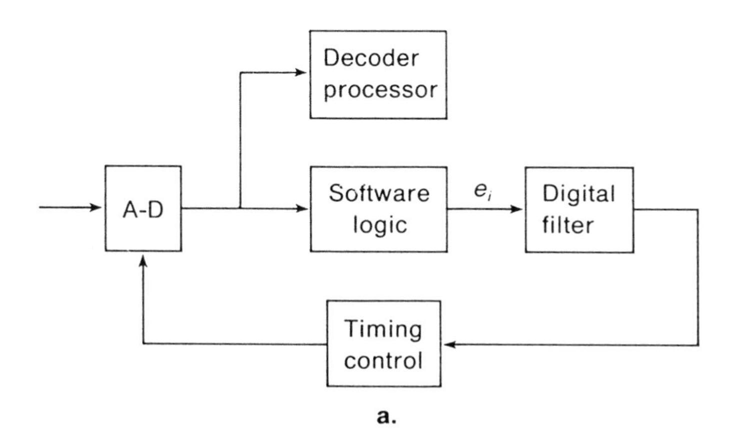

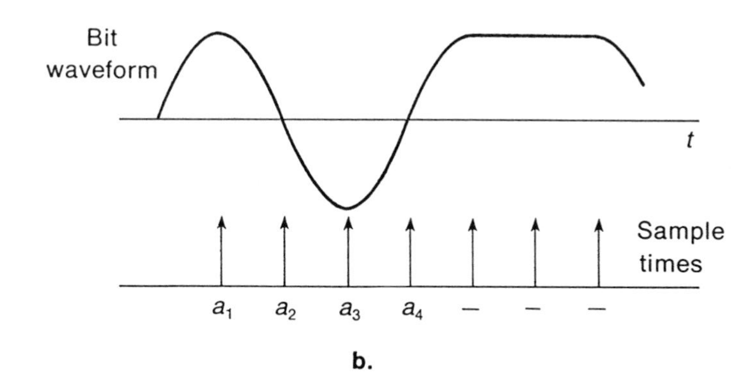

FIGURE 9.5 The digital transition tracking loop (DTTL). (a) Block diagram. (b) Time line.

where $\{a_i\}$ are the samples in Figure 9.5b. The error samples are then digitally accumulated and the result is used to step the clocking of the sampler. When in synchronization, the edge samples will be at bit crossover, and zero correction voltages are generated. Bit timing can then be extracted from all odd (or all even) sample times. As the timing slides off, the voltages in Eq. (9.2.8) will be collected and the clock stepped in the proper direction to pull back into synchronization.

A DTTL has the advantage of high-speed digital processing and being chip implementable. At high SNR its operation is identical to analog bit timing, but performance degrades rapidly at low SNR. Because of its discrete nature, acquisition and pull-in times may be slow, and rigorous analysis requires state transition formulations (Holmes, 1982; Lindsey and Anderson, 1968; Simon, 1970). Baseband filtering prior to the A-D conversion is important for shaping properly the bit waveforms in Figure 9.5b for linear zero crossing response.

9.3 DATA RECLOCKING, ROUTING, AND MULTIPLEXING

With the digital decoding of the previous section, decoded data bits are regenerated with the timing of the DTTL timing subsystem. Since this bit clock tracks the data bit edges, its timing follows the timing of the uplink bits. Thus, any bit timing variations in the received data, owing to uplink oscillator jitter, Doppler variations, etc. (along with time variations caused by DTTL noise), will be transferred directly to the decoded bit stream.

To remove these timing variations for subsequent data processing, the decoded bit stream is often reclocked by a local, stabilized clock. A typical reclocking system is shown in Figure 9.6. Midbit sampling is timed by a local bit clock, which also generates a clock pulse each bit time to present the new bit waveforms. The input digital waveform is sampled, and the bit polarity is used to multiply the clock pulse, resulting in a pulse digital waveform at the output having the bits sequence of the input. If the decoded digital waveform contains clock jitter as shown in Figure 9.6b (i.e., the bit periods are not all exactly T_b s long), the reclocking transfers the bits to the more stable local clock. Hence, reclocking "cleans up" bit jitter, as long as the jitter does not exceed half a bit period.

Reclocking, however, cannot compensate for continual clock drifting. If the input bit clock tends to drift in frequency relative to the local bit clock, the input bit periods will continually decrease or increase relative to the local clock bit periods. The midbit sampling may eventually produce two samples from the same bit, or may skip over a bit. To compensate for this, an *elastic buffer* (Williams, 1987) circuit is inserted,

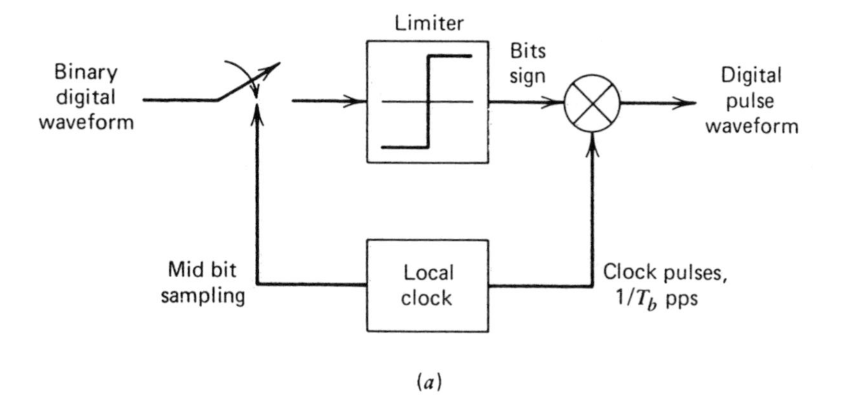

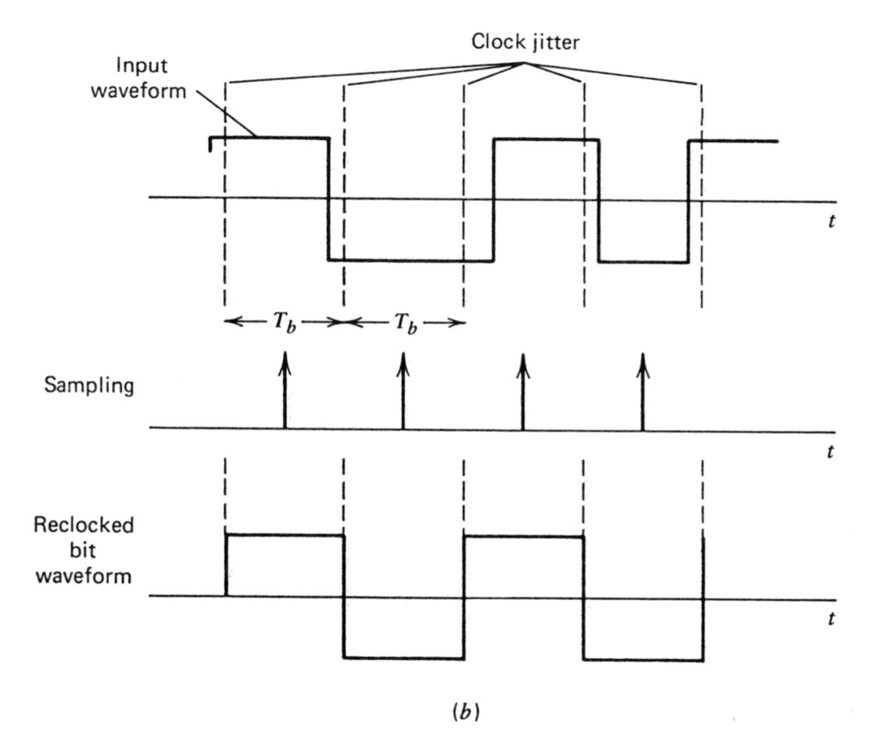

FIGURE 9.6 Reclocking circuit. (a) Block diagram. (b) Waveform diagram.

in addition to the reclocking, to stabilize the frequency drift. An elastic buffer operates as shown in Figure 9.7a. The input bits at rate r_{in} are sampled and clocked out at rate r_o by the local clock. The clock rates are monitored via a set/reset switch in which the input bit edge starts a pulse generator and the local clock samples stop the pulse. If perfect timing is maintained ($r_{\text{in}} = r_o$), the generated pulses are always of the same width. As the source clock drifts relative to the buffer output clock, the pulse width is altered accordingly (Figure 9.7b). A short-term integrating filter (effectively measuring the short-term dc value of the pulses) will indicate the changing pulse width. When the pulse width becomes too narrow or

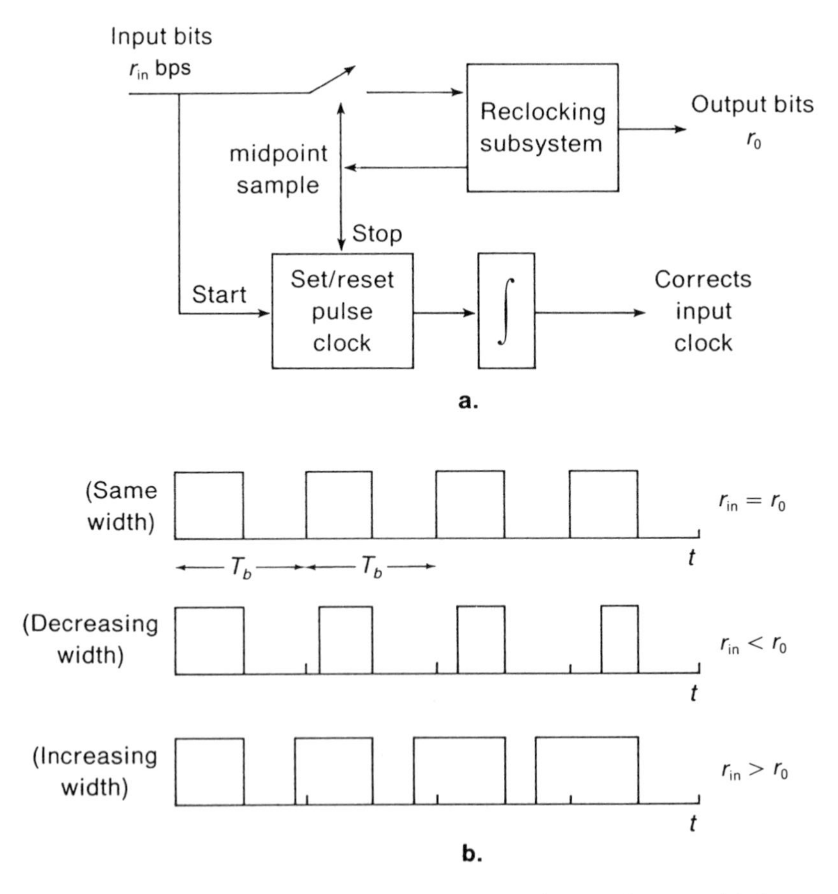

FIGURE 9.7 Elastic buffer for monitoring input bit rate. (a) System diagram. (b) Pulse width waveforms.

too wide, the input clock can be corrected. If the source clock is not accessible, the buffer can insert extra bits (*bit stuffing*) or delete bits to maintain the proper timing.

The presence of a stable decoded bit stream at the satellite can now be used for subsequent processing. It can be directly remodulated for downlink transmission, or it can be multiplexed with other decoded bit streams to form a common time division multiplexed (TDM) bit stream. This multiplexing is achieved by a *parallel–serial converter* (PSC) circuit, using the implementations shown in Figure 9.8. When all bit streams produce bits at the same rate, we need only commutate over the bit set with a high-speed rotating switch, reading out one bit at a time from each bit stream in sequence and continually recycling (Figure 9.8a). When the rates are unequal, however, a slightly more complicated system must be implemented, making use of the digital shift registers in Figure 7.3a. Input bits to the register can be made to shift through the register at a prescribed clock rate. A PSC is constructed by using unequal shift registers placed

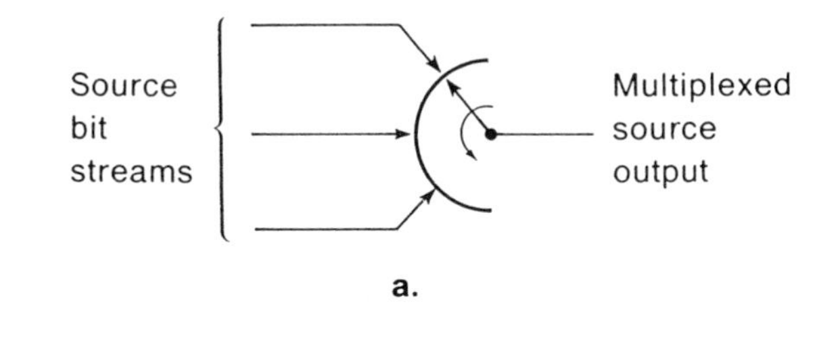

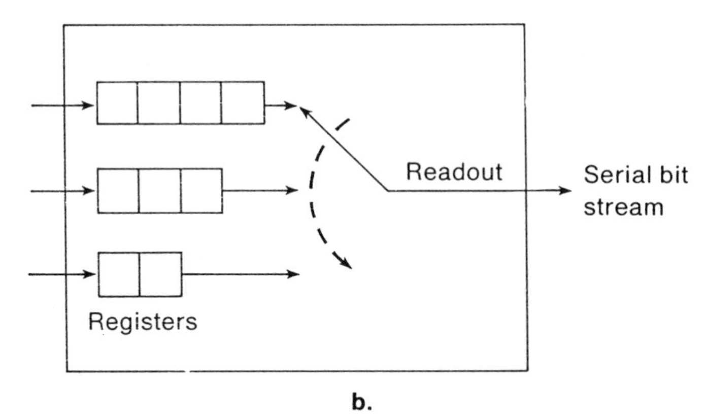

FIGURE 9.8 Parallel-serial converter (PSC). (a) Equal input bit rates. (b) Unequal input bit rates with shift registers.

in parallel with a single communication switch (Figure 9.8b). A register must be available for each different bit stream being multiplexed. The different bit streams are simultaneously loaded into the appropriate registers and the entire register contents are read out in sequence at fixed time intervals to accomplish the multiplexing.

Assume that there are N digital sequences, having rates R_i bits/s, $i = 1, 2, \ldots, N$, fed into a specific register of a PSC at its own rate. The register lengths (number of register stages) must be selected such that after some interval in time all the registers will fill simultaneously. If h_i is the register length of the ith such register of a particular PSC, this loading condition requires that a time τ exists such that the number of bits of rate R_i occurring in τ s is exactly h_i. Hence, τ must be such that

$$R_i \tau = h_i \tag{9.3.1}$$

for all inputs to the PSC. This immediately implies that for any two input rates, say, R_i and R_j, the corresponding register lengths must be related by

$$\frac{h_i}{h_j} = \frac{R_i}{R_j} \tag{9.3.2}$$

For smallest register lengths (simplest PSC), it is necessary that the h_i have the smallest possible integer value. Thus, each h_i should be the smallest factor of the set of input rates (R_i). Equivalently, the set of input rates should be related to the set of register lengths by

$$R_i = \alpha h_i \tag{9.3.3}$$

where α is the largest common multiple of all rates. We see that a set of PSC input rates specifies the required register lengths, whereas a set of registers specifies the ratios of input rates. The minimal register size will occur if the h_i are the smallest prime factors, and the rates R_i are selected to be multiples of these prime numbers. As an example, suppose that we wish to multiplex three digital sources having the rates $R_1 = 72$ bps, $R_2 = 48$ bps, and $R_3 = 60$ bps. The smallest integer factors of these rates are then $h_1 = 6$, $h_2 = 4$, and $h_3 = 5$, respectively, and the common factor is 12 (i.e., $R_i = 12h_i$). A PSC as in Figure 9.8b would therefore require three parallel registers with six, four, and five stages, respectively.

After the registers of the PSC are filled, the bits can be read out from each register in sequence at a constant readout rate. That is, the commutating switch in Figure 9.8b moves to the first register, reads out all stored bits in sequence, then moves to the second register and reads out all bits, and so on. One complete cycle of register readout must occur in τ s, the time it takes to fill each, with each register refilling immediately after emptying. Hence, the output serial bit rate of the PSC is

$$R_o = \frac{1}{\tau} \sum_{i=1}^{N} h_i = \sum_{i=1}^{N} R_i \text{ bits/s} \tag{9.3.4}$$

The multiplexed output rate is therefore always equal to the sum of the individual source rates.

Whenever several data sources are multiplexed, it is generally necessary to insert synchronization bits periodically to aid in the receiver demultiplexing. If K_s synchronization bits are inserted after every K_b data bits at the output of each PSC, then the data rate of the PSC output bit stream is increased by the factor

$$R_s = \frac{K_s}{K_b/R_o} = R_o\left(\frac{K_s}{K_b}\right) \tag{9.3.5}$$

The total effective PSC multiplexed output bit rate is then

$$R_o + R_s = R_o\left(1 + \frac{K_b}{K_s}\right) \tag{9.3.6}$$

Thus, the total multiplexed bit rate is not increased significantly if the synchronization bits are a small fraction of K_b frame bits. The required number of synchronization bits depends on the accuracy to which this synchronization word must be recovered at the receiver. The exact time of arrival of the synchronization word at the receiver is needed to demultiplex properly the bit stream and synchronize the data frame. This is referred to as *frame* synchronization.

Frame synchronization is obtained at the receiver by having the receiver "look" in each data frame for the synchronization word, just as in unique word detection in Section 6.3. It does this by having the receiver baseband processor use a digital word correlator that stores the known synchronization word.

As the recovered TDM bits are shifted through the register bit by bit, a word correlation is made with the stored word. When the transmitted word is received, its correlation will produce a large signal output that can be noted as a threshold crossing. This crossing can be used to generate a time marker that marks all subsequent data words of the TDM frame. During the next frame, the synchronization word is reinserted, the word

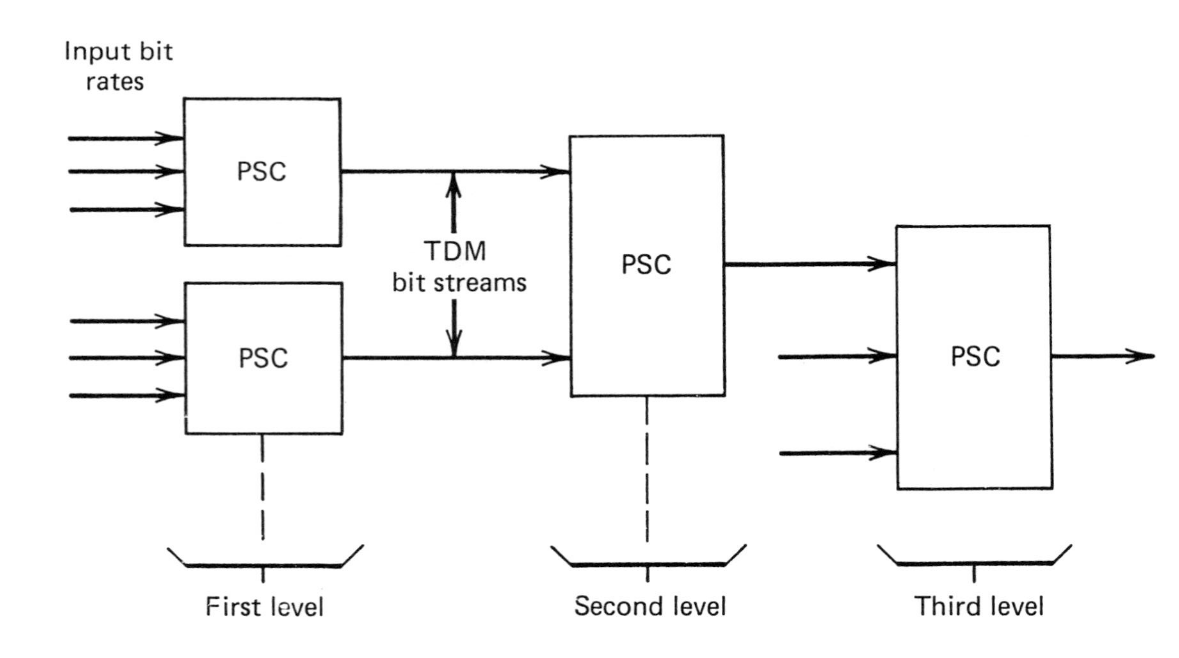

FIGURE 9.9 Higher level multiplexing with PSC.

is detected, and the word timing is again established for the new frame.

Deinterleaving of the multiplexed bit stream is accomplished by a commutator operating in synchronism with the arriving bit stream. The synchronized commutator taps off the group of bits of each source in sequence, feeding them to the proper destination. The synchronization bits, inserted at the PSC, maintain the required commutation synchronism for this deinterleaving.

The output bit stream of a single PSC can be combined with other PSC outputs to produce an entire multiplex hierarchy, as shown in Figure 9.9. The PSC outputs due to multiplexing source bit streams at level 1 are similarly combined to form a bit stream at level 2, which, in turn, can be further combined at level 3, and so on. The result is a structured multiplexing format in which a large number of digital sources can be combined into a single bit stream.

Standardized PSC hardware has been established for this purpose. For example, a PSC has been developed for operating on 24 digital channels, each operating at 64 kbps (voice A-D systems). This produces a multiplexed bit rate of 24×64 kbps $= 1.536$ Mbps. With eight synchronization bits inserted every millisecond, a resulting multiplexed bit rate of 1.536 Mbps $+ 8$ kbps $= 1.544$ Mbps is created. This is referred to as a standard *T_1 digital multiplexer*, and is commonly used in digital subscriber service. Higher-order multiplexers have also been standardized at the rates 6.312 Mbps (T_2 multiplexing, for video telephone), 44.736 Mbps (T_3 multiplexing, for 600 voice channels), and 274.176 Mbps (T_4 multiplexing, for 3600 voice channels).

Unique word detection can also be used to recognize address words embedded in the individual bit streams, with the correlation pulse used to route data via bit switching. The switched bits can then be routed to a specific memory buffer cell for latter transmission to a specific destination. This permits an entire complex of parallel decoded data at the satellite to be separated into specified route directions, as shown in Figure 9.10. The PSC generates the TDM data with the embedded routing addresses. The parallel memory banks recognize the addresses, reading the data into memory or buffer registers. The registers are then read-out, or "dumped," periodically to route the data in the proper direction.

9.4 TDM–FDM CONVERSION

Data routing, as discussed in the previous section, is a basic part of modern on-board processing. In Figure 9.10 it was shown how data bits from parallel channels could be placed in a sequential (TDM) format,

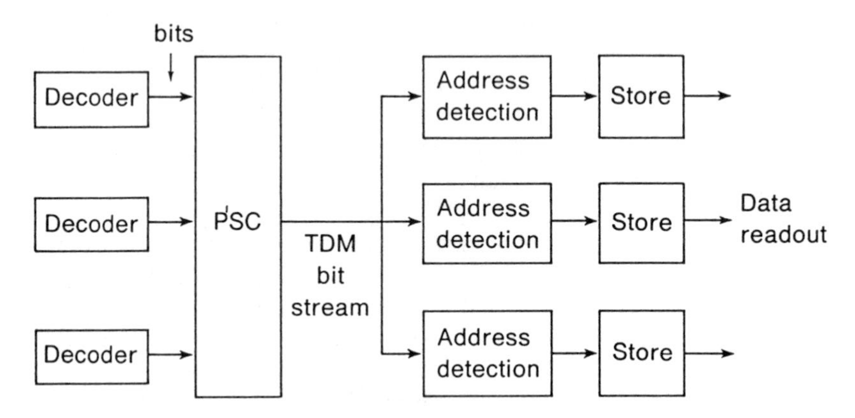

FIGURE 9.10 Data routing via parallel-series conversion and address recognition.

then later could be separated out into parallel memory banks by address recognition. Another closely related operation in parallel–serial on-board processing is the conversion between an FDM format and a TDM format. This permits parallel, frequency-separated channels to be serially multiplexed, or vice versa. This FDM/TDM conversion is rapidly becoming a necessary and important operation in processing satellites.

Direct FDM/TDM conversion can be done, of course, with parallel oscillator banks and commutating switches, as shown in Figure 9.11. Such standard procedures, however, require an excessive number of duplicate components and would use primarily analog hardware. These disadvantages have led to the development of "bulk" digital processors that perform the FDM–TDM conversions directly with digital hardware. Such devices are referred to as *transmultiplexers* (Bellanger and Daguet, 1974; Takahuta et al., 1978).

The basic operation of a digital TDM-to-FDM transmultiplexer can be described from Figure 9.12. As shown in Figure 9.12a a waveform of bandwidth B_m Hz is sampled at the rate $f_s = 2B_m$ samples/s to produce the sampled spectrum shown. These samples are then quantized and processed in the digital filter such that each input sample is converted to N output samples, with the digital filtering designed to have the frequency transfer function shown. As a result, the output sequence occurs at a rate Nf_s samples/s, and corresponds to a sampled version of the same source at this higher rate.

Now repeat for a second waveform with same bandwidth, but use a digital filter with response shown in Figure 9.12b. This filter can be

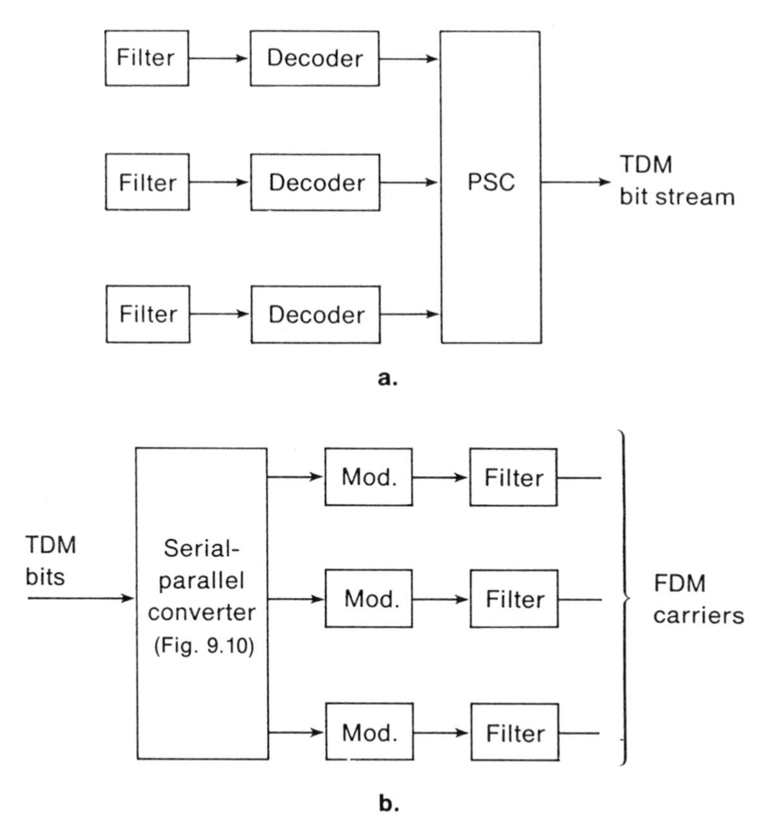

FIGURE 9.11 Direct conversion. (a) FDM to TDM. (b) TDM to FDM.

obtained from that in Figure 9.12a by inserting an additional frequency shift (which translates to an appropriate phase shift in the digital domain). We now combine the two systems by interlacing the input samples and directly summing the output samples, producing a resulting output sequence with the combined spectrum in Figure 9.12c. The latter has separated the individual source information, converting source A into bands A and source B into bands B. Note that the input baseband sampling for each source is physically done at the minimal possible rate f_s, but the output sequence is created at the higher rate. Thus, the frequency translation is accomplished entirely in software, instead of by analog oscillators. A bandpass filter located along the frequency axis at any multiple of Nf_s can then be used to generate an analog FDM waveform from the FDM sample sequence.

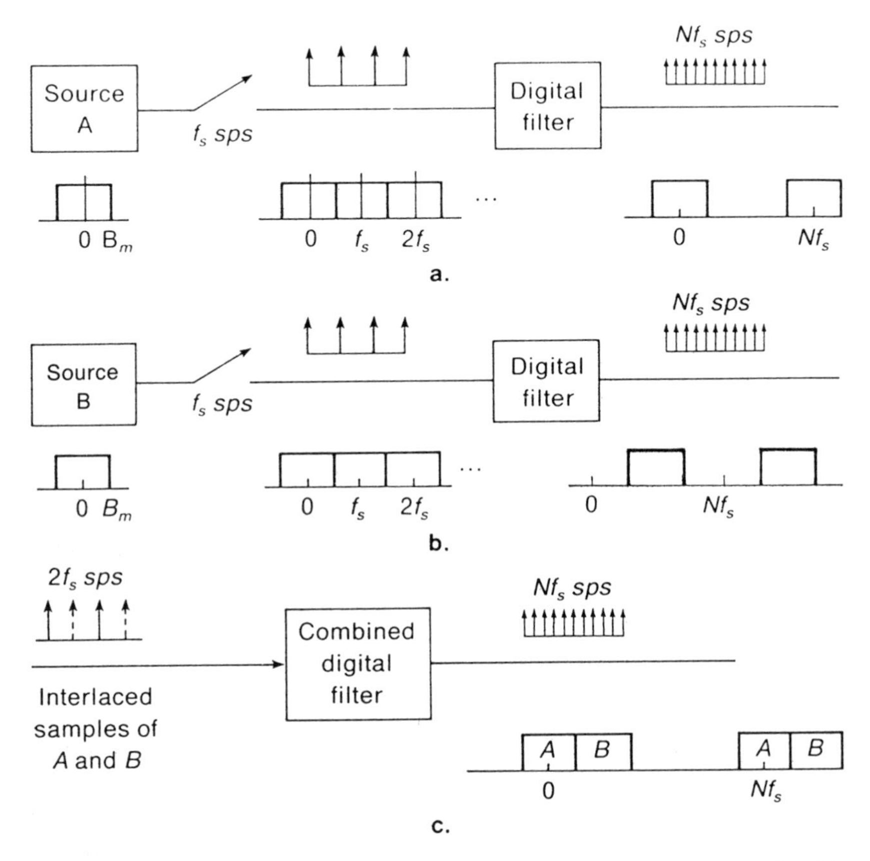

FIGURE 9.12 Digital TDM-FDM converter. (a) For source A. (b) For source B. (c) Combined for A and B. Interlaced samples converted to FDM output samples.

The concept can now be extended to M waveforms, each sampled at rate f_s and having their samples interleaved to produce the TDM input format at rate Mf_s. The subsequent digital filtering bank produces the output sample sum sequence with the FDM repeated spectrum. A band-pass filter located at any Nf_s band along the frequency axis then converts the output sequence to a basic FDM waveform with each source information occupying a disjoint frequency band.

The parallel bank of digital filters in Figure 9.12 can be reduced to a single bulk filter by taking account of the redundancy in filter shapes. This conversion was shown (Maruto and Tomozowa, 1978) to produce the equivalent processor shown in Figure 9.13. The latter is composed of

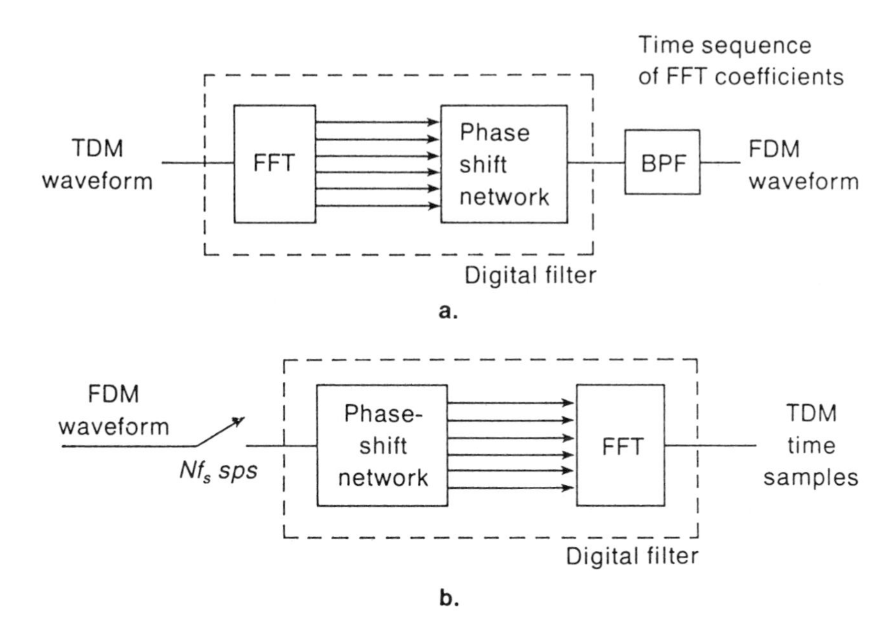

FIGURE 9.13 Transmultiplexors. (a) TDM to FDM. (b) FDM to TDM.

a fast Fourier transformer (FFT) that converts each TDM frame of N input samples into an output sequence of Fourier coefficients, followed by a phase shifting network that produces the equivalent output sequence to that generated in Figure 9.12. The well-known technology for digital FFT processing has made it possible to package bulk transmultiplexers into standard solid-state hardware for satellite applications.

The system can be reversed to permit FDM-to-TDM transmultiplexing. The latter system is shown in Figure 9.13b. The input is the FDM bandpass waveform, which is undersampled at the rate $2NB_m$, where B_m is the individual channel bandwidth. This produces the sampled version of the input FDM spectra. The samples are phase shifted and processed in a FFT, which computes sequentially in time the N Fourier coefficients of each group of N input samples. The resulting FFT output sequence is then a TDM sequence in which every Nth sample is the sample value of the baseband information associated with one particular input FDM frequency band. Thus, a direct conversion of the FDM input to TDM sample sequence is produced by the transmultiplexer. It should be pointed out that analog transmultiplexers can also be designed using SAW filters and chirp carrier multiplication (Hays and Hartmann, 1976).

9.5 ON-BOARD REMODULATION

On-board data decoding and routing will require remodulation at the satellite prior to retransmission. Spaceborne digital modulation invariably produces degraded carrier signal quality, owing to the variations in temperature, vibrations, and aging under which the modulator must operate. If the modulation is first done at lower frequencies (MHz) then upconverted to the higher transmission bands (GHz), the modulation bandwidth and data rates will be limited severely. This might preclude, for example, the retransmission of high-rate TDM data from the satellite. If the modulation encoding is accomplished directly at the transmission frequency, higher data rates can be achieved, but the resulting carrier waveform is often nonperfect. This leads to degraded decoding performance at the receiver.

To assess the expected degradation due to nonperfect spacecraft bit modulation, it is first necessary to identify the prime waveform distortions, then evaluate their effect on the eventual receiver decoding. Consider the standard BPSK carrier in Section 2.3, having the form

$$c(t) = A \cos[\omega_c t + \theta(t) + \psi] \tag{9.5.1}$$

where $\theta(t)$ accounts for the data phase modulation. Ideally, $\theta(t)$ shifts between 0 and π rad every T_b s according to the data bits, and the receiver decodes with the bit energy $E_b = A^2 T_b/2$. In modern modulation hardware in which gigahertz carriers are directly phase shifted at high bit rates, the resulting BPSK carrier is often generated with the following specific anomalies:

- Amplitude imbalance. The amplitude A in Eq. (9.5.1) may be different for a one bit than for a zero bit.
- Phase imbalance. The phase switching will occur between phase angles that are not exactly π rad apart.
- Transient phase time. The BPSK phase shifting will have a nonzero transient time while switching between a zero and a one bit.
- Bit time imbalance. The bit intervals for a one bit and zero bit are different.

The combined effect is to produce a phase modulation function that only approximates the ideal, as shown, for example, in Figure 9.14a. The result of ideal phase-coherent decoding, as in Figure 9.14b, is to produce during any bit a decoding bit energy of

$$E_b = \frac{1}{2T_b}\left[\int_0^{T_b} A(t) \cos[\theta(t)]\, dt\right]^2 \tag{9.5.2}$$

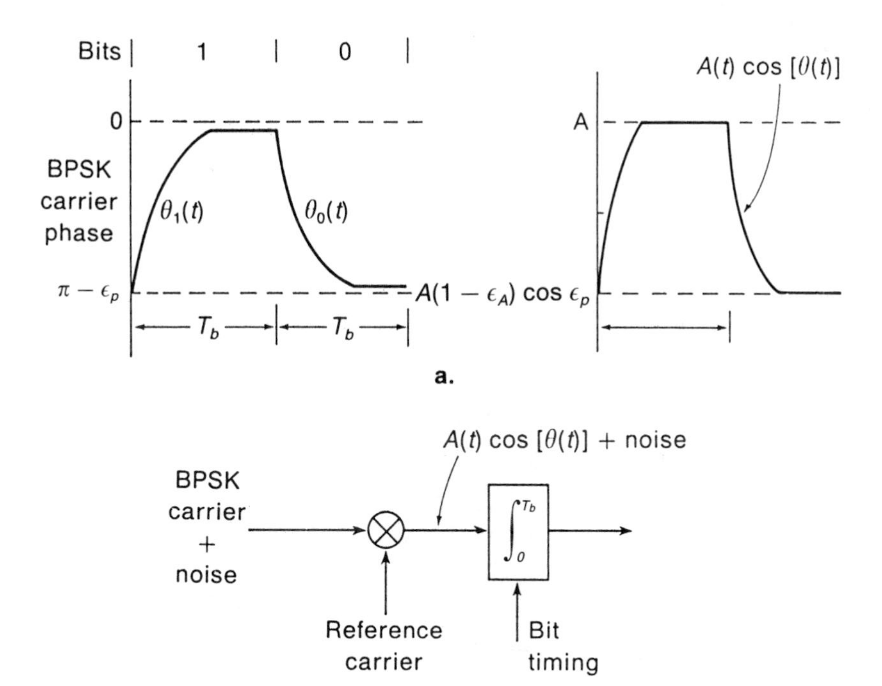

FIGURE 9.14 Nonperfect BPSK carrier modulation. (a) Carrier phase and baseband waveforms. (b) BPSK decoder.

where T_b is the bit time. The form of $A(t)$ and $\theta(t)$ will be dependent on the particular bit being decoded and on the previous bit. Referring to Figure 9.14a and Eq. (9.5.2) we can write

$$E_b = \begin{cases} \dfrac{A^2}{2T_b}\left[\displaystyle\int_0^{T_b} \cos[\theta_1(t)]\,dt\right]^2, & 0 \to 1 \\ \dfrac{A^2 T_b}{2}, & 1 \to 1 \\ \dfrac{A^2 T_b}{2}\left[(1-\epsilon_A)\cos(\pi - \epsilon_p)\right]^2, & 0 \to 0 \\ \dfrac{A^2(1-\epsilon_A)^2}{2T_b}\left[\displaystyle\int_0^{T_b} \cos[\theta_0(t)]\,dt\right]^2, & 1 \to 0 \end{cases} \tag{9.5.3}$$

where ϵ_A and ϵ_p are the fractional amplitude and phase imbalances. The resulting bit-error probability is determined by averaging over these bit

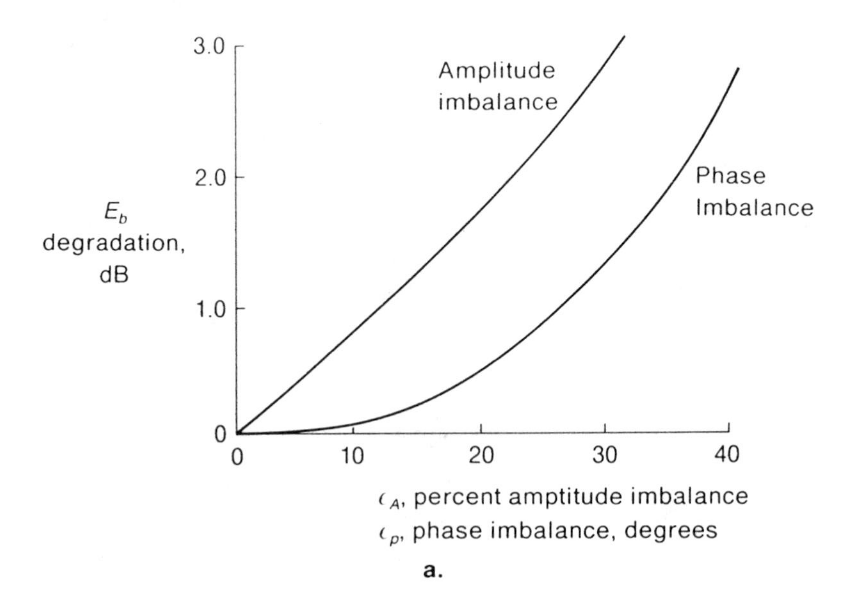

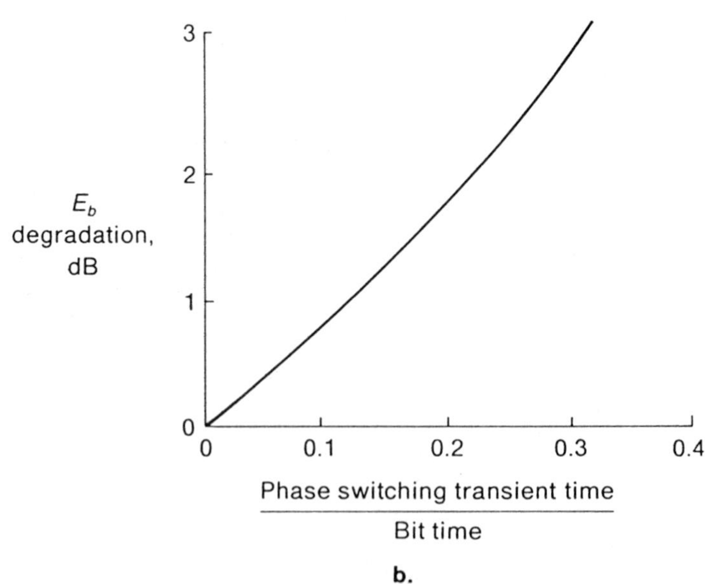

FIGURE 9.15 Decoding energy degradation due to imperfect BPSK. (a) Amplitude and phase imbalance. (b) Phase transient degradation.

possibilities. Hence, the ideal BPSK PE in Table 2.4 is modified to

$$\begin{aligned}
\mathrm{PE} = {} & \tfrac{1}{4}Q\left[\left(\frac{A^2 T_b}{N_0}\right)^{1/2}\left[\frac{1}{T_b}\int_0^{T_b}\cos[\theta_1(t)]\,dt\right]\right] + \tfrac{1}{4}Q\left[\left(\frac{A^2 T_b}{N_0}\right)^{1/2}\right] \\
& + \tfrac{1}{4}Q\left[\left(\frac{A^2 T_b}{N_0}\right)^{1/2}[(1-\epsilon_A)(\cos\epsilon_p)]\right] \\
& + \tfrac{1}{4}Q\left[\left(\frac{A^2 T_b}{N_0}\right)^{1/2}(1-\epsilon_A)\frac{1}{T_b}\int_0^{T_b}\cos[\theta_0(t)]\,dt\right] \qquad (9.5.4)
\end{aligned}$$

To evaluate Eq. (9.5.4) and assess the performance degradation requires an accurate model of the waveform imbalances and the switching phase transients $\theta_1(t)$ and $\theta_0(t)$. Some reported studies (Ziemer and Peterson, 1985) have attempted to evaluate these effects. Figure 9.15 shows some estimated E_b degradations for individual imbalances. Figure 9.15a shows the effect of amplitude and phase imbalances, assuming no other imperfections. Figure 9.15b estimates the effect of the bit switching phase transient by assuming no decoding energy during the phase transient time between the bit changes. The exhibited degradations are somewhat similar in their behavior, and each separately inserts implementation losses to the overall link. Curves of this type permit operational specifications to be placed on space-qualified modulation hardware during design. The anomaly most difficult to account for is the bit asymmetry [not included in Eqs. (9.5.3) and (9.5.4)], since the bit integration over unequal bit times will involve the effect of both adjacent bits (preceding and following bit), and will yield a more complicated result than Eq. (9.5.4).

9.6 ON-BOARD BASEBAND PROCESSING WITH BEAM HOPPING

An area of particular interest is on-board baseband processing used in conjunction with hopping beams to provide a highly interconnected satellite network with wide area coverage. As discussed in Section 3.9, the use of hopping beams produce spot beam coverage without having to generate large numbers of multiple beams when wide coverage is needed. However, the desire to maintain connectivity between earth stations in different hops require that on-board processors be present to properly redistribute data from the uplink hops to the downlink hops.

A diagram of a hopping beam system is shown in Figure 9.16. The uplink beam hops throughout its pattern, collecting RF transmission at each hop from all transmitters in its beam. A similar hopping beam is

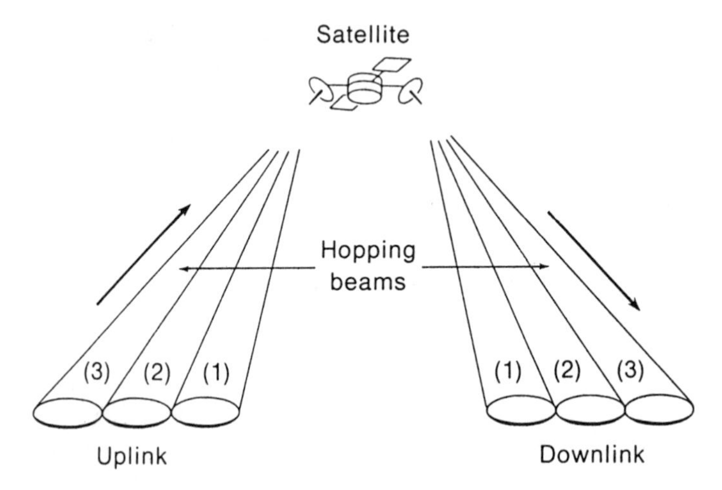

FIGURE 9.16 Hopping beam system.

also present for the downlink, and can be generated from a completely separate antenna, or can be provided by the same uplink beam (i.e., the same hopping beam provides an uplink and downlink frequency band at each hop position). Since an uplink station may wish to communicate with receiving stations in different hops, its uplink data must be addressed, and the satellite must provide the processing to deliver the data to the correct downlink hop. This forces the on-board processor to have both demod-remod and store-and-forward routing capability.

Spatial beam hopping can be achieved with mechanically steered antennas or with phased arrays using adjustable, digitally controlled, phase shift networks, which can cycle through a family of phase shifts to form each scanning beam. The procedure can be extended to commandable hopping in which beam location is selected by data words sent to the satellite to elect the proper antenna phasing during each hopping interval.

The modulation format within an uplink beam can be either FDMA or TDMA. With FDMA a parallel set of bandpass demodulators must operate at every hop to decode the data, tending to produce bulky and redundant satellite hardware. The preferred format is TDMA, in which uplink data is collected in serial slots, and a single demodulator can produce a frame of baseband data in each hop. As the uplink beam hops, it collects and decodes sequences of data frames, forming an uplink TDMA bit stream at the satellite. This bit stream must then be realigned for the downlink transmission by the baseband on-board processor.

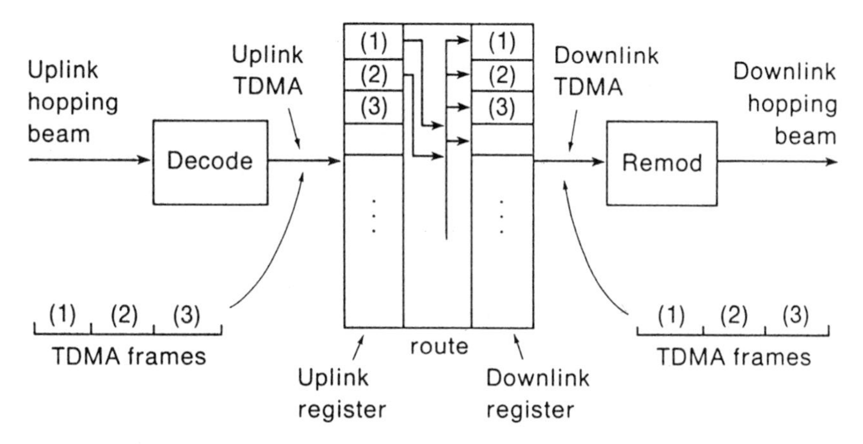

FIGURE 9.17 Satellite baseband processor for beam-hopping TDMA.

The TDMA on-board processor would appear as in Figure 9.17. Each uplink data frame from each hop is serially shifted into one stage of a parallel shift register, the next hop frame is placed into a second register, and so on. The data of each shift register is then clocked through digital routing subsystems with address recognition circuitry used to place the slots of the TDMA frame into a parallel set of output registers, using the serial-parallel conversion circuits considered in Section 9.3. As the uplink registers are clocked out and cleared, the next cycle of hops is used to refill the uplink register. The output registers are then read out one frame at a time producing the downlink TDMA bit stream, as shown. The slot data of each uplink frame is properly aligned with each downlink hop. As the downlink beam hops (either separately or in conjunction with the uplink hops), the data is distributed to the receivers in the proper hop footprint. Thus each uplink slot of TDMA data is repositioned in the proper downlink frame, and the on-board processing has converted an uplink TDMA bit stream to a downlink bit stream with relocated slots. This operation is referred to as *slot interchanging*. High speed slot-interchanging digital hardware is therefore an important element for this particular type of on-board processor.

9.7 MULTIPLE SPOT BEAMING

Closely associated with the present emphasis on on-board processing is the use of more antenna spot beaming from the satellite. The rapid development of phased-array antenna technology, using lightweight, accurate waveguide phase shifters in the feed assembly (recall Figure 3.7), has made it possible to produce easily multiple beams from a single antenna reflector with minimal cross effects. By using large enough reflectors, the resulting beams can each be made to produce small area spots in the downlink pattern. As was pointed out in Section 3.8, smaller spot sizes increase beam EIRP, reduce downlink interference, and provide better control of coverage contours. In addition, the various spot beams can be coupled with the ability to address and route downlink data on the satellite to provide complete interconnectivity of all uplink and downlink spots.

Another important advantage of multiple beaming is its ability to increase the information transmission capability of a band limited satellite through frequency reuse. If the downlink beams have no overlapping patterns, then the satellite can (theoretically) transmit the same frequency band over disjoint beams, each with independent data modulation, with no mutual interference. By dividing a desired coverage area into multiple spots, a limited frequency band can be reused to the extent by which the footprints do not overlap. By additionally inserting dual polarization within each beam, the amount of bandwidth reuse can be then doubled. Thus, a b beam downlink with bandwidth B_{RF}, can theoretically achieve a reuse factor of

$$\text{Reuse factor} = \frac{2bB_{\mathrm{RF}}}{B_{\mathrm{RF}}} = 2b \tag{9.7.1}$$

The difficulty is that complete separation of antenna patterns is impossible if an entire area must be covered with equal downlink EIRP. Consider the diagram in Figure 9.18a, showing a one-dimensional view of the received field intensity from two separate beams, plotted along the earth's surface. If the beams were widely separated, some receiving areas will be at low intensity levels. If the beams are spaced closer, so that no area will receive less than a 3-dB degradation from full intensity (Figure 9.18b), then clearly there could be significant interference from the adjacent beam if both used the same frequency band. It is therefore necessary in this case to use separate frequency bands in each beam so that receiver frequency filtering will reduce or eliminate the interference. A third beam (Figure 9.18c) can use the same band as beam 1, owing to the larger beam separation from the latter. As a result, adjacent antenna beams generally use separate frequency bands. Figure 9.19 shows how six elongated beams can be selected to span the continental United States, while reusing the same two frequency bands. Multibeaming would also permit adjustment

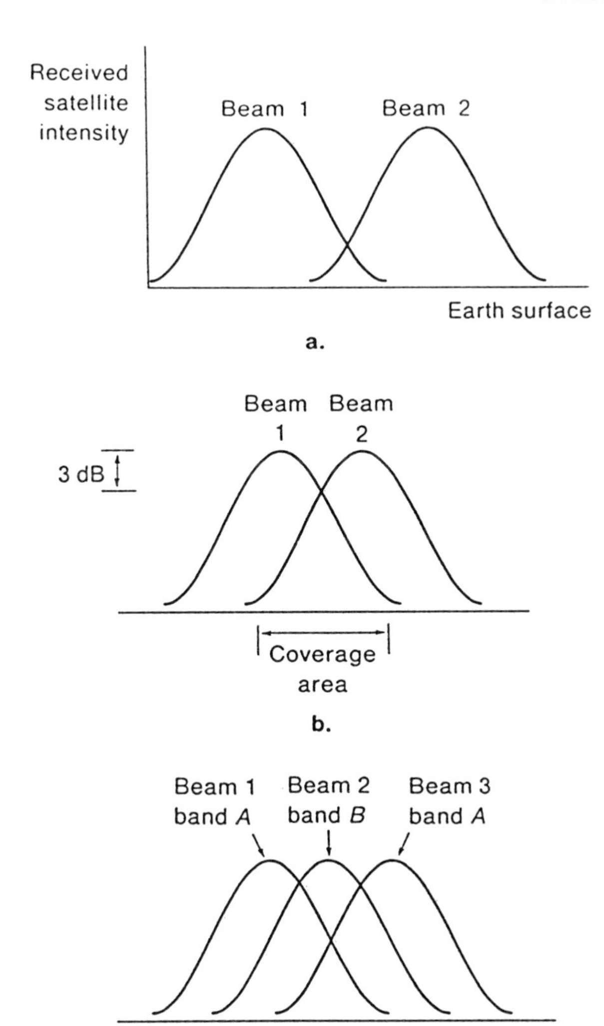

FIGURE 9.18 Multiple beam patterns. (a) Wide separation. (b) Closely spaced patterns. (c) Frequency band allocation.

of the bandwidths (data rates) over the coverage areas, with some spots transmitting higher rates to match the traffic flow for that area.

The trend in future satellites is toward increasing numbers of beams. The beam-location problem becomes more complicated, however, when Figure 9.19 is extended to two dimensions, where a given beam footprint can be surrounded on all sides by adjacent beams, as shown in Figure

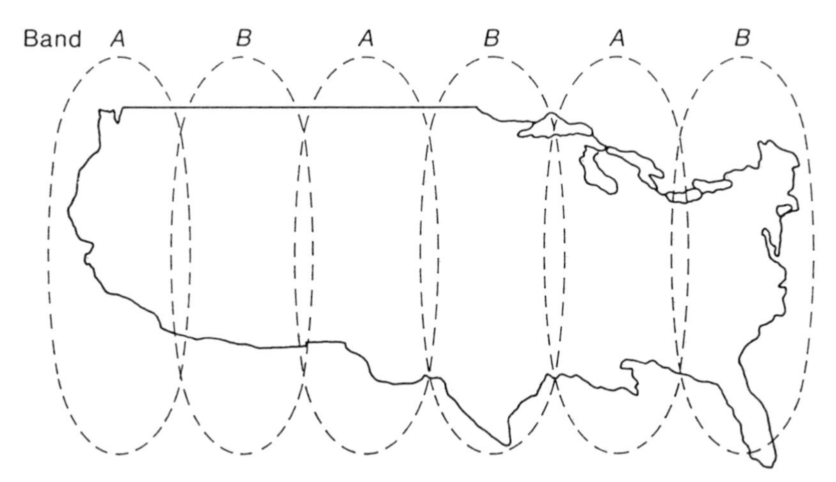

FIGURE 9.19 An example of 6-beam coverage of continental US (CONUS) sharing two frequency bands.

9.20. It is evident that more than two frequency bands must be used, and these bands must be assigned carefully over all beams. As the beam patterns become tighter, more frequency bands must be used to control the interference. The combined crosstalk from all other beams onto a particular beam depends on beam separation, pattern shape, and the frequency bands allocated. Various studies have been reported (Salmesi, 1980) for assessing beam crosstalk in terms of beam allocation and number

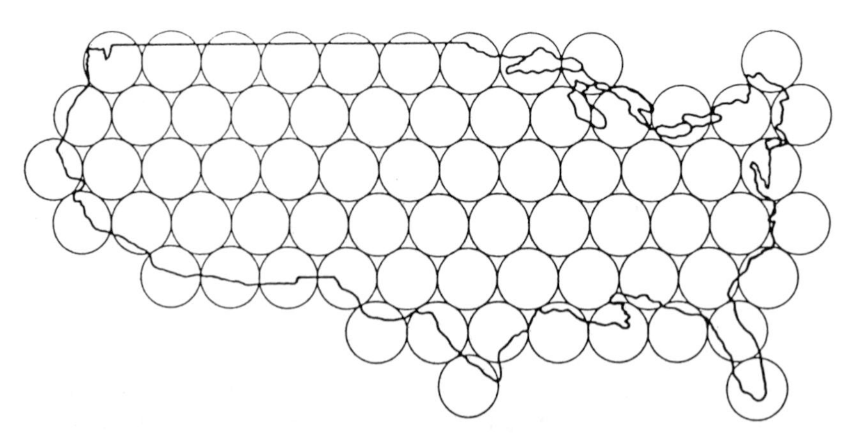

FIGURE 9.20 Multibeam coverage of CONUS with multiple frequency bands.

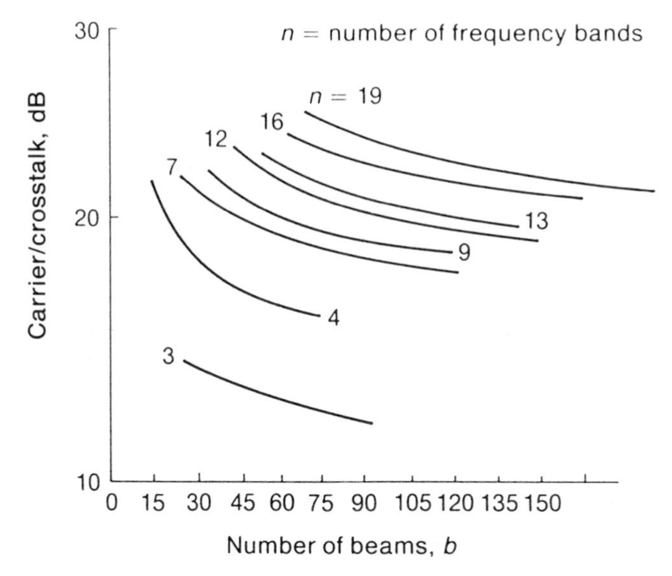

FIGURE 9.21 Carrier power-to-antenna crosstalk ratio vs. number of antenna beams and frequency bands. Beams separated at 3-dB points Crosstalk computed at 3-dB point of each beam. Parabolic antennas. [From Salmesi (1980).]

of distinct frequency bands. One such result is shown in Figure 9.21 for a specific beam pattern, and shows the desired signal power to crosstalk ratio as a function of the number of beams and frequency bands. As more bands are used, the frequency reuse factor in Eq. (9.6.1) decreases. With b dual polarized beams and n bands, each of width B_{RF} Hz, the reuse factor now becomes

$$\text{Reuse factor} = 2b/n \tag{9.7.2}$$

Thus, the data transmission capability of the satellite is reduced as more bands are needed to reduce the beam crosstalk. Figure 9.22 plots the reuse factor in Eq. (9.7.2) as a function of the number of beams, with n adjusted to maintain a 20-dB crosstalk ratio.

The multibeaming advantage in a satellite system is one of the prime motivations for pursuing the concept of space platforms, where extremely large (hundreds of feet) orbiting stations, with correspondingly large antenna dishes, would be conceived to take full advantage of the frequency reuse. Such platforms, constructed in stages in space, would provide the

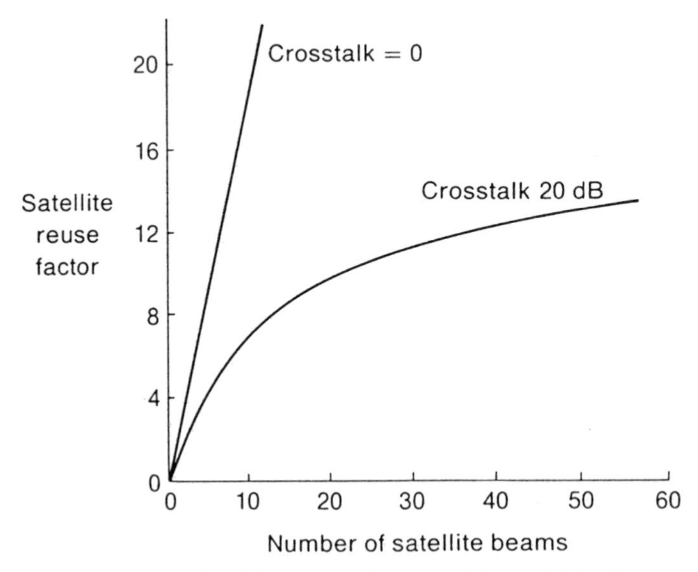

FIGURE 9.22 Satellite reuse factor with multiple beams, with beam crosstalk.

basis for "antenna farms," in which the operation of many individual multibeam satellites could be condensed theoretically to one orbit location. By internal interconnection of antennas using on-board processing (rather than interconnecting satellites via crosslinks), the data efficiency of a specific orbit slot is greatly increased.

9.8 PHOTONIC ON-BOARD PROCESSING

One of the most promising areas of research and development for future satellites is in the use of photonic (electro-optical) hardware to aid on-board processing. Optical devices require less power to operate, are lighter in weight, smaller in size, provide larger bandwidths, and are impervious to electromagnetic interference. These properties can be converted directly to savings in satellite power, weight, and size for future satellite payloads, and can aid in offsetting the expected hardware growth as higher demands for on-board processing are required. Although questions of space qualification, ruggedness, and feasibility must still be fully addressed before lightwave processing becomes a viable satellite subsystem, the vast potential for photonics in future satellites cannot be ignored.

The basic objective of satellite photonics is to make use of optical interfaces for interconnecting and cross-strapping the RF uplink and downlink (or crosslink) ports of a satellite. Thus the photonic hardware would replace part or all of the bulky microwave circuitry of typical on-board processors. The system would appear as in Figure 9.23a. The uplinks refer to the collected RF waveforms, which can be a set of parallel spot beams, or a sequential readout of programmed hopping beams. The

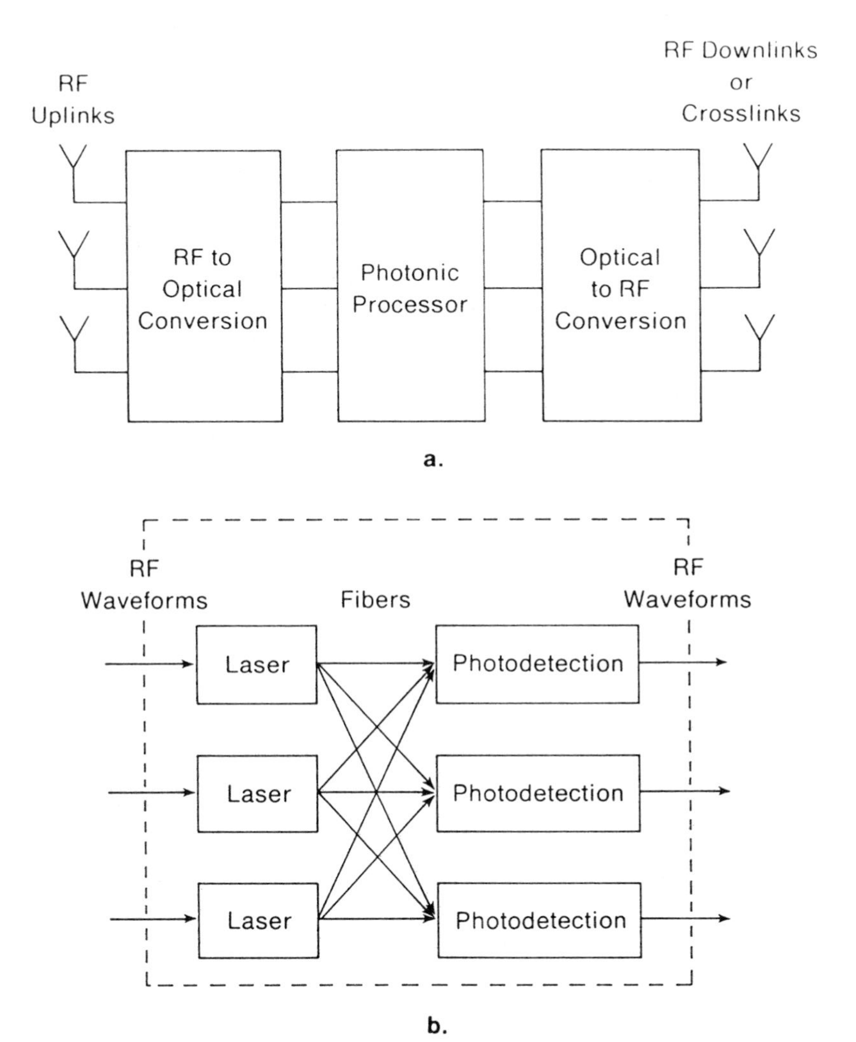

FIGURE 9.23 Photonic on-board processor. (a) Block diagram. (b) Photonic interconnects.

outputs refer to the antenna or diplexor ports which deliver the RF waveforms to the specific downlink or crosslink receivers. The photonic subsystem provides the processing interface between these sets of ports. It can be as simple as optically aided cross-connection of the RF fields and can involve either fixed or reconfigureable fiberoptic connections. If a photonic system can exactly replace an existing microwave processor, it should immediately pay dividends in satellite power, weight, packaging, and electromagnetic rejection, all factors that will have significant influence on payload cost and applications.

An example of a carrier coupled photonic subsystem is shown in Figure 9.23b. A set of RF carrier waveforms are individually modulated on to laser sources, coupled by fiber to the proper output port, and photodetected to remove each RF waveform. The latter is then frequency translated and power amplified for transmission. This represents the simplest photonic interconnect, and basically replace the modern day microwave gate matrix. The optical modulators and photodetectors can be implemented in integrated circuitry, reducing its packaging to centimeters. The fibers can be shared star-couplers, fiber bundles, or fiber arrays. The key issue is not implementability, but rather its degradation on the overall RF link. The details of link SNR, linearity of modulation, crosstalk, isolation, and subsequent contributions to satellite amplifier nonlinearities are beyond our scope here, but system studies in these areas have been reported (Simons, 1990).

Figure 9.24a shows how a photonic gate matrix can be redesigned as a dynamic switch matrix for SS-TDMA operation. The fiber interconnects are switched at programmed rates to reconnect the ports. Present technology favors lithium niobate switching subsystems (IEEE, 1988) into which the fibers are imbedded. Switching is achieved electronically by applying external voltages that alter the velocity (or phase) of the optical fields, causing a controlled coupling to the alternate outputs. Figure 9.24b shows how the 2×2 switch of Figure 9.24a can be interconnected to form higher order gates. The technology of today's switches may already be suitable for satellite applications.

Rather than interconnecting RF fields, the photonics can be used to introduce and route optical pulses. The optical pulses can be generated from the RF fields either by direct time sampling or following data demodulation, as in baseband demod-remod operation. In either case the generated light pulses are routed to output terminals from which the RF data or waveforms are reconstructed. The fact that light pulses can be produced and transmitted at significantly higher rates than the RF data samples permits the photonics to accomplish multiplexing, distribution, and addressing of the bits during routing. These potential capabilities

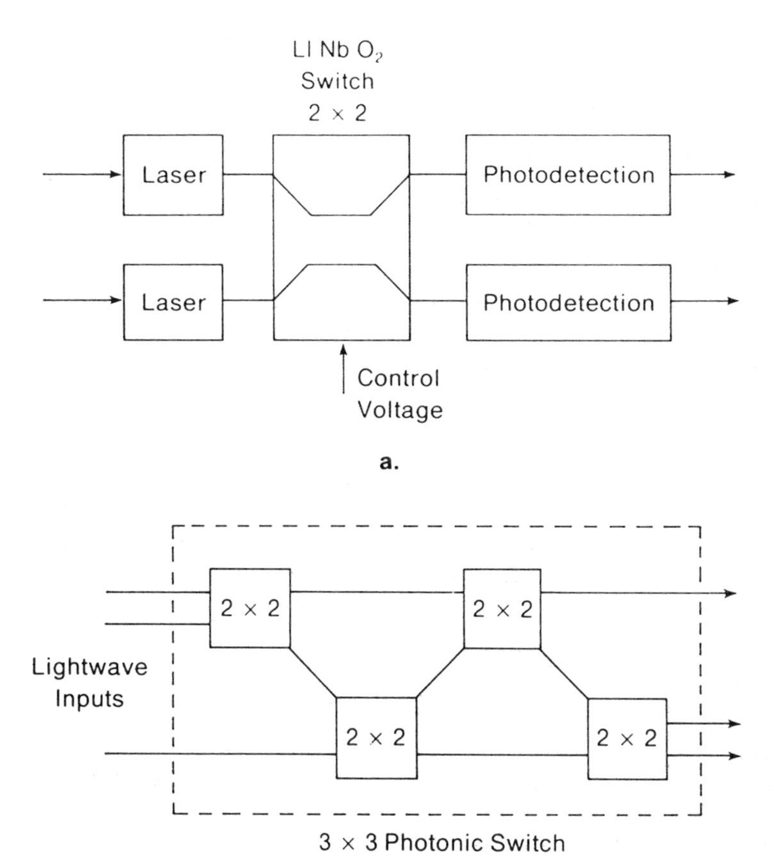

FIGURE 9.24 Photonic gate with switching. (a) 2 × 2 switch. (b) 3 × 3 switch.

have not yet been fully explored, and research toward innovative and novel uses of these advantages in satellite processing must still be carried out.

One primary advantage of pulsed photonics is that parallel sources can easily be multiplexed into optical TDM formats. Figure 9.25 shows a block diagram of four 250 Mb/s RF data streams multiplexed into a 1 Gb/s optical stream. The latter can be carried by fiber to an output where, for example, the photodetected output is directly modulated on to an EHF crosslink. The photonics therefore provides the physical interconnect from the low rate uplink RF to the high rate crosslinks.

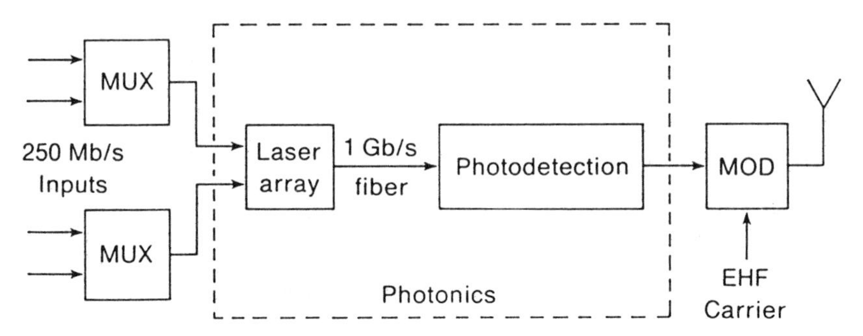

FIGURE 9.25 Photonic multiplexor.

While the achievable capabilities of fiberoptic systems continue to advance at a rapid pace, its application to the satellite environment is still primarily a research and development area. Its possibilities of playing a principal role in future satellite systems must be explored rigorously.

REFERENCES

Bellanger, M., and Daguet, J. (1974), "TDM-FDM Transmultiplexors—Digital Polyphase and FFT," *IEEE Trans. Commun.*, **COM-22** (9), 1189–1204.

Hays, R., and Hartmann, C. (1976), "SAW Devices for Communications," *Proc. the IEEE*, **64**, pp. 856–871.

Holmes, J. K. (1982), *Coherent Spread Spectrum Systems*, Wiley, New York.

IEEE, (1988), Special Issue on "Photonic Switching" *Journal on Selected Areas in Comm.*, Vol. 6 (August).

Lindsey, W., and Anderson, T. (1968), "Digital Data Transition Tracking Loops," *Proc. Int. Tele. Conf.*, Los Angeles, pp. 259–271.

Maruto, R., and Tomozowa, A. (1978), "An Improved Method For SSB-FDM Modulation and Demodulation," *IEEE Trans. Commun.*, **COM-26** (5), 720–725.

Salmesi, A. (1980), "A Parametric Analysis of Satellite-Borne Multiple Beam Antennas," *JPL Report* 80–52.

Simon, M. (1970), "Optimization of the Performance of Digital Data Tracking Loop," *IEEE Trans. Comm. Tech.*, **COM-18**, 686–690.

Simons, R. (1990), *Optical Control of Microwave Devices*, Artech House, Norwood, MA.

Takahuta, F., Hirata, Y., and Ogawa, A. (1978), "Development of a TDM-FDM Transmultiplexor," *IEEE Trans. Commun.*, **COM-26** (5), 726–733.

Williams, R. (1987), *Communication System Analysis and Design*, Prentice-Hall, Englewood Cliffs, NJ.

Ziemer, R., and Peterson, R. (1985), *Digital Communication and Spread Spectrum Systems*, Macmillan, New York, Chapter 3.

PROBLEMS

9.1. Sampling theory requires that a signal $s(t)$ of bandwidth B Hz be sampled at the rate $2B$ samples/s to ensure exact reconstruction. Prove this fact by deriving the spectrum of a sampled signal [transform of the product of $s(t)$ and a periodic delta function pulse train at the sampling rate] and showing there is no spectral distortion of $s(t)$.

9.2. The mean squared error (MSE) of a quantizer is the average squared error due to converting all voltage samples in a quantization interval to a single value. If this quantized value is taken as the midpoint of the interval having width Δ, determine the MSE for a voltage sample uniformly distributed over the interval. Rewrite this MSE in terms of the number of quantization intervals q, assuming that the quantizer covers the range $(-v, v)$ volts.

9.3. An A-D converter is used to digitize a source having a uniform amplitude distribution over $(-A, A)$ volts, and a flat frequency spectrum out to 10 kHz. A four-bit quantizer is used exactly covering $(-A, A)$ volts. Sketch a plot showing how the mean square digitizing error, as a fraction of the source waveform power, varies with the output bit rate.

9.4. Show that transforming a voltage sample that has probability density $p_d(x)$ by the transformation $y = \int_{-\infty}^{x} p_d(u)\, du$ will generate a uniformly distributed sample over (0, 1).

9.5. A signal $d(t)$ is time limited to Ts (but is not band-limited). A communicator wishes to digitize the waveform [send samples of some type to represent $d(t)$]. Using the sampling theorem *in the frequency domain*, explain (a) what type samples should be sent, (b) the reconstruction process at the receiver.

9.6. Three digital sources, having the bit rates $R_1 = 44$ bps, $R_2 = 96$ bps, $R_3 = 120$ bps, are to be time multiplexed. (a) Show that a PSC for this set could be designed with three parallel registers of lengths 12,

8, and 10 stages. (b) Show that these same rates can be multiplexed with fewer total register stages by first multiplexing two sources in one PSC, and then multiplexing the output with the third source in another PSC.

9.7. Design a parallel-series converter for sources with rates $R_1 = 100$ kbps, $R_2 = 45$ kbps, $R_3 = 6$ kbps, and $R_4 = 3$ kbps. (a) What will be the bit rate at the PSC output? (b) What will the bit rate be if a synchronization bit is inserted every 200 bits?

9.8. A false frame synchronization word detection will occur when random bits fill the correlator and happen to match the frame synchronization word. Show that if there are N bits in the synchronization word, and M of the N bits are needed for detection, then the probability of false synchronization is

$$\text{Prob. false synchronization} = \frac{1}{2^N} \sum_{i=M}^{N} \binom{N}{i}$$

9.9. Show that a complex analog Fourier transformer (output time function equal to the input transform) can be constructed with the SAW filter system in Figure P.9.9. [*Hint*: use the fact that $2xy = x^2 + y^2 - (x - y)^2$.]

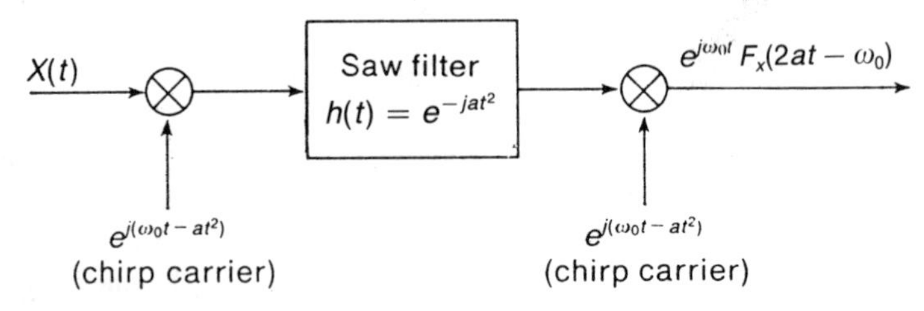

FIGURE P9.9

9.10. Derive the modified expression for PE in Eq. (9.5.4) for a BPSK bit waveform with data asymmetry (unequal data bit lengths). Assume the timing subsystem in Figure 9.14b integrates over the average bit time of the waveform. Let $\theta_1(t)$, $\theta_0(t)$, ϵ_A, and ϵ_p all be zero (no other imperfections).

CHAPTER 10
Satellite Crosslinks

It has been pointed out that insertion of a satellite data crosslink, or intersatellite link (ISL), has important advantages in the operation of an overall satellite system. These advantages, however, must be carefully assessed against the weight and power penalty imposed on the satellite, owing to the additional crosslink equipment and the possible downlink performance degradation with a crosslink transmission inserted. As the potential for larger satellites and payloads increases, these disadvantages are somewhat diluted, and the insertion of crosslinks is rapidly becoming an important aspect of satellite systems. In this chapter some design considerations for satellite crosslinking are presented.

10.1 THE CROSSLINK SYSTEM

The specific frequency bands allocated for satellite crosslinks were given in Chapters 1 and are summarized in Table 10.1. Frequency bands have been allocated at the K-band and EHF frequencies shown. K-band crosslinks have been designated primarily for low-data-rate systems, and therefore it is expected that much of the future crosslink design activity will be concentrated in the EHF bands. The allocation of high-frequency bands for crosslinks have the advantages of permitting smaller hardware and wider crosslink bandwidths, and they occur at frequencies that are shielded by the atmosphere from terrestrial and downlink interference.

TABLE 10.1 Allocated crosslink frequency bands

	Band (GHz)	*Bandwidth (GHz)*
K-band	22.55–23.55	1
	32.0–34.0	2
EHF	54.25–58.25	4
	59.0–64.0	5

Figure 10.1 sketches a satellite system with an ISL. The advantages of the crosslink are immediately obvious. Earth stations can be interconnected that are further apart (e.g., different hemispheres), total Earth coverage is larger, and satellite elevation angles are improved (stations in the coverage area of satellite A would have had to view satellite B at lower angles). On the other hand, crosslinks present some basic design difficulties. Besides the obvious problem of maintaining a communication link in which both transmitter and receiver are spaceborne, crosslink satellites must invariably utilize on-board processing to separate out downlink and crosslink data. For a crosslink that serves primarily as a relay network, transmissions must be transparent to (not degrade) the crosslink data. This often requires crosslink operation at higher CNR than in the downlinks. In addition, crosslinks insert additional propagation delay that may become intolerable to a round trip message. The crosslink itself will most likely require significantly higher data rates, since it will generally involve the simultaneous relaying of multiple uplinks. Hence, ISL bandwidths will be larger than the individual downlink bandwidths. Since

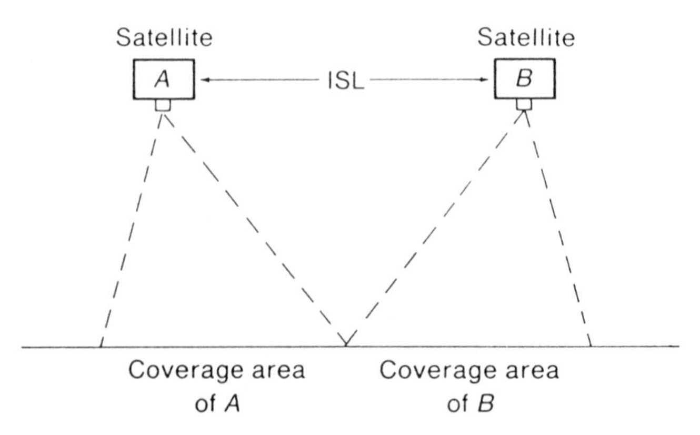

FIGURE 10.1 Intersatellite links (ISL).

on-board remodulation and demodulation are often required to complete the crosslink, modulation formats for crosslinks are generally restricted to basic techniques, such as coherent BPSK or QPSK, or low-order noncoherent modulation. These formats preclude the opportunity to achieve the power benefits of more sophisticated waveform encoding on the crosslinks.

Satellite crosslink distances will depend on the orientation of the satellite constellation involved. Two satellites in geostationary orbits will be physically separated by a distance dependent on their longitude difference angle, as was shown in Figure 3.23. Figure 10.2 plots crosslink distances as a function of separation angle. Satellites at 2° separation, for example, are about 1000 miles apart, while maximum distance (occurring when two synchronous satellites have separation angle approaching 160° while maintaining line-of-sight above the Earth's atmosphere) is approximately twice the synchronous altitude ($\approx$45,000 miles).

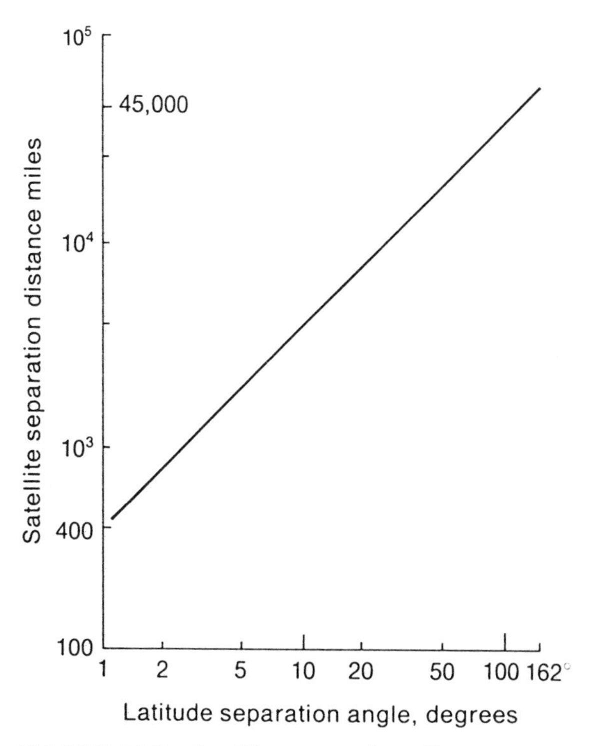

FIGURE 10.2 Satellite separation distance versus latitude separation angle in synchronous orbit.

The successful insertion of K-band or EHF crosslinks will depend on the component development of spaceborne hardware at these bands. The two key elements are the EHF crosslink transmitter power sources and the low-noise amplifier (LNA) for the crosslink receivers. Figure 10.3a shows reported results (Park et al., 1973; Schroeder and Haddad, 1979) of power output capability of both TWTA and IMPATT diode amplifiers in the EHF range. While the larger wave tubes have potential for tens of

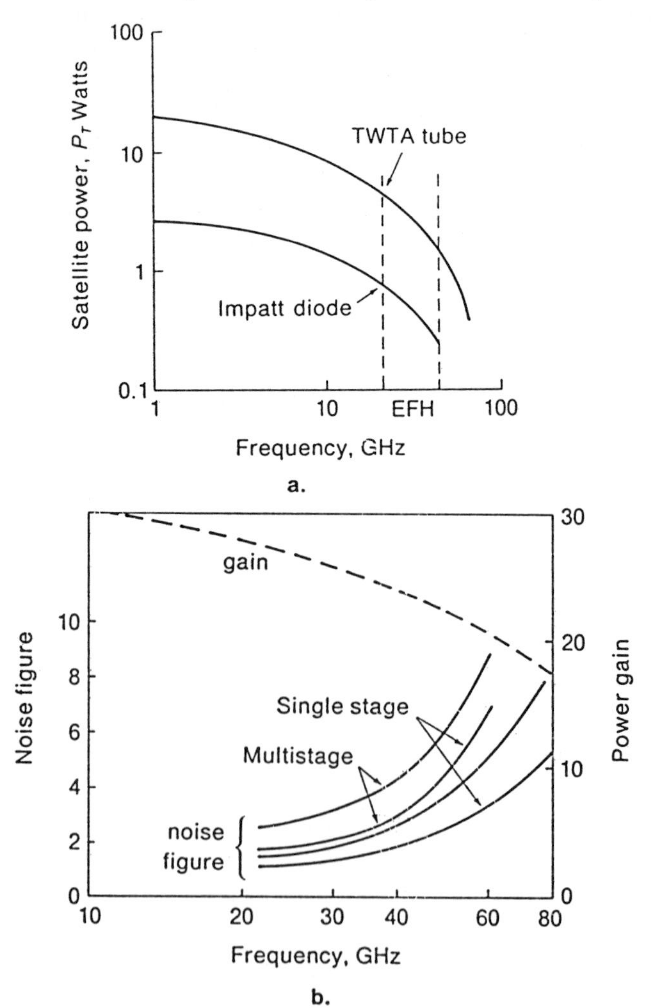

FIGURE 10.3 EHF components. (a) power capability. (b) FET LNA amplifier. Solid line = noise figure, dashed lines = gain. From Watkin, 1985.

watts capability, the more compact solid-state diodes will be limited to only a few watts. Figure 10.3b shows LNA performance at EHF for use in receiver front ends. Recall from our discussion in Section 3.4 that the front-end LNA noise figure will determine the receiver noise figure if its power gain is suitably high. EHF receiver noise figures are reported to be in the 4–8-dB range.

While noise temperatures of up/down links tend to be somewhat stationary with geostationary satellites, crosslinks may exhibit considerable variations in background temperatures. The changing orientation of one satellite relative to another may alter the background, possibly producing a significant temperature change. If the background is clear sky, its temperature contribution may be an order of 50–100 K, and a 7-dB receiver noise figure will produce a receiver noise temperature of about 1200 K. However, if a portion of the Sun wanders near the field of view, T_{eq} may increase to the 3000 K range. A satellite receiver with the Sun entirely filling the field-of-view may have 6000 K background temperatures.

10.2 CROSSLINK POWER BUDGETS

A crosslink receiver carrier to noise ratio was derived in Section 3.6. We rewrite the corresponding C/N_0 value here as simply

$$\frac{C}{N_0} = \frac{P_T g_t g_r L_p}{kT^\circ_{eq}} \tag{10.2.1}$$

where P_T is the transmitter power, g_t and g_r are the antenna gains, L_p is the propagation loss over the ISL distance Z in Figure 10.2, k is Boltzman's constant, and T°_{eq} is the receiver noise temperature. Figure 10.4 plots the key parameters of this equation at the crosslink frequencies, showing the antenna gains and beamwidths as a function of antenna size, and the space loss at various satellite crosslink distances. Note that 3-ft antennas at EHF will produce gains of about 52 dB and beamwidths less than one-half of a degree.

The C/N_0 expression can also be written directly in terms of the transmit satellite EIRP, or the receive satellite RIP, as

$$\begin{aligned}\frac{C}{N_0} &= \frac{(\text{EIRP})L_p g_r}{kT^\circ_{eq}} \\ &= \frac{(\text{RIP})g_r}{kT^\circ_{eq}}\end{aligned} \tag{10.2.2}$$

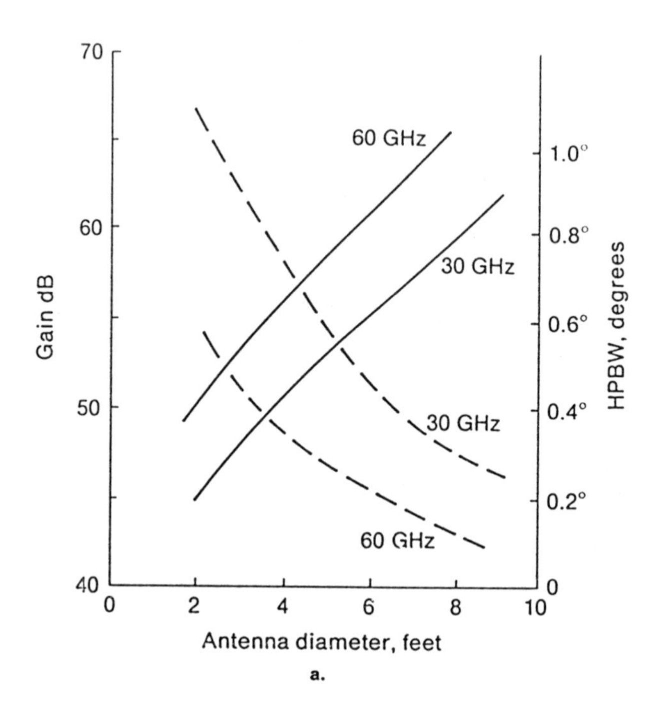

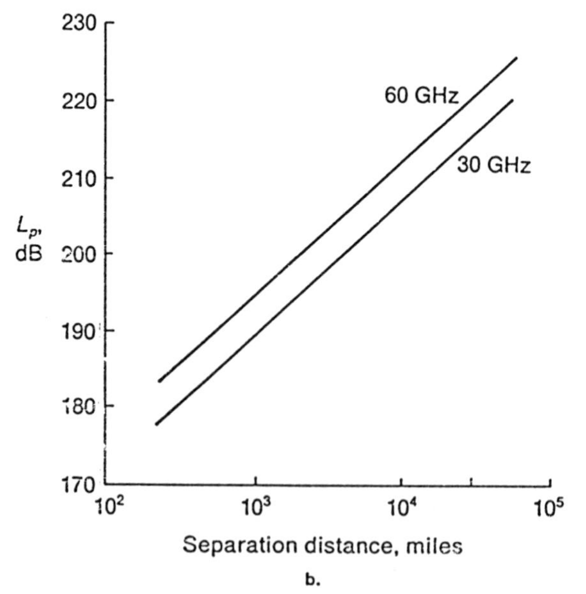

FIGURE 10.4 (a) Link parameters. Antenna gain and half-power beamwidths. (b) Propagation loss L_p versus separation distance in miles, at crosslink frequencies.

Equation (10.2.2) separates out the contributions of the transmitting satellite, in terms of the power it can deliver to the receiving satellite, and the receiving satellite, in terms of its receive g_r/T°_{eq} capability.

The crosslink parameters in Figure 10.4 will generate the resulting link performance. For a BPSK link, the E_b/N_0 value must provide the required bit error probability to sustain the data rate. The decoding E_b/N_0 is obtained from Eq. (10.2.1) as

$$\frac{E_b}{N_0} = \frac{C/N_0}{R_b} \tag{10.2.3}$$

where R_b is the bit rate. Figure 10.5 plots the available crosslink BPSK data rate at PE $= 10^{-5}$ ($E_b/N_0 = 9.6$ dB) for several operating parameters. In Figure 10.5a data rate is plotted versus antenna size, assuming identical antennas at each satellite, at maximum range, with several transmitter power levels. In Figure 10.5b data rate is shown versus satellite distance and antenna size.

The results in Figure 10.5 assume perfect beam pointing so that the maximum gain of the antennas is available. However, maintaining accurate pointing with the narrow EHF beamwidths in Figure 10.4a is difficult, owing to the uncertainty in true satellite location and inaccuracies in satellite attitude (orientation in space). With antenna-beam-pointing errors inserted, the CNR at the receiver will be reduced by any off-axis antenna gain values due to mispointing at the transmitter or receiver or both. If the antenna gains in Eq. (10.2.1) are now written in terms of the antenna patterns $g(\phi)$, to indicate its variation with spatial pointing angle ϕ, then Eq. (10.2.2) becomes

$$\frac{E_b}{N_0} = \left(\frac{P_T L_p}{k T^{\circ}_{eq} R_b}\right) g_t(\phi_{et}) g_r(\phi_{er}) \tag{10.2.4}$$

where ϕ_{et}, ϕ_{er} are the pointing errors at the transmitter and receiver, respectively. For a parabolic dish with half-power beamwidth $\phi_b \cong \lambda/d$ rad, the decibel loss from peak gain due to pointing error ϕ_e rad can be estimated from Eq. (3.2.5) as

$$\begin{aligned} \text{Decibel pointing loss} &= 11.1(\phi_e/\phi_b)^2 \\ &= (123.5)\phi_e^2 d^2 \text{ (m)} f^2 \text{ (GHz) dB} \end{aligned} \tag{10.2.5}$$

Figure 10.6 shows how the crosslink E_b/N_0 degrades for circular dish antennas as the diameter is increased, assuming constant and equal

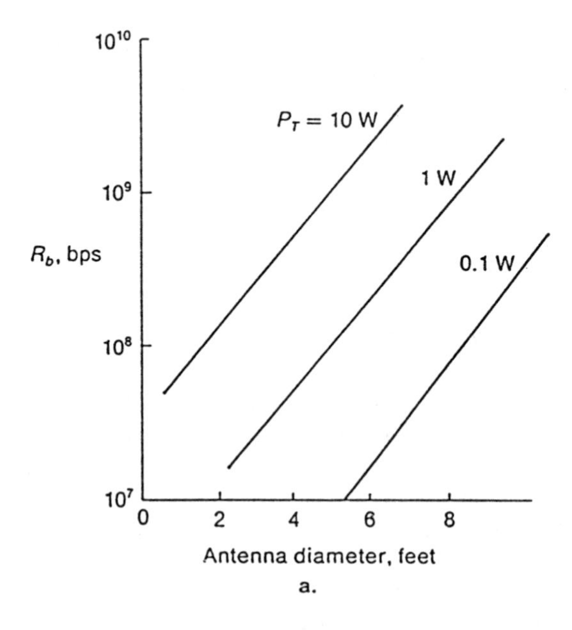

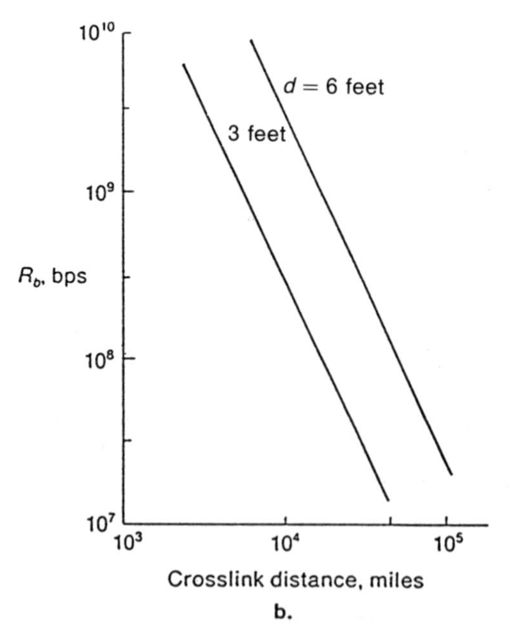

FIGURE 10.5 Achievable crosslink data rates. $f = 60$ GHz, $T^\circ_{eq} = 1000$, $E_b/N_o = 9.6$ dB. Equal transmit and receive antennas. (a) Versus antenna diameter at $z = 45{,}000$ miles. (b) Versus crosslink distance at $P_T = 1$ W.

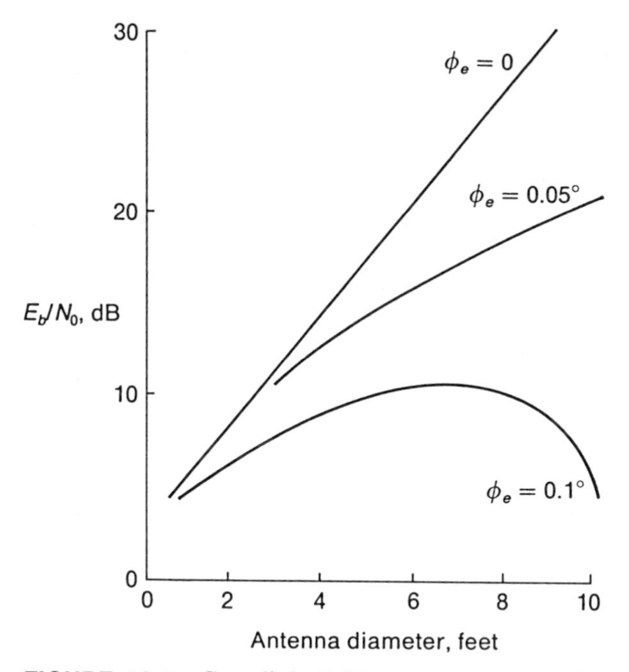

FIGURE 10.6 Crosslink E_b/N_o versus antenna diameter. ϕ_e = pointing error at transmitter and receiver. z = 45,000 miles, f = 60 GHz, T°_{eq} = 1000°, R_b = 50 Mbps.

pointing errors at each end ($\phi_{et} = \phi_{er} = \phi_e$). Increasing the antenna size produces increasing E_b/N_0 until the corresponding narrowing beamwidths begin to cause a rapid antenna gain fall-off for a particular pointing error. From these curves the link loss due to mispointing can be estimated. Alternatively stated, Figure 10.6 shows how the link can be improved if antenna mispointing can be removed. Since the pointing errors are limited by the inherent uncertainty (that may in fact drift with time), they generally cannot be overcome by any fixed open-loop pointing from either end. The only way to compensate adequately at the receiver is to make use of the arriving transmissions from the transmitting satellite to point back with the receiver antenna. This automatic pointing of the receiver antenna via tracking of the arrival transmitter beam is referred to as *autotracking*. Curves similar to Figure 10.6 become instrumental in determining whether the insertion of autotrack hardware to remove pointing loss is a viable trade-off against open-loop mispointing. The autotrack subsystems are considered in Section 10.5.

10.3 COHERENT AND NONCOHERENT COMMUNICATIONS AT EHF

Satellite crosslinks will be primarily digital links in order to interface with data routing and on-board processing. Moderate data (data rates on the order of 10–100 Mbps) will most likely be first modulated onto C-band carriers (3–6 GHz) and upconverted to EHF to generate the crosslink carrier. Higher data rates (100–500 Mbps) must be modulated directly on the EHF carriers to take advantage of the higher modulation bandwidths. Modulation formats can be either coherent (BPSK or QPSK) or non-coherent (FSK or DPSK). Coherent systems provide a power savings in achieving a desired PE, but require additional receiver hardware to provide the necessary carrier referencing for decoding.

While accurate phase referencing is common circuitry in modern C-band and X-band links, the extension of coherent communications to EHF frequencies presents phasing problems that can be significantly more difficult. This is due to the fact that frequency instabilities, Doppler shifts, and phase noise spectra are proportionally higher at the crosslink bands. These increases, coupled with the common $\times 2$ and $\times 4$ multiplication used for modulation wipe-off during coherent phase referencing, require wider tracking bandwidths, and hence more severe thermal noise effects.

Frequency phase referencing procedures for BPSK and QPSK are discussed in Appendix B. The basic decoding subsystem is shown in Figure 10.7. The modulated carrier, with phase modulation $\theta(t)$ and carrier phase $\psi(t)$, is either directly decoded at the crosslink frequency, or down-converted to an appropriate C-band or lower IF frequency. Initial EHF carrier frequency uncertainty (that must be acquired prior to any data decoding) could be significant. For example, at 60 GHz a frequency uncertainty of one part in 10^6 is about 60 kHz. Since this generally exceeds the typical phase tracking loop bandwidths, frequency searching may have to be inserted to establish initial frequency lock-up. The time to search

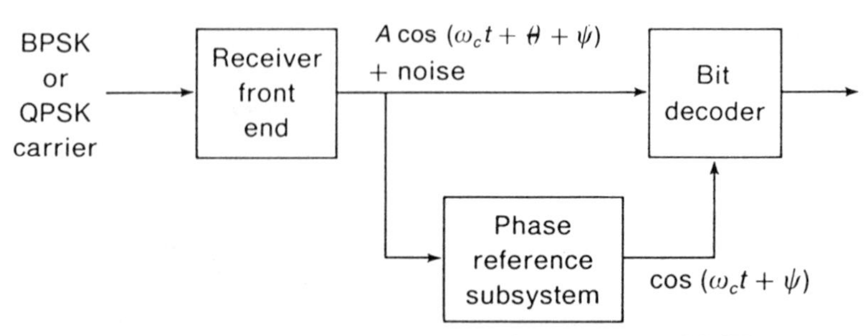

FIGURE 10.7 Bit decoding subsystem for BPSK or QPSK.

an initial uncertainty of $\pm\Delta f$ Hz, with a BPSK tracking loop of bandwidth B_L and damping factor of 0.707, is approximately

$$T_{ac} = \frac{2(2\Delta f)}{(2B_L)^2/2} = \frac{2\Delta f}{B_L^2} \text{ s} \tag{10.3.1}$$

Figure 10.8 plots the acquisition time to acquire a crosslink frequency having a specific fractional uncertainty, as a function of B_L. Acquisition time may be considered as part of the initial "start-up" time of a satellite brought into operation in a particular constellation, and its actual time may not be critical. However, acquisition time can become important during outage or blackout periods during normal satellite operation, since crosslink data could be lost while reacquiring. A 10-Mbps crosslink that requires 2 s to reacquire, for example, could lose about 20×10^6 consecutive data bits.

After carrier frequency acquisition, a coherent phase reference at the decoding frequency must be extracted by the referencing subsystem for data decoding. As was pointed out in Chapter 2, phase referencing errors

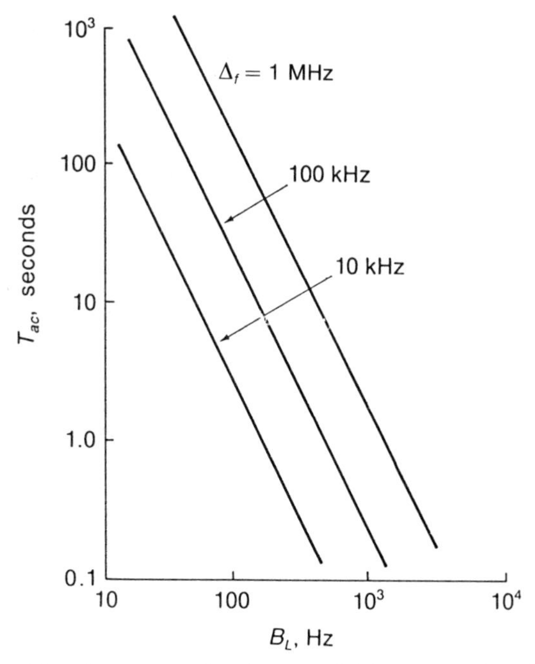

FIGURE 10.8 Carrier acquisition time for frequency offset Δ_f versus phase tracking loop bandwidth.

ψ_e should be restricted to about 10–15° for BPSK and about 4–7° for QPSK. For phase tracking systems, the reference phase error will generally have a bias (constant) and a random component. The bias component will be due to Doppler rates (linearly changing carrier frequencies) occurring during relative satellite acceleration, and phase bias values will be

$$\psi_{eb} = \frac{2(\dot{\Delta f})}{B_L} \text{ rad}, \qquad \text{for BPSK} \tag{10.3.2a}$$

$$= \frac{4(\dot{\Delta f})}{B_L} \text{ rad}, \qquad \text{for QPSK} \tag{10.3.2b}$$

where $\dot{\Delta f}$ is the frequency rate in Hz/s. Again, low acceleration rates could still produce significant Doppler rates at EHF relative to typical loop bandwidths. This may be a prime motivation for decoding at lower frequencies (C-band or less), where the corresponding $\dot{\Delta f}$ is less.

The random component of the phase error is due to thermal noise and oscillator phase noise entering the phase tracking loop. The thermal noise produces a BPSK tracking variance of

$$\sigma_n^2 = \frac{B_L T_b}{E_b/N_0}\left(1 + \frac{2}{E_b/N_0}\right) \tag{10.3.3}$$

where T_b is the bit time and E_b/N_0 is adjusted for the desired PE. Phase noise on the received crosslink carrier and on the local VCO produces an additional tracking variance of

$$\sigma_p^2 = \frac{1}{2\pi}\int_{-\infty}^{\infty} S_\psi(\omega)|1 - H(\omega)|^2 \, d\omega \tag{10.3.4}$$

where $S_\psi(\omega)$ is the combined phase noise spectra at the decoding frequency, and $H(\omega)$ is the phase tracking loop gain function. For a second-order loop, with a damping factor ξ and loop noise bandwidth B_L,

$$|1 - H(\omega)|^2 = \frac{(\omega/2B_L)^4}{[1 - (\omega/2B_L)^2]^2 + 4\xi^2(\omega/2B_L)^2} \tag{10.3.5}$$

Figure 10.9a shows typical oscillator phase noise spectra (Hewlett, 1985) referred to a 60-GHz crosslink carrier frequency. These phase noise spectral curves are then used to evaluate Eq. (10.3.4), with Eq. (10.3.5)

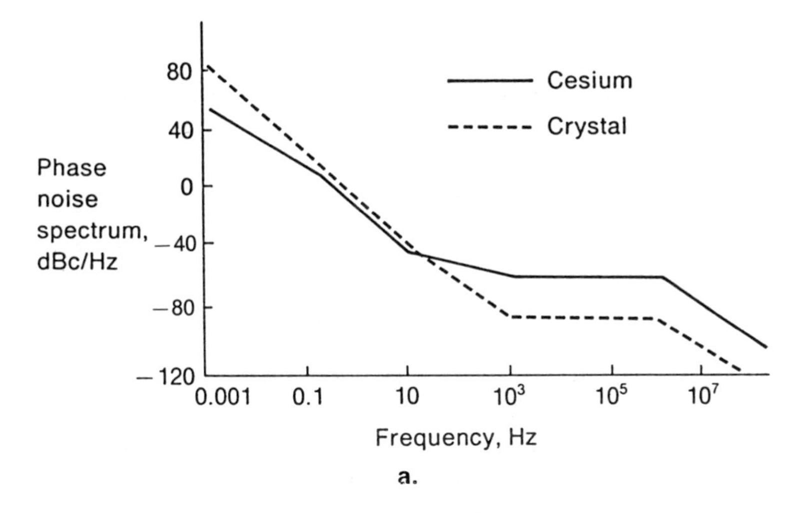

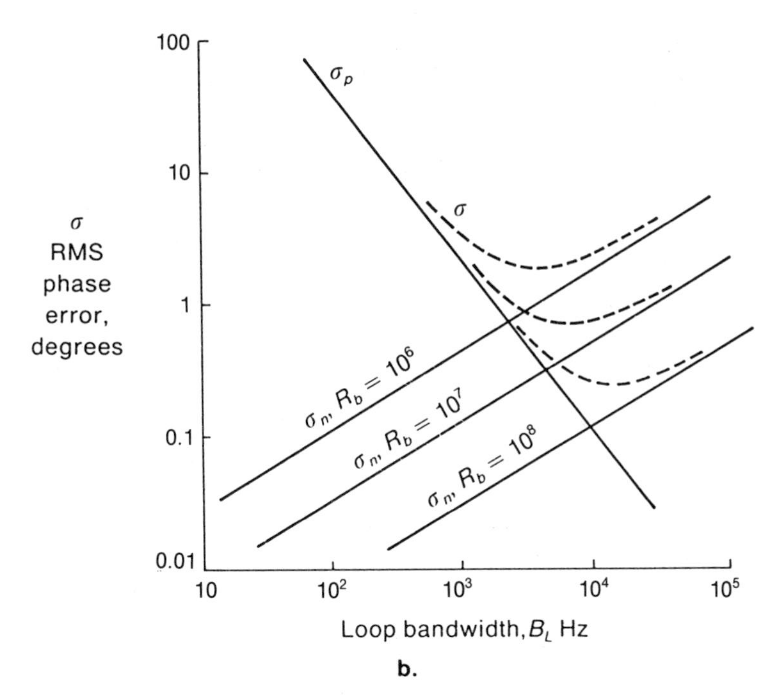

FIGURE 10.9 (a) Phase noise spectra of typical satellite oscillators, referenced to 60 GHz. (b) Total RMS phase error, $\sigma = (\sigma_p^2 + \sigma_n^2)^{1/2}$: σ_p = error due to phase noise spectrum. σ_n = error due to thermal noise, E_b/N_o = 10 dB.

inserted. A plot of the resulting σ_p versus loop bandwidth is shown in Figure 10.9b. Also included in the thermal noise effect in Eq. (10.3.3), again exhibiting the basic trade-off of phase error variation with loop bandwidth. Increasing B_L increases the thermal noise contribution, but reduces the phase noise effect. The combined phase error variance $(\sigma_n^2 + \sigma_p^2)^{1/2}$ can then be superimposed to determine the loop bandwidth yielding minimum phase error. This choice of bandwidth must be contrasted to the bandwidth needed for satisfactory acquisition time in Figure 10.8. This comparison will determine whether dual bandwidth loops (wide band for acquisition sweeping and narrow band for phase tracking) need be implemented in the crosslink receiver. Also note that the achievable phase accuracy depends on the received E_b/N_0 and bit rate, and therefore is directly related to the link budget.

Coherent bit decoding directly at EHF is also hampered by other phasing effects besides the phase referencing accuracy. One is the possible phase differential occurring during the transfer of equal phased waveforms to different circuit points. Since a carrier wavelength at 60 GHz is about 0.5 cm, path distance differentials of even tenths of a centimeter can translate to phase differentials of 90° or more. This becomes important in a decoding subsystem as in Figure 10.7. While the phase referencing system may produce a perfectly referenced carrier relative to the input data carrier, the transfer of this reference to the decoder multiplier may introduce a phase differential that would appear much like an equivalent phase reference error. Phase composition circuitry would have to be inserted to alleviate this imbalance.

Another effect is caused by the difficulty in maintaining exact orthogonality in EHF decoders using quadratic correlators, as in QPSK. Figure 10.10 shows a QPSK decoder generating both a referenced carrier and its 90°-shifted version, each using its own separate path for decoder multiplication. Any inaccuracy in the path lengths will lead to a relative phase shift that will degrade the orthogonality of the referencing carriers. This nonorthogonality converts to both a power degradation and the insertion of crosstalk in the quadrature arm of the decoder. Note that the orthogonality imbalance effect ψ_s is slightly different from a phase reference error, ψ_e. The imbalance ψ_s can be either positive or negative, and can either add to or subtract from the phase reference error.

In a noncoherent crosslink system, the phase referencing subsystem is no longer required, but the carrier phase noise still will degrade the detection performance, as discussed in Appendix A.9. For a noncoherent binary FSK signaling format, the bit-error probability can be determined as

$$\mathrm{PE}_2 = \tfrac{1}{2} e^{-\bar{D}(E_b/2N_0)} \tag{10.3.6}$$

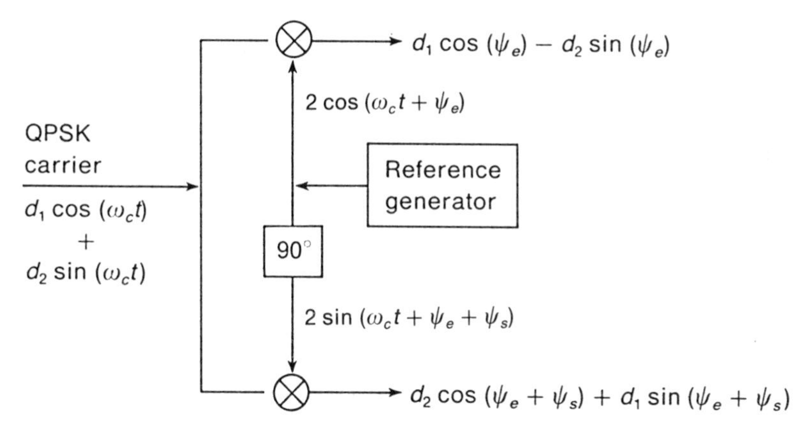

FIGURE 10.10 Effect of phase referencing error (ψ_e) and phase imbalance (ψ_s) on QPSK decoding.

where $\bar{D}$ is the effective energy degradation

$$\bar{D} = \frac{2}{T_b} \int_0^{T_b} \left(1 - \frac{\tau}{T_b}\right) e^{-\sigma^2(\tau)/2} \, d\tau \tag{10.3.7}$$

T_b is the bit time, and

$$\sigma^2(\tau) \triangleq \frac{4}{\pi} \int_{-\infty}^{\infty} S_\psi(\omega) \sin^2(\omega T_b/2) \, d\omega \tag{10.3.8}$$

Thus, the spectrum $S_\psi(\omega)$ of the phase noise on the crosslink carrier tones generates the phase structure function $\sigma^2(\tau)$, which must be integrated in Eq. (10.3.7) to determine the bit energy degradation, $\bar{D}$. Note that $\bar{D}$ depends on both the spectrum itself and the bit rate. Again, using the phase noise spectra in Figure 10.9a, the resulting $\sigma(\tau)$ and $\bar{D}$ are shown in Figures 10.11a and 10.11b, the latter as a function of crosslink frequency and bit rate. The value of $\sigma^2(\tau)$ scales directly with carrier frequency so that phase noise degradation is most severe at EHF.

For an M-ary FSK system, using M orthogonal tones, the bit-error probability can be approximated as

$$\text{PE}_M \cong (\tfrac{1}{2}M)\text{PE}_2 \tag{10.3.9}$$

now with

$$\text{PE}_2 = \tfrac{1}{2} e^{-\bar{D}(E_b \log_2 M)/2N_0)} \tag{10.3.10}$$

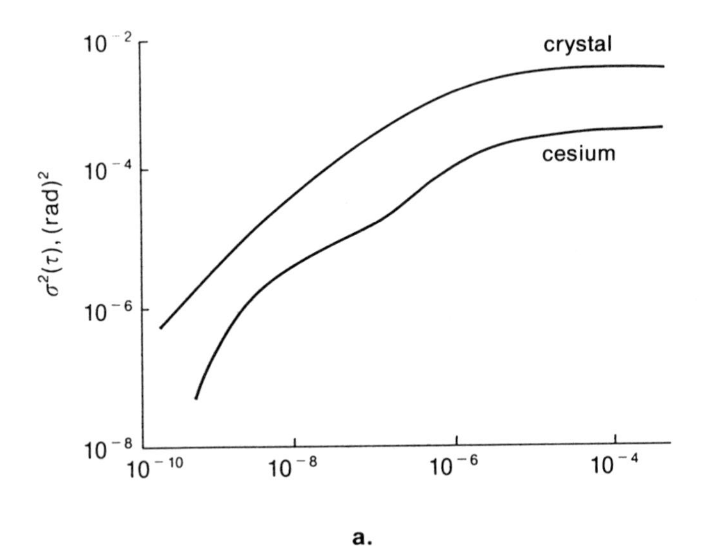

a.

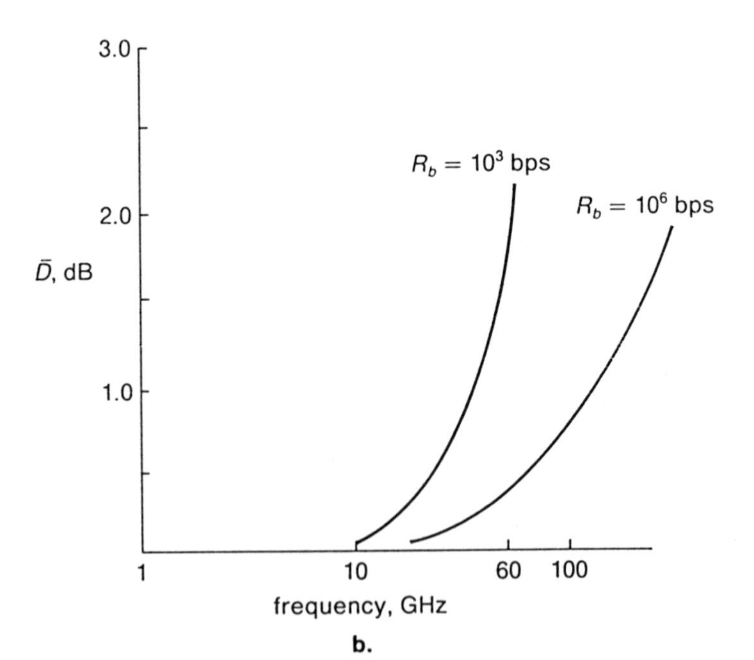

b.

FIGURE 10.11 (a) Differential phase noise variance function $\sigma^2(\tau)$. [Phase noise structure function in (10.3.8).] Frequency = 1.0 GHz. (b) Bit energy degradation due to phase noise for non-coherent binary FSK. Phase noise spectrum in Figure 10.9a. From Alexovich (1990).

The degradation is $\bar{D}$ is again given by (10.3.7) with T_b replaced by the symbol time T_w- $= T_b \log_2 M$.

In a DPSK crosslink, excessive phase noise can cause a random phase differential between one bit time and the next, even though the phase may be relatively constant over each bit time T_b. The resulting bit-error probability must then be averaged over this phase differential. For a DPSK decoder with phase differential $\Delta = \psi_1 - \psi_2$ between two adjacent bit intervals, the DPSK bit error probability is given by (Alexovich and Gagliardi, 1990)

$$\text{PE}(\Delta) = \int_0^{\infty} x e^{-(x^2+a^2)/2} I_0(ax) \int_x^{\infty} y e^{-(y^2+b^2)/2} I_0(yb)\, dy\, dx \quad (10.3.11)$$

where

$$a = \left[\frac{2E_b}{N_0}(1 + \cos\Delta)\right]^{1/2} \quad (10.3.12a)$$

$$b = \left[\frac{2E_b}{N_0}(1 - \cos\Delta)\right]^{1/2} \quad (10.3.12b)$$

The average DPSK bit-error probability is then obtained as

$$\text{PE} = \int_{-\infty}^{\infty} \text{PE}(\Delta) p(\Delta)\, d\Delta \quad (10.3.13)$$

where $p(\Delta)$ is the probability density of the phase differential Δ. When $\psi(t)$ is a Gaussian phase noise process, the phase differential evolves as a Gaussian variable with zero mean and variance $\sigma_\Delta^2 = \mathscr{E}(\psi_1 - \psi_2)^2 = \sigma^2(T_b)$ where $\sigma^2(\tau)$ is given in Eq. (10.3.8). For the phase noise spectra in Figure 10.9a, producing the differential variance function in Figure 10.11a, the variance can be read directly at the particular bit time. Figure 10.12 shows the degradation in DPSK PE by integrating Eq. (10.3.13) with a Gaussian density, at several values of σ_Δ^2. These figures provide a direct comparison of DPSK and binary FSK noncoherent performance in the presence of carrier phase noise. Note that the phase noise effect in noncoherent FSK produces a direct degradation to the decoding symbol energy E_b at all PE, while the effect in DPSK is to produce a PE floor at the higher values of E_b.

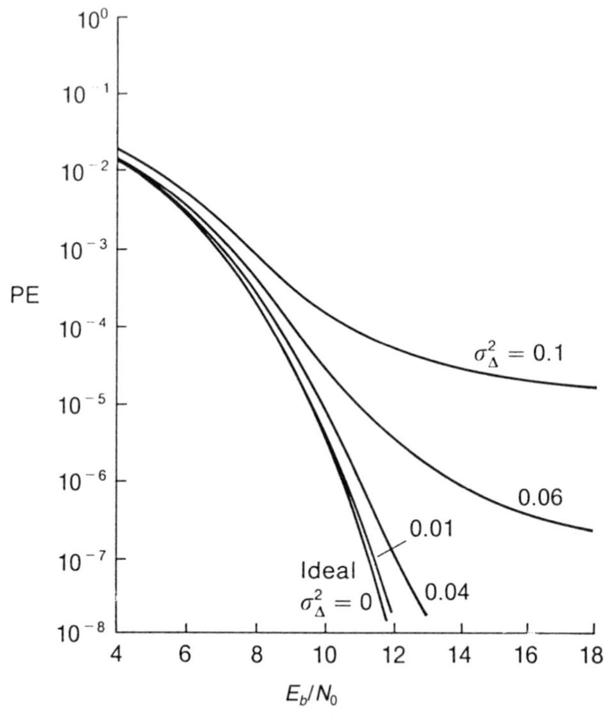

FIGURE 10.12 Bit error probability PE for DPSK with Gaussian phase noise having differential phase variance σ_Δ^2.

10.4 AUTOTRACKING

In Section 10.2 it was shown how the communication performance of a crosslink can be improved by using automatic antenna pointing (autotracking). In autotracking the RF receiver antenna is designed to point continually toward the transmitter by tracking the arrival angle of the received beam. This can be implemented by the system shown in Figure 10.13. An error voltage proportional to the pointing error between the antenna boresight and the arriving beam is generated in the autotrack subsystem. This error voltage is then used to control the gimbal axis in a feedback loop that repoints the antenna so as to reduce the error. Thus, the receiver antenna undergoes a continual tracking operation on the arriving transmitter beam. Since the arrival beam angle relative to the receiver is described by both an azimuth and elevation angle, a pointing error voltage must be generated simultaneously in both coordinates to control the antenna properly in both axes. In a two-way system a receiver

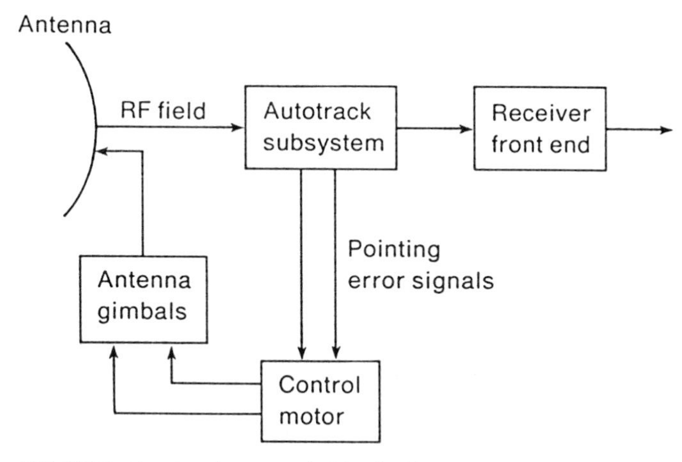

FIGURE 10.13 Autotrack block diagram.

at the transmit satellite would likewise autotrack on the crosslink return beam.

Autotrack pointing error voltages are generated by standard monopulse techniques (King, 1988). Figure 10.14a shows a four-feed monopulse antenna operating with a common reflector dish. Each feed produces its own antenna pattern slightly offset from boresight, with all patterns overlapping in the forward direction, as shown. By taking sums and differences of the RF carrier at each feed output, combinations of the individual beam patterns can be achieved. To see this, consider the simplified diagram in Figure 10.14b showing a one-dimension (azimuth or elevation) version, with two beams having the gain patterns shown. If the outputs of the feeds producing these beams are subtracted, the resulting combined RF carrier waveform is equivalent to receiving the arriving RF carrier with the difference pattern shown in Figure 10.14c. If the feed outputs are summed, the equivalent pattern is given by the sum of the two beam patterns. Hence, the sum and difference patterns of the beams represent the effective gain patterns that generate the sum and difference carrier waveforms from the antenna feeds. The sum beam has its maximum gain on boresight and therefore represents the effective receive antenna pattern. The difference beam, on the other hand, produces a carrier with a unique amplitude and sign, dependent on the offset angle of the arriving transmit beam. (The sign can be interpreted as a phase shift in the difference carrier relative to the phase of the sum carrier.) The amplitude and phase of the difference carrier therefore gives a measurement of the pointing error of the sum beam relative to the arriving carrier direction. An instantaneous measurement of the difference amplitude and

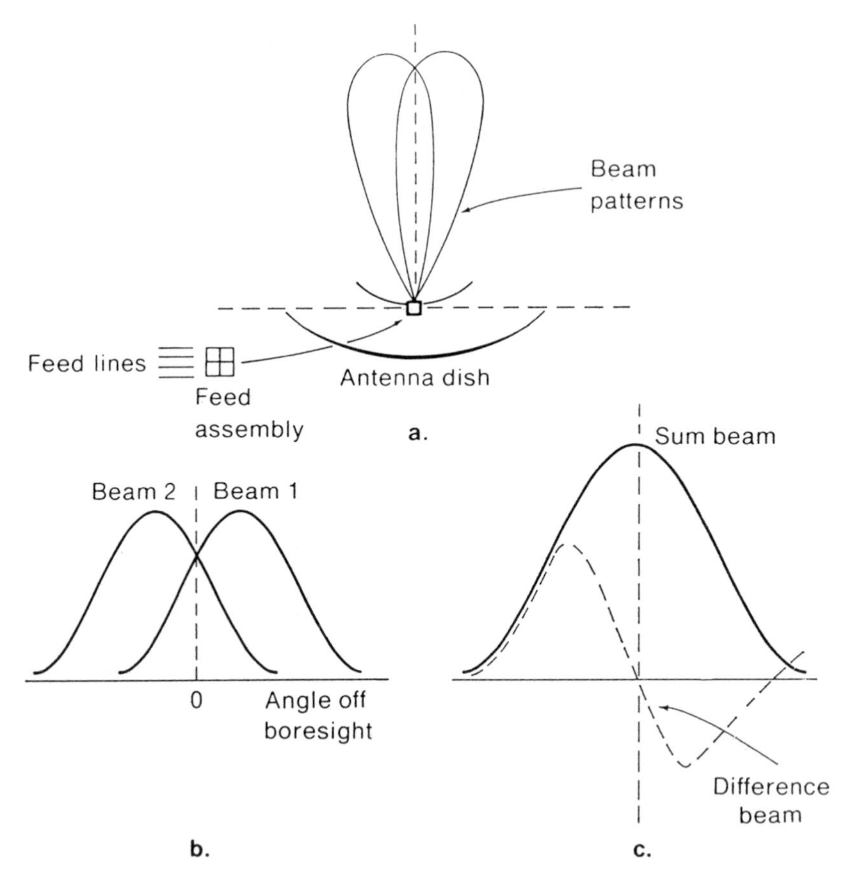

FIGURE 10.14 Autotrack beams. (a) Formation of beams from feed assembly (only 2 beams shown). (b) One dimensional diagram of 2 beams. (c) Resulting sum and difference pattern from beams in (b).

phase can then be used to provide the necessary pointing error voltages for the autotrack operation. In essence, the difference carrier can be used to keep the sum beam pointed toward the transmitter.

The actual monopulse system would be formed as in Figure 10.15a, with the four feed beams combined and subtracted to give instantaneous azimuth and elevation error signals. (Note beams are first summed in pairs, then subtracted, to give the individual two-dimensional errors.) The sum (of all four beams) then represents the total receiver antenna pattern for carrier processing. The three carrier waveforms, two differences and one sum, can then be translated to IF, with the difference carriers each

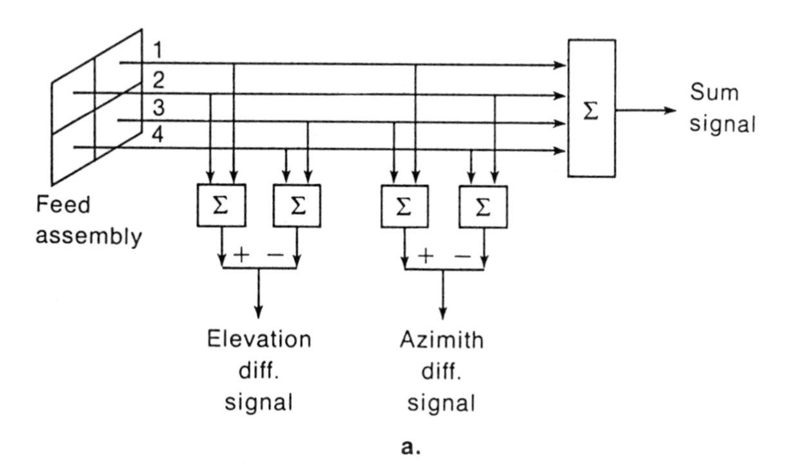

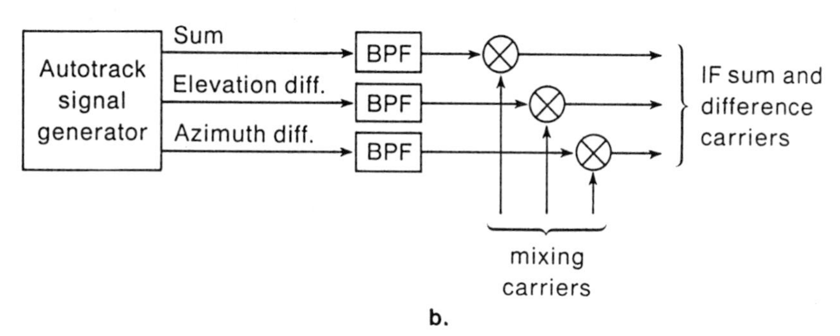

FIGURE 10.15 Autotrack front end. (a) Signal generator. (b) Front end RF-to-IF conversion.

amplitude detected and phase-compared to the sum carrier to extract the autotrack error signals, as shown in Figure 10.15b.

The disadvantage of the system in Figure 10.15 is the duplication of the RF and IF channels for all these carriers. It would be beneficial, from a satellite weight and power point of view, to be able to process the sum and difference carriers in a common RF channel. This can be achieved by modulating the pointing error signals directly on to the sum carrier. The simplest method for the modulation is to add the sum carrier to phase-shifted versions of the difference carriers. Consider again the one-dimensional beam version in Figures 10.14b and 10.14c. Suppose an arrival angle offset of ϕ occurs, producing feed output carrier amplitudes of A_1 and A_2, respectively (Figure 10.16a). If the difference carrier

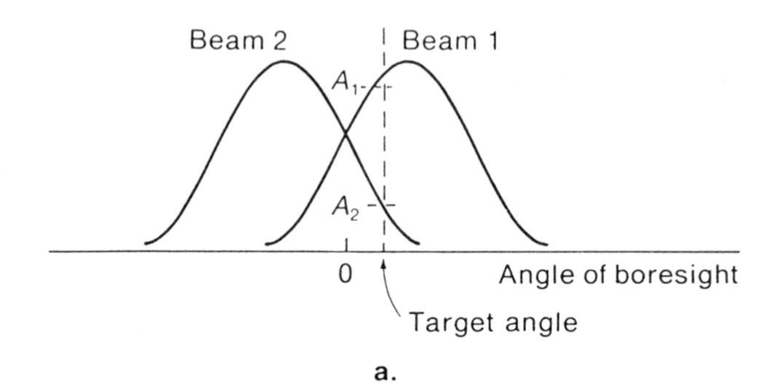

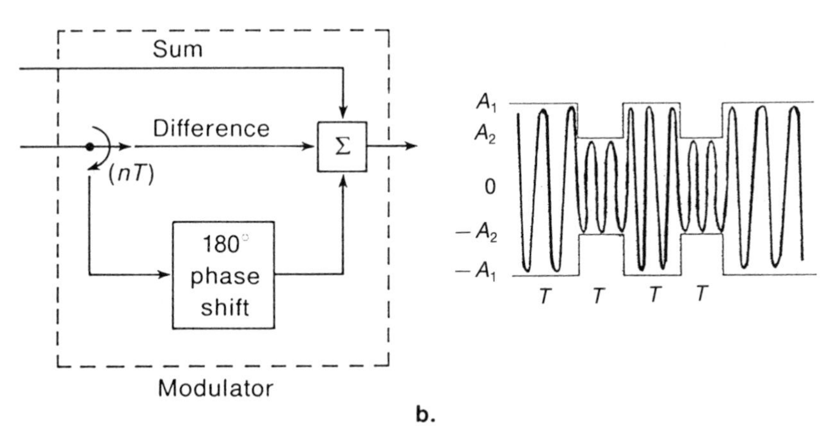

FIGURE 10.16 Generation of modulated autotrack signal. (a) Beam amplitudes due to target offset. (b) Autotrack modulator.

waveform is added to the sum carrier, it will produce a resulting RF carrier with amplitude

$$\tfrac{1}{2}[(A_1 + A_2) + (A_1 - A_2)] = A_1 \tag{10.4.1}$$

This combined carrier has the amplitude produced by receiving with beam 1 alone. Now phase shift the difference carrier by 180° and again add to the sum carrier, producing the amplitude

$$\tfrac{1}{2}[(A_1 + A_2) - (A_1 - A_2)] = A_2 \tag{10.4.2}$$

The carrier now has the amplitude of beam 2. As the 180° phase shift is removed and applied, the carrier amplitude shifts from A_1 to A_2 and back, as shown in Figure 10.16b. This produces an amplitude modulation equivalent to "lobing" from beam 1 to beam 2, with the lobing rate equal to the phase-shifting rate. The depth of the modulation on the carrier is equal to the difference beam amplitude shown earlier. When the antenna is perfectly aligned, the amplitude modulation is zero, and the carrier is equal to the sum beam carrier. As the antenna moves off-axis, amplitude modulation proportional to the offset appears on the sum carrier. The modulated sum carrier can then be translated to IF in a single RF channel, and the amplitude modulation can be retrieved for autotracking, while the sum signal is processed for data.

A separate RF channel must still be provided for both azimuth and elevation errors. Further simplification can occur by time sharing a single RF channel. The elevation error is first amplitude-modulated onto the sum carrier with difference carrier phase shifting, for a fixed time period, then the azimuth modulation is inserted. By using a periodic gating signal in the autotrack modulator, the two error signals are sequentially modulated onto the sum carrier, with the same gating signal used at the IF detector to separate the individual autotrack errors.

The amplitude-modulated autotrack signals are stripped from the sum carrier and used for gimbal control. The azimuth and elevation error signals are separated and filtered to drive the individual gimbals, as shown in Figure 10.17. The gimbal movement is usually in the form of stepping motors (King, 1988) that rotate the antenna axis in discrete steps. Noise added to the RF signals during error detection adds directly to the autotrack signals, and therefore contributes a random component to the instantaneous error signals. In addition, satellite mechanical vibrations,

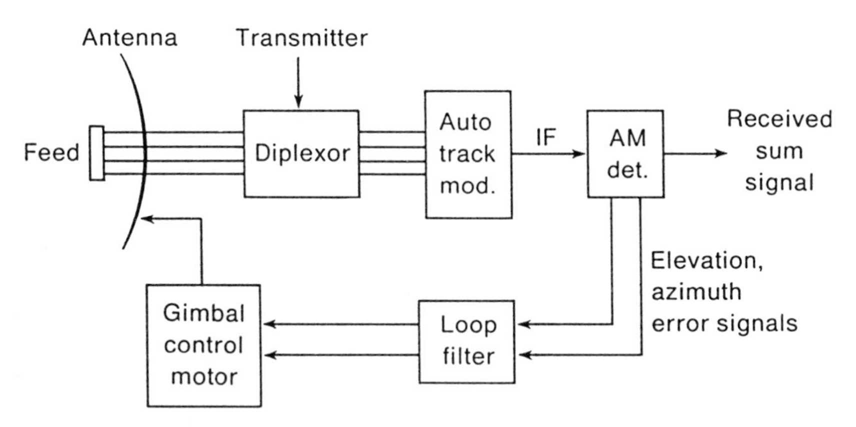

FIGURE 10.17 Autotrack subsystem with modulator and AM error recovery.

metallic bending, and stress perturbations appear as disturbances to the antenna pointing, and contribute to the true pointing errors being generated at any time. These random pointing errors, due to both noise and vibrations, can cause the stepping motors to chatter back and forth as it attempts to correct for the instantaneous pointing offsets.

10.5 AUTOTRACK LOOP ANALYSIS

The monopulse sum and difference carrier waveforms generate the error voltages for autotracking. Thus, a direct conversion occurs between the antenna pointing error $\phi_e(t)$ and the error voltage $e(t)$ at the autotrack detector in Figure 10.17. Assuming no crosstalk between the azimuth and elevation errors, either voltage error can be related to its pointing error by

$$e(t) = S[\phi_e(t)] \tag{10.5.1}$$

where S represents the functional mapping of the monopulse system. This error voltage, for both the azimuth and elevation channels, is then filtered for gimbal control. The dynamical behavior of the autotracking operation can therefore be modulated, just as for the phase-tracking loops in Appendix B. The error voltage $e(t)$ in Eq. (10.5.1) is filtered by the control loop and used to drive the gimbal motors in the appropriate axis. The motor converts the feedback signal at its input to position error at its output so as to reduce the pointing error $\phi_e(t)$. The overall tracking operation can therefore be modeled by the equivalent autotrack loop in Figure 10.18. The input to the loop is the dynamical motion of the line-of-sight to the transmitter, relative to the instantaneous receiver antenna boresight. The loop filter $F(s)$ represents the combined filtering applied to the error voltage, which drives the motors. Since line-of-sight angle variations between synchronous satellites will generally be slow, the filtering bandwidths tend to be narrow, on the order of 1–100 Hz .

The noise entering the loop is due to the receiver noise that appears with the detected amplitude signals. Vibration, bending, and stress motions in the gimbaling hardware appear as an inserted random angle displacement to the pointing of the antenna, and also appears as an equivalent loop input. The dynamical performance of the autotracking, in either azimuth or elevation, can then be predicted from equivalent loops of this type by accurately modeling their input disturbances. Note the direct comparison of the autotrack loop and the phase-lock loops described in Appendix B. In the former case, the variable in the loop

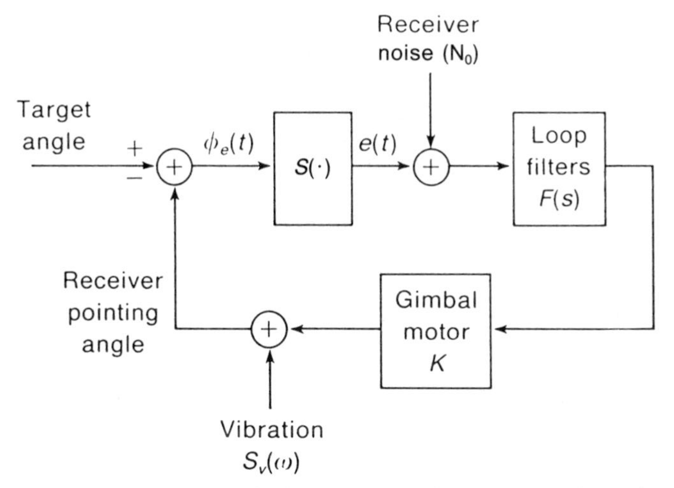

FIGURE 10.18 Equivalent autotrack loop model. Noise is white with level N_0. Vibration spectrum is $S_v(\omega)$.

represents pointing angles, while the latter case has carrier phase angle as its variable.

In typical tracking operations the autotrack will operate in a near-linear mode, so that Eq. (10.5.1) can be approximated as

$$e(t) = \left[\frac{A_s b_0}{2\phi_b}\right]\phi_e(t) \tag{10.5.2}$$

where b_0 is the fractional amplitude modulation index on the sum carrier, ϕ_b is the beamwidth of the sum beam, and A_s is the sum carrier amplitude with perfect pointing. The tracking loop in Figure 10.18 can be replaced by a linear equivalent feedback loop, and analyzed by the standard linear tracking theory, as presented in Appendix B.1. The loop will have the closed loop gain function

$$H_L(\omega) = \frac{KF(\omega)}{1 + KF(\omega)} \tag{10.5.3}$$

where $K = A_s b_0 K_M/2$, and the linearity of the loop permits the error contribution from each of the inputs to be evaluated and summed separately in Figure 10.18.

The total mean squared random pointing error can be obtained by determining the transfer from each input disturbance to the error point.

During accurate line-of-sight pointing the principal disturbance will come from the receiver noise, and will produce a mean squared pointing jitter of

$$\sigma_n^2 = \frac{2N_0 B_L}{(A_s b_0/2\phi_b)^2} \text{ rad}^2 \tag{10.5.4}$$

where $N_0 B_L$ is the RF noise power in the autotrack loop bandwidth B_L, and

$$B_L = \frac{1}{2\pi} \int_0^{\infty} |H_L(\omega)|^2 \, d\omega \tag{10.5.5}$$

This jitter can be written as a fractional error relative to the sum pattern beamwidth as

$$\left(\frac{\sigma_n}{\phi_b}\right)^2 = \frac{4}{b_0^2 (P_c/N_0 B_L)} \tag{10.5.6}$$

where $P_c = A_s^2/2$ is the sum carrier unmodulated power at the AM detector input. Equation (10.5.6) is plotted in Figure 10.19 with modulation index b_0 as a parameter. Note that pointing jitter decreases as a higher index is used for the error modulation, but increased modulation on the sum carrier may degrade the data processing in the sum channel.

Random gimbal vibrations and bending, defined by a power spectrum $S_v(\omega)$, are filtered by the loop transfer function $H_L(\omega)$ in Figure 10.18, to produce its contribution to the mean squared pointing error. This becomes

$$\sigma_v^2 = \frac{1}{2\pi} \int_{-\infty}^{\infty} S_v(\omega) |1 - H_L(\omega)|^2 \, d\omega \tag{10.5.7}$$

Note that reduction of σ_v^2 requires increasing the autotrack loop bandwidth to cover the spectral extent of $S_v(\omega)$. The autotrack bandwidth must be wide enough to track out the significant frequency components of any vibration. Of course, increasing the loop bandwidth produces more pointing jitter by the receiver noise, as indicated by Eq. (10.5.4).

The total mean squared autotrack pointing jitter is then obtained from Eqs. (10.5.4) and (10.5.7) as

$$\sigma_e^2 = \sigma_n^2 + \sigma_v^2 \tag{10.5.8}$$

This indicates the ultimate accuracy in the ability to point the receiving antenna toward the transmitter.

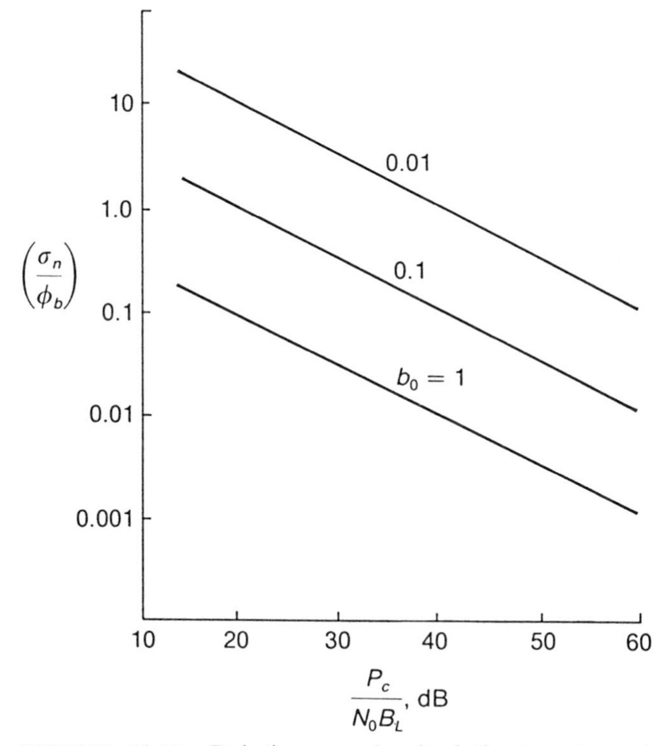

FIGURE 10.19 Pointing error (rms) relative to antenna beam half-power beamwidth. P_c = received carrier power from transmitter. b_0 = autotrack modulator modulation index.

With a Gaussian assumption on the noise and vibration processes, the pointing error in the linear autotrack loop, for both the azimuth and elevation subsystem, evolves as a Gaussian jitter process, each with the variance value in Eq. (10.5.8). For circular symmetric antenna sum patterns, the important parameter is the magnitude of the combining pointing error due to both azimuth and elevation errors, and is given by

$$|\phi_e(t)| = [\phi_L^2(t) + \phi_A^2(t)]^{1/2} \tag{10.5.9}$$

When $\phi_L(t)$ and $\phi_A(t)$ are Gaussian, $\phi_e(t)$ becomes a Rayleigh variable, with its probability density at any t being

$$p_{|\phi_e|}(\phi) = \frac{\phi}{\sigma_e^2} e^{-\phi^2/2\sigma_e^2}, \qquad \phi \geq 0 \tag{10.5.10}$$

The above describes statistically the instantaneous pointing error magnitude during autotrack operation.

10.6 EFFECT OF AUTOTRACK POINTING ERRORS ON PE

Perfect autotracking will remove the pointing error and thereby avoid the antenna pointing loss in Section 10.2. However, practical autotracking accuracy will be limited to the pointing jitter in Eq. (10.5.8). The actual improvement achieved by autotracking will require a performance evaluation taking this jitter into account.

Since the autotrack error will vary relatively slowly with respect to the typical crosslink data rates (the autotrack bandwidth is much smaller than the bit rate), the pointing error appears as a stationary random variable during each bit time. This means bit-error-probability (PE) performance for a digital crosslink will require an average over the PE caused by mispointing. Given a random receiver pointing error $|\phi_e|$, the decoding E_b/N_0 is obtained from Eq. (10.2.2) as

$$\frac{E_b(\phi_e)}{N_0} = \left[\frac{(\text{RIP})}{N_0 R_b}\right] g_r(\phi_e) \tag{10.6.1}$$

where $g_r(\phi_e)$ is the receiver gain pattern at angle ϕ_e. Assuming a parabolic dish with diameter d_r, and using the small-angle approximation in Eq. (3.3.6), Eq. (10.6.1) can be written as

$$\frac{E_b(\phi_e)}{N_0} = \left(\frac{E_b}{N_0}\right)_{\text{I}} c_1^2 d_r^2 e^{-c_2(\phi_e d_r)^2} \tag{10.6.2}$$

where $c_1 = (\pi/\lambda)$, $c_2 = 2.6/\lambda^2$, and $(E_b/N_0)_{\text{I}}$ is the ratio value obtained from the isotropic received power. For a BPSK or QPSK crosslink, the resulting PE is given by

$$\text{PE}(\phi_e) = Q\left[\left(\frac{2E_b(\phi_e)}{N_0}\right)^{1/2}\right]$$

$$= Q\left[\left(\frac{2E_b}{N_0}\right)_{\text{I}}^{1/2} c_1 d_r e^{-(c_2/2)(\phi_e d_r)^2}\right] \tag{10.6.3}$$

For Gaussian loop pointing errors, the combined autotrack error has the

Rayleigh density in Eq. (10.5.10). The average bit-error probability is then

$$\mathrm{PE} = \int_0^\infty \mathrm{PE}(\phi)\left[\left(\frac{\phi}{\sigma_e}\right)e^{-\phi^2/2\sigma_e^2}\right]d\phi$$

$$= \int_0^\infty Q\left[\left(\frac{2E_b}{N_0}\right)_{\mathrm{I}}^{1/2} c_1 d_r e^{-c_2(\phi d_r)^2/2}\right]\left(\frac{\phi}{\sigma_e}\right)e^{-\phi^2/2\sigma_e^2}\,d\phi \qquad (10.6.4)$$

where σ_e^2 is the mean squared loop pointing error in Eq. (10.5.8). Letting $u = \phi/\sigma_e$, PE can be simplified to

$$\mathrm{PE} = \int_0^\infty Q\left[\left(\frac{2E_b}{N_0}\right)_{\mathrm{I}}^{1/2} c_1 d_r e^{-c_2(u d_r \sigma_e)^2/2}\right]u e^{-u^2/2}\,du \qquad (10.6.5)$$

The above can be directly integrated to determine the resulting PE at specific values of $(E_b/N_0)_{\mathrm{I}}$, d_r, and σ_e. Closer inspection, however, reveals that Eq. (10.6.5) depends on σ_e only through its multiplication and division with E_b and d_r. That is, if we denote

$$\gamma_b = \frac{2E_b c_1^2/N_0}{\sigma_e^2} \qquad (10.6.6a)$$

$$\gamma_r = d_r \sigma_e \qquad (10.6.6b)$$

then Eq. (10.6.5) is

$$\mathrm{PE}(\gamma_b, \gamma_r) = \int_0^\infty Q\left[(\gamma_b)^{1/2}\gamma_r e^{-c_2(u\gamma_r)^2/2}\right]u e^{-u^2/2}\,du \qquad (10.6.7)$$

and PE depends on only the normalized parameters γ_b and γ_r in Eqs. (10.6.6), when operated at a fixed frequency. Equation (10.6.7) is plotted in Figure 10.20a as a function of these parameters. It is immediately obvious that a specific link PE can be maintained at various combinations of γ_p and γ_r. With a particular value of each selected, the required isotropic E_b/N_0 and receive antenna diameter d_r can then be obtained from Eq. (10.6.6) for the operating σ_e in Eq. (10.5.8). Note that as σ_e increases (higher rms pointing errors), the antenna must be reduced in size, and the transmit EIRP increased, to maintain the selected values of γ_b and γ_r.

It would be expected that preferred performance is with minimal transmit power (γ_p) while achieving a given PE. Hence, the set of (γ_p, γ_r) combinations in Figure 10.20a that produce the desired PE while having

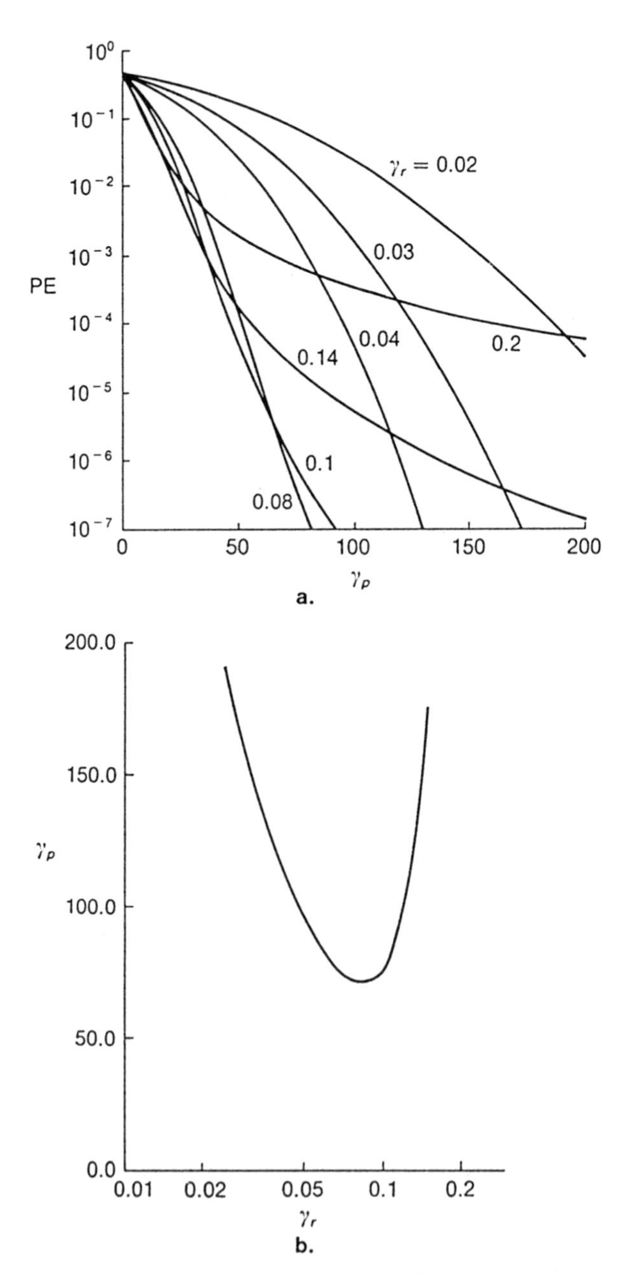

FIGURE 10.20 Bit error probability with autotrack pointing errors. (a) PE vs γ_p and γ_r. (γ_p = normalized power, γ_r = normalized antenna diameter.) (b) γ_p and γ_r values producing PE $= 10^{-6}$.

the smallest γ_p value will constitute an optimal solution. For example, Figure 10.20b shows the (γ_p, γ_r) combinations producing a PE $= 10^{-6}$, as read from Figure 10.20a. From this one can set the optimal receive antenna size for permitting the minimal required transmitter power when operating in the presence of a given pointing variance σ_e^2. The results can be easily extended to account for pointing errors at both ends of the crosslink.

10.7 OPTICAL CROSSLINKS

There is considerable interest in using optical lightwaves for satellite crosslinks, as an alternative to the K-band and EHF crosslinking. With the availability of feasible light sources, and the existence of efficient optical modulators, communication links with optical beams are presently being given serious consideration in intersatellite links. Furthermore, laser operation (at frequencies of about 10^{14} Hz) provide a 10^5 factor over C-band frequencies, which theoretically can now produce about a 90-dB power advantage, as well as a 10^5 advantage in information bandwidth. Although the power advantage aids in overcoming the weaker link, a portion of this improvement can be used to reduce component size (and therefore spacecraft cost and weight). Hence, optical transmitters and receiver packages are usually smaller and lighter than the equivalent RF or microwave subsystems.

The introduction of laser communications, however, requires some modification to the design, implementation, and analysis of the crosslinks. These modifications are caused by the different characteristics of optical devices as opposed to their RF counterpart. In particular, there are: (1) a different technology used for component and device development (quantum-mechanical principles instead of electronics), (2) a laser source instead of electronic oscillators and amplifiers, (3) narrow light beamwidths (fractions of a degree) instead of antenna gain patterns of several degrees, (4) lensing systems for beam transmission and focusing, instead of RF antennas, and (5) photodetectors for light reception instead of the usual field-current conversion in RF reception. Although these differences primarily pertain to device description (all electromagnetic waves obey the same laws of radiation propagation, whether at RF or optical frequencies), they do tend to introduce new nomenclature, terminology, and parameters from those in RF links. These in turn alter link analysis and performance evaluation.

In an optical communication link a laser is used as the transmitting source, and optical fields are generated at carrier frequencies in the optical portion of the frequency spectrum. Figure 10.21 shows this part of the electromagnetic spectrum (recall Figure 1.13) along with the corresponding wavelengths associated with each frequency. Note that an optical system operates at wavelengths on the order of microns (10^{-6} m), which is several orders of magnitude smaller yet than those of millimeter waves (K-band). Since the atmosphere will have extreme deleterious effects at these wavelengths, optical links are usually designed to avoid propagation through the atmosphere. Thus, a satellite crosslink, involving transmissions above the Earth's atmosphere, is a natural candidate for optical wavelength propagation.

A laser optical link model is shown in Figure 10.22. A laser source generates a light wave, which is focused into an optical beam and propagates a distance Z over a free-space path to a collecting area A_r. Focused laser beams are typically modeled as having circular symmetric beam patterns with beamwidth

$$\phi_b = \frac{\lambda}{d_t} \tag{10.7.1}$$

where λ is the wavelength of the laser light and d_t is the diameter of the transmitting optics of the laser. It is important to comprehend the

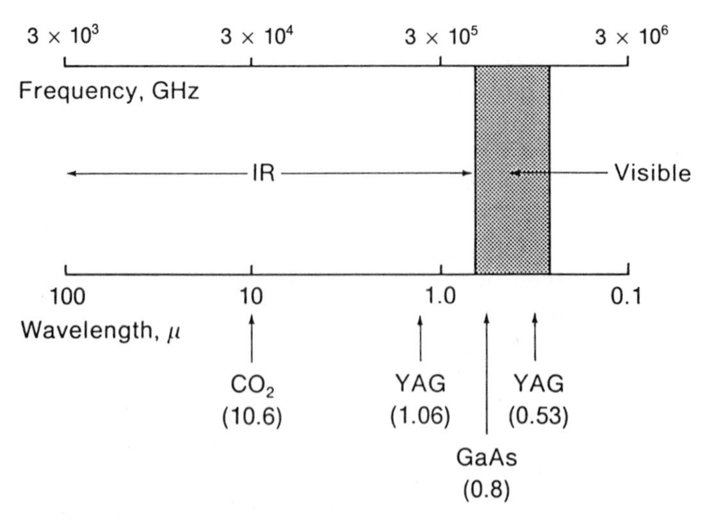

FIGURE 10.21 Upper end of the electromagnetic frequency chart (Wavelength = 3×10^8/frequency). Location of some optical sources shown.

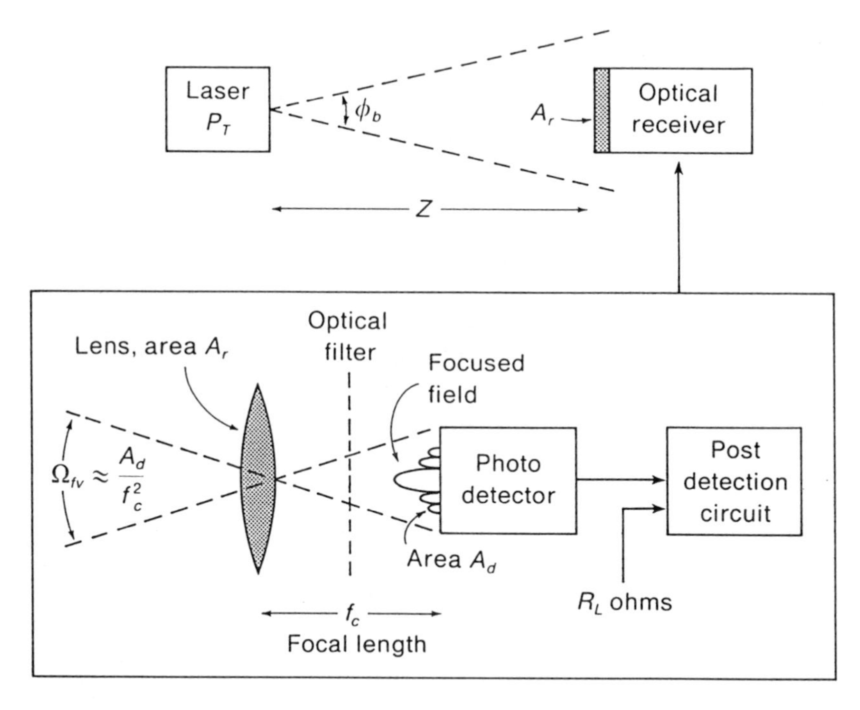

FIGURE 10.22 Laser link model and optical receiver.

order of the dimensions involved. In the visible optical range ($\lambda \approx 1\ \mu m = 10^{-6}$ m) a transmitting lens of only 6 in. (0.15 m) will produce a beamwidth of about 10 μrad (recall 17.5 mrad $\cong 1°$). Hence, a laser transmitter produces beamwidths several orders of magnitude narrower than beamwidths used in RF systems. In optical links, therefore, we deal with beamwidths measured in microradians or milliradians, instead of degrees.

The laser source power P_T is distributed over the beamfront at the distance Z. The optical power density incident over a collecting area A_r in Figure 10.22 is then computed as in an RF system. Hence, the power collected over A_r is

$$P_r = \frac{P_T A_r}{\phi_b^2 Z^2} \tag{10.7.2}$$

A natural tendency is to attempt to rewrite Eq. (10.7.2) in terms of antenna gain values and propagation losses, as we did in earlier RF link analysis.

This indeed can be done for the transmitting end by again defining $g_t \cong 4\pi/\phi_b^2$, as in Eq. (3.2.1). At optical wavelengths, this will produce extremely high values of g_t (e.g., a 6-in. optical transmitter will have a gain of 115.6 dB). Of course, this will be largely overcome by the corresponding large space loss $[L_P = (\lambda/4\pi Z)^2]$ at these wavelengths. The introduction of gain values to replace receiving area A_r in Eq. (10.7.2) in the same way, however, can be misleading in the optical system. This is due to the fact that receiver field of view (which determines receiver gain) does *not* depend only on receiver area A_r but on other receiver focusing parameters as well. (We shall explore this point later.) For the present it is perhaps best to simply deal with received optical power directly, as in Eq. (10.7.2).

When Eq. (10.7.1) is used in Eq. (10.7.2), we have instead

$$P_r = \frac{P_T(d_t d_r)^2}{\lambda^2 Z^2} \tag{10.7.3}$$

which depends only on the squared product of the transmit and receive diameters. Figure 10.23 plots P_r as a function of optic size for a propagation length $Z = 45{,}000$ miles at wavelength $\lambda = 1$ micron, and assuming identical transmit–receive optics. We see that power levels comparable to those received in RF links are obtainable with only milliwatts of source power and with optic diameters on the order of a foot or less. This significant reduction in source power values and optic sizes is a key driving force for continued interest in laser links.

In optical link analyses it is common to convert power values P_r to equivalent *photoelectron count rates*. This is accomplished by dividing P_r by the energy value hf_0, where h is Planck's constant ($h = 6.6 \times 10^{-34}$ W/Hz) and f_0 is the optical frequency. Hence, we denote the number of received photoelectrons per second (usually called simply *counts* per second), corresponding to P_r in Eq. (10.7.3), as

$$n_r = \frac{P_r}{hf_0} \text{ counts/sec} \tag{10.7.4}$$

The number of received counts per sec is, therefore, simply a normalization of received power. At optical frequencies (say $\lambda \cong 1\ \mu$), Eq. (10.7.4) becomes $n_r = (5.1)10^{18} P_r$, and it is evident that count rates are usually extremely large numbers (relative to the fractional watt values for P_r). Figure 10.23 is also labeled in counts per sec obtained by using this conversion factor at 1 μ.

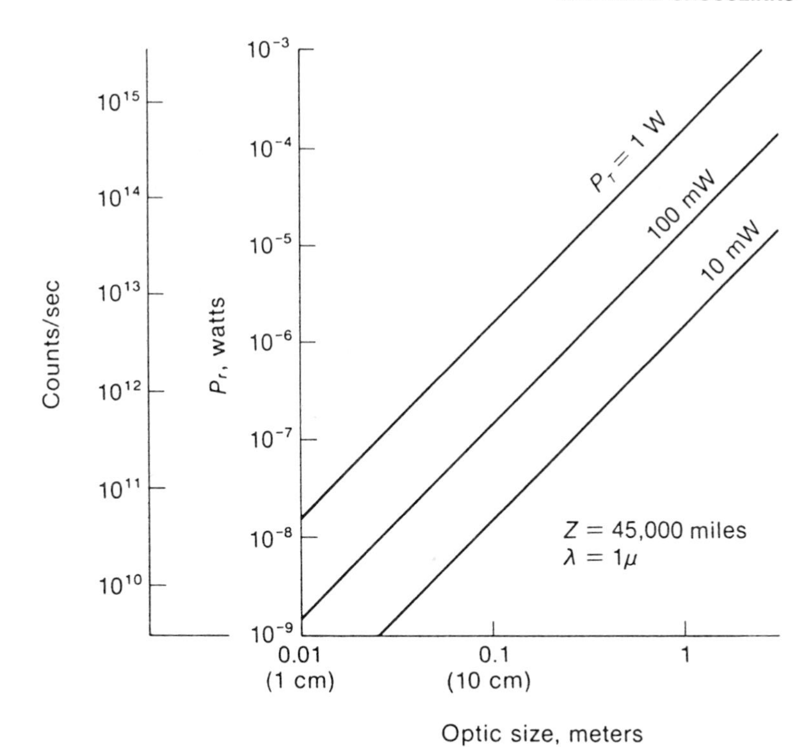

FIGURE 10.23 Received optical power vs. optic size. Transmit and receiver optics are the same size. P_T is the laser source power. Counts per sec refers to received photoelectron count rate.

Optical Components

Laser sources exist at specific wavelengths, depending on the material achieving the lasing. Materials of various types (gases, liquids, semiconductors) have been used to produce lasers. The material determines the physical properties, power conversion efficiency, and required auxiliary hardware of the laser. Table 10.2 summarizes the characteristics of common laser sources projected for use in space communication links. The lasers extend from high-powered, low-efficiency, bulky devices to the smaller, lightweight GaAs (gallium arsenide) solid-state diodes. The latter devices are inherently low powered (output power on the order of tens of milliwatts), and it is generally necessary to combine diodes into arrays to increase source power outputs.

The receiver optical filter in Figure 10.22 limits the range of wavelengths that pass through it. It therefore plays the same role as an RF front-end filter, with its prime function being to reduce the amount of

TABLE 10.2 Summary of laser sources

Laser Type	*Wavelength (μm)*	*Average Power Output*	*Efficiency*	*Characteristics*
Nd–YAG (neodymium–yittrium) aluminium garnet) crystal	1.06 0.532	0.5–1 W 100 MW	0.5–1% 0.5–1%	• Requires elaborate modulation equipment • Requires diode or solar pumping • 10,000 life hours • Frequency doubling loses efficiency
GaAs (solid-state diode)	0.8–0.9	40 MW	5–10%	• Small, rugged, compact • Directly and easily modulated • Easily combined into arrays • Nanosec and pulsing • 50,000 life hours, reliable
CO_2 (gas laser) carbon dioxide	10.6	1–2W	10–15%	• In IR range (poor detectors) • Discharge tube • Modulation difficult • 20,000 life hours, existing technology
HeNe (helium neon)	0.63	10 MW	1%	• Gas tube, power limited, inefficient • Requires external modulation • 50,000 life hours

unwanted radiation entering the photodetector. The range of wavelengths around the laser wavelength allowed by the optical filter is called the *optical bandwidth.* Typical optical filter bandwidths at 1 micron generally range from 10 to 100 angstroms (1 angstrom $= 10^{-4}$ micron), corresponding to an equivalent frequency bandwidth of about $B_0 \approx 10^{11}$–10^{12} Hz (100–1000 GHz). Hence, front-end optical bandwidths are several orders larger than those encountered in microwave systems.

The primary element for receiving an optical field is a focusing lens and a photodetecting surface. In RF systems an antenna dish concentrates the impinging field at the feed point, which electronically converts the field at the feed directly to an electronic signal current. In optics, the focusing lens (or perhaps a series of lens and mirrors) likewise focuses the field onto the photodetector. However, since the optical field is at such a high frequency, it cannot be directly detected. Instead the photodetector responds quantum mechanically by using the radiation to excite its photoemissive surface and produce a photoelectron (current) flow. The latter depends on the instantaneous *field intensity* over the detector surface (*field intensity* is the square of the field envelope; the integral of intensity over the detector area is the *detected field power*). Hence, a photodetector detects the instantaneous field power (instead of the field itself) of the focused field on the detector. This means that all modulation must be imposed on field power (called *intensity modulation*) or must be inserted in such a way that a power detector will recover it. The amount of field power focused onto the detector is equal to the amount of field power incident over the focusing lens. Hence, photodetected power can be equivalently computed by determining receiving field power over the lens area, instead of by focused field power over the detector area. The former is obtained directly from Eq. (10.7.2) with A_r corresponding to the receiver lens aperture area. Hence, Eq. (10.7.2) is indeed valid for computing detector field power even though it does not explicitly involve the actual photodetecting area.

A photodetector will respond to all radiation focused on its photoemissive surface. This therefore defines the optical receiver field of view as the field arrival angles over which the lens will focus the impinging field onto the photodetector surface. From Figure 10.22 we see that this will depend on the detector area A_d and the optical focal length rather than on the size of the receiving lens. Recall that in RF systems, receiver field of view depended only on the size of the receiving antenna A_r, and was given as the *diffraction-limited* field of view, λ^2/A_r. Thus, optical systems have a field of view that can be adjusted independently of the receiver lens area. Typically, the focal length is set to approximately the aperture diameter ($f_c \approx \sqrt{A_r}$), and the optical field of view is generally stated as

$$\Omega_{fv} \cong \frac{A_d}{f_c^2} \approx \frac{A_d}{A_r} = \left(\frac{A_d}{\lambda^2}\right)\left(\frac{\lambda^2}{A_r}\right) \tag{10.7.5}$$

Since detector area is on the order of millimeters or centimeters, while λ is on the order of microns, we see that the optical field of view is usually many times larger than the diffraction-limited field of view (λ^2/A_r) of the same aperture area. This point becomes important in assessing optical visibility and optical receiver noise.

Photodetectors can be of several types (Yariv, 1975) and have specific external characteristics of interest to communication analyses. The most important are listed below.

Detector Efficiency. Only a portion of the incident power may be detected by a photoemissive surface. The detector efficiency η indicates the fraction of received power that is actually detected. Detector efficiency is wavelength dependent, and depends on the material used in the photoemissive surface (silicon, germanium, etc.). Efficiency values are usually in the range (0.15–0.90) for visible frequencies, but fall off rapidly at the higher wavelengths (lower frequencies). The detector efficiency allows us to convert received power P_r in Eq. (10.7.2) to effective detected power ηP_r.

Gain. Photodetectors may be relatively simple photoemissive structures (e.g., PIN diodes) or may be designed as photomultipliers, such that a single emitted photoelectron from its primary surface, owing to incident radiation, may eventually produce multiple photoelectrons at its output. This increases the output current flow for a given incident power, thereby achieving an effective gain during the detection process. This gain may be achieved by cascaded emissive surfaces (as in photomultiplier tubes) or by inherent avalanche mechanisms, as in *avalanche photodetectors* (APD). Photomultiplication gain can be interpreted as either a multiplication of the number of electrons emitted during reception, or as an increase in the current output of the detector. Unfortunately, the multiplication factor achieved by the gain mechanisms is often random in nature, producing a random gain distributed about some mean gain $\bar{G}$. This causes a statistical dependence of the output current on this gain, and introduces an additional degree of randomness to the detector output model. The gain variance, σ_α^2, or photomultiplier *spread*, is an indication of the extent of this randomness. The photomultiplier *excess noise factor* is defined as

$$F \triangleq 1 + \frac{\sigma_d^2}{(\bar{G})^2} \tag{10.7.6}$$

Typically, vacuum tube photomultipliers have mean gains of approximately $\bar{G} = 10^4$–10^6, and excess noise factors between 1 and 2. An APD, being a smaller, lighter device, is limited to gains of about 50–300. Its excess noise factor increases with $\bar{G}$ and is usually in the range of 2 to 5 for the above gain values.

Responsivity. A parameter closely related to efficiency is the detector responsivity, labeled in amps/W. The responsivity simply indicates how much current will be produced for a given power input. Responsivity values for detectors account not only for their efficiency factors, but also for any inherent photomultiplication mean gain. Formally, the responsivity u is given by

$$u = \frac{e\eta\bar{G}}{hf_0} \text{ amps/W} \tag{10.7.7}$$

where e is the charge of an electron ($e = 1.97 \times 10^{-19}$ coulomb). Hence, responsivity is also frequency-dependent, and indicates the average photodetecting capability of the device.

Bandwidth. The photodetector bandwidth determines the rate of power variation that can be detected. It therefore indicates the highest frequency at which the power can be varied and have the variation detected by the output current. As such, it constrains the maximum modulation bandwidth used with intensity modulation. Typical detector bandwidths are usually 1–10 GHz. This bandwidth should not be confused with the optical bandwidth of the receiver filter.

Optical Receiver Noise

An anomaly in optical detection is that a photodetector does not respond perfectly to impinging field power. Instead, the output current observed will appear as a random variation around the true input power value. This random variation is called detector *shot noise*, and is caused by the statistical nature of the emission of photoelectrons during radiation reception. The actual shot noise spectral level is known to be (Gagliardi and Karp, 1975)

$$N_{sn}(\omega) = \bar{G}^2 F e u \bar{P} \text{ amps}^2\text{/Hz} \tag{10.7.8}$$

where $\bar{G}$, F, and u are the detector gain, noise factor, and responsivity parameters previously discussed; e is the electron charge; and $\bar{P}$ is the time-averaged mean received power.

In addition to shot noise, a photodetector produces *dark current*. Dark current is output current that appears even with no input radiation, and is caused by the random thermal emission of photoelectrons due to the inherent temperature of the device. Dark current is a random process, and must be considered an additional noise in the receiving operation. It also is described as a shot noise process, with a spectral level proportional to the average output dark current. Thus, its spectral level is

$$N_{dc}(\omega) = eI_{dc} \text{ amps}^2/\text{Hz} \qquad (10.7.9)$$

where I_{dc} is the detector mean dark current. Typical detector dark current values are in the range 10^{-16}–10^{-12} ampere, and can be further reduced by detector cooling.

To the output detector current must be added the thermal noise current of the postdetection circuitry. This introduces a noise current spectral level of (Van der Ziel, 1954)

$$N_t(\omega) = \frac{4kT^\circ_{\text{eq}}}{R_L} \text{ amp}^2/\text{Hz} \qquad (10.7.10)$$

where k is Boltzmann's constant, R_L is the impedance loading the photodetector, and T°_{eq} is its equivalent noise temperature. The last depends on the noise figure of the loading circuitry (amplifiers, networks, etc.). Note that the thermal noise spectrum is reduced by increasing the load impedance R_L, but this, in conjunction with output capacitance, may tend to reduce the bandwidth of the loading circuitry.

The preceding noise processes can be combined into a total receiver current noise process with the combined two-sided spectral level

$$N_0 \triangleq N_{sn} + N_{dc} + N_t \text{ amps}^2/\text{Hz} \qquad (10.7.11)$$

where the right-hand terms are given in Eqs. (10.7.8), (10.7.9), and (10.7.10).

An optical receiver collects background noise just as an RF antenna. The amount of background power at wavelength λ impinging on an optical receiver having optical bandwidth $\Delta\lambda$, collecting area A_r, and field of view Ω_{fv} is given by Kapeika and Bordogna (1970); Karp, 1988)

$$P_b = H(\lambda)\Delta\lambda A_r \Omega_{fv} \qquad (10.7.12)$$

where $H(\lambda)$ is the background spectral radiance at wavelength λ. Table 10.3 lists values of $H(\lambda)$ at several wavelengths for common background sources. Total background power is obtained by multiplying the appropri-

TABLE 10.3 Tabulation of background radiance values

	$H(\lambda)$, $\frac{watts}{cm^2\text{–}ster\text{–}micron}$ $\lambda = 1$ *micron*	$\lambda = 10$ *microns*
Sun	2×10^3	3×10^2
Sunlit sky (observed from Earth)	2×10^1	1.7×10^0
Moon	3×10^{-3}	4×10^{-5}
Venus	10^{-3}	10^{-4}
Star	10^{-9}	2×10^{-11}
Earth (viewed from sync orbit)	10^{-3}	6×10^{-5}

ate radiance values by the proper values of optical bandwidth, field of view, and receiver collecting area. Contributions from extended backgrounds (e.g., sky as viewed from Earth) increase continually with field of view, while localized sources (stars, planets, etc.) contribute only if they appear in the field of view. Their power contribution increases only until the receiver field of view encompasses the entire source, after which the power remains constant. Figure 10.24 plots typical background power levels as a function of receiver beamwidth for specific optical background noise sources. As in Eq. (10.7.4) background power can be converted to equivalent receiver background count rates as

$$n_b = \left(\frac{\eta}{hf_0}\right) P_b \tag{10.7.13}$$

where again h is Planck's constant and f_0 is the optical frequency. This conversion is included in Figure 10.24. The primary effect of background noise is to add to the total receiver average power, and therefore increases the shot noise spectral level in Eq. (10.7.8).

Optical Link Analysis

The results of the previous equations can now be applied to the analysis of an optical communication link. With intensity modulation, an informa-

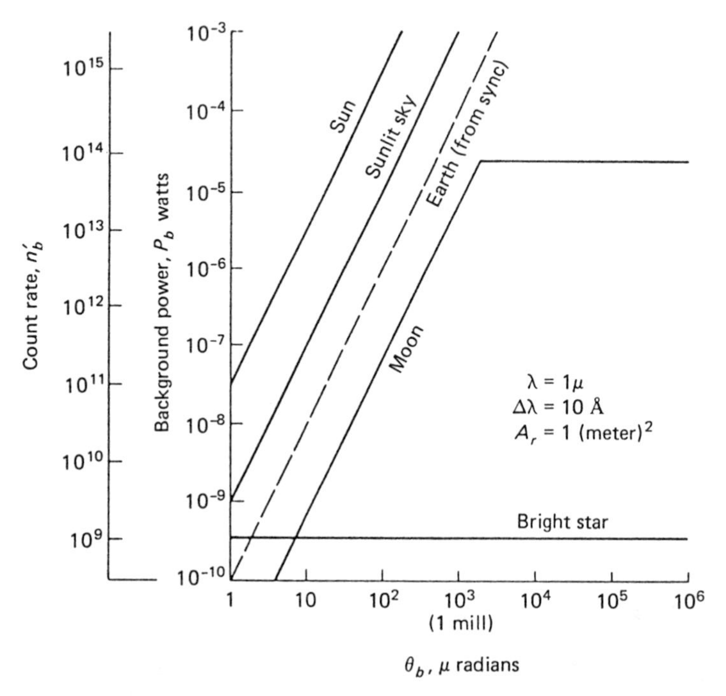

FIGURE 10.24 Background power and count rate due to stated sources.

tion waveform $m(t)$, having bandwidth B_m Hz, is modulated onto the laser field power. The instantaneous received field power collected over the receiver lens area in Figure 10.22 is then

$$P_r(t) = P_r[1 + \beta m(t)] \tag{10.7.14}$$

where P_r is the average received power in Eq. (10.7.3) and β is the intensity modulation index. The background adds a noise power of P_b in Eq. (10.7.12). After photodetection, a current process, $u[P_r(t) + P_b]$, is generated along with the detector noise. A filter for $m(t)$, with bandwidth B_m, produces a signal power

$$P_s = (uP_r\beta)^2 P_m \tag{10.7.15}$$

and a total noise power

$$P_n = N_0(2B_m) \tag{10.7.16}$$

Using Eqs. (10.7.15)–(10.7.16) with $P_m = 1$ we therefore have a photo-detected SNR of

$$\mathrm{SNR}_s = \frac{P_s}{P_n}$$

$$= \frac{(uP_r\beta)^2}{[eu\bar{G}^2F(P_r + P_b) + eI_{dc} + 2kT^\circ_{\mathrm{eq}}/R_L]2B_m} \qquad (10.7.17)$$

The preceding is often called the *electronic*, or *postdetection*, SNR, as opposed to the *optical* SNR at the input (P_r/P_b). The SNR_s can be written in several ways for clarity. Dividing through by $(\bar{G}e)^2$ allows us to rewrite as

$$\mathrm{SNR}_s = \beta^2\left[\frac{(\eta P_r/hf_0)^2}{[(\eta/hf_0)F(P_r + P_b) + I_{dc}/\bar{G}^2e + 2kT^\circ_{\mathrm{eq}}/\bar{G}^2e^2R_L]2B_m}\right] \qquad (10.7.18)$$

Note that the detected SNR_s is nonlinearly related to the optical power P_r, and depends on both P_r and P_b, and *not* simply on their ratio. That is, the electronic SNR is not a simple scaling of the optical SNR. This means that front-end optical losses (in the front-end mirrors, lens, and optical filters) must be correctly accounted for in computing both P_r and P_b. (In RF analysis, all front-end gains and losses cancel out, since only their ratio is needed for performance evaluation.) Note further that both the dark current and thermal noise are reduced by the square of the photodetector mean gain. Hence, high-gain photomultipliers aid in reducing the effect of postdetection noise, although the detector noise factor F also tends to increase with higher gain values.

To further simplify, we introduce the use of count rates, defined in Eqs. (10.7.4) and (10.7.13), and write

$$\mathrm{SNR}_s = \beta^2\left[\frac{n_s^2}{[F(n_s + n_b) + n_d + n_t]2B_m}\right] \qquad (10.7.19)$$

where we have denoted

$$n_d \triangleq \frac{I_{dc}}{\bar{G}^2e} = \left[\begin{array}{l}\text{Average number of}\\ \text{dark current photoelectrons}\\ \text{per second}\end{array}\right] \qquad (10.7.20)$$

$$n_t = \frac{2kT^\circ_{\mathrm{eq}}}{\bar{G}^2e^2R_L} = \left[\begin{array}{l}\text{Average number of}\\ \text{thermal photoelectrons}\\ \text{per second}\end{array}\right] \qquad (10.7.21)$$

Hence, SNR_s appears in a more compact form when written in terms of equivalent count rates instead of power values. Note that thermal noise counts are usually more significant, and we again see the advantage of high photodetection gains in reducing these rates. It is convenient to rewrite Eq. (10.7.19) in the form

$$SNR_s = \beta^2\left(\frac{n_s}{2B_m}\right)\left[\frac{n_s}{Fn_s + (Fn_b + n_d + n_t)}\right] \tag{10.7.22}$$

Since n_s is a count rate, the term $(n_s/2B_m)$ can be interpreted as the number of signal counts in a time $1/2B_m$ s and is often referred to as the *quantum-limited* SNR for a bandwidth B_m. This connotation follows since $n_s/2B_m = \eta P_r/hf_0 2B_m$, and can be equivalently interpreted as a signal power-to-quantum noise power ratio, the latter having an effective noise spectrum $N_0 \triangleq hf_0$. The first term in Eq. (10.7.22) represents a modulation loss, while the last bracket represents a reduction, or squaring loss, on this quantum-limited SNR. Note that when the signal count rate n_s exceeds the combined noise count rate, SNR_s approaches the quantum-limited bound. Interestingly, even if the total noise count is zero, SNR_s is not infinite but is bounded by the presence of this quantum noise. When in quantum-limited operation, we see that SNR_s varies linearly with n_s, but when the noise count dominates, SNR_s varies as the square of n_s. Thus, at high SNR_s, performance is linear in laser power, but at low SNR_s it improves as its square.

An optical crosslink between two satellites can be established in several ways. Figure 10.25 shows one concept, which uses direct intensity modulation of the entire RF uplink onto the laser field for the crosslink. We assume a C-band RF carrier with a 500-MHz bandwidth for the uplinks. The uplink satellite uses a direct frequency translation in which the uplink carrier bandwidth is directly intensity-modulated onto the optical carrier for the crosslink. It must be assumed that the upper RF frequency at C-band is within the receiving photodetector bandwidth. If not, the uplink bandwidth must first be down-connected before laser modulation. The uplink produces an RF carrier to noise ratio CNR_u. The photodetected waveform after the crosslink transmission, which exists at the C-band uplink frequency, is then translated to the downlink frequency, power amplified, and retransmitted. The total downlink retransmitted noise power (uplink plus total detector noise) is received at the ground station. In addition the downlink receiver adds in its thermal noise. The resulting downlink CNR, after uplink, optical crosslink, and downlink transmission, can be written in the form

$$CNR_d = [(CNR_u)^{-1} + (SNR_s)^{-1} + (CNR_r)^{-1}]^{-1} \tag{10.7.23}$$

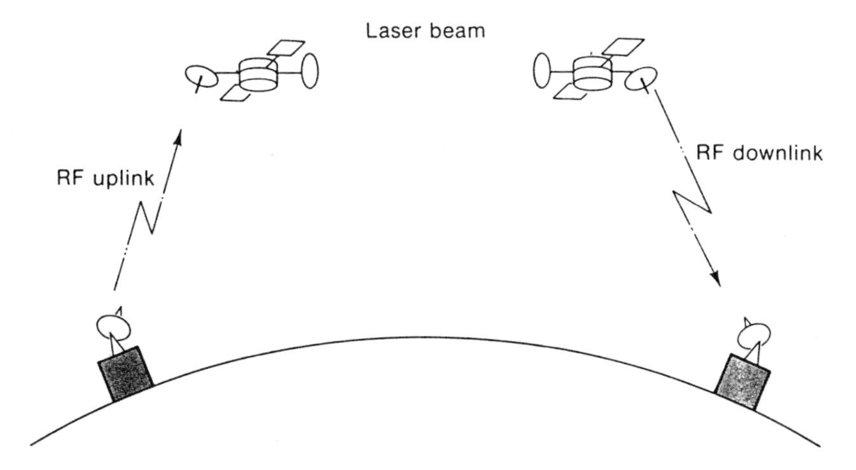

FIGURE 10.25 Laser-satellite crosslink model.

where SNR_s is given in Eq. (10.7.18) and all CNR refer to the uplink satellite carrier bandwidth. Comparison to Eq. (3.6.10) shows that the presence of the optical crosslink simply adds a term to the basic uplink-downlink equation, while modifying the usual downlink CNR_r term. This latter term is due to the satellite amplifier carrier suppression effects, which are now determined by the optical link. We again see that overall performance is determined by the weaker of the three links. The system is *transparent* to the optics if the optical SNR_s is maintained at a larger value than the individual RF uplink and downlink CNR in the satellite bandwidth.

Digital Optical Crosslinks

In some cases it may be necessary to establish a laser crosslink to transmit digital data at a prescribed performance level, independent of the uplink and downlink. This may occur, for example, in a decode–encode system where data on the uplink is decoded at the satellite prior to retransmission on the crosslink. In other cases, the digital data may be generated directly at one satellite, and is to be transmitted to the other satellite via a direct optical link. The crosslink design then requires relating bit rates and error probabilities to the optical link parameters, rather than merely computing CNR values.

Figure 10.26 shows a block diagram of the system. The data bit stream is modulated onto the laser, and the photodetected output is decoded back to bits. The easiest method is simply to intensity-modulate bit waveforms directly onto the laser, and decode the photodetector output

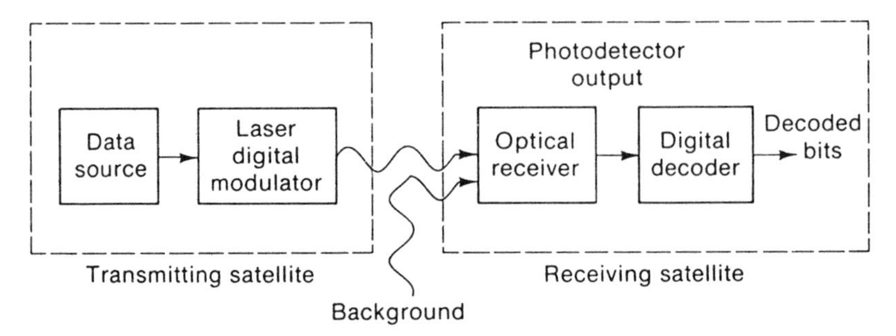

FIGURE 10.26 Optical digital satellite link.

by standard baseband circuitry. The bit waveforms can be NRZ, Manchester, or BPSK, using a data subcarrier. NRZ systems utilize the least intensity bandwidth, while Manchester and subcarrier waveforms avoid the low-frequency (dc) optical interference. Performance of these systems can be determined by using Gaussian* error probability curves, with an E_b/N_0 determined from the detected SNR_s derived earlier. Specifically, we evaluate the detected SNR_s at a bandwidth equal to the bit rate. Hence,

$$\left(\frac{E_b}{N_0}\right) = \text{SNR}_s|_{B_s = R_b = 1/T_b} \tag{10.7.24}$$

where R_b is the bit rate, T_b is the bit time, and SNR_s is obtained from either Eq. (10.7.17) or (10.7.19). Table 10.4 lists the resulting bit-error probability for several encoding schemes. Hence, performance can be estimated via standard Gaussian curves using the appropriate values of E_b/N_0. Note that the latter will depend primarily on the parameters:

$$K_s \triangleq n_s T_b = \text{Number of signal counts in a bit time}$$

$$K_b \triangleq n_b T_b = \text{Number of background noise counts per bit time}$$

$$K_n \triangleq (n_d + n_t)T_b = \text{Number of effective detector noise counts per bit time} \tag{10.7.25}$$

---

* The use of Gaussian receiver statistics in computing error probabilities is only an approximation. Photodetected noise processes are shot noise processes (Gagliardi and Karp, 1975) with counting statistics evolving as classes of Poisson processes. It has been shown that these processes approach Gaussian statistics at high count values (number counts per bit $\gtrsim 50$), and PE can be adequately approximated from Gaussian analysis. For further discussions of these models see Gagliardi and Karp, (1975).

TABLE 10.4 Tabulation of digital error probabilities PE for continuous carrier modulation

	Signal Format	*PE*	E_b/N_0	*Bit rate, (bps)*
Direct intensity modulation	NRZ Manchester PSK	$Q\left(\sqrt{\frac{2E_b}{N_0}}\right)$	$\beta^2 \frac{K_s^2}{K_s + K_b + K_n}$ $K_s = n_s T_b$ $K_b = n_b T_b$ $K_n = (n_d + n_t)T_b$	$\frac{1}{T_b}$
	Noncoherent FSK	$\frac{1}{2}e^{-E_b/2N_0}$		
Pulsed digital	On–off keyed (OOK)	$\frac{1}{2}Q\left(\sqrt{\left(\frac{E_b}{N_0}\right)_0}\right) + \frac{1}{2}Q\left(\sqrt{\left(\frac{E_b}{N_0}\right)_1}\right)$	$\left(\frac{E_b}{N_0}\right)_0 = \left(\frac{1}{4}\right)\frac{K_s^2}{K_s + 2K_b + 2K_n}$ $\left(\frac{E_b}{N_0}\right)_1 = \left(\frac{1}{4}\right)\frac{K_s^2}{2K_b + 2K_n}$	PRF
	Pulse modulation (PPM)	$\frac{M}{2}Q\left(\sqrt{\frac{E_b}{N_0}}\right)$	$\frac{K_s^2}{K_s + 2K_b + 2K_n}$	$(\log_2 M)$PRF

Thus, in laser-encoded digital systems, error probabilities can be determined directly from the number of detected signal and noise counts occurring in a bit time.

When either the background or receiver noise counts are high relative to the signal count, E_b/N_0 may be low, and performance for any of the schemes in Table 10.4 will be poor. This can often be improved by pulsing the laser, and encoding the resulting light pulses (Figure 10.27a). The laser source is pulsed on and off at a prescribed *pulse repetition frequency* (PRF), producing a light pulse with fixed pulse width (τ) and peak pulse power (P_p) every 1/PRF s. The laser, therefore, operates at an average power P_r (referred to the receiver) satisfying

$$P_p \tau = P_r/\text{PRF} \tag{10.7.26}$$

At the receiver the pulsed laser field is photodetected (Figure 10.27b), and its output integrated over the τ-s pulse time, producing a signal count

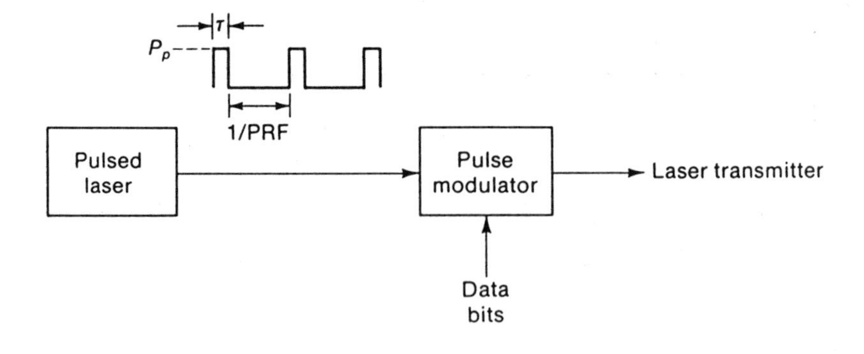

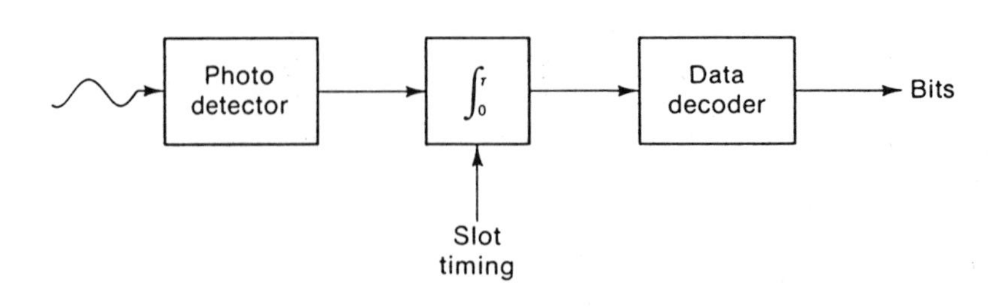

FIGURE 10.27 Pulsed-laser digital modulation system. (a) Pulsed transmitter. (b) Receiver decoder.

$K_s = n_p\tau$, where n_p is the detected count rate produced during the peak power P_p. The detected pulse SNR after photodetection is now

$$\mathrm{SNR}_s = \frac{K_s^2}{K_s + K_b + K_n} = K_s\left[\frac{P_p}{P_p + P_b + P_n}\right] \tag{10.7.27}$$

where K_s, K_b, and K_n are detected pulse counts over a pulse time.

With average laser power constrained, SNR_s can be increased with higher peak power P_p. Hence, performance is improved by using higher and narrower laser pulses, while still maintaining the same average laser power. The noise terms in the bracket in Eq. (10.7.27) are now offset by the peak laser power, instead of by its average power. Note from Eq. (10.7.26) that peak power is increased by either decreasing PRF (slower pulse rate) or decreasing τ (narrower pulse width, with a corresponding increase in laser modulation bandwidth). The minimal value of τ is determined by the maximum photodetector bandwidth. The maximum value for P_p is determined by the largest peak power that the laser source can generate. Laser pulsing is achieved by two basic source modifications—*mode locking* and *Q-switching*. Mode locking allows laser pulses to be formed at a high PRF (hundreds of megapulses per second). Q-switching involves a cavity dumping procedure that allows extremely high pulses to be generated at a low PRF (tens of kilopulses per second).

With pulsed lasers, digital data is transmitted by encoding each of the individual pulses. In *on–off keying* (OOK) a laser pulse (or no pulse) is transmitted to represent each bit. In *pulse position modulation* (PPM) the laser pulse is delayed into one of M possible pulse locations during each pulse period. In this case $\log_2 M$ bits is sent with each pulse. In each system the corresponding error probability depends on the ability of the receiver to detect the true pulse states, which are related to the detected pulse counts. Table 10.4 lists the performance characteristics of these pulse-encoding formats. Note the data rate is directly related to PRF, while error probability depends on both signal and noise counts occurring in a pulse time τ.

Optical Beam Pointing

An optical crosslink will require precise pointing and beam tracking, just as for the EHF link. In fact the optical pointing is even more severe due to the narrower beams (μrad instead of mrad) that occur with lightwaves.

However, the basic optical pointing procedures are carried out in much the same way as with RF autotracking. Each satellite points its optical beam back toward the other satellite by tracking the arriving optical beam line of sight from that satellite. The standard procedure for tracking an arriving optical field direction is by the use of quadrant detectors for the photodetection (Figure 10.28). The quadrant detector uses an array of individual photodetectors instead of one large detecting surface. By focusing the received field to a spot on the quadrant array, pointing-error voltages can be generated in azimuth and elevation by properly combining the quadrant outputs. The error voltages are proportional to the displacement of the focused spot from the array center. In essence, the quadrant outputs operate much like the separate feeds of the RF monopulse tracker. The sum of the quadrant outputs form the optical signal used for the link communications, while the error signals are fedback and used to control the pointing gimbals by keeping the focused spot on the quadrant crosshairs. If there is significant relative motion between the satellites during the two-way transit time, it may be necessary to point the return beam slightly ahead to ensure illumination of the receiving satellite when it arrives.

Tracking-error studies associated with gimbal-controlled positioning

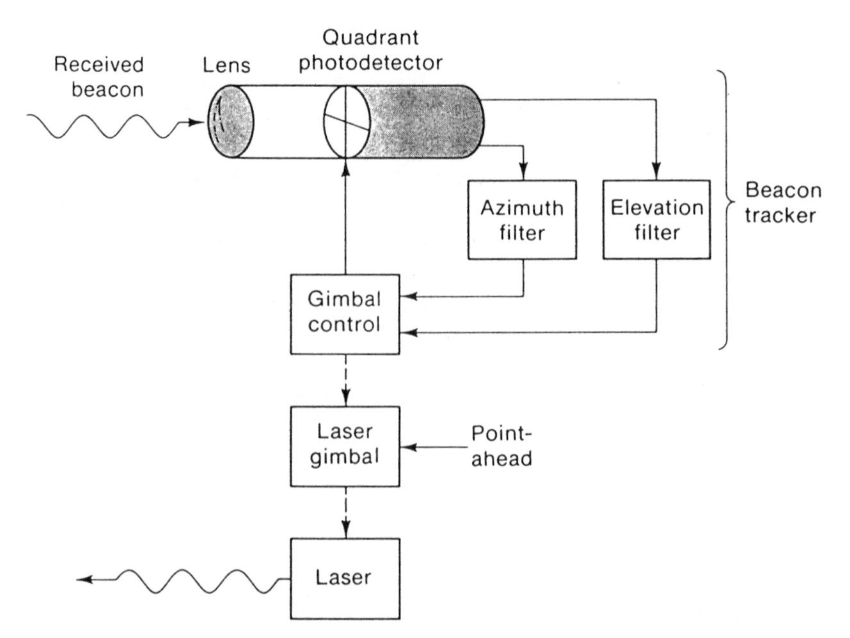

FIGURE 10.28 Optical beacon-tracking subsystem.

loops have been previously reported (Gagliardi and Karp, 1975; Gagliardi and Sheikh, 1980). The primary tracking errors are caused by motion of the LOS, noise, clutter, and internal mechanical vibrations. LOS motion should be minimal when both satellites are in synchronous orbits. If satellites are in near-earth orbits, the tracking loop bandwidth B_L must be large enough to cover the LOS rotation. If Ω_L is the angular rotation rate of the LOS in rad/s, we require approximately

$$B_L \geq \Omega_L/4 \tag{10.7.28}$$

for accurate LOS tracking. The rms tracking error due to quadrant detector noise is given by

$$\theta_{\text{rms}} = \frac{1.4\lambda/d}{\sqrt{\text{SNR}_s}} \tag{10.7.29}$$

where λ is the optical beacon wavelength, d is the beacon receiver diameter, and SNR_s is the quadrant detector SNR given by

$$\text{SNR}_s = \left(\frac{16n_s}{3\pi B_L}\right)\left[\frac{n_s}{n_s + 4n_n}\right] \tag{10.7.30}$$

Here n_s is the beacon count rate and n_n is the total noise count per each quadrant detector. The analysis of a specific optical beam tracking loop can be carried out similar to the RF analysis in Section 10.6, and the effect on link PE performance can be also estimated (Chen and Gardner, 1989).

REFERENCES

Aerospace Corp. (1983), "Communication Satellite Technology," Aerospace Corp. Report TOR-0083(3417-03)-02.

Agarwal, V., and Taggart, D. (1979), *60 GHz Crosslink Technology Study*, Aerospace Corp. Report ATM 79(4417-01)-8.

Alexovich, J. (1988), "Frequency Synthesizer Phase Distortion Effects in Frequency Hopped Communications," *PhD Dissertation presented to the Dept. of Electrical Engineering*, University of Southern California.

Alexovich, J., and Gagliardi, R. (1990), "Effect of Phase Noise on Noncoherent Communications," *Trans. IEEE on Comm.*, **COM-22** (September), pp 921–933.

Chen, C. and Gardner, C. (1989), "Impact of Random Pointing Errors on the Design of Optical Intersatellite Communication Links, *IEEE Trans. on Comm.*, **COM-37** (March,) pp. 252–260.
Gagliardi, R. and Karp, S. (1975), *Optical Communications*, Wiley, New York, Chapter 4.
Gagliardi, R. and Sheikh, M. (1980), "Pointing Error Statistics in Optical Beam Tracking," *IEEE Trans. Aerospace and Electronic Systems*, **AES-16** (5) (September) 674–682.
Hewlett-Packard (1985), "Phase Noise Characteristics of Microwave Oscillators," Hewlett-Packard Product Note 11729B-1.
Kapeika, N. and Bordogna, J. (1970), "Background Noise in Optical Systems," *Proc. IEEE*, pp. 1571–1578.
Karp, S. et al. (1988) *Optical Channels*, Plenum Press, New York.
King, M. (1988) "RF Crosslink System Design—Appendix A" in *Laser Satellite Communications*, M. Katzman, ed. Prentice-Hall, Englewood Cliffs, NJ.
Park, S., Tanzi, P., and Kelley, D. (1973), "IMPATT Power Amplifiers For Digital Communications," *IEEE Trans. Microwave Theory and Techniques*, **21**, 716–720.
Schroeder, W., and Haddad, C. (1973), "Nonlinear Properties of IMPATT Devices," *Proc. of the IEEE*, **61** (2), 153–181.
Van der Ziel, A. (1954), *Noise*, Prentice-Hall, Englewood Cliffs, NJ.
Watkin, E. (1985), "LNA Designed for Advanced EHF Systems," *Microwave System News*, **12**. pp. 1127–1141.
Yariv, A. (1975), *Quantum Electronics*, 2d ed., Wiley, New York.

PROBLEMS

10.1. A crosslink can only be pointed (open loop) to within 50% of a beamwidth at each end of a crosslink. How much transmitter power can be saved by inserting a perfect autotrack subsystem at each end?

10.2. Derive Eq. (10.2.5) from Eq. (3.2.5).

10.3. Determine the time it will take to acquire a 60-GHz carrier having a frequency uncertainty of 10 parts per million, using a carrier phase tracking loop of bandwidth 500 Hz.

10.4. The Doppler rate on a 60-GHz crosslink is 0.01 parts per million. What carrier tracking bandwidth B_L is needed to maintain no more than a 12° phase-reference error bias?

10.5. Assume a BPSK carrier with a phase noise spectrum $S_\psi(\omega) = S_0$. Evaluate the degradation D that it will cause in a binary FSK noncoherent link.

10.6. A first-order linear tracking loop with a noise bandwidth B_L is used in Figure 10.18. Assume a gimbal vibration enters as a sine wave at frequency f_m. What B_L is needed to produce a 20-dB rms vibration reduction in pointing with this loop?

10.7. An autotrack subsystem produces a 30% modulation on the sum carrier, and operates with a beamwidth of 0.1° and a tracking bandwidth of 10 Hz. What received P_c/N_0 is needed to maintain an rms pointing error no larger than 10 millidegrees?

10.8. After autotrack, the sum RF signal in Figure 10.17 may be used for data decoding. Assume the data bit rate is much larger than the autotrack bandwidth. Estimate the worst case degradation to the decoding E_b/N_0 due to autotrack using an AM index b_0 in Eq. (10.5.2). *Hint*: decoding can occur at any point on the signal waveform.

10.9. (a) Show that with no noise and an integrating loop filter in Figure 10.18 $[F(\omega) = 1/j\omega]$ the pointing error $\phi_e(t)$ satisfies the differential equation

$$\frac{d\phi_e}{dt} + KS(\phi_e) = \frac{d\theta}{dt}$$

where $\theta(t)$ is the target angle motion. (b) Show that the pointing errors $\phi_1(t)$ and $\phi_2(t)$ at both ends of a crosslink, using the same loop as in (a), must satisfy the joint pair of differential equations

$$\frac{d\phi_1(t)}{dt} + K_1 g[\phi_2(t - t_d)]S(\phi_1) = \frac{d\theta_1}{dt}$$

$$\frac{d\phi_2(t)}{dt} + K_2 g[\phi_1(t - t_d)]S(\phi_2) = \frac{d\theta_2}{dt}$$

where $g(\phi)$ is the antenna gain patterns and t_d is the one-way propagation time between satellites.

10.10. The count rate n_r defined in Eq. (10.7.4) actually refers to the number of photons collected from a quantum-mechanical field. (For this reason, an optical receiver is often called a *photon bucket*.) Prove that $n_r T$ is the number of field photons occurring in a time period T. [*Hint*: The energy of a photon of frequency f_0 is hf_0.]

10.11. (a) Plot the gain and beamwidth of an optical ($\lambda = 1\ \mu$) antenna as a function of aperture diameter d. (b) Plot the propagation loss at the same λ as a function of distance.

10.12 In a properly designed lensing system with focal length f_c, the field imaged on the detector $f_d(x, y)$ is the two-dimensional Fourier transform of the field passing through the lens aperture, $f_l(x, y)$. That is,

$$f_d(x, y) = \int_{\text{plane}} f_l(u, v) \exp\left[-j\frac{2\pi}{\lambda f_c}(xu + yv)\right] du\, dv$$

Assume an arriving plane wave (field constant over an infinite plane), and a square lens aperture of width d. Compute the imaged field.

10.13. Given a blackbody background at temperature 300 K. How much noise per Hz would be collected at wavelength of 0.5 μm in an optical receiver area of 2 m^2 and a 1-deg field of view?

10.14. In certain optical detection models, the count detected over a time interval has a Poisson distribution, in which the probability that the count is n (an integer) is given by

$$P(n) = \frac{K^n}{n!} e^{-K}$$

where K is the mean interval count. Use this to write the Poisson error probability of an M-slot PPM system, assuming the pulsed slot has mean count $K_s + K_b$, and all other slots have mean count K_b. Pulse decoding is based on selecting the signal slot as that with the largest count.

10.15. The *noise equivalent power* (NEP) of an optical receiver is defined as the value of received power P_r such that $\sqrt{\text{SNR}_s} = 1$ in a 1 Hz noise bandwidth. Derive an equation for the NEP of an optical receiver with shot noise, background noise, dark current, and thermal noise.

10.16. It is desired to send 10^5 bps with OOK encoding over a quantum-limited optical (1 μm) channel from Jupiter (10^{10} km) and achieve a PE $= 10^{-4}$. How many joules (watt-sec) are required from a source having a 6-in. transmitting lens, assuming the receiver uses a 2 m^2 area?

10.17. A four-bit PPM direct detection optical communication system is desired with a word-error probability of 10^{-4}, and has a background count energy of $K_b = 3$ in each slot. Approximately how many joules per pulse must be received to operate the system. (1 joule = 1 watt-s).

10.18. An optical PPM system uses a source of average power P and pulse repetition rate of r pulses per sec. Derive the equation for the peak pulse power in a system sending R_b bits per second.

10.19. An optical tracking system operating at a 1-μ wavelength uses a 1-kHz tracking bandwidth in both elevation and azimuth. The receiver has a 6-in. lens and operates with a total noise count rate of 10^4 cps. What received signal count rate n_s is needed to track (in either elevation or azimuth) within an rms pointing error of 0.5 μrad?

CHAPTER 11
VSAT and Mobile-Satellite Systems

The satellite systems considered in the previous chapters basically involved satellite transmissions to earth stations with moderate or large size antennas. One of the future challenges is the application of satellites to communicate directly with small, low-cost terminal stations. These stations have been given the acronym VSAT—for *very-small-aperture terminals*—and refer to ground stations that operate with antenna apertures around 1 m or less. VSATs can be fixed terminals with a specific location and orientation relative to the satellite, or can be associated with mobile terminals placed on moving vehicles, such as autos, trucks, ships, or planes. Since the communication characteristics of a fixed-point VSAT differs from those associated with mobile terminals, the former will be simply referred to as VSATs, while the mobile terminals will be separately addressed.

The VSAT/mobile-satellite concept has already been mandated for service by the FCC, with allocated frequency bands for precisely this purpose. It is expected that successful development of VSAT–satellite links will evolve into widely diverse applications when eventually implemented. The system requires the development of some hardware somewhat different from today's technology, as we shall see in this chapter. Preliminary construction and testing with experimental satellites has begun, and the hardware development is underway for several of the proposed concepts being pursued.

11.1 THE VSAT–SATELLITE-SYSTEM CONCEPT

The concept of a VSAT–satellite network system is shown in Figure 11.1. The satellite (or a network of satellites) is to provide the necessary interconnection for transmission between fixed or mobile VSATs. The VSAT terminals are projected to be compact, lightweight, solid-state packages with small antennas, that can be placed on rooftops and building facades, or can be surface-mounted on moving vehicles. The intention is for each such terminal to operate independently, and communicate directly with the satellite. The satellite may provide simple bent-pipe relaying through a single transponder, or may insert various levels of on-board processing to complete the links. The individual links may be straightforward VSAT-VSAT transmissions, or involve communications between fixed and mobile terminals.

The frequency bands presently allocated for VSAT service are listed in Table 11.1. They permit VSAT operation at UHF, L-band, and K-band, with specific allocations for both uplink and downlink service. UHF and L-band assignments are primarily for mobile systems, while the K-band is designated for fixed-point systems. It is expected that the preferred use of the UHF allocated bands will be based on an FDMA concept, in which the available bandwidths are partitioned into 5-KHz subbands, and assigned on a per channel basis. Thus, a 5-MHz total UHF band can conceivably support about 1000 separate channels. However, other multi-

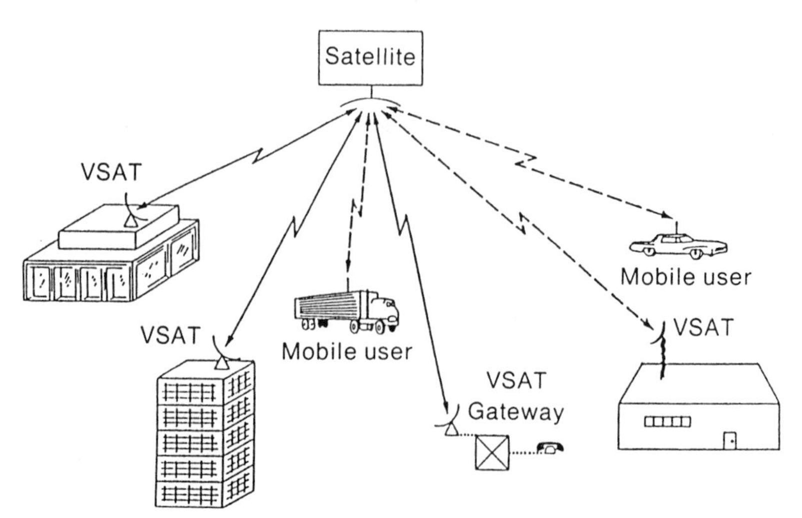

FIGURE 11.1 VSAT and Mobile Satellite System.

TABLE 11.1 Allocation frequencies for mobile satellites

Band	*Frequency*	*Use*
UHF	821–825 MHz	Uplink
UHF	845–851 MHz	Uplink
UHF	866–870 MHz	Downlink
UHF	890–896 MHz	Downlink
L-band	1.631–1.634 GHz	Uplink
L-band	1.530–1.533 GHz	Downlink
K-band	14–14.5 GHz	Uplink
K-band	11.7–12.2 GHz	Downlink

ple-access formats are still being considered for more efficient use of the available K-band bandwidths. With the limited band size, and the expected proliferation in requests for VSAT services, it is expected that viable frequency reuse techniques will be an important part of system development.

The use of small terminals for the transmitter and/or receiver stations immediately indicate that the up and down satellite links will be limited in power. Not only is a small compact solid-state transmitter restricted in its power generation capability, but its small-aperture antenna gain is likewise limited. In the mobile environment the problem is further compounded by the difficulty in controlling antenna pointing during vehicle motion (without a significant increase in equipment complexity), which means near omnidirectional antenna patterns must be used to ensure satellite visibility. This restricts most mobile antenna gains to only a few decibels. This combination of limited link power and restricted channel bandwidth with the FDMA format presents a challenge to the communication engineer to maintain a quality UHF link. On the other hand, a fixed-point VSAT terminal at K-band using only a 1-ft antenna pointed directly at the satellite could produce as much as 30 dB of gain, simplifying the link design.

The satellite segment of a small-aperture network will most likely require wide-coverage downlink patterns to ensure operation with geographically scattered terminals. To accomplish this with a single downlink antenna pattern will greatly restrict satellite antenna gain levels, which can further complicate the link power capability. The use of spot-beam antennas increases gain, but may force multiple-beam or hopping-beam coverage, thereby complicating the satellite electronics. These trade-offs will be examined in the next section.

While a fixed-point VSAT can be pointed directly at a satellite, the mobile VSAT system must contend with the continual vehicle motion that changes orientation relative to satellite line of sight. Several concepts proposed for vehicle antenna are shown in Figure 11.2. In Figure 11.2a is a simple roof-mounted, omni-directional flat antenna to provide satellite coverage in all directions. In Figure 11.2b is a planar array or verticle dipole to provide some degree of antenna pointing by restricting coverage to about a 30° elevation angle (the approximate angle at which a point in the United States observes a synchronous satellite at the Equator). Figure 11.2c is a possible form of steerable antenna designed to produce

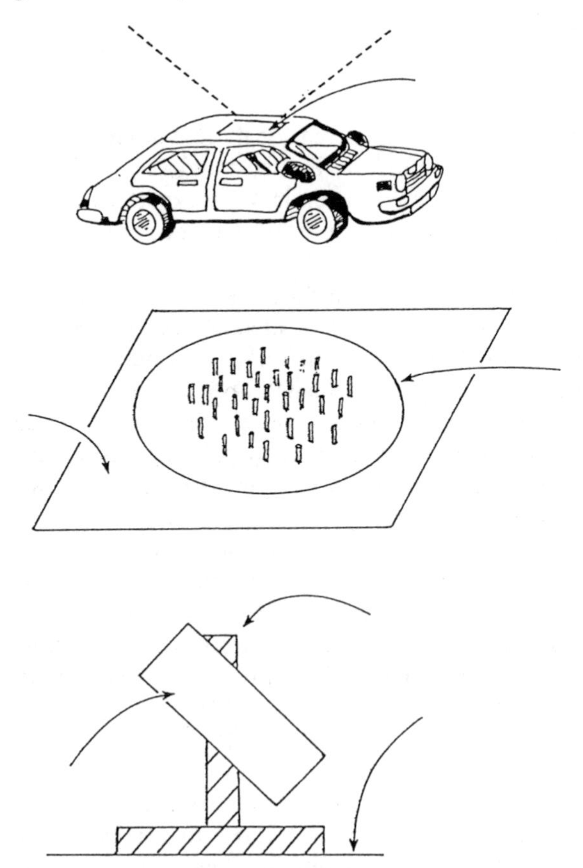

FIGURE 11.2 Vehicle mounted antennas. (a) Flat, roof-mounted. (b) Phase array. (c) Steerable flat.

a focused pattern that continually tracks the satellite throughout vehicle movement. The beam steering can be done electronically via array processing, or mechanically by physically rotating a gimballed flat plate or dish antenna. A 30-in. antenna at UHF will provide about 12 dB of gain if pointed at the satellite. Details of the steerable antenna construction and the generation of steering signals can be found in Bell and Naderi (1986), Huang (1988), JPL (1988a) and will not be considered here.

The modulation formats to be used in VSAT digital transmissions will depend on the data rate desired. For the mobile systems with 5-kHz channel bandwidths, high throughput modulation using constant envelope signaling (see Section 2.7) will be the most likely candidates. The class of MPSK and M-ary CPFSK produce the higher throughput, but consideration must be given to both spectral overlap in adjacent bands (channel crosstalk) and performance degradations during decoding. In VSAT uplinks with relatively wide beamwidths, care in signal selection is needed to avoid excessive adjacent satellite interference.

It is expected that the transmission of voice (audio) over mobile links will be a basic requirement. Direct PCM of voice (see Table 2.2) will not be possible over a 5-kHz channel, and forms of encoded voice must be used. Reported results on voice encoding using vector quantized linear predictive coding (Chen and Gersho, 1987; Davidson et al., 1987) have indicated good quality voice transmission can be achieved with about 4800 bps. Transmission of this rate over a 5-kHz channel specifies that a throughput of about 1 bps/Hz is required. This means standard BPSK (with a throughput of 0.5) will not be satisfactory, and some form of M-ary signaling, with $M \geq 4$, is needed. These higher-M systems, however, generally require increased E_b/N_0 values and, more important, require more accurate phase referencing. It is again apparent that trying simultaneously to satisfy power limitations (low E_b/N_0 values) and bandwidth limitations (high throughput) in the VSAT environment will require careful system design and accurate tradeoff analyses. In the following sections we examine some of these alternatives.

11.2 SMALL-TERMINAL–SATELLITE LINK ANALYSIS

Transmission requirements for establishing a small terminal-to-satellite system can be grossly determined by first reviewing the link analysis associated with such systems. The key concern of the uplink is the ability of a small terminal transmitter to generate enough uplink EIRP satisfactorily to complete a satellite uplink. Similarly, the receiving capability (g/T°) of a small terminal must be sufficient for the downlink. It must also

TABLE 11.2 Typical VSAT uplink parameters

	UHF	*K-band*
Frequency	823 MHz	14 GHz
Path loss	182.8 dB	206.7 dB
Atmosphere loss	0.3 dB	4.0 dB
Bandwidth	5 KHz	5 KHz
T°_{eq} of receiver	1000 K	1000 K
Satellite antenna efficiency	55%	55%

be remembered that in a bent-pipe satellite system, the SNR of the up and down links will combine to produce an even smaller receiver SNR, effectively making the weak link even weaker. To understand the level of these effects, we examine some basic UHF and K-band link budgets.

Table 11.2 lists typical parameters of a UHF and K-band satellite uplink, designed to operate over a 5-KHz receiver bandwidth. Figure 11.3 then uses these values to plot transmitter EIRP and satellite antenna size

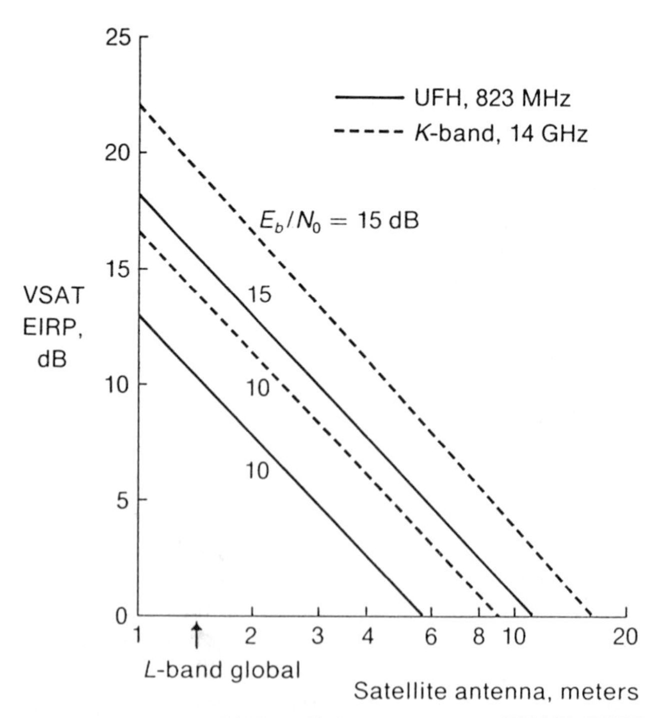

FIGURE 11.3 VSAT uplink parameters. VSAT EIRP versus satellite antenna diameter for given E_b/N_0. Link parameters in Table 11.2.

needed to produce the corresponding E_b/N_0 in the uplinks. A small transmitting UHF terminal with a 1 W amplifier and operating through an omnidirectional antenna produces an EIRP of about 3 dBW. Figure 11.3 shows that for an $E_b/N_0 = 10$ dB the satellite antenna must be about 3.7 m, somewhat larger than those being used on current satellites. However, a UHF antenna of this size has a beamwidth of only 6.5°, and does not produce global coverage. (It would illuminate only about two-thirds of CONUS.) Hence, the UHF uplink must operate with a satellite spot beam, and cannot achieve full CONUS coverage while obtaining the uplink E_b/N_0. A 1.5-m UHF satellite global antenna would have a reduced gain, and the transmitter EIRP must be increased to about 12 dBW to maintain the link. This would require some form of vehicle antenna pointing to achieve the additional gain.

An alternative is to lower the data rate of the mobile transmitter, and use only a portion of the available 5-kHz bandwidth. For example, at UHF a global satellite antenna could operate with a 2 W, omnidirectional vehicle antenna if the data rate was reduced to about 1250 bps.

These uplink budget parameters also explain the increasing interest in developing a VSAT uplink network based upon low earth-orbiting (LEO) satellites, instead of synchronous satellites. A satellite at an altitude of 1000 miles would have a space loss reduced by $(22{,}300/1000)^2 = 497.3 = 26.9$ dB. This space loss reduction can be used to reduce the EIRP of the VSAT and the size of the satellite antenna. For example, the one watt VSAT transmitter can be reduced to 100 mW (which can be provided by hand-held equipment), and the satellite antenna can be reduced from 3.7 m to about 1.5 ft. (17 dB less gain). The LEO satellite will be in orbit (nonstationary) and the system must be implemented so as to provide continuous earth coverage and satellite handoff. Constellations of numerous (50–100) LEO satellites continuously circling the earth have been proposed for this concept. It is expected that VSAT-LEO networks will be rigorously explored in the next generation.

A K-band VSAT uplink, although having slightly more atmospheric loss, can operate at $E_b/N_0 = 10$ dB with a 1 m satellite antenna provided the VSAT EIRP is about 16.5 dBW. A 2-W VSAT transmitter must therefore have an antenna gain of about 13 dB, which can easily be attained with a K-band aperture of only a few inches if pointed directly at the satellite. A 1-m K-band VSAT antenna will produce an $E_b/N_0 = 10$ dB over a bandwidth of about 5 MHz with the same satellite antenna. Hence, K-band VSAT systems can provide significantly higher data rates than UHF systems.

A similar analysis can be carried out for a VSAT downlink. Table 11.3 lists the downlink parameters and Figure 11.4 plots the required satellite EIRP and VSAT antenna size needed to achieve the stated E_b/N_0 in a

TABLE 11.3 Typical VSAT downlink parameters

	UHF	*K-band*
Frequency	868 MHz	12 GHz
Path loss	182.2 dB	206 dB
Atmosphere loss	0.1 dB	5.0 dB
Bandwidth	5 KHz	5 KHz
T°_{eq} of receiver	1200 K	1200 K
E_b/N_0	10 dB	10 dB

5-kHz bandwidth. An omniantenna requires a satellite EIRP of about 28 dBW per channel for a link $E_b/N_0 = 10$ dB. The transmission of a band of 1000 such channels would therefore require a satellite transponder capable of producing about 58 dBW of EIRP. This is much higher than currently available from commercial synchronous satellites. A UHF

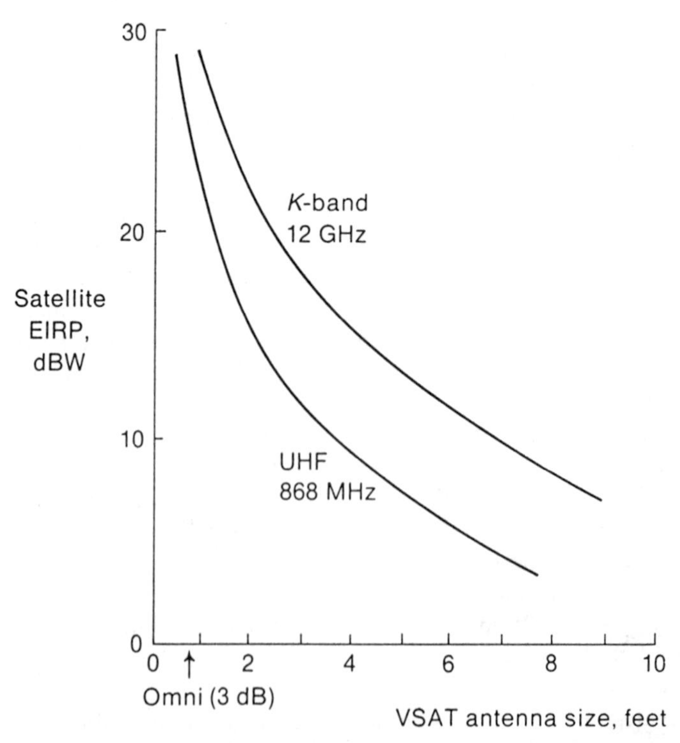

FIGURE 11.4 VSAT downlink parameters. VSAT antenna size versus satellite EIRP per 5 KHz channel. Link parameters in Table 11.3.

global antenna has only about 19 dB of gain, implying the increased satellite EIRP must be developed from a higher payload power capability. This fact entails a new generation of synchronous satellite design in order to provide this additional power requirement for VSAT networks.

It should be emphasized again that designing the VSAT uplink and downlink each for a particular E_b/N_0 value may produce a much lower decoding E_b/N_0 in the receiver. This is due to the E_b/N_0 loss from intermodulation in transponding the FDMA channels, and the loss from the up–down combination of the two equivalent links. For example, two equal E_b/N_0 links will combine to produce at least a 3-dB loss at the receiver. Thus, each VSAT up and down link must be individually designed for performance well above the desired decoding performance, further complicating the link design in Figures 11.3 and 11.4.

The necessity of using spot beams in UHF downlinks to achieve the desired EIRP in VSAT links can be used to develop a complete multiple beam cellular concept for establishing a VSAT network. The desired coverage area is divided into segments covered by a set of nonoverlapping spot beam, as was shown in Figure 9.20. The footprints of each beam over the coverage area are referred to as *cells*. Each cell is produced by a downlink beam that establishes the necessary EIRP for all receivers in its footprint. In addition, the multiple cells allow for frequency reuse, as described in Section 9.7, which increases the number of terminals that can be accommodated in the network. The cell concept requires the satellite to have both multibeam capability and on-board processing to separately switch the uplink information to the receivers in the appropriate cells. The smaller the cell size used, the higher the EIRP per cell and the greater the frequency reuse, but the beamwidth of each cell must be decreased via larger satellite antennas. For example, Table 11.4 lists the sizes of UHF satellite antennas to produce cells of specific dimensions, and the number of such cell beams needed to cover CONUS. We again see the

TABLE 11.4 Multiple beam coverage of contintental USA at UHF

Satellite Antenna Size (m)	*Cell Footprint Diameter (miles)*	*Number Cells to Cover CONUS*
3	3000	1
4.9	1500	2
5.3	1000	3
9.14	430	7
20	125	24

direct relation between the development of large, more sophisticated synchronous satellites and the establishment of extensive VSAT networks.

11.3 VSAT AND HUB NETWORKS

The use of multiple VSAT in point-to-point satellite systems has numerous potential applications in future communication networking. The real-time daily transmission of business information, banking data, stock reports, news services, polling operations, etc., are all examples of important applications for the development of a unified VSAT network. Such systems will operate at K-band with low-cost, small pointable antennas that can be located conveniently on rooftops, building surfaces, or conceivably can even be reduced to portable or manpack operation. By operating at K-band, higher antenna gains are available, and bandwidths will correspond to those of present-day transponders, being in the range 30–500 MHz. Each terminal will operate independently, and transmit directly through a satellite to a receiving destination. Transmission activity will most often include multiplexing of data and voice, and will be usually limited to low duty cycles (few hours per day). Data accuracy requirements will invariably be stringent, requiring overall link bit error probabilities on the order of 10^{-6} or less.

The choice of accessing format to accommodate a complete network of VSAT over a particular satellite bandwidth may not be a simple choice among the basic formats discussed in Chapters 5–7. The use of FDMA, in which the available band is subdivided (as projected for the UHF mobile bands) and a subband assigned to a particular VSAT, is the simplest to implement, but could produce excessive intermodulation and possible backoff losses. In addition, assigning dedicated frequency bands to low-duty-cycle terminals can be wasteful of the available bandwidth, since the FDMA system can not directly take advantage of the low activity factor. This can be seen from Eq. (5.1.4). If each individual FDMA channel uses error correcting coding with code rate r (to lower the required VSAT EIRP), and uses constant envelope MPSK modulation with appropriate guard band spacing, the required channel bandwidth to transmit a data rate of R_b bps is approximately $B = (2/r \log_2 M)\epsilon R_b$ Hz. Here ϵ (≥ 1) accounts for the channel widening factor of the guard bands. The number of available channels for a specified RF bandwidth is then

$$K = \frac{B_{RF}}{2\epsilon R_b/r \log_2 M} = \frac{B_{RF}/2R_b}{\epsilon/r \log_2 M} \quad \text{for FDMA} \tag{11.3.1}$$

The numerator is the link spreading ratio defined in Eq. (7.1.3). The number of FDMA channels is therefore reduced by the code rate inserted and the guard spacing needed to remove the adjacent channel crosstalk. On the other hand, a CDMA system with the same spreading ratio can theoretically have, from Eq. (7.3.22),

$$K = \frac{B_{RF}/2R_b}{v\Upsilon} \quad \text{for CDMA} \tag{11.3.2}$$

where v is the fractional activity factor of the VSAT terminals, and Υ is the required user E_b/N_0 with coding. Thus the VSAT activity factor and the coding threshold reduction can directly increase the CDMA network size, producing an important CDMA advantage. In addition, the development of high-speed microchip versions of the code generators will simplify the VSAT transmitter package. However, CDMA networks must still combat the intermodulation problem, and require receiving terminals that must perform the necessary code acquisition and tracking prior to decoding. The power-efficient TDMA format makes maximal use of the satellite power, but requires more complicated transmitter packages for carrying out the necessary uplink network synchronization. Thus, the mere selection of a common accessing format for an entire VSAT network involves a careful trade-off of transmitter, satellite, and receiver complexities.

One concept used today to improve the overall operation of a VSAT network is the insertion of a common hub terminal, as shown in Figure 11.5. The hub is designed to be a large earth terminal, centrally located to provide an intermediate facility to complete the VSAT-to-VSAT transmissions. A transmitting VSAT would transmit through the satellite to the hub, where the signals from all active transmitters are received and separated. The hub station then retransmits the entire data field from all VSAT transmitters back through the satellite to the receiving VSAT terminals to complete the link. Uplinks from VSATs and from the hub, and downlinks to the VSATs and to the hub, can be easily separated by dual polarizations. By having a large hub terminal (high transmitting EIRP and high receiving $g/T°$) several of the disadvantages of the direct VSAT–satellite–VSAT link can be overcome. With a large $g/T°$, the satellite–hub downlink can be completed with a sufficiently high SNR, even with satellite back-off. This means each VSAT–satellite–hub link performance will be determined solely by the VSAT uplink, and is not degraded by the hub. The presence of the fixed-point hub also permits spot beaming from the satellite to further improve the quality of the hub downlink.

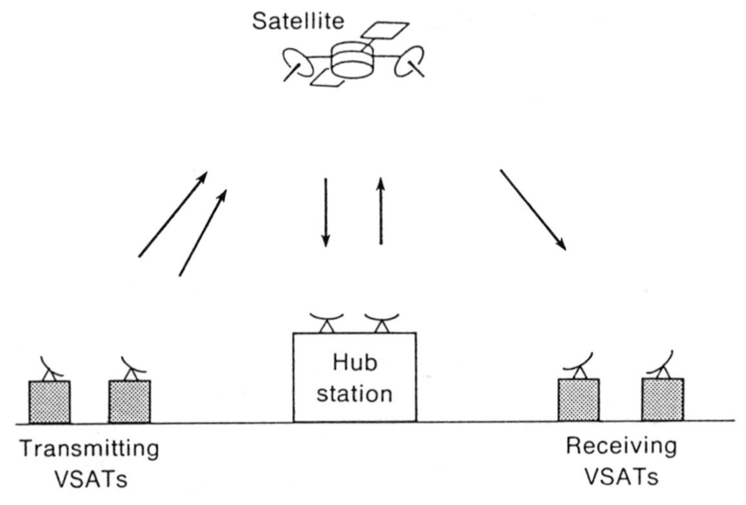

FIGURE 11.5 VSAT-hub network.

Likewise, the hub uplink, using its higher EIRP, provides a strong uplink for returning the data through the satellite to the receiving VSAT terminals. The performance of the final VSAT downlink is then determined primarily by the latter link alone. Hence, the insertion of the hub permits the overall VSAT–VSAT transmission to be replaced by the cascade of the VSAT uplink followed by the VSAT downlink, rather than the degraded up–down combination of the two relatively weak links that would occur with no hub. As was shown in Section 3.4, cascaded digital links always have better decoding performance than combined links.

The use of the hub also permits better matching of the multiple-accessing format to the specific link. In particular, the hub can be used to provide a format conversion from its downlink to its uplink. In this way the VSAT uplinks, being most favorable to FDMA or CDMA operation with the simplest transmitters, can be designed as such, with the hub alleviating the intermodulation back-off problem. The hub collects the active VSAT carriers from the satellite, decodes the data, and can reformat its uplink into TDMA frames, with each uplink terminal allocated a specific interval slot. The TDMA bit stream is then transmitted back through the satellite, obtaining the power savings advantages of TDMA during satellite relaying. The data from all VSAT uplinks is then contained in a common downlink transmitted to all receiving VSATs either in a global beam or in multiple or hopped spot beams. The data of any transmitting VSAT terminal can be obtained at any receiving terminal by time gating to the proper time slot, with frame synchronization easily acquired from the satellite downlink itself. Note that the hub must provide the necessary processing for converting the accessing modes, but can also insert control, command, and monitoring functions for all VSAT terminals. If the networks involve transmission between fixed-point K-band stations, and mobile L-band terminals, the hub can also provide the necessary frequency conversion during the retransmission.

The disadvantage of the hub system is that two extra satellite links have been added to the VSAT–VSAT link due to the double hop. This adds additional time delay to the data transmission, which may become prohibitive in voice links. For example, since it takes about 0.25 s for satellite link propagation, the transmission delay with a hub system will be at least 0.5 s (plus added delay due to the hub processing) longer than a standard relay link. When the delay begins to approach 1 s or more, real-time voice links may become unmanagable.

Many of the advantages of the hub concept can be retained with the insertion of on-board processing, as was discussed in Chapter 9. The ultimate objective is to simply replace the hub with an equivalent on-board processor; i.e., move the hub to the satellite. The processor would have to perform the necessary FDMA–TDMA conversion via the bulk demodulators discussed in Section 9.6. This would also permit additional advantages by interfacing with multiple antenna beams using routing algorithms, producing the most power-efficient VSAT network. It is expected that future satellite payloads will continually strive toward this goal, with the VSAT concept being a driving force for its development.

In VSAT–hub networks the important parameter becomes the operating data rate that the VSAT terminals can support. With the excessive K-band satellite bandwidths, this data rate will most likely be determined primarily by the power capabilities of the links, rather than by bandwidth restrictions. For the VSAT uplinks, this is a direct function of the achievable transmitter EIRP. Figure 11.6a shows the VSAT uplink data rates that can be supported at K-band with an $E_b/N_0 = 10$ dB, with different VSAT power levels and antenna sizes. A wide beam satellite is assumed with a receiving $g/T^\circ = -5$ dB.

With the hub–VSAT downlink operating in a TDM mode, each VSAT receiver must have the capability of decoding the entire TDMA bit stream. This occurs at a higher rate than the individual terminal data rates (recall from Chapter 6 that N sources at rate R_b produces a TDMA bit rate of about NR_b). Assuming a sufficient transponder bandwidth, the TDM data rate will be determined solely by the satellite EIRP and the VSAT antenna size. Figure 11.6b plots the achievable downlink data rate at $E_b/N_0 =$ 10 dB in terms of EIRP and aperture size.

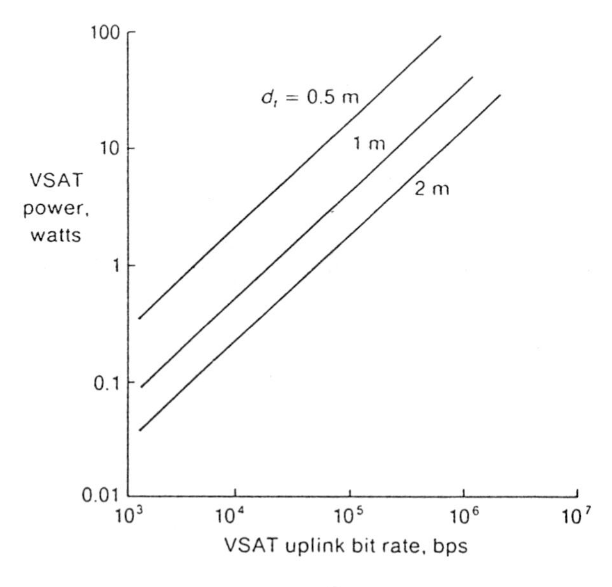

a.

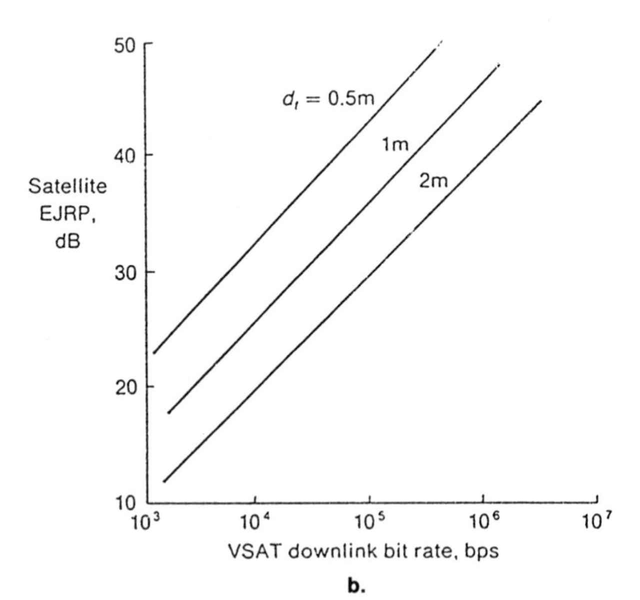

b.

FIGURE 11.6 (a) VSAT uplink data rate at K = band versus VSAT transmitter power. d_t = VSAT antenna diameter. $f = 14$ GHz, satellite $g/T^\circ = -5$ dB. (b) VSAT downlink data rate at K-band versus satellite EIRP. d_t = VSAT antenna diameter. $f = 12$ GHz.

11.4 DIRECT HOME TV BROADCASTING

A special type of VSAT–satellite system of considerable interest is the direct broadcasting of television from satellite to home from synchronous orbit (Figure 11.7). These systems require VSAT terminals composed of small rooftop antennas to be placed at the home and receive relayed TV transmissions directly from the satellite rather than from the local TV microwave networks. Satellites used for this type of video transmissions are referred to as *direct broadcast satellites* (DBS). Currently, direct satellite reception requires fairly large and expensive antennas (6–10 ft), precluding simple rooftop operation. The future generation of DBS is planned for the VSAT environment, making direct satellite reception simpler and less costly.

A DBS is planned to operate at K-band with large RF bandwidths, and fairly high demodulated SNR for the commercial TV receivers. In addition, the amplitude modulation (AM) formats of the commercial terrestrial TV system must be matched to the constant-envelope FM transmissions of the satellite. Although this requires a carrier-waveform converter (FM to AM) as additional external home circuitry, it allows increases in AM CNR to be made through the FM improvement operation at the expense of downlink bandwidth. By increasing FM deviation of the video carrier in the satellite uplink, improved AM CNR is achievable at the home receiver after satellite retransmission. This permits a

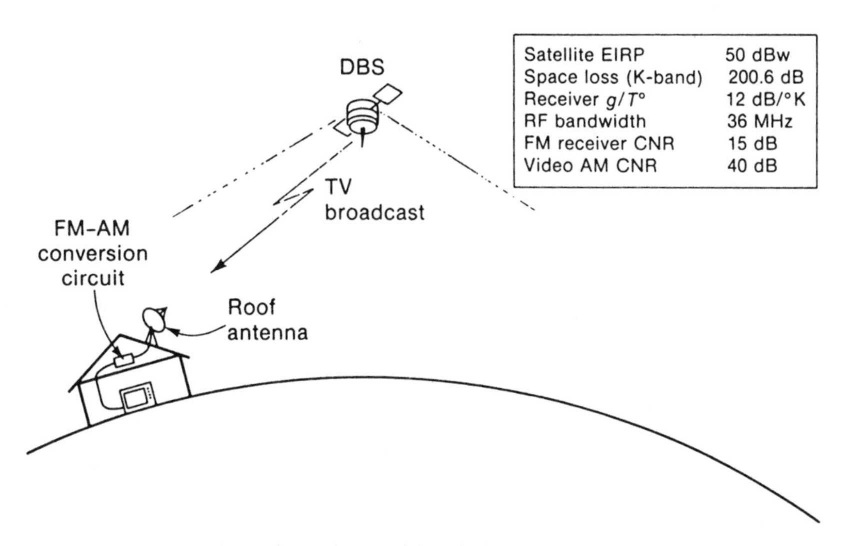

FIGURE 11.7 Direct broadcast video link.

relatively low FM CNR for the VSAT link, while still generating a suitable video AM CNR at the TV set.

A link budget for a downlink home broadcast system at 12 GHz is included in Figure 11.7. By using an RF bandwidth of 36 MHz, an FM improvement of 26 dB allows an AM CNR of 40 dB to be achieved with a satellite downlink EIRP of 50 dBw and a rooftop receiver (g/T°) of 12 dB/K. To achieve the last, careful control of the noise figure and efficiency of the rooftop antenna and front-end electronics is needed. Figure 11.8 plots the required roof antenna dish diameter as a function of receiver noise figure F, and antenna efficiency to achieve this g/T° at 12 GHz. To limit antenna size to about a 1-m, 60% dish, a receiver noise figure of about 4.5 dB is needed in the front-end electronics. Hence, the success of direct video broadcasting is inherently related to the development of both low-cost, low-noise electronics (Douville, 1977; Konishi, 1978; Pritchard and Kase, 1981), and high EIRP satellites.

One area where DBS are expected to play a key role is in *high definition television* (HDTV). The latter system uses wideband video signals ($\approx$30 MHz) to produce a sharper, high-quality television picture. The wider band cannot be easily accommodated in present terrestrial microwave or cable links without sacrificing existing formats or reducing picture quality. One solution is to transmit HDTV via direct broadcast satellites to homes equipped with appropriate wideband electronics and television sets. The availability of wide satellite bandwidths to handle the extended RF HDTV bandwidths ($\approx$200 MHz), along with the power efficiency of

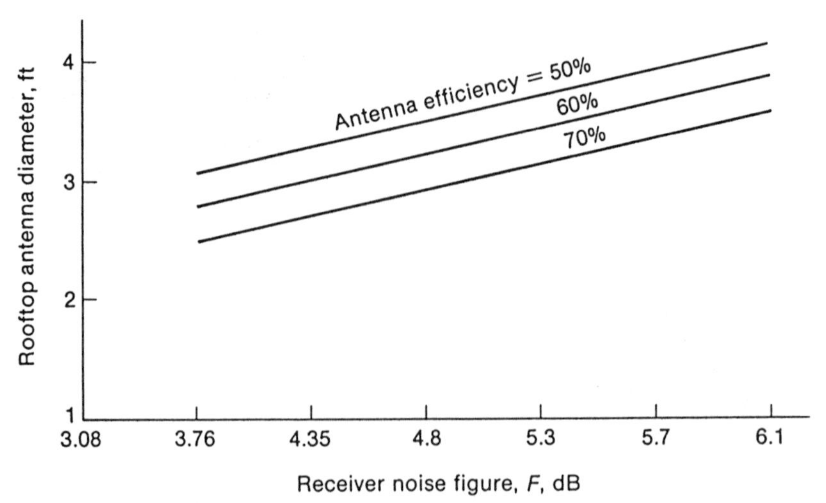

FIGURE 11.8 Roof antenna size vs. receiver noise figure and antenna efficiency to achieve $g/T^\circ = 12$ dB/K ($T_b^\circ = 50$ K, frequency = 12 GHz).

K-band transmission from high EIRP satellites, makes this form of VSAT network an extremely important commercial entity for the near future.

11.5 THE MOBILE-SATELLITE CHANNEL

A satellite system for communicating directly to mobile terminals requires the basic link analysis of Figures 11.3 and 11.4, but will have additional channel characteristics due to the terminal motion. This motion can introduce spectral changes in transmitted waveforms that can cause additional degradations relative to those of fixed-point VSATs. The effects are primarily caused by the time-varying multipaths that will appear in trying to transmit an electromagnetic field from a fixed orbital point (satellite) to a moving terminal (vehicle). The multipaths act to produce fading and frequency shifting that invariably lower the link performance. By understanding their generation, and establishing plausible channel models, these mobile degradations can often be combated, or at least reduced.

The basic UHF mobile downlink channel is shown in Figure 11.9. The transmitted field from the satellite is received by the moving vehicle in three basic ways: (1) the direct component, which is the line-of-sight modulated field collected by the terminal directly from the satellite; (2) the specular component, that represents the field collected from a single reflection (usually from the nearby Earth surface) that arrives as a delayed version of the direct component; and (3) the diffuse component, which is

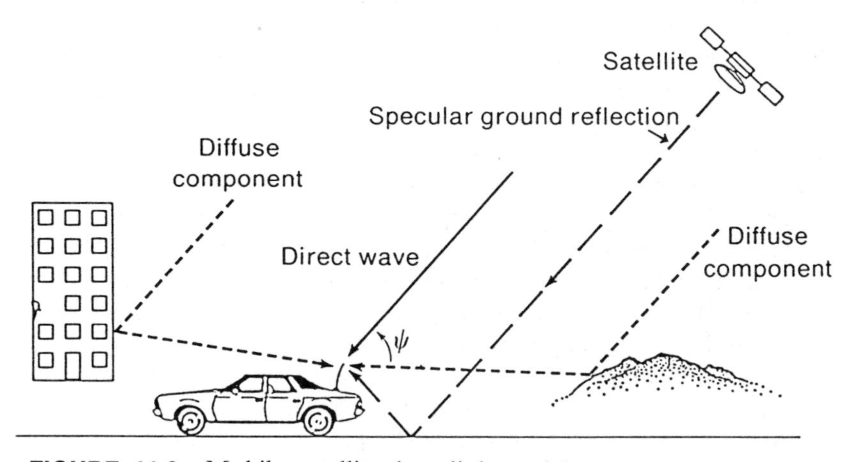

FIGURE 11.9 Mobile–satellite downlink model.

the combined effect of the multipath (multiple reflections) from all directions collected by the vehicle antenna. With wide angle or omnidirectional reception at the vehicle, much of the scattered multipath will be received and contribute to the diffuse noise.

The direct component generally represents the strongest component, and is composed of the modulated satellite carrier with the insertion of a possible Doppler frequency shift due to the terminal motion. This frequency shift is given by

$$f_d = (v/c) f_0 \cos \psi_l \tag{11.5.1}$$

where f_0 is the transmitted carrier frequency, v is the vehicle velocity, c is the speed of light, and ψ_l is the elevation angle of the line-of-sight in Figure 11.9. For low angles, a vehicle speed of 60 mph can produce a Doppler shift at UHF of about $f_d \simeq \pm 100$ Hz. This shift, plus any inherent oscillator offset, represents the carrier bandwidth shift when received at the vehicle. Maritime vessels will have Doppler shifts of only a few hertz.

The power level of the received direct component may be further reduced by "shadowing" (due to trees or terrain) as in Figure 11.10a or may have complete outages (due to blockage by buildings and mountains) as in Figure 11.10b. Shadowing and outages are generally temporally nonstationary, and only occur intermittently during the vehicle motion.

The specular field component is a delayed, phase-shifted replica of the direct field, and is generally received at a significantly lower power level due to the reflection loss during its propagation. Specular components can augment or interfere with the direct component at the receiver. The strongest specular components are those due to nearby ground reflections, but are generally received at angles outside the upward-looking gain pattern of the vehicle antenna. As a general result, specular fields can often be neglected in multipath analysis. Even if the direct component has been blocked completely, the specular component is generally well below the receiver noise floor to provide any useful data from the satellite.

The diffuse component is composed of the random accumulation of all field reflections from all horizontal directions (Figure 11.11). Although each individual reflection may be insignificantly small, the totality of all such reflections can combine to produce a nonnegligible interference signal to the direct component. Each such multipath component arrives with a random amplitude, random phase, and random horizontal angle, dependent on the number and type of reflections. An arrival angle of ϕ_n in Figure 11.11 produces a resulting Doppler shift of $f_d \cos(\phi_n)$, where f_d is given in Eq. (11.5.1). The accumulation of a large number of such

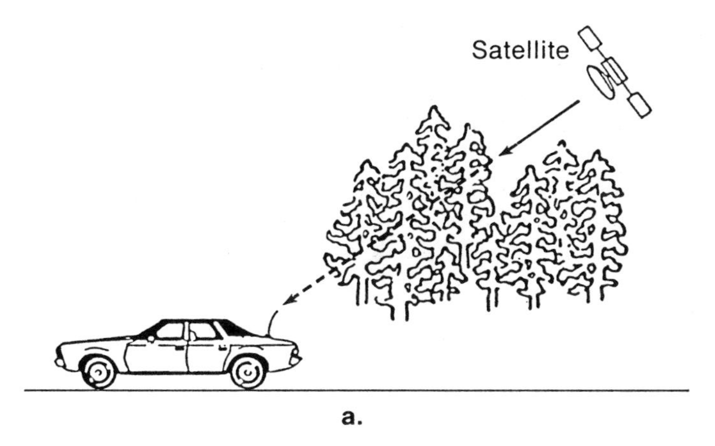

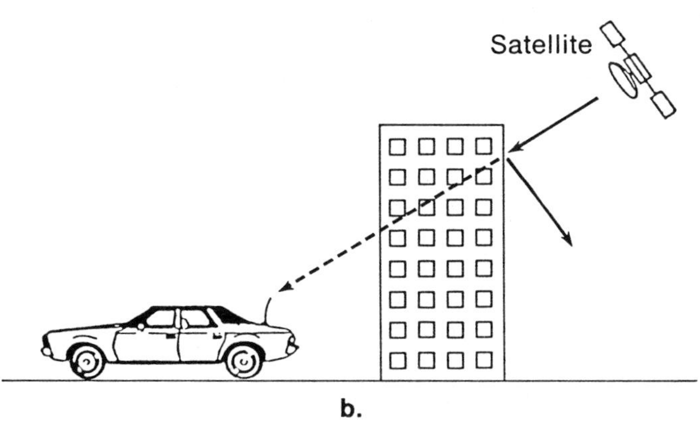

FIGURE 11.10 Power losses in mobile–satellite links. (a) Due to shadowing. (b) Due to blockage.

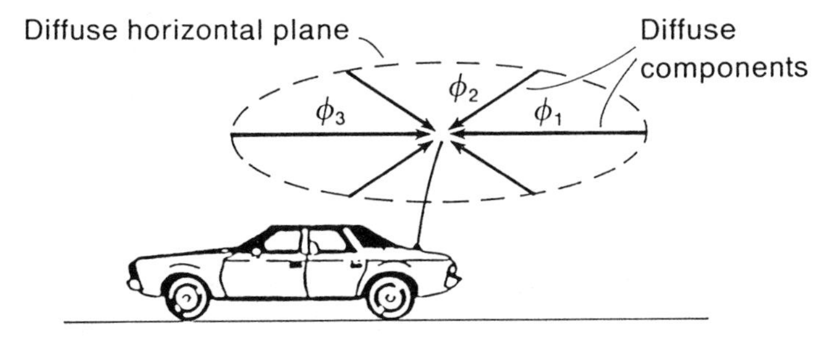

FIGURE 11.11 Arrival directions of diffuse components.

components, with random angles ϕ_n, sum to produce an effective additive Gaussian bandpass noise process, having a frequency spectrum caused by the spread of the Doppler shifts. Thus, for land vehicles the diffuse noise appears with a bandwidth of about 200 Hz centered around the carrier frequency.

The overall result of the downlink mobile multipath channel is to produce a received satellite signal composed of the sum of the direct modulated carrier component and the diffuse multipath noise. The total power collected by the mobile receiver is then the sum of the direct and diffuse power,

$$P_c = A^2/2 + \sigma_d^2 \tag{11.5.2}$$

where A is the direct component amplitude and σ_d^2 is the diffuse power over its Doppler spread bandwidth. The power P_c is the total power available from the satellite at the receiver (neglecting specular and scattered field power), and is the power determined by the link analysis in Section 11.2. The combined multipath received waveform can then be written as

$$c(t) = \alpha(t) \cos[2\pi(f_0 + f_d)t + \theta(t) + v(t)] \tag{11.5.3}$$

where $\alpha(t)$ is the instantaneous amplitude, $\theta(t)$ is the phase modulation, and $v(t)$ is an effective phase noise caused by the diffuse noise. Thus, the presence of the diffuse component is to insert randomness into the amplitude and phase of the modulated satellite carrier signal. It is well known that the envelope $\alpha(t)$ is a Ricean random process, having the probability density at any time t of

$$p(\alpha) = \frac{\alpha}{\sigma_d^2} e^{-(\alpha^2 + A^2)/2\sigma_d^2} I_0\left(\frac{\alpha A}{\sigma_d^2}\right), \qquad \alpha \geq 0 \tag{11.5.4}$$

It is convenient to define the normalized carrier power

$$s = \frac{\alpha^2/2}{P_c} \tag{11.5.5}$$

so that the power in $\alpha(t)$ is then $P_\alpha = sP_c$. The parameter therefore indicates the portion of the available satellite power that is contained in the multipath carrier. Its probability density can be obtained from Eq.

(11.5.4) by direct transformation,

$$p(S) = \left.\frac{p(\alpha)}{\alpha/P_c}\right|_{\alpha=\sqrt{2SP_c}}$$

$$= (1 + r)e^{-s(1+r)-r}I_0(2\sqrt{sr(1 + r)}), \qquad s \geq 0 \qquad (11.5.6)$$

where

$$r = \frac{\text{Power in direct component}}{\text{Power in diffuse component}} \qquad (11.5.7)$$

The parameter r in Eq. (11.5.7) is referred to as the *Rice parameter*, and indicates the relative power contribution of the direct and diffuse components. Note that the probability density of s in Eq. (11.5.6) depends only on the Rice parameter r. As $r \to \infty$ all power is in the direct component, indicating reception is via a direct carrier line-of-sight transmission from the satellite, and the diffuse component is negligible. As $r \to 0$, the received power is all diffuse. In the latter case, the envelope density in Eq. (11.5.4) collapses to a Rayleigh density, and the channel is referred to as a *Rayleigh channel*. Hence, as the parameter r increases, the mobile channel transforms from a Rayleigh channel to a Rice channel, and finally to a direct line of sight (no multipath) channel.

The value of the Rice parameter is a function of the reflective terrain in the vicinity of the vehicle terminal, and is strongly dependent on the elevation angle of the vehicle–satellite line-of-sight. Figure 11.12 plots some simulated and measured values (JPL, 1988b) for r for operation in urban, rural, and maritime environments, as a function of the elevation angle ψ_l. The parameter increases significantly as the satellite is observed at higher angles, where more of the horizontal multipath is rejected. Likewise, a dense collection of reflectors, as in metropolitan environments, tends to produce lower values of r, and operation closer to that of a Rayleigh channel. The rural environment is the more benign, whereas the maritime involves primarily long-range sea reflections whose severity is strongly dependent on the ocean wave structure. During periods of shadowing (due to trees, foliage, and terrain) the Rice parameter may be reduced by several (3–10) decibels from the average values stated here.

As a random variable, s in Eq. (11.5.6) has the higher probability of having a value less than 1. This means that most of the time the multipath carrier is received with less available power than the total power predicted from the satellite link budget. This power loss is referred to as fading, and translates directly to a degradation in processing the satellite signal. The

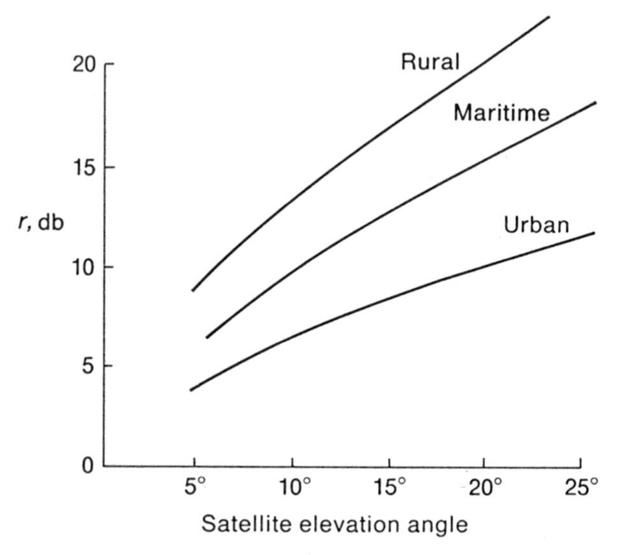

FIGURE 11.12 Rice parameter (r) vs elevation to satellite for several channel environments.

amount of additional satellite power needed to overcome the fade in achieving a specified receiver performance level is called the *fade margin.*

While Eq. (11.5.6) indicates instantaneous power statistics, the temporal behavior of the envelope variation $\alpha(t)$ are also of importance. Of particular interest are the fade duration (the length of time a specific fade lasts) and the fade rate (how often a specific fade occurs). These data are obtained primarily from reported field measurements observed over a specific channel for long time periods. Figure 11.13a shows some reported data (Hong, 1988; JPL, 1988b; Schmier and Bostian, 1987; Vogel and Goldhirch, 1988) indicating the average time periods that a particular fade depth was observed. For example, a fade of -5 dB typically lasts about 20–100 ms at the stated elevation angle. Fade rates are obtained by attempting to measure the bandwidths of the time variations of $\alpha(t)$ in Eq. (11.5.3). Reported bandwidth (Figure 11.13b) are about 0.2–1.0 Hz, indicating fade cycles of about 1–5 s. These channel statistics become increasingly important in the ultimate design of the communication link.

11.6 COMMUNICATING OVER THE MOBILE-SATELLITE CHANNEL

The presence of multipath in the mobile-satellite channel causes the downlink phase-modulated carrier at the mobile receiver to have the form

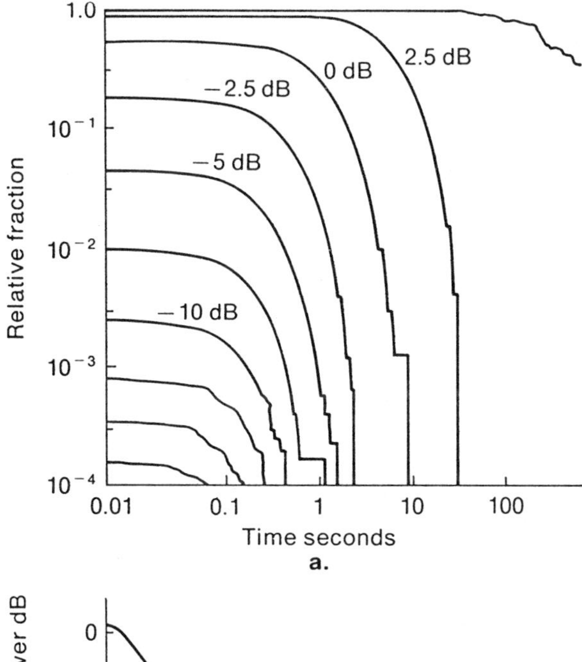

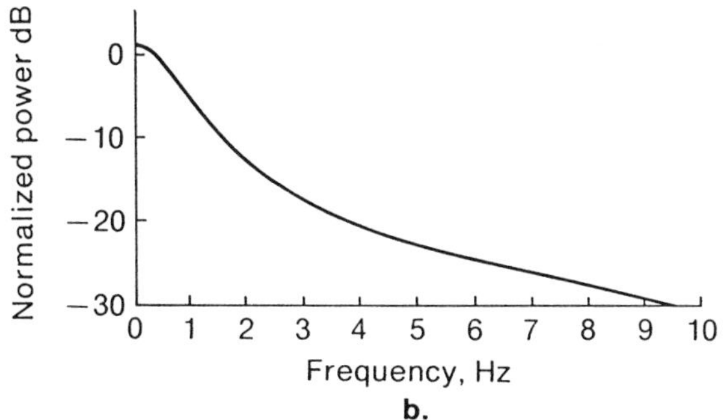

FIGURE 11.13 Fade statistics for mobile channels. (a) Relative time length of each fade level, (b) spectral density fade variations.

of Eq. (11.5.3), with the envelope $\alpha(t)$ having the normalized power distribution in Eq. (11.5.6). The latter indicates that the carrier power is most likely to be received with a power fade, relative to the total power available from the satellite. This power fade leads directly to degraded performance.

In a digital link the multipath effect is to cause the bit energy to depend on $\alpha^2(t)$. If we assume that the envelope $\alpha(t)$ changes slowly with respect to the bit time T_b, then we can consider α to be a fixed random variable during each bit decoding interval. For a given value of s in Eq. (11.5.5) the binary decoding error probability, for additive white Gaussian receiver noise, is then

$$\begin{aligned} \mathrm{PE}(s) &= Q(\sqrt{2sE_b/N_0}) \quad \text{for BPSK, QPSK} && (11.6.1a) \\ &= \tfrac{1}{2}e^{-sE_b/N_0} \quad \text{for DPSK} && (11.6.1b) \\ &= \tfrac{1}{2}e^{-sE_b/2N_0} \quad \text{for FSK} && (11.6.1c) \end{aligned}$$

where $E_b = P_c T_b$ is the bit energy based on receiving the total power P_c in Eq. (11.5.2). The average bit error probability PE for the multipath channel is then obtained by averaging PE(s) over the fade variable s. Hence

$$\mathrm{PE} = \int_0^\infty \mathrm{PE}(s)p(s)ds \qquad (11.6.2)$$

with $p(s)$ given in Eq. (11.5.6). The integral can be directly evaluated numerically. Figure 11.14 shows the result for the BPSK or QPSK coherent link operating over the Ricean channel with several parameter

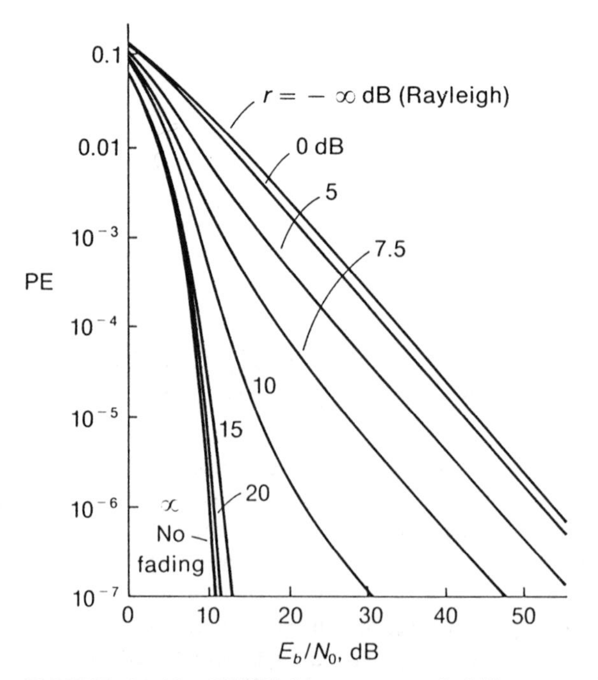

FIGURE 11.14 BPSK bit error probability vs unfaded E_b/N_o for the Rician channel. r = Rice parameter.

values r. The PE performance degrades substantially from the ideal $r = \infty$ case (no multipath) through increasing multipath to the Rayleigh fade ($r = 0$) case. Note the severe penalty in E_b/N_0 needed to combat the fade, approximately 7 dB at PE $= 10^{-5}$ with a 10-dB Rice parameter, and approaching almost 30 dB for the Rayleigh channel. This means link analysis based purely on line-of-sight transmission with ideal Gaussian channel decoding can be completely misleading in estimating performance for the faded channel.

Figure 11.15 compares the systems in Eq. (11.6.1) for the Rayleigh fade (worst case) channel. In the later case Eq. (11.6.2) can be integrated directly to produce

$$\mathrm{PE} = \frac{1}{2}\left[1 - \sqrt{\frac{E_b/N_0}{1 + E_b/N_0}}\right], \quad \text{BPSK, QPSK} \tag{11.6.3a}$$

$$= \frac{1}{2 + 2(E_b/N_0)}, \quad \text{DPSK} \tag{11.6.3b}$$

$$= \frac{1}{2 + (E_b/N_0)}, \quad \text{FSK} \tag{11.6.3c}$$

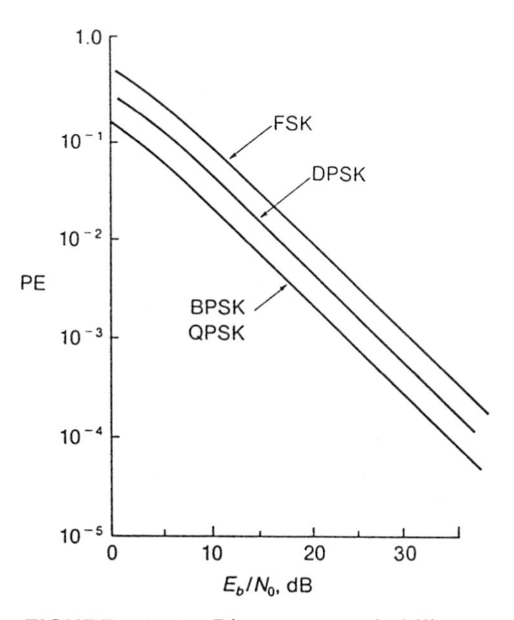

FIGURE 11.15 Bit error probability vs E_b/N_o for Rayleigh fading channel.

We see that Rayleigh fading has caused each system to have PE performance that improves only linearly with E_b/N_0, instead of exponentially as in Eq. (11.6.1). This results in performance much worse than for the nonfading channel, and again indicates that the satellite transmit power must be significantly increased to account for this fade loss. From Figure 11.15 we also observe that the coherent (BPSK) system still performs better than the noncoherent systems (about 3 dB) even in the presence of the Rayleigh fade.

The plots of the coherent performance curves in Figure 11.15, however, assumes that perfect phase-coherent decoding can be achieved, and the only effect of the multipath is the power fade. This means that during bit decoding of the phase modulation in Eq. (11.5.3) a phase-tracking loop is available to track the phase noise $v(t)$ imposed by the diffuse noise component. The discussion in Appendix B indicates that phase referencing requires a loop tracking bandwidth B_L that satisfies $B_L T_b \leq 10^{-2}$, or equivalently $B_L \leq R_b/100$. For 5 kbps data, this means $B_L \leq 50$ Hz. Since phase spectral spreading due to Doppler can be as high as 100 Hz in some mobile environments, suitable tracking bandwidths may be insufficient to reduce the phase reference error to a negligible level. The result is an inherent phase-tracking error that produces an irreducible bit-error probability, as was indicated in Figure 2.20. This flattening of the PE curves at high E_b/N_0 due to nonideal phase referencing therefore acts as a bit-error floor that exists even if enough satellite power was available to overcome the fade. Thus, in addition to the power fade, the multipath can insert a phase noise floor to the coherent PE curves. With this taken into account, the noncoherent formats such as DPSK and FSK that require no phase referencing may in fact perform better than the coherent systems at the higher E_b/N_0. In cases where the Doppler is low (≤ 10 Hz, for example, as in the maritime systems), the coherent systems with phase-tracking decoding will always be the most advantageous.

Coherent digital operation without the phase-referencing floor can be achieved by avoiding the use of phase-tracking loops. One alternative is the use of pilot tones for the referencing. In this system a carrier pilot tone that is phase coherent with the modulated carrier is inserted in the satellite downlink, as shown in Figure 11.16a. The pilot tone can be located anywhere in the 5-kHz spectrum, but the most advantageous location is at the spectral nulls. In BPSK or QPSK systems this would occur at band edge (Figure 11.16b), or a null can be created at band center by modifying the baseband modulation waveform (e.g., with Manchester encoding, as in Figure 11.16c). At the receiver the coherent pilot tone, having the same phase-noise variation as the modulated carrier, can be directly filtered from the spectrum and used to provide the phase reference

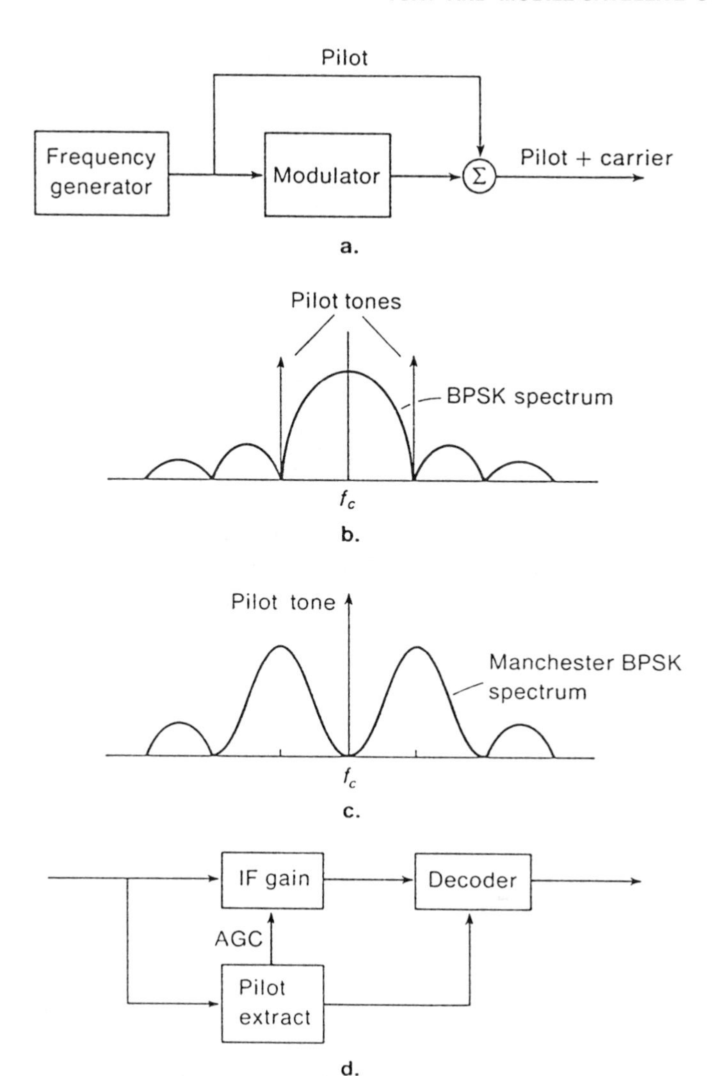

FIGURE 11.16 Pilot tone system. (a) Encoder block diagram. (b) BPSK spectrum. (c) Manchester BPSK spectrum. (d) Decoder block diagram.

for coherent decoding (Figure 11.16d). The pilot tone filter can be narrow (order of the Doppler shift) so that pilot extraction is obtained with minimal noise. The power in the pilot tone need only be sufficient to generate a suitable pilot SNR. No phase locking is utilized so the limitation on loop B_L is avoided. Schemes using dual coherent pilot tones that can be combined after receiver filtering to improve SNR have also been proposed (Simon, 1986).

Pilot tone operation has the added advantage that channel fading over the entire 5-kHz band can be monitored during the pilot extraction in Figure 11.16d. The fade estimate can then be used to provide gain control in the decoding channel. When the pilot tone amplitude decreases during a fade, the decoding gain can be proportionally increased, effectively dividing out the fade effect during decoding. Alternative techniques involve using the pilot amplitude estimate in a threshold test to detect the presence of a deep fade and simply terminate decoding during that period. These uses of the pilot tone, aside from generating the phase reference, can be considered forms of channel "probing" and channel "state" measurements used to aid the decoding.

Pilot tone systems also have some basic disadvantages. The insertion of a pilot no longer produces a constant-envelope carrier. In addition, the nonlinearities of the satellite transponder in bent-pipe operation can cause power robbing and intermodulation, and a subsequent reduction in both carrier and pilot power. (This is why it is advisable to maintain the pilot power as low as possible, using narrow filters and spectral null locations.) Pilot tones at band edge will be susceptible to adjacent channel interference, while those at band center require bit reshaping that generally increases spectral tails.

The phase-coherency problem can also be avoided by use of differential phase modulation. The class of M-ary DPSK, where the M phases of the present digital symbol, is referenced to the phase transmitted on the previous symbol, are especially advantageous for the mobile channel. As long as the residual carrier phase remains approximately constant over adjacent symbols, it need not be estimated to decode the M-ary differential phase states. Although a slight degradation in performance occurs, the necessity of phase tracking or pilot tone subsystems is avoided. Frequency tracking may still be required but this occurs at a much lower rate than the phase variations.

Degradation in PE performance due to fading suggests improvement via some form of channel coding and error correction. As discussed in Chapter 2, error correction can be achieved by first converting the data bits to channel chips and transmitting chips over the mobile channel. At the receiver the decoded chips are then processed to reconstruct the data bits, achieving some degree of PE improvement. Figure 11.17 shows the result of the expected gain in using a rate $\frac{1}{2}$, constraint length 7 convolution code over a Rayleigh fading DPSK channel. Also shown is the result of a (15, 9) Reed–Solomon block code. Note that a significant portion of the fade loss is reduced by insertion of the coding, with the fade margin at PE $= 10^{-5}$ reduced to about 7 dB.

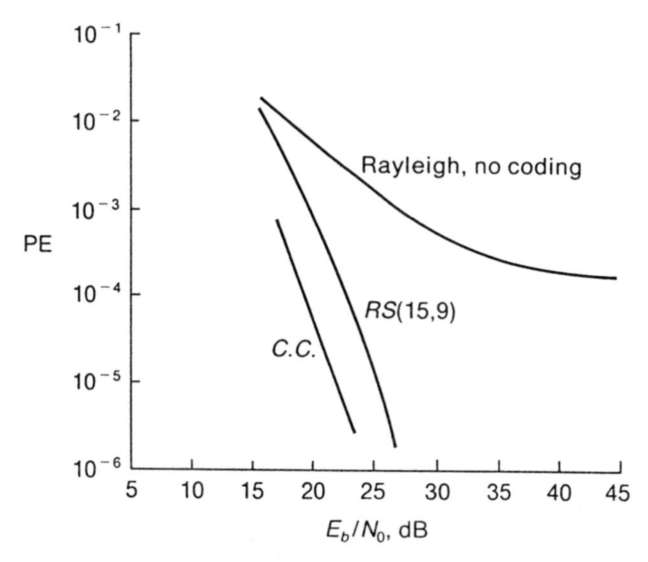

FIGURE 11.17 PE improvement with coding over DPSK Rayleigh channel. *RS* = Reed–Solomon block code. C.C. = convolution code, rate $\frac{1}{2}$, length 7.

The insertion of channel coding, however, slows up the data rate, since the transmitted chip spectra must be accommodated within the allowable mobile channel bandwidth. Thus a $\frac{1}{2}$ rate code over a 5-kHz DPSK channel bandwidth would reduce the data rate to only about 1.25 kbps. Higher rate codes would support higher data rates, but with less coding improvement in Figure 11.17. The loss of data rate over a fixed bandwidth can be equalized with use of a modulation format with a higher throughput. If a total data throughput of 1 bps/Hz is desired (5 kbps over the 5-kHz channel) it is necessary to have a chip throughput that makes up for the code rate loss. To see this we note

$$\begin{aligned}\text{Data throughput} &= \frac{\text{bps}}{\text{Hz}} \\ &= \left[\frac{\text{bits}}{\text{chip}}\right]\left[\frac{\text{chips/s}}{\text{Hz}}\right] \\ &= [\text{code rate}][\text{chip throughput}] \qquad (11.6.4)\end{aligned}$$

The first term is the coding rate inserted, whereas the second is the throughput that the modulation transmission allows for the coded chips.

For a rate $\frac{1}{2}$ code, a chip throughput of 2 is required, the latter achievable, for example, with a fairly complex 16-ary PSK signal (see Figure 2.34). In general, a code rate of $1/r$ will require a 2^{2r}-PSK modulation to restore the data throughput to 1. The code inserts the error correction capability, while the modulation encoding increases the data throughput. In fact, it has been found that the coding can be optimally adjusted to match the modulation for the most efficient PE performance. This topic is discussed in detail in Section 11.8.

11.7 INTERLEAVING TO COMBAT DEEP FADING

The predicted coding gain improvements for the fading channel were based on the assumption of random chip errors that were effectively corrected by the inserted coding. In the mobile channel the presence of shadowing and long-time deep fades can produce long strings of transmission error bursts that cannot be connected by the inserted codes. For example, a deep fade lasting 100 ms with a transmitted chip rate of 4.8 kcps means that a string of 480 consecutive chips will be transmitted during the fade, and all most likely will be decoded in error. The insertion of coding, designed to have a correction capability of only a few errors per chip word, would be useless during the fade. To use a more powerful code having the correction capability to cover the expected burst length would require a prohibitively long block length and extensive decoder processing. Furthermore, for the majority of the time (when the fade does not occur) the excessive coding hardware is unnecessary and inefficient.

A solution to this burst error problem is the use of *interleaving* and *deinterleaving*. The latter corresponds to an additional level of relatively simple digital processing inserted at both the transmitter and receiver to combat the error bursts. An interleaver and deinterleaver is simply a matrix of read-in/read-out shift registers in which chips can be sequentially read in to fill rows or columns, then read out sequentially by either rows or columns.

An interleaved digital system is shown in Figure 11.18a and operates as follows. The source data bits are first coded into n-chip words by standard block coding operations. The sequence of chip words then fill the *columns* of transmit interleaver (Figure 11.18b). When the interleaver matrix is filled, its *rows* are read-out in order, and the resulting chip sequence transmitted to the receiver over the binary communication link. At the receiver the decoded chips are used to fill the *rows* of the deinterleaver. When the latter is filled, its *columns* are read out in sequence to regenerate the original coded chip sequence generated at the transmit-

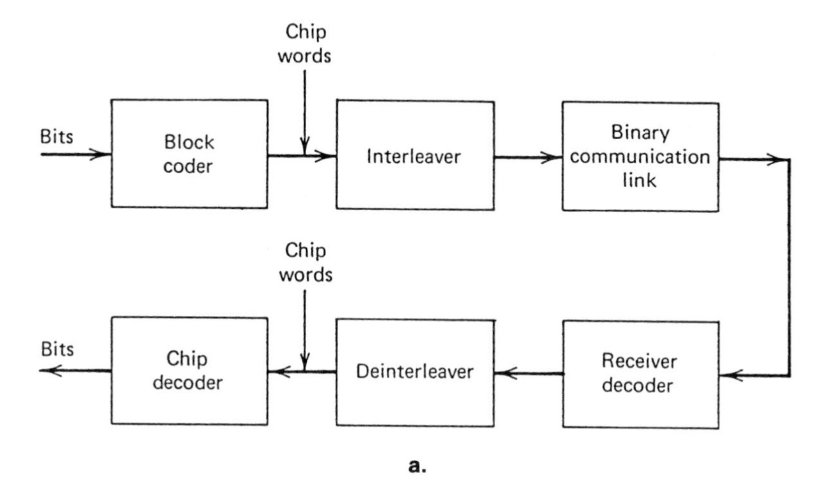

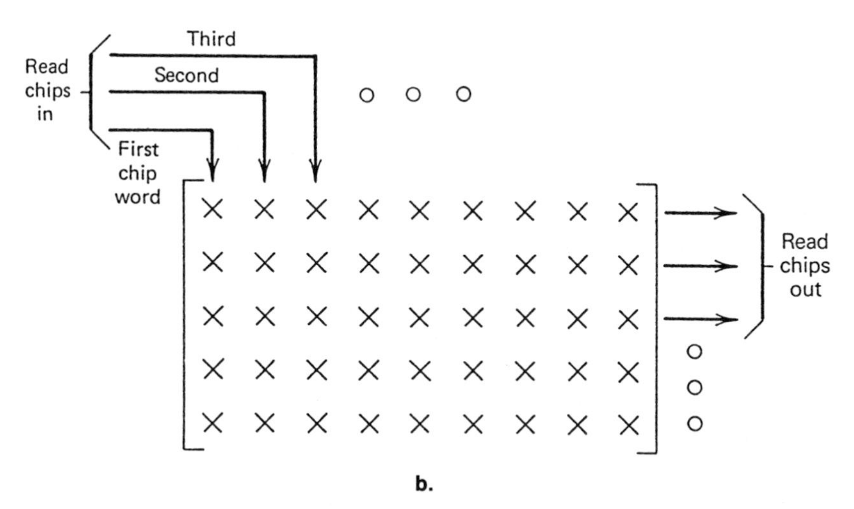

FIGURE 11.18 Channel coding with interleaving and deinterleaving. (a) Block diagram. (b) Interleaver matrix for read-in and read-out. ($\times$ denotes chips.)

ter. Except for the additional time to fill and read out the interleaver and deinterleaver, the overall link is theoretically transparent to their presence.

Now assume a fade occurs during chip transmission, lasting over N consecutive chips. This means that N consecutive entries of some row of the deinterleaver are incorrect (assuming no other chip error occur), as

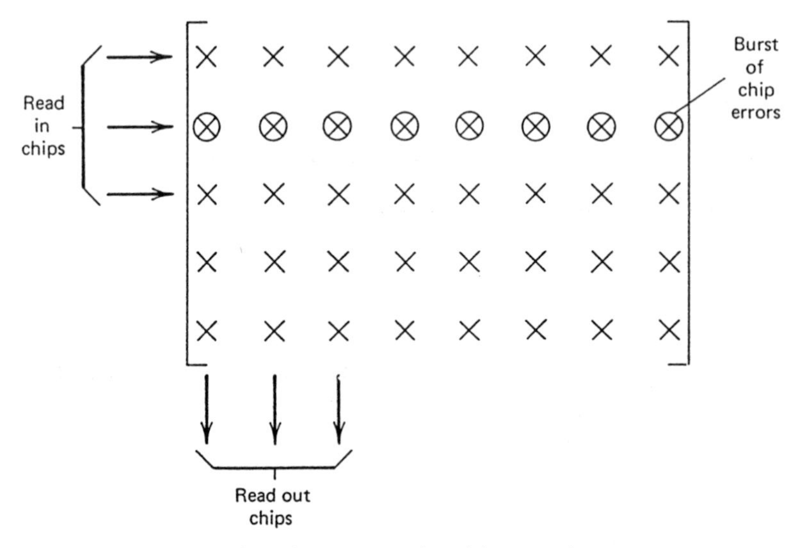

FIGURE 11.19 Deinterleaver matrix with error burst.

shown in Figure 11.19, for example. When the chips are read-out by columns, however, the resultant chip words will have at most one error. Hence, a relatively simple error correcting code used to form the transmitter chip words can correct for the entire N chip fade. (If the N chips straddled two rows, then a slightly stronger code is needed, but this event will occur with low probability if the row length greatly exceeded the fade length.) This interleaving permits simple codes to combat long fades.

It is clear from Figure 11.19 that the height (number of rows) of the interleaving matrix must accommodate the coded chip word size, while the width (number of columns) should exceed the expected fade duration. As a rough rule, an interleaver with width w and height h storing code words with correction capability t, can correct for fades of length wt. However, the interleaving matrices cannot be too large since it must fill in a time period shorter than the time between fades (so only one fade will occur during each interleaving time). In addition, large interleavers increase the delay time required to read in and read out the registers. As an example, a (15,9) block code, having a correction capability of three chips per word, operating with a 15×60 interleaver, can handle fade bursts of up to $3 \times 60/4800 = 37$ ms. The interleaver however will require $15 \times 60/4800 = 0.18$ s to fill and empty, thereby adding an additional $2 \times 0.18 = 0.36$ s to the total system delay. A larger interleaver will correct for longer fade bursts, but of course will add additional delay.

Hence, interleaver design requires careful balance of fade duration and fade rate, and generally reduces to a trade-off of burst length correction versus overall link delay.

11.8 COMBINED CODING AND MODULATION FOR THE MOBILE CHANNEL

In Section 11.6 it was pointed out that the use of channel-coding (converting bits to chips) can improve the error probability of a faded link. The primary disadvantage was that the bit rate had to decrease to maintain the desired channel bandwidth. In Chapter 2 it was shown that waveform encoding of blocks of bits can also improve a communication line. While some waveform signaling formats improved bit-error probability, others permitted increasingly larger block sizes to be transmitted over band-limited channels. The most important of this latter type were the class of MPSK and CPFSK waveforms, which directly improved throughput, but required significant power increases to maintain PE performance. A question then arises as to whether some form of channel coding can possibly be combined with bandwidth-efficient waveform encoding to produce a communication link that is advantageous for the mobile channel.

In order to achieve this, the channel coding must provide for the coding of sequences of data words into sequences of waveform encoded signals. Decoding is then based on the entire sequence rather than deciding one word at a time. Sequence coding allows the insertion of coding memory, so that signal selection for a particular data word depends on the past words (that is, on the previous "states" of the coder). This memory helps to increase the separability of the sequences. The design objective then is to find systematic ways of deriving the necessary coding and the choice of signals so that this sequence decoding improves error probability without expanding bandwidth.

A way to accomplish this was proposed by Ungerboeck (1982, 1987), using the concept of signal set partitioning in conjunction with convolutional trellis coding. The details will not be covered here, but the method can be summarized as follows. The idea is to begin with a bandwidth-efficient signal set larger than that needed to encode each data word alone. This set is then partitioned into subsets, and only signal combinations within each subset that have largest minimum distance (vector separation) are used. The coding operation then selects sequences of the candidate signals from the family of subsets. Decoding is achieved in two steps: (1) by making preliminary decisions among the candidate signals within each subset, then (2) deciding among the sequence of subsets that could have occurred via the coding. This latter operation is carried out similar to the Viterbi search algorithms, and eventually decide the most likely sequence based on the preliminary decisions. The procedure results in sequence decisions that are made from signals having a greater minimum distance than if independent block waveforms were transmitted.

The design of the required convolutional code is not easily described. Most of the usable codes that have been reported have been derived by heuristic code selection, using some basic rule for optimally selecting signal paths while transiting state diagrams. The interested reader may wish to pursue this topic in Ungerboeck (1982, 1987) and Wei (1984).

A simple example of the set of partitioning is shown in Figure 11.20. Two-bit words are to be transmitted in sequence. If a direct word-by-word waveform encoding is used, then each word will be coded into a QPSK signal set as in Figure 11.20a (four signals in two-dimensional space). A decision would then be made among the QPSK signal vectors each word time, with a minimum signal distance of $\sqrt{2E_w}$. Now consider beginning with the 8 PSK signal set of Figure 11.20b and partitioning into the two subsets of Figure 11.20c. The four signals to be used during any word time will be an antipodal signal from each subset. For example, the signal vector points (1, 5, 2, 6) represent one set, while (3, 7, 4, 8) would represent another. Other combinations of the same signal pairs can be denoted. Within each subset a maximal distance (separation of $2\sqrt{E_w}$) will be made. The coding operation at the transmitter determines which signal from which subset should be selected during each word time, based on the data bit sequence. All of the 8 PSK signals will eventually be used, when considering all the possible sequences that can occur.

The transmitter encoder has the overall block diagram shown in Figure 11.21a. The sequence of two-bit words are convolutionally coded into a corresponding sequence of three-chip words via the convolutional coder with rate $\frac{2}{3}$. Each chip word is then waveform encoded into one of the eight phase states (Figure 11.21b), and transmitted as an 8 PSK carrier waveform each word time. The receiver decoder makes the antipodal phase comparison from the known candidate pairs in Figure 11.20c, and stores the resulting signal "score" concerning the possible order in which the phase subsets occurred. A sequence decision concerning the most likely phase states is then made. The final decision is made from among sequence of signals having average distance greater than the $\sqrt{2E_w}$ if independent word coding was used. This translates directly to improved decoding performance. Note that the combination of a rate $\frac{2}{3}$ coder with an 8 PSK modulation format (the latter having a chip throughput of $(\log 8)/2 = \frac{3}{2}$) produces an overall bit throughput of one in Eq. (11.6.4), as desired for the mobile 5-kHz channel.

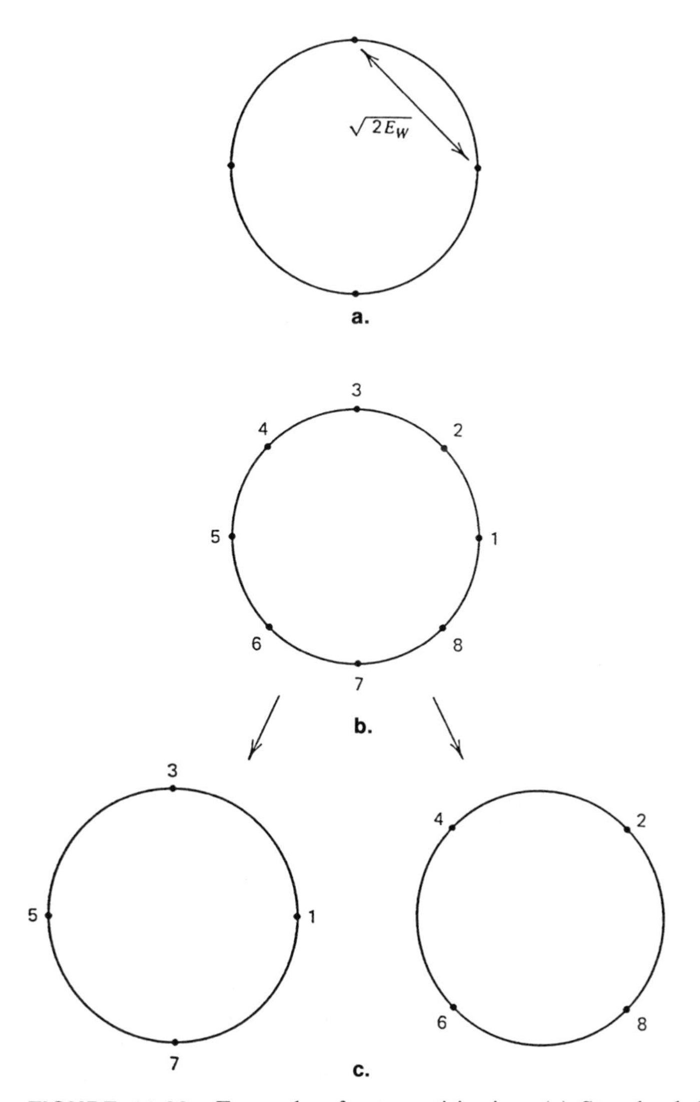

FIGURE 11.20 Example of set partitioning. (a) Standard QPSK. (b) 8 PSK. (c) Partitioned version of (b).

The system in Figure 11.21 can be extended in complexity by using a higher number of phases (more signal vector points) and by coding sequences of longer words (for example, four-bit words instead of two-bit words). The basic concept remains the same, with the convolutional coder mapping the bit words onto the partitioned phase states. Classes of QASK

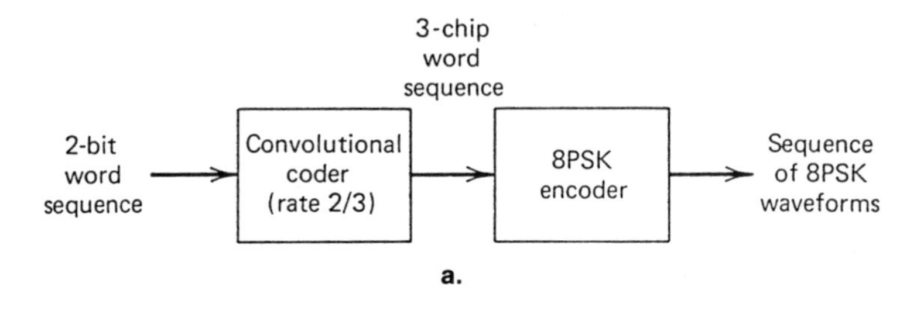

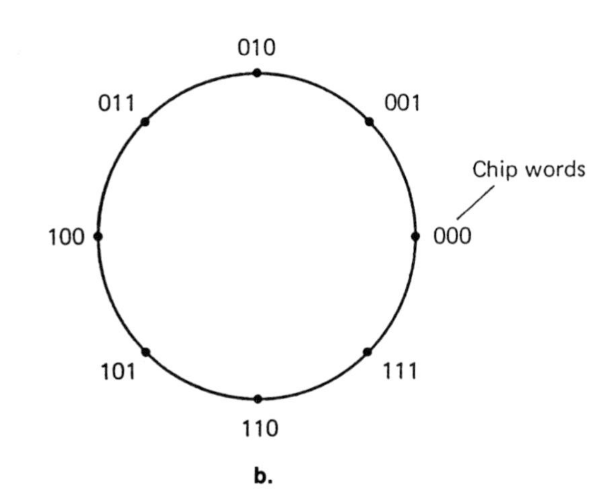

FIGURE 11.21 Coder for 2-bit, 8 PSK transmission. (a) Block diagram. (b) Phase mapping.

or ASK/PSK signal constellations can be used as alternatives to the MPSK signal sets.

Figure 11.22 compares the resulting bit error probability for sequence coding in a nonfading, Gaussian noise channel, predicted from simulation and theoretical bounds, as reported in the literature (Ungerboeck, 1982, 1987). The uncoded QPSK curve refers to the PE obtained if independent two-bit word-by-word encoding is used. The result of the sequence coding is shown for a rate $\frac{2}{3}$ convolutional coder using a two-bit 8 PSK system (Figure 11.21) and a more complex four-bit 16 PSK system. Note that at $PE = 10^{-5}$, approximately 3 dB improvement is gained over uncoded QPSK with 8 PSK, and this can be extended to about 5 dB with 16 PSK.

Figure 11.23 extends the results of coding to a fading Gaussian channel. The curves show the simulated improvement obtained in a Rayleigh and Rice ($r = 10$) fading channel for the case of the rate $\frac{2}{3}$ coder with 8 PSK modulation.

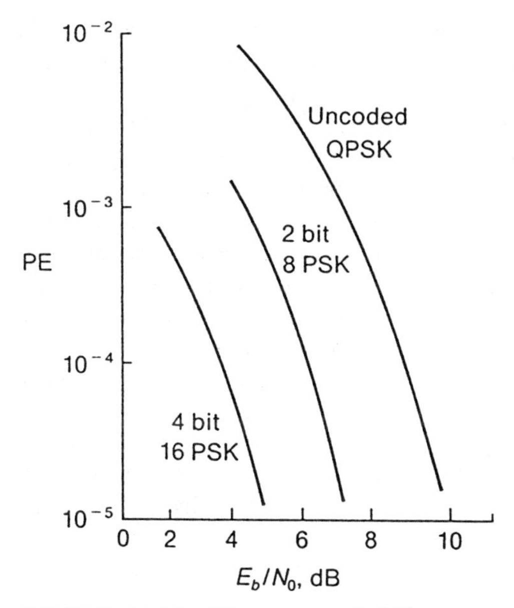

FIGURE 11.22 Bit error probability versus E_b/N_b.

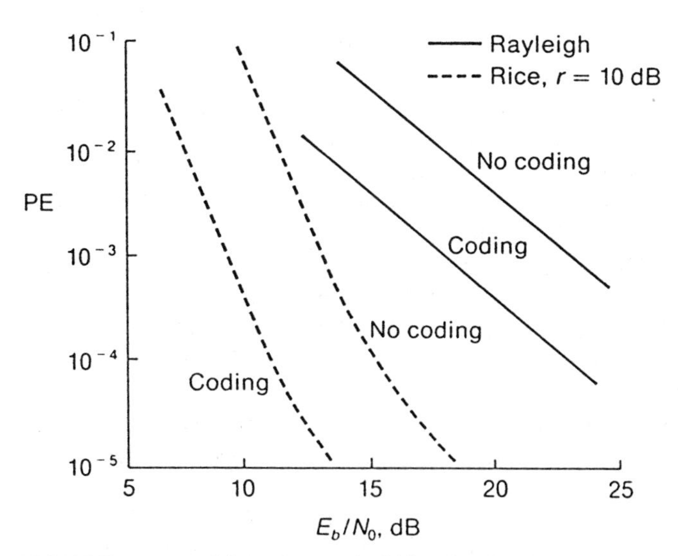

FIGURE 11.23 Bit error probability for fading Gaussian channel. Rate $\frac{2}{3}$, BPSK sequence coding.

It is to be emphasized that the power improvement as indicated in Figure 11.23 is achieved without increasing bandwidth or reducing data rate, as was necessary to obtain the results in Figure 11.17, since the transmission of 8 PSK and 16 PSK signals require the same bandwidth as uncoded QPSK. Instead, the improvement is obtained by using a signal set larger than that needed. That is, the improvement is obtained by inserting redundancy in the signal set, and only using a fraction of the available signals each word time. For this reason, this combination of sequence coding and set partitioning is referred to as *signal space coding*. Contrast this with the error correction described in Section 11.6, where the redundancy appeared in the extra chips that were sent. While the latter required excess bandwidth, the receiver still required only a binary decoder. Signal space coding requires no increase in bandwidth, but complicates the decoder design, since an increasingly larger number of signal waveforms must be handled by the receiver decoder. This again illustrates the primary technological direction of modern digital communications, in which extensive decoder processing is accepted as a viable tradeoff for bandwidth and power efficiency of the communication channel.

REFERENCES

Bell, D., and Naderi, F. (1986), "Mobile Satellite Communications—Vehicle Antenna Update," *AIAA Space System Technology Conference*, June 9–12.

Chen, J., and Gersho, A. (1987), "Real Time Vector Speech Coding at 4800 bps with Adaptive Filtering," *Proceedings of ICASSP*, 51.3–51.8.

Davidson, G., Chen, J., and Gersho, A. (1987), "Speech Coding For Mobile Satellite Experiment," *Proceeding of the International Communication Conference* (ICC), (June).

Douville, R. (1977), "A 12 GHz Low Cost Earth Terminal for Direct TV Reception," *IEEE Trans. MIT*, **MTT-25**. pp. 1045–1054.

Hong, U. (1988), "Measurement and Modeling of Land Mobile Satellite Propagation at UHF," *IEEE Trans. Antennas and Propagation*, **AP-36**. pp. 733–740.

Huang, J. (1988), "Circularly Polarized Conical Patterns From Microstrip Antennas," *IEEE Trans. Antennas and Propagation*, **AP-32** (9), 991–994.

JPL (1988a), *Proceedings of the Mobile Satellite Conference*, JPL Publication 88-9, May, 1988, Session E, Vehicle Antennas, pp. 209–255.

JPL (1988b), *Proceedings of the Mobile Satellite Conference*, JPL Publication 88-9, Section F, Propagation, pp. 87–145.

Konishi, Y. (1978), "12 GHZ FM Receiver for Satellite Broadcasting," *IEEE Trans. MIT*, **MIT-26**, 830–839.

Pritchard, W., and Kase, C. (1981), "Getting Set for Direct Broadcast Satellites," *IEEE Spectrum*, pp. 716-717.

Schmier, R., and Bostian, C. (1987), "Fade Durations in Mobile Radio Propagation," *IEEE Trans. Vehicle Technology*, **VT-21**, 971–981.

Simon, M. (1986), "Dual Pilot Tone Calibration Techniques," *IEEE Trans. Vehicle Technology*, **VT-35**, 63–70.

Ungerboeck, G. (1982), "Channel coding with multilevel/phase signals," *IEEE Trans. Information Theory*, **IT-28**, 55–67.

Ungerboeck, G. (1987), "Trellis coded modulation with redundant signal sets—Part I and Part II," *IEEE Communication Magazine*, **25** (2), 5–22.

Vogel, W., and Goldhirch, J. (1988), "Fade Measurements at L-band and UHF in Mobile Systems," *IEEE Trans. Antennas and Propagation*, **AP-36**, 32–43.

Wei, L. F. (1984), "Rotationally invariant convolutional channel coding with expanded signal space—Part I: 180 degrees," *IEEE Trans. Selected Areas in Comm.*, **SAC-2**, 659–672.

PROBLEMS

11.1. Consider the antenna patterns produced by a dipole (whip antenna), as shown in Table 3.1. Show how this may be designed as an auto-mounted roof-top antenna that will have a field of view covering a satellite at 30° elevation, no matter what the direction of the auto. Estimate the dipole length for a UHF and K-band system.

11.2. Use the results of Chapter 7 to design a CDMA system for 100 orthogonal users, each transmitting 100 kbps in a VSAT–satellite network. Estimate the code length, code rate, and carrier bandwidth needed by each transmiting VSAT.

11.3. A satellite is observed at an elevation angle of 30° by a vehicle moving at 10 mph. Determine the Doppler shift at UHF in the received carrier. Repeat for a velocity of 100 mph.

11.4. A Ts burst of a sine wave of amplitude A and frequency ω_c is transmitted to a receiver. Determine the shortest multipath delay that can be tolerated to ensure that the cross-correlation of direct and multipath signal is less than 10% of the direct signal power. Assume $\omega_c T \gg 1$, and the multipath signal has the same amplitude as the direct signal when received.

11.5. Show that a differential phase shift between the direct and specular components of a multipath signal will effectively "filter" the waveform. Estimate the filter bandwidth.

11.6. Write the probability that s in Eq. (11.5.5) is less than one (i.e., that a fade occurs) in terms of the Q-function in Eq. (A.8.5). *Hint*: convert the condition that $s \leq 1$ to an equivalent condition on the amplitude $\alpha(t)$.

11.7. Formally integrate Eq. (11.6.2) and prove Eqs. (11.6.3b) and (11.6.3c).

11.8. A BPSK system operates with $E_b/N_0 = 10$ dB. A pilot tone system is used to extract the carrier reference. The noise $n(t)$ at the pilot filter output produces a phase variation on the pilot of $\psi(t) = n(t)/A$, where A is the pilot amplitude. The pilot has a Doppler shift that may be as large as 1/1000 of the bit rate. What is the rms phase error that can be expected in using the filtered pilot as a carrier reference?

APPENDIX A
Review of Digital Communications

This appendix is a detailed backup for some of the results presented in the text. The objective is to guide the interested reader through some of the mathematical steps leading to equations, figures, and tabulations presented in Chapter 2. Further detail can be found in the references given at the end of this appendix.

A.1 BASEBAND DIGITAL WAVEFORMS

Let a data modulated baseband waveform be written as

$$m(t) = \sum_{k=-\infty}^{\infty} d_k p(t - kT_b + \epsilon) \qquad \text{(A.1.1)}$$

where $\{d_k\}$ is a sequence of independent, binary random variables, each ± 1 and equal likely, and ϵ is a uniformly distributed location variable over $(0, T_b)$. Define the waveform correlation as

$$R_m(\tau) = \mathscr{E}[m(t)m(t + \tau)] \qquad \text{(A.1.2)}$$

where $\mathscr{E}$ is the expectation operator. It therefore follows that $m(t)$ in Eq.

(A.1.1) has the correlation

$$
\begin{aligned}
R_m(\tau) &= \sum_k \sum_q \mathscr{E}[d_k d_q] \mathscr{E}[p(t - kT_b + \epsilon) \times p(t + \tau - qT_b + \epsilon)] \\
&= \sum_k \frac{1}{T_b} \int_{t+kT_b}^{t+(k+1)T_b} p(u)p(u+\tau)du \\
&= R_{pp}(\tau)
\end{aligned} \tag{A.1.3}
$$

where

$$
R_{pp}(\tau) \triangleq \frac{1}{T_b} \int_{-\infty}^{\infty} p(t)p(t+\tau)dt \tag{A.1.4}
$$

The spectral density of $m(t)$ is the Fourier transform of $R_m(\tau)$, and is then

$$
\begin{aligned}
S_m(\omega) &= \text{Fourier transform of } R_{pp}(\tau) \\
&= \frac{1}{T_b} |F_p(\omega)|^2
\end{aligned}
$$

where

$$
F_p(\omega) = \int_{-\infty}^{\infty} p(t) e^{-j\omega t} \, dt \tag{A.1.5}
$$

Hence, the Fourier transform of the bit waveform in Eq. (A.1.5) determines the spectral density of the baseband sequence. Some specific examples of baseband waveforms are: (1) the NRZ waveform,

$$
\begin{aligned}
p(t) &= 1 \qquad 0 \le t \le T_b \\
&= 0 \qquad \text{elsewhere}
\end{aligned} \tag{A.1.6}
$$

for which the Fourier transform produces

$$
|F_p(\omega)|^2 = T_b^2 \left| \frac{\sin(\omega T_b/2)}{(\omega T_b/2)} \right|^2 \tag{A.1.7}
$$

and (2) the Manchester waveform,

$$\begin{aligned} p(t) &= 1 && 0 \le t \le T_b/2 \\ &= -1 && T_b/2 \le t \le T_b \\ &= 0 && \text{elsewhere} \end{aligned} \tag{A.1.8}$$

for which

$$\begin{aligned} F_p(\omega) &= \int_0^{T_b/2} e^{-j\omega t}\,dt - \int_{T_b/2}^{T_b} e^{-j\omega t}\,dt \\ &= \frac{2\sin(\omega T_b/2) - \sin(\omega T_b)}{\omega} + j\frac{2\cos(\omega T_b/2) - (1 + \cos\omega T_b)}{\omega} \end{aligned} \tag{A.1.9}$$

Using

$$2\sin^2\omega T_b = 1 - \cos 2\omega T_b$$

$$4\sin^4(\omega T_b/4) = 1 - 2\cos(\omega T_b/2) + \cos^2(\omega T_b/2)$$

allows us to write

$$|F_p(\omega)| = \frac{4}{\omega^2}\,[\sin^4(\omega T_b/4)] \tag{A.1.10}$$

These spectra are plotted in Figure 2.6 in Chapter 2.

A.2 BPSK SYSTEMS

BPSK carriers are obtained by modulating baseband sequences onto carriers. The waveform is then

$$c(t) = Am(t)\cos(\omega_c t + \psi) \tag{A.2.1}$$

where $m(t)$ is of the form of Eq. (A.1.1) and ψ is taken as a random phase, uniformly distributed over $(0, 2\pi)$. The carrier has correlation function

$$\begin{aligned} R_c(\tau) &= A^2\mathscr{E}[m(t)m(t+\tau)]\mathscr{E}[\cos(\omega_c t + \psi)\cos[\omega_c(t+\tau) + \psi]] \\ &= \frac{A^2}{2}R_m(\tau)\cos(\omega_c\tau) \end{aligned} \tag{A.2.2}$$

Its spectral density is then

$$S_c(\omega) = \frac{A^2}{2}[S_m(\omega) \otimes][\tfrac{1}{2}\delta(\omega + \omega_c) + \tfrac{1}{2}\delta(\omega - \omega_c)]$$

$$= \frac{A^2}{4}[S_m(\omega + \omega_c) + S_m(\omega - \omega_c)] \tag{A.2.3}$$

where $\otimes$ denotes frequency convolution. Hence, the BPSK carrier has a spectrum obtained by shifting the spectrum of $m(t)$ to $\pm\omega_c$.

Decoding of BPSK in the presence of Gaussian noise is achieved by coherent phase correlation, followed by bit integration (see Figure 2.16). Letting

$$r(t) = 2p(t)\cos(\omega_c t + \hat{\psi}) \tag{A.2.4}$$

be the local reference, the decoder therefore forms over each bit interval the decoding variable

$$z \triangleq \int_0^{T_b} [c(t) + n(t)]r(t)dt \tag{A.2.5}$$

where $n(t)$ is a white Gaussian noise process of one-sided spectral level N_0. Bit decisioning is then made on the sign of z. For the data sequence in Eq. (A.1.1), with $p(t)$ normalized to unit power, we have

$$z = Ad_0 T_b \cos\psi_e + AT_b \sum_{k \neq 0} d_k r_{pp}(k)\cos\psi_e + n \tag{A.2.6}$$

where d_0 is the present bit, n is a Gaussian variable with variance $N_0 T_b/2$, ψ_e is the phase-reference error, and

$$r_{pp}(k) = \frac{1}{T_b}\int_0^{T_b} p(t)p(t - kT_b)dt \tag{A.2.7}$$

The first term is the data bit being decoded and the second term represents intersymbol interference (II) from other bits. If the data pulse $p(t)$ is restricted to a bit time $[p(t) = 0,\ t$ outside $(0, T_b)]$, we have $r_{pp}(k) = 0$, $k \neq 0$, and there is no II. The probability that a bit error will occur (z

has the opposite sign from d_0) in the absence of II is then

$$\mathrm{PE} = Q\left[\left(\frac{2E_b}{N_0}\right)^{1/2} \cos \psi_e\right] \tag{A.2.8}$$

where

$$\frac{E_b}{N_0} = \frac{(A^2/2)T_b}{N_0} \tag{A.2.9}$$

and

$$Q[x] = \frac{1}{2\pi}\int_x^{\infty} e^{-t^2/2}\, dt \tag{A.2.10}$$

The $Q(x)$ function is the well-known Gaussian tail integral (Abramowitz and Stegun, 1965), and is shown in Figure A.1 along with some often used approximations. When used in the form of Eq. (A.2.8) with $\psi_e = 0$, the BPSK curve in Figure 2.17 results.

If the phase error ψ_e is a random variable, then PE in Eq. (A.2.8) must be averaged over the statistics of ψ_e. Hence we formally write

$$\mathrm{PE} = \int_{-\infty}^{\infty} \mathrm{PE}(\psi_e)p(\psi_e)d\psi_e \tag{A.2.11}$$

For a Gaussian phase error having zero mean and variance σ_e^2, this becomes

$$\mathrm{PE} = \int_{-\infty}^{\infty} Q\left[\left(\frac{2E_b}{N_0}\right)^{1/2} \cos x\right] \frac{e^{-x^2/2\sigma_e^2}}{\sqrt{2\pi}\,\sigma_e}\, dx \tag{A.2.12}$$

The resulting integration is shown in Figure 2.19, for several σ_e values. The result shows how BPSK performance is degraded with nonperfect phase referencing.

If the system is phase coherent ($\psi_e = 0$) and II occurs, the statistics of the interference must be incorporated. Formally, it would be necessary to interpret the II sum in Eq. (A.2.6) as a constant, for each data sequence (d_k), that either adds to or subtracts from the signal term. PE would then be obtained by averaging the resulting conditional PE over all possible data sequences. Alternatively, the exact statistics of the II can be computed

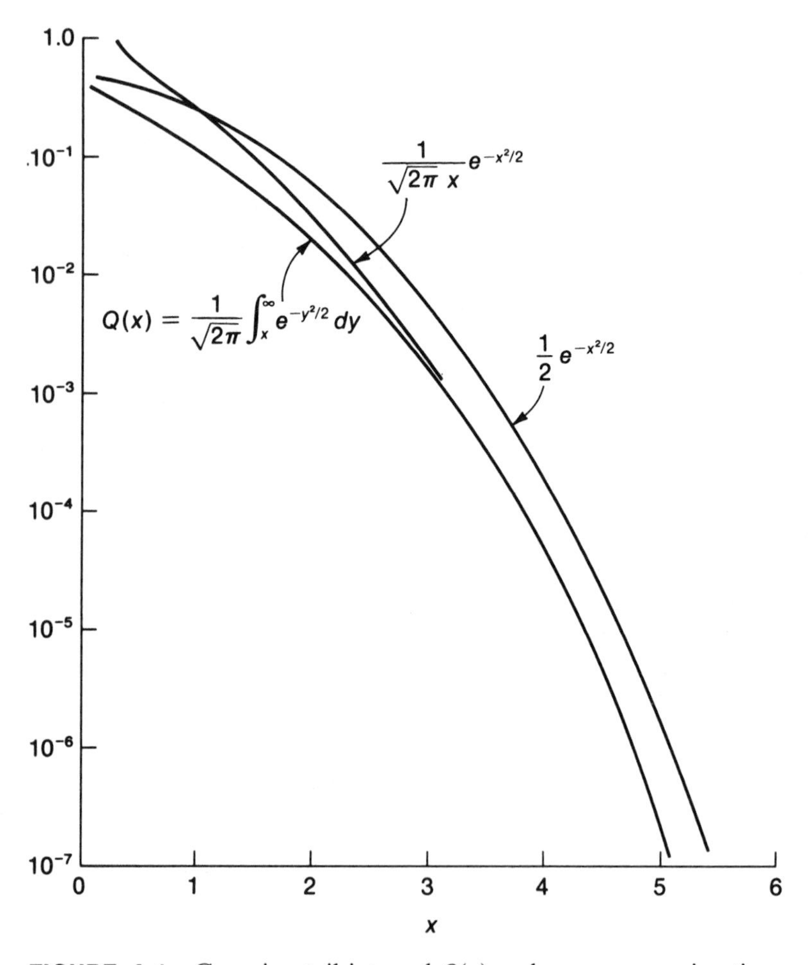

FIGURE A.1 Gaussian tail integral $Q(x)$ and some approximations.

(Shimbo, 1988), or the averaged PE can be estimated via moment analysis (Benedetto et al., 1987). The simplest approximation is to assume the II sum is Gaussian with zero mean and variance

$$\sigma_I^2 = \mathscr{E}\left[\sum_k d_k r_{pp}(k)\right]^2$$

$$= \sum_k r_{pp}^2(k) \qquad \text{(A.2.13)}$$

Then PE becomes

$$\text{PE} = Q\left[\sqrt{2}\left(\frac{E_b}{N_0 + (\sigma_I^2/T_b)}\right)^{1/2}\right] \tag{A.2.14}$$

This shows that II degrades performance from that of an ideal BPSK decoder. This analysis approach is used in Chapters 4 and 5.

A.3 QPSK CARRIER WAVEFORMS

QPSK carriers are formed by simultaneous modulation of two separate baseband waveforms onto the quadrature components of a common carrier. Hence

$$c(t) = Am_c(t)\cos(\omega_c t + \psi) + Am_s(t)\sin(\omega_c t + \psi) \tag{A.3.1}$$

where

$$m_c(t) = \sum_k a_k p(t - kT_s) \tag{A.3.2a}$$

$$m_s(t) = \sum_k b_k q(t - kT_s) \tag{A.3.2b}$$

and a_k, b_k are ± 1. The correlation of $c(t)$ is then

$$\begin{aligned} R_c(\tau) &= \mathscr{E}[c(t)c(t+\tau)] \\ &= \frac{A^2}{2}\{R_{m_c}(\tau)\cos(\omega_c\tau) \\ &\quad + R_{m_s}(\tau)\cos(\omega_c\tau) \\ &\quad + \mathscr{E}[m_c(t)m_s(t+\tau)]\sin(\omega_c\tau) \\ &\quad + \mathscr{E}[m_s(t)m_c(t+\tau)]\sin(\omega_c\tau)\} \end{aligned} \tag{A.3.3}$$

If the data waveforms $m_c(t)$ and $m_s(t)$ contain uncorrelated data bits, this reduces to

$$R_c(\tau) = \frac{A^2}{2}[R_{m_c}(\tau) + R_{m_s}(\tau)]\cos(\omega_c\tau) \tag{A.3.4}$$

The corresponding spectral density is then

$$S_c(\omega) = \frac{A^2}{4}[S_{m_c}(\omega + \omega_c) + S_{m_s}(\omega + \omega_c)] + \frac{A^2}{4}[S_{m_c}(\omega - \omega_c) + S_{m_s}(\omega - \omega_c)] \quad \text{(A.3.5)}$$

Hence, the QPSK carrier spectrum is the combined baseband spectra of each quadrature component superimposed at $\pm\omega_c$.

Decoding is achieved via the quadrature cross-correlator in Figure 2.16. The decoding voltages generated in each arm for a given symbol interval $(0, T_s)$ are

$$I = \int_0^{T_s} [c(t) + n(t)]p(t) \cos(\omega_c t + \hat{\psi})dt \quad \text{(A.3.6a)}$$

$$Q = \int_0^{T_s} [c(t) + n(t)]q(t) \sin(\omega_c t + \hat{\psi})dt \quad \text{(A.3.6b)}$$

Using the QPSK waveform in Eq. (A.3.2) and assuming that the bit waveforms $p(t)$ and $q(t)$ are each T_s seconds long and unit power, Eq. (A.3.6) becomes

$$I = AT_s\left(a_0 + \sum_{k \neq 0} a_k r_{pp}(k)\right) \cos \psi_e + AT_s \sum_k b_k r_{pq}(k) \sin \psi_e + n_c \quad \text{(A.3.7a)}$$

$$Q = AT_s\left(b_0 + \sum_{k \neq 0} b_k r_{qq}(k)\right) \cos \psi_e + AT_s \sum_k a_k r_{qp}(k) \sin \psi_e + n_s \quad \text{(A.3.7b)}$$

where now

$$r_{pq}(k) = \frac{1}{T_s} \int_0^{T_s} p(t)q(t - kT_s)dt \quad \text{(A.3.8)}$$

and n_c and n_s are independent Gaussian variables with the variance $N_0 T_s/2$. In Eq. (A.3.7) the first term represents the present data symbol being decoded. The first summation represents II, the second summation represents quadrature channel interference, and the n terms are the noise interference. If the decoder is perfectly coherent ($\psi_e = 0$), the quadrature interference is zero, and both I and Q reduce to the BPSK decisioning variables in Eq. (A.2.6). Hence, each quadrature channel can be separately decoded as a BPSK link, and the bit error probability of each is given in Eq. (A.2.8) or (A.2.14), with T_s replacing T_b. We point out that Eq. (A.3.7) is valid for any symbol waveform $p(t)$ and $q(t)$, and therefore includes both OQPSK and MSK, described in Chapter 2.

If the phase error ψ_e is not zero, both II and quadrature channel interference must be included. It is no longer obvious that the bit integrator is the preferable decoder. A sampling decoder (phase coherent demodulation followed by a symbol sampler) would produce the decoding variable for I of

$$I = AT_s\left(a_0 + \sum_{k \neq 0} a_k p(k)\right) \cos \psi_e + AT_s \sum_k b_k q(k) \sin \psi_e + n_c \tag{A.3.9}$$

where $p(k)$ and $q(k)$ are the quadrature samples of the waveforms $p(t - kT_s)$ and $q(t - kT_s)$. If the symbol waveform $q(t)$ was designed such that all the samples $q(k)$ during the present symbol sampling of $p(t)$ were zero, no quadrature interference would occur. This can occur for both OQPSK and MSK, and shows the advantage of waveform shaping in QPSK decoding.

If $p(t)$ and $q(t)$ were restricted to one symbol time $(0, T_s)$, then Eq. (A.3.9) simplifies to

$$I = AT_s(a_0 \cos \psi_e + b_0 \sin \psi_e) + n_c \tag{A.3.10}$$

When $\psi_e \neq 0$, the quadrature channel symbol b_0 will either add to or subtract from the voltage of the correct symbol. The probability of error is then the average over these two possibilities. Hence

$$\text{PE} = \frac{1}{2} Q\left[\left(\frac{2E_b}{N_0}\right)^{1/2} (\cos \psi_e + \sin \psi_e)\right] + \frac{1}{2} Q\left[\left(\frac{2E_b}{N_0}\right)^{1/2} (\cos \psi_e - \sin \psi_e)\right] \tag{A.3.11}$$

A given phase error ψ_e will always act to increase PE from the ideal case.

If the phase error ψ_e is random, PE must be further averaged over its probability density. In OQPSK, symbol transitions in the other channel occur at the midpoint of each symbol. If no transition occurs, PE is identical to the QPSK result in Eq. (A.3.11). When a transition occurs, the cross-coupled term involving $\sin \psi_e$ changes polarity at midpoint, and the integrated quadrature interference during the first half of the symbol is canceled by that during the second half. Hence no interference appears, and PE is identical to that of a BPSK link with the same phase error. Since a symbol transition occurs with probability one-half, the OQPSK PE is then the average of the PE for QPSK and BPSK with the same phase error ψ_e.

A.4 FSK

In frequency shift keyed systems, one of the two waveforms:

$$c(t) = A \cos[(\omega_c + \omega_d)t + \psi], \qquad t \in (0, T_b) \tag{A.4.1a}$$

or

$$c(t) = A \cos[(\omega_c - \omega_d)t + \psi], \qquad t \in (0, T_b) \tag{A.4.1b}$$

is transmitted, depending on the data bit. Each waveform occupies a bandwidth of $2/T_b$ Hz around each carrier frequency. Decoding is achieved using a two-channel decoder, one for each waveform in Eq. (A.4.1). In a coherent decoder, a phase coherent reference at each frequency is used to bit-correlate in each channel. The channel with the largest correlator output is selected as the transmitted bit. The correct channel correlates to

$$z_c = \frac{AT_b}{2} + n_c \tag{A.4.2a}$$

whereas the incorrect channel correlates to

$$z_{\text{in}} = n_s \tag{A.4.2b}$$

For Gaussian noise, the probability of deciding correctly is then

$$\text{PE} = Q\left[\left(\frac{E_b}{N_0}\right)^{1/2}\right] \tag{A.4.3}$$

Note that E_b is effectively reduced by one-half from coherent BPSK. This result is plotted as coherent FSK in Figure 2.17.

If noncoherent FSK decoding is used (Figure 2.20), bit decisioning is based on the largest *envelope* sample in the two channels, and knowledge of carrier phase ψ is not required. The correct channel contains the bit tone in Eq. (A.4.1) plus Gaussian bandpass noise. Hence the BPF output is

$$y(t) = A\cos(\omega_0 t + \psi) + n_c(t)\cos(\omega_0 t + \psi) + n_s(t)\sin(\omega_0 t + \psi) \quad \text{(A.4.4)}$$

where ω_0 is either $(\omega_c + \omega_d)$ or $\omega_c - \omega_s)$ and $n_c(t)$ and $n_s(t)$ are the noise quadrature components in Eq. (4.2.5). The envelope of this waveform is

$$\alpha(t) = [(A + n_c(t))^2 + (n_s(t))^2]^{1/2} \quad \text{(A.4.5)}$$

If the BPF bandwidth is matched to the bit time, a sample α_c of this envelope is known to have the probability density (Papoulis, 1965)

$$p(\alpha_c) = \frac{\alpha_c}{\rho^2} e^{-(\alpha_c^2 + \rho^4)/2\rho^2} I_0(\alpha_c), \qquad \alpha_c \geq 0 \quad \text{(A.4.6)}$$

where $I_0(x)$ is the imaginary Bessel function and

$$\rho^2 = \frac{2E_b}{N_0} = \frac{A^2 T_b}{N_0} \quad \text{(A.4.7)}$$

If the frequency separation $2\omega_d$ is wide enough ($\omega_d \gg 1/T_b$), the incorrect bandpass channel contains only noise, and a sample α_n of its envelope has density

$$p(\alpha_n) = \frac{\alpha_n}{\rho^2} e^{-\alpha_n^2/2\rho^2}, \qquad \alpha_n \geq 0 \quad \text{(A.4.8)}$$

The probability of a bit error is the probability that the incorrect envelope sample α_n will exceed the correct sample α_c. Using Eqs. (A.4.6) and (A.4.8) this becomes

$$\begin{aligned} \text{PE} &= \int_0^\infty p(\alpha_c) \int_{\alpha_c}^\infty p(\alpha_n) d\alpha_n \, d\alpha_c \\ &= \tfrac{1}{2} e^{-\rho^2/4} \end{aligned} \quad \text{(A.4.9)}$$

The result is plotted in Figure 2.17 as noncoherent FSK.

A.5 DPSK

Another method of decoding binary data without the need for a coherent reference is by the use of differential phase shift keying (DPSK). Here BPSK waveforms are used, but the previous bit waveform serves as the phase coherent reference for the present bit. A one or zero bit is sent by transmitting the same phase or the opposite phase of the previous bit. The bits are decoded by the processing system in Figure 2.23a. The system performs the bit test

$$y_c(0)y_c(T_b) + y_s(0)y_s(T_s) \lesseqgtr 0 \tag{A.5.1}$$

where

$$\begin{pmatrix} y_c(t) \\ y_s(t) \end{pmatrix} = \int_{t-T_b}^{t} x(t) \begin{pmatrix} \cos(\omega_c t) \\ \sin(\omega_c t) \end{pmatrix} dt \tag{A.5.2}$$

and $x(t)$ is the input. This is equivalent to the test $E_1 \gtrless E_0$, where

$$E_1^2 = [y_c(0) + y_c(T_b)]^2 + [y_s(0) + y_s(T_b)]^2 \tag{A.5.3a}$$

$$E_0 = [y_c(0) - y_c(T_b)]^2 + [y_s(0) - y_s(T_b)]^2 \tag{A.5.3b}$$

When a one bit is sent, the first bracket of E_1^2 is a Gaussian variable with mean $\sqrt{2E_b T_{T_b}} \cos \psi$ and variance $(N_0/2)(2T_b/2) = N_0 T_b/2$. Likewise the second bracket is Gaussian with mean $\sqrt{2E_b T_b} \sin \psi$ and the same variance. Combining and following the development in Eq. (A.4.5) and (A.4.6) we see that E_1 is a Rician variable with noncentrality parameter $\sqrt{2E_b T_b}$. The variables in the first and second bracket of E_0^2 have zero mean, and each has the variance $N_0 T_b/2$, so that E_0 in Eq. (A.5.3b) has a Rayleigh density. The probability that $E_1 < E_0$ when a one bit is sent during $(0, T_b)$ will integrate as in Eq. (A.4.9) to produce $\text{PE} = (\frac{1}{2}) \exp(-\rho^2/4)$, where now $\rho^2 = (\sqrt{2E_b T_{T_b}})^2/(N_0 T_b/2) = 4E_b/N_0$. The resulting DPSK bit error probability is then

$$\text{PE} = \tfrac{1}{2} e^{-E_b/N_0} \tag{A.5.4}$$

where E_b is the bit energy in T_b s. This result is included in Table 2.3 and Figure 2.17.

A DPSK carrier can also be decoded by the simpler, but suboptimal, delay-and-correlate system in Figure 2.23b. The multiplier output for any bit is

$$z = \int_0^{T_b} [Ad_0 \cos(\omega_c t + \psi) + n_0(t)][Ad_{-1} \cos(\omega_c t + \psi) + n_{-1}(t)]dt \tag{A.5.5}$$

where $n_0(t)$ and $n_{-1}(t)$ are the noise processes during the present and past bit times. The variable z expands as

$$z = \frac{A^2 T_b}{2} (d_0 d_{-1}) + n + \int_0^{T_b} n_0(t) n_{-1}(t)dt \tag{A.5.6}$$

where n is Gaussian variable with variance $2A^2 N_0 T_b/4$. Since $d_0 d_{-1} = \pm 1$, corresponding to the present bit, the correlation has recovered the present bit without use of a coherent reference. However, the decoding variable z contains the noise–noise crossproduct term due to the use of a noisy reference from the past bit interval. If we argue that the integration variable is approximately Gaussian (since the integration bandwidth is generally much less than the noise bandwidth), we can estimate PE performance by computing the mean and variance of z. If this is done the PE is approximately

$$\begin{aligned} \text{PE} &= Q\left[\frac{E_b^2}{E_b N_0 + N_0^2/4}\right]^{1/2} \\ &= Q\left\{\left(\frac{E_b}{N_0}\right)^{1/2}\left[\frac{1}{1 + (N_0/4E_b)}\right]^{1/2}\right\} \end{aligned} \tag{A.5.7}$$

The inner bracket term therefore accounts for the degradation due to the noise correlation. When $E_b/N_0 \gg 1$,

$$\begin{aligned} \text{PE} &\approx Q\left[\left(\frac{E_b}{N_0}\right)^{1/2}\right] \\ &\approx \frac{1}{\sqrt{2\pi}(E_b/N_0)^{1/2}} e^{-(E_b/2N_0)} \end{aligned} \tag{A.5.8}$$

At high energy levels ($E_b/N_0 \approx 10$) the delay-and-correlate DPSK decoder in Figure 2.23b is about 2 dB worse than the optimal DPSK decoder in Figure 2.23a.

A.6 MPSK

In MPSK, a carrier with one of M distinct phase shifts $\theta_i = i(360°)/M$, $i = 0, 1, \ldots, M-1$, is transmitted. The modulated carrier therefore appears as

$$c(t) = A\cos(\omega_c t + \theta(t) + \psi) \tag{A.6.1}$$

where $\theta(t)$ shifts to one of the M phases θ_i every symbol time T_s. The spectrum of the MPSK carrier therefore corresponds to a BPSK carrier with symbol width T_s, and therefore occupies a main-hump bandwidth of $2/T_s$ Hz around ω_c.

Decoding is achieved by detecting the phase of the received noisy waveform and determining to which phase states θ_i the detected phase is closest. A phase detector requires a phase referenced carrier in order to distinguish the modulation phase θ_i from the carrier phase ψ. The probability of a symbol error is obtained by determining the probability that the detected phase is not closer to the true phase. For additive Gaussian noise this has been determined to be (Lindsey and Simon, 1973)

$$\begin{aligned}\text{PWE} &= \frac{M-1}{M} - \tfrac{1}{2}\,\text{Erf}\left[\left(\frac{E_w}{N_0}\right)^{1/2}\sin\left(\frac{\pi}{M}\right)\right] \\ &\quad - \frac{1}{\sqrt{\pi}}\int_0^{\sqrt{E_w/N_0}\sin(\pi/M)} e^{-y^2}\,\text{Erf}\left[y\cot\left(\frac{\pi}{M}\right)\right]dy\end{aligned} \tag{A.6.2}$$

where E_w is the word energy ($E_w = A^2 T_w/2$), and

$$\text{Erf}(x) = \frac{2}{\sqrt{\pi}}\int_0^x e^{-y^2}\,dy \tag{A.6.3}$$

For $M = 2$ and $M = 4$, Eq. (A.6.2) reduces to the previous BPSK and QPSK results. When $M \gg 1$, the integral in Eq. (A.6.2) can be neglected, and PWE is adequately approximated by Eq. (2.7.3) in Chapter 2. The resulting word error probabilities are shown in Figure 2.31 for various M and E_w/N_0.

A.7 CORRELATION DETECTION OF ORTHOGONAL BPSK WAVEFORMS AND PHASE-SEQUENCE TRACKING

In Section 2.7, we considered the mapping of blocks of data bits into orthogonal sequences of BPSK carrier waveforms. Here we consider the decoding of these carriers.

Let a_i be a binary sequence $a_i = \{a_{iq}\}$, $a_{iq} = \pm 1$ of length v. Let $c_i(t)$ be a BPSK carrier formed from this sequence as

$$c_i(t) = A \sum_{q=1}^{v} a_{iq} p(t - qT) \cos(\omega_c t + \psi) \qquad 0 \leq t \leq T_w = vT \quad \text{(A.7.1)}$$

where $p(t)$ is the modulating waveform. Let a_j and $c_j(t)$, $j = 1, \ldots, M$ be a similar sequence and carrier formed in the same way, constituting a set of M such carriers. Let each carrier have energy E_w over $(0, T_w)$. Assume the ith carrier is transmitted, and received with additive white Gaussian noise $n(t)$, as in Eq. (A.2.5), forming

$$x(t) = c_i(t) + n(t) \tag{A.7.2}$$

Consider a decoding bank of M phase coherent correlators, as shown in Figure 2.29b, each matched to one of the possible carriers $c_j(t)$. The jth correlator computes the correlation

$$\begin{aligned} y_j &= \int_0^{T_w} x(t) c_j(t) dt \\ &= \int_0^{T_w} c_i(t) c_j(t) dt + n \end{aligned} \tag{A.7.3}$$

where n is a Gaussian random variable with variance $N_0 E_w/2$. Under the condition $\omega_c \gg 2/T_w$, the first term integrates to

$$\frac{A^2}{2} \int_0^{T_w} \sum_{q=1}^{v} (a_{iq} a_{jq}) p^2(t - qT) dt = (E_w T/T_w) \sum_{q=1}^{v} a_{iq} a_{jq} \tag{A.7.4}$$

Orthogonal sequences are those for which

$$\begin{aligned} \sum_{q=1}^{v} a_{iq} a_{jq} &= 0 \qquad i \neq j \\ &= v \qquad i = j \end{aligned} \tag{A.7.5}$$

Orthogonal sequences of binary variables can be formed from Hadamard matrices (Brennan and Reed, 1965), and M orthogonal sequences require a length $v \geq M$. With orthogonal sequences, y_j in Eq. (A.7.3) is a Gaussian variable with mean zero if $j \neq i$, and with mean E_w for $i = j$. The probability that the transmitted sequence is correctly decoded is the probability that y_i exceeds all other y_j. Thus, if y_j has probability density

$p_{y_j}(y)$, then the probability of decoding the sequence in error is then

$$\text{PWE} = 1 - \int_{-\infty}^{\infty} p_{y_i}(y)\left[\int_{-\infty}^{y} p_{y_j}(x)dx\right]^{M-1} dy \tag{A.7.6}$$

When the Gaussian densities are substituted for each y_j, we obtain the entry in Table 2.4.

Now assume the bit sequence a_i phase modulates the carrier, with an arbitrary pulse shape $p(t)$, forming instead

$$c_i(t) = A \cos\left(\omega_c t + \psi + 2\pi\Delta_f \sum_{q=1}^{v} a_{iq}p(t - qT)\right) \tag{A.7.7}$$

The binary sequence now produces a carrier phase function that traces out a trajectory dependent on the bit sequence and waveform $p(t)$ (see Figure 2.12). Two different sequences therefore trace out two different trajectories.

Phase tracking decoding is obtained by attempting to determine which phase trajectory is being received during an interval $(0, T_w)$ by observing the noise carrier. This is achieved by correlating the latter with a phase coherent carrier having each possible trajectory, and deciding which has the largest correlation. The jth correlator output now has the mean value

$$\begin{aligned} m_{ij} &= \int_0^{T_w} c_i(t)c_j(t)dt \\ &= E_w \int_0^{T_w} \cos\left[2\pi\Delta_f \sum_{q=1}^{v} (a_{iq} - a_{jq})p(t - qT)dt\right] \end{aligned} \tag{A.7.8}$$

Note that the mean signal again depends on the bit sequence, but it is effectively observed through the cosine function.

The probability of erring in decoding between the correct y_i and an incorrect y_j will depend on the *distance* between the carrier waveforms, given by

$$\begin{aligned} d_{ij}^2 &= \int_0^{T_w} (c_i(t) - c_j(t))^2 dt \\ &= 2(E_w - m_{ij}) \\ &= 2E_w\left[1 - \int_0^{T_w} \cos\left[2\pi\Delta_f \sum_{q} (a_{iq} - a_{jq})p(t - qT)dt\right]\right] \end{aligned} \tag{A.7.9}$$

Performance of the entire system (assuming all possible transmitted sequences) can be union bounded by

$$\text{PWE} \leq (M-1)Q\left[\frac{d_{\min}^2 E_N}{2N_0}\right]^{1/2} \tag{A.7.10}$$

where $d_{\min}$ is the *minimum* distance d_{ij} over all i and j, $i \neq j$. Thus, performance bounds can be determined by estimating the $d_{\min}$ over all sequences in Eq. (A.7.9). For given phase functions $p(t)$ and deviations Δ_f, this generally requires numerical calculations or simulation. This was the procedure used in performance evaluation for CPFSK carriers reported in Chapter 2.

A.8 MFSK

In MFSK one of M distinct frequencies is used to transmit data words. Again, each transmitted word appears as a frequency burst of T_s s and therefore occupies a bandwidth of approximately $2/T_s$ Hz around each frequency. Noncoherent decoding is achieved by bandpass filtering at each frequency and sampling the envelope of each filter output. The transmitted frequency is decided by the one with the largest sample value. Following the discussion of binary FSK, the envelope sample value of the correct filter has the probability density in Eq. (A.4.6), while all other incorrect samples have the density in Eq. (A.4.8). The probability of a word error is then

$$\text{PWE} = 1 - \int_0^\infty p(\alpha_c)\left[\int_0^{\alpha_c} p(\alpha_n)d\alpha_n\right]^{M-1} d\alpha_c \tag{A.8.1}$$

In evaluating Eq. (A.8.1) we use the facts:

$$\left[\int_0^{\alpha_c} xe^{-x^2/2}\,dx\right]^{M-1} = [1 - e^{-\alpha_0^2/2}]^{M-1}$$

$$= \sum_{q=0}^{M-1}(-1)^q\binom{M-1}{q}e^{-q\alpha_c^2/2} \tag{A.8.2}$$

and

$$\int_0^\infty ue^{-u^2(q+1)/2}I_0(pu)du = \frac{\exp[(p^2/2)/1+q]}{1+q} \tag{A.8.3}$$

When Eqs. (A.8.2) and (A.8.3) are used in Eq. (A.8.1), PWE integrates to the function given in Table 6.2. This rather complicated function is straightforward to compute, and can be adequately approximated by Eq. (2.7.4).

A function closely related to PWE in Eq. (A.8.1) is the probability that a noncoherent envelope variable α_c or α_n will not exceed a level λ. This can be written as

$$\text{Prob}(\alpha_c < \lambda) = \int_0^{\lambda} p(\alpha_c) d\alpha_c \tag{A.8.4a}$$

and

$$\text{Prob}(\alpha_n > \lambda) = \int_{\lambda}^{\infty} p(\alpha_n) d\alpha_n \tag{A.8.4b}$$

These can be conveniently written in terms of the *Marcum Q function*, defined as

$$Q(a, b) \triangleq \int_b^{\infty} \exp\left[-\frac{a^2 + x^2}{2}\right] I_0(ax) x \, dx \tag{A.8.5}$$

From Eqs. (A.4.6) and (A.4.8) it then follows that

$$\text{Prob}(\alpha_c < \lambda) = 1 - Q[\rho, \lambda] \tag{A.8.6a}$$

$$\text{Prob}(\alpha_n > \lambda) = Q[0, \lambda] \tag{A.8.6b}$$

The Q function in Eq. (A.8.5) has been tabulated (Marcum, 1950), and recursive computational methods have been developed for its evaluation (Brennan and Reed, 1965). It occurs in noncoherent threshold detection, as when determining code acquisition probabilities in Chapter 7.

A.9 EFFECT OF PHASE NOISE ON NONCOHERENT FSK AND DPSK SYSTEMS

In coherent BPSK and QPSK systems, any carrier phase noise adds to the phase reference error (see Appendix B) that degrades performance according to Eqs. (A.2.8) and (A.3.11). In noncoherent systems carrier phase noise also degrades performance by altering the statistics of the decoding variables.

In MFSK the received carrier during any word time will be one of the frequency shifted waveforms

$$c(t) = A \cos[(\omega_c + 2\pi i/T_w)t + \psi(t)] \qquad i = 1, 2, \ldots, M \tag{A.9.1}$$

where $\psi(t)$ is the carrier phase noise. The FSK decoder is a bank of bandpass envelope detectors tuned to each of the frequencies. The envelope sample of each channel is a Rician variable with noncentrality parameter dependent on the phase noise $\psi(t)$. Hence Eq. (A.8.1) becomes

$$\mathrm{PWE}(\psi) = 1 - \frac{1}{M} \sum_{i=1}^{M} \int_0^\infty x \exp[-\tfrac{1}{2}(x^2 + C_{ii})] I_o(x\sqrt{C_{ii}})$$

$$\times \left\{ \prod_{\substack{k=1 \\ k \neq i}}^{M} \int_0^x y \exp[-\tfrac{1}{2}(y^2 + C_{ik})] I_o(y\sqrt{C_{ik}}) dy \right\} dx \tag{A.9.2}$$

where I_0 is the modified Bessel function and

$$C_{ik} = 2 \frac{E_w}{N_0} D_{ik}$$

$$D_{ik} \triangleq \frac{1}{T_w^2} \left| \int_0^{T_w} \exp\left(-\frac{j2\pi(k-i)}{T_w} t + j\psi(t) \right) dt \right|^2 \tag{A.9.3}$$

Equation (A.9.2) corresponds to a conditional M-ary error probability among independent Rician variates, each with a noncentrality parameter C which depends on the phase noise through D_{ik}. Specifically, the correct signal variable depends on D_{ii}, which effectively reduces the available word energy E_w, whereas all incorrect (noise-only) variables depend on D_{ik}. The latter accounts for the fraction of the available word energy detected in the noise-only detectors, and may be considered a measure of the nonorthogonality caused by the phase noise. The phase noise can be integrated to determine the $\{D_{ik}\}$ set, from which Eq. (A.9.2) can be evaluated to determine the PWE. Since $\psi(t)$ is a random process the $\{D_{ik}\}$ are a set of random variables, and the conditional PWE must be averaged as

$$\mathrm{PWE} = \mathscr{E}[\mathrm{PWE}(\psi)] \tag{A.9.4}$$

where $\mathscr{E}$ is the expectation operator. Hence, Eq. (A.9.2) must be averaged over the joint statistics of the $\{D_{ik}\}$ random variables. Note they are not independent. Typically, however, $D_{ik} \ll D_{ii}$ and PWE (ψ) in Eq. (A.9.2) then simplifies to the result in Table 2.6 with E_w replaced by DE_w, $D = D_{ii}$.

The PWE can then be estimated by replacing D by its mean value

$$\bar{D} = \mathscr{E}[D] = \frac{1}{T_w^2} \int_0^{T_w} \int_0^{T_w} \mathscr{E}[e^{j(\psi(t) - \psi(s))}] dt ds \tag{A.9.5}$$

When $\psi(t)$ is a stationary random phase process, the expectation in the integrand defines the characteristic function of the Gaussian variable $[\psi(t) - \psi(s)]$. The resulting expression for $\bar{D}$ is then

$$\bar{D} = \frac{1}{T_w^2} \int_0^{T_w} \int_0^{T_w} e^{-\sigma^2(t-s)/2} \, dt ds \tag{A.9.6}$$

where $\sigma^2(\tau) = \mathscr{E}[\psi(t) - \psi(t + \tau)]^2$. From properties of structure functions (Gagliardi, 1984) associated with stationary processes, this can be written as

$$\sigma^2(\tau) = \frac{8}{2\pi} \int_0^{\infty} S_\psi(\omega) \sin^2(\omega\tau/2) d\omega \tag{A.9.7}$$

where $S_\psi(\omega)$ is the two-sided spectral density of the phase noise $\psi(t)$. The double integral in Eq. (A.9.6) can be simplified to a single integral by a simple change of variable, producing

$$\bar{D} = \frac{2}{T_w} \int_0^{T_w} \left(1 - \frac{\tau}{T_w}\right) e^{-\sigma^2(\tau)/2} \, d\tau \tag{A.9.8}$$

The result is used in Chapter 10.

In DPSK the effect of the phase noise $\psi(t)$ is to alter the noncentrality parameter of the variables E_1 and E_0 in Eq. (A.5.3). The resulting conditional PE for DPSK with phase noise is then

$$\begin{aligned} \mathrm{PE}(\psi) = 1 - \int_0^{\infty} & x \exp[-\tfrac{1}{2}(x^2 + D4E_b/N_0)] I_0(2x\sqrt{DE_b/N_0}) \\ & \times [Q'(2\sqrt{D_n(E_b/N_0)}, x) dx \end{aligned} \tag{A.9.9}$$

where $Q'(a, b) = 1 - Q(a, b)$, with $Q(a, b)$ defined in Eq. (A.8.5), and

$$D = \frac{1}{4T_b^2}\left|\int_0^{2T_b} e^{j\psi(t)}\,dt\right|^2 \tag{A.9.10a}$$

$$D_n = \frac{1}{4T_b^2}\left(\left\{\int_0^{T_b} \cos[\psi(t+T_b)] - \cos[\psi(t)]dt\right\}^2 + \left[\left\{\int_0^{T_b} \sin[\psi(t+T_b)] - \sin[\psi(t)]^2 dt\right\}^2\right)\right. \tag{A.9.10b}$$

The PE can then be computed by the average

$$\text{PE} = \int_0^{\infty}\int_0^{\infty} \text{PE}(D, D_n)p(D, D_n)dD\,dD_n \tag{A.9.11}$$

where $p(D, D_n)$ is the joint probability density of the random variables D and D_n induced by the phase noise $\psi(t)$.

In the typical noncoherent link the bandwidth of the phase noise is much less than the bit time T_b of the data. In this case it can be assumed that $\cos\psi(t)$ and $\sin\psi(t)$ are constant over each T_b interval. Letting $\psi_2 = \psi(t + T_b)$ and $\psi_1 = \psi(t)$ during any $(0, T_b)$, the D variables reduce to

$$D = (\cos\psi_1 + \cos\psi_2)^2 + (\sin\psi_1 + \sin\psi_2)^2 = 2(1 + \cos(\Delta\psi)) \tag{A.9.12}$$

and

$$D_n = (\cos\psi_2 - \cos\psi_1)^2 + (\sin\psi_2 - \sin\psi_1)^2 = 2[1 - \cos(\Delta\psi)] \tag{A.9.13}$$

and depend only on the phase differential $\Delta\psi = \psi_2 - \psi_1$ of the phase noise. This fact is used in Chapter 10.

A.10 MASK

In MASK the data words are encoded into carriers with one of M amplitude levels $\{A_i\}$. The carriers therefore have the form

$$c(t) = A_i\cos(\omega_c t + \phi) \qquad 0 \le t \le T_w \tag{A.10.1}$$

For any set $\{A_i\}$ the carrier $c(t)$ corresponds to a T_w s carrier burst, and therefore always occupies a bandwidth of $2/T_w$ Hz. The average power of the signal set is

$$\bar{P}_c = \frac{1}{M} \sum_{i=1}^{M} (A_i^2/2) \tag{A.10.2}$$

If the amplitude levels A_i are all evenly spaced over positive and negative amplitude values, each separated by distance Δ volts, we have $A_i = (2i - 1)\ (\Delta/2)$, $1 \leq i \leq M/2$, and $A_i = A_{i-(M/2)}$, $M/2 \leq i \leq M$. The average power of this set is then

$$\bar{P}_c = \frac{\Delta^2(M^2 - 1)}{12} \tag{A.10.3}$$

Decoding in MASK is achieved by phase-coherent carrier referencing, followed by word integration, to produce the variable

$$z = A_j T_w + n \tag{A.10.4}$$

where n has variance $N_0 T_w/2$, and A_j is the transmitted amplitude during $(0, T_w)$. A word decision is based on which $A_i T_w$ product z is closest. An error occurs if n causes z to be outside the intervals $(A_j + \Delta/2, A_j - \Delta/2)$. For Gaussian noise this occurs with probability

$$\begin{aligned} \text{PWE} &= \frac{1}{M}\left[(M-2)2Q\left(\frac{\Delta T_w}{\sqrt{2N_0 T_w}}\right) + 2Q\left(\frac{\Delta T_w}{\sqrt{2N_0 T_w}}\right)\right] \\ &= \frac{M-1}{M}\left[2Q\left(\frac{\Delta T_w}{\sqrt{2N_0 T_w}}\right)\right] \end{aligned} \tag{A.10.5}$$

When written in terms of the average power in Eq. (A.10.3) we obtain

$$\text{PWE} = \left(\frac{M-1}{M}\right)2Q\left[\left(\frac{6\bar{E}_w}{(M^2-1)N_0}\right)^{1/2}\right] \tag{A.10.6}$$

where $\bar{E}_w = \bar{P}_c T_w$ is the average word energy over the signal set. This result is included in Table 2.6 in Chapter 2.

REFERENCES

Abramowitz, A., and Stegun, I. (1965), *Handbook of Mathematical Functions*, National Bureau of Standards, Washington, DC., chapter 26.

Benedetto, S., Biglieri, E., and Castellani, V. (1987), *Digital Transmission Theory*, Prentice-Hall, Englewood Cliffs, NJ.

Brennan, L., and Reed, I. (1965), "A Recursive Method for Computing the Q-Functions," *IEEE Trans. Info. Theory*, **IT-11**.

Gagliardi, R. (1984), *Satellite Communications*, 1st ed., Van Nostrand Reinhold, New York.

Lindsey, W., and Simon, M. (1973), *Telecommunication System Design*, Prentice-Hall, Englewood Cliffs, NJ.

Marcum, C. (1950), *Tables of the Q-function*, Rand Corp. Research Memo RN 399.

Papoulis, A. (1965), *Probability, Random Variables, and Random Processes*, McGraw-Hill, New York.

Shimbo, O. (1988), *Transmission Analysis in Communication Systems —Volumes 1 and 2.* Computer Science Press, Rockville, MD.

APPENDIX B
Carrier Recovery and Bit Timing

Carrier recovery (sometimes referred to as *phase referencing*) is the operation of extracting a phase coherent reference carrier from an observed noise received carrier. *Bit timing* is the operation of extracting a time coherent bit rate clock from an observed noisy data-modulated waveform. Both of these operations are fundamental to digital communications. In this appendix we review the basic analysis associated with both these operations.

B.1 CARRIER RECOVERY

Carrier recovery is achieved by forcing a local voltage controlled oscillator (VCO) to track the phase of a received carrier. The VCO output can then be used as a receiver reference source for decoding operations. The VCO and the received carrier are said to be phase locked, or phase coherent, if the phase error (phase difference) of the two carriers, when referred to the received carrier frequency, is small ($\leq 20^\circ$).

Phase tracking of an unmodulated carrier is achieved via the *phase locked loop* (PLL) shown in Figure B.1a. Let the loop input be written as

$$x(t) = A\cos(\omega_c t + \psi(t)) + n(t) \tag{B.1.1}$$

where $\psi(t)$ is the phase to be tracked, and $n(t)$ is again white noise of

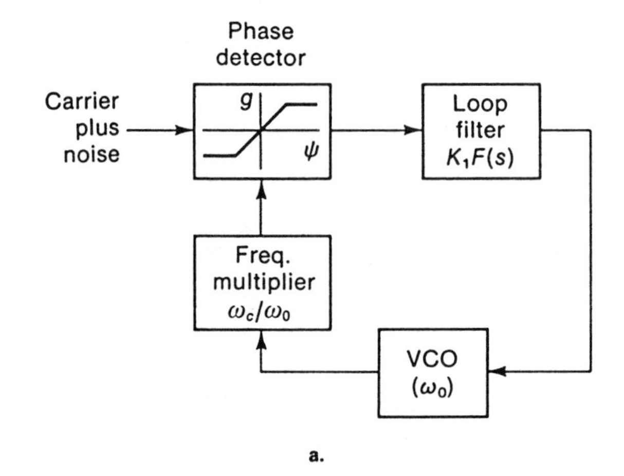

a.

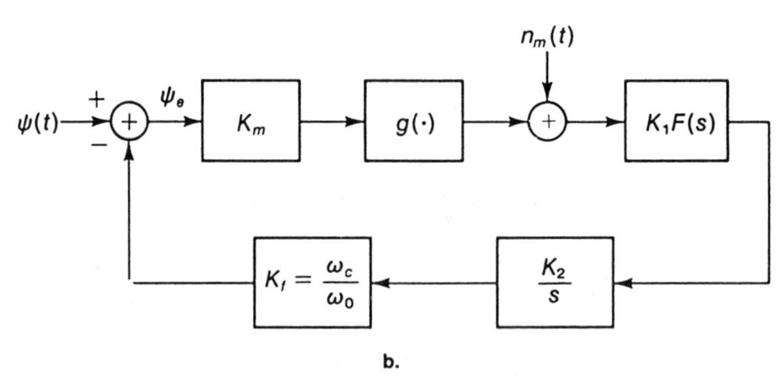

b.

FIGURE B.1 Phase lock tracking loop for unmodulated carrier. (a) Block diagram. (b) Equivalent phase model.

one-sided spectral level N_0 in a wide bandwidth around the carrier frequency ω_c. The VCO, at rest frequency ω_0, is frequency-multiplied to ω_c to track the phase of the received carrier. A phase comparator determines the instantaneous phase error between the two carriers. This error voltage is then filtered and used to correct the phase of the VCO. If the phase comparator is a phase detector with gain $K_m g(\psi)$, the phase error voltage appears as*

$$e(t) = K_m g[\psi(t) - \hat{\psi}(t)] + n_m(t) \quad \text{(B.1.2)}$$

---

* If the phase comparator is a phase *mixer*, then $g(\psi)$ is replaced by $\sin(\psi)$, and $e(t)$ is multiplied by the carrier amplitude A. This multiplies K_m by A and scales the spectrum of $n_m(t)$. These conversions are easily accounted for in all subsequent equations.

where $g(\psi)$ is the nonlinear phase conversion function, $\hat{\psi}(t)$ is the VCO phase, and $n_m(t)$ is the mixer output noise. The latter is a baseband white noise process of two-sided spectral level $K_m^2 N_0/A^2$. The voltage $e(t)$ is filtered by the loop filter function $K_1 F(s)$ and used to adjust the frequency of the VCO. The PLL therefore has the equivalent phase model shown in Figure B.1b. Note that the VCO is effectively replaced by an integrator to account for the conversion from frequency control at its input to phase variations at its output. If we define the tracking phase error at frequency ω_c as

$$\psi_e(t) = \psi(t) - \hat{\psi}(t) \tag{B.1.3}$$

it follows that

$$\frac{d\psi_e}{dt} = \frac{d\psi}{dt} - \frac{d\hat{\psi}}{dt} \tag{B.1.4}$$

From Figure B.1b it is evident that

$$\begin{aligned} \frac{d\hat{\psi}}{dt} &= K_f K_2 K_1 F\{K_m g(\psi_e) + n_m(t)\} \\ &= KF\left\{g(\psi_e) + \frac{n_m(t)}{K_m}\right\} \end{aligned} \tag{B.1.5}$$

where K is the total loop coefficient gain

$$K = K_m K_1 K_2 K_f \tag{B.1.6}$$

and $F\{\cdot\}$ represents the filter operation of $F(s)$ on the time function in $\{\cdot\}$. Substitution of Eq. (B.1.5) into Eq. (B.1.4) generates the system differential equation describing the phase referencing operation of the phase lock loop,

$$\frac{d\psi_e}{dt} + KF\{g(\psi_e)\} = \frac{d\psi}{dt} - KF\left\{\frac{n_m(t)}{K_m}\right\} \tag{B.1.7}$$

Note that the equation is a nonlinear, stochastic differential equation forced by the phase dynamics $d\psi/dt$ of the received carrier and the filtered random noise $n_m(t)$. If the phase error is small ($\psi_e \gtrsim 20°$), the loop is said to be *in-lock*, and the system can be linearized by assuming $g(\psi_e) \approx \psi_e$.

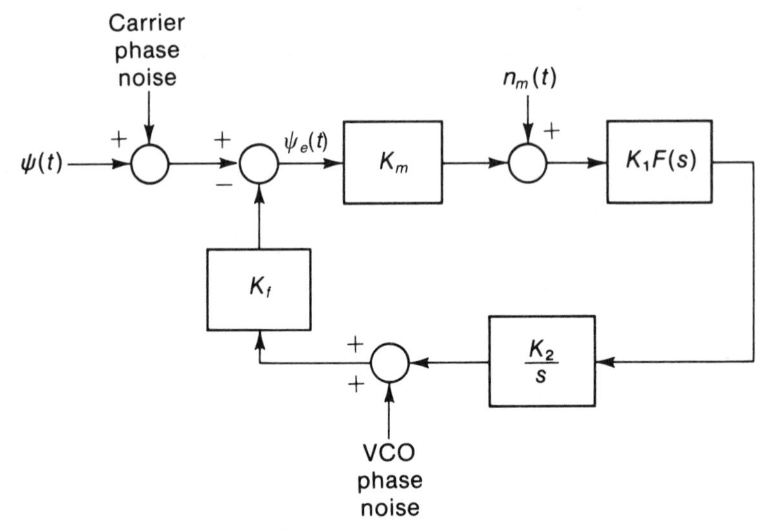

FIGURE B.2 Linear phase model with phase noise.

The nonlinear phase system in Figure B.1b then becomes a linear feedback system, as shown in Figure B.2. The inputs to the loop are the input phase process $\psi(t)$, the input noise $n_m(t)$, and any other phase variations superimposed on the received carrier or VCO, such as oscillator phase noise. We see that the phase error $\psi_e(t)$ of the PLL evolves as a superposition of the contributions from all phase sources. Figure B.2 includes the addition of the phase noises of the carrier and the VCO, each adding directly at the points shown. Phase error can be computed by determining the response at the error point due to each input. This error will have a deterministic part [due to tracking phase dynamics in $\psi(t)$] and a random part (due to tracking the phase noise and thermal noise inputs). Each of these error responses will depend on the loop gain function defined as

$$H(\omega) = \frac{KF(\omega)/j\omega}{1 + KF(\omega)/j\omega} \tag{B.1.8}$$

Table B.1 summarizes the resulting form for $H(\omega)$ for some common loop filter functions $F(\omega)$.

The important dynamic variations in the received carrier phase $\psi(t)$ are phase offsets, frequency offsets (Doppler), and linear frequency changes. These lead to constant phase errors depending on the type of loop. Table B.2 summarizes the resulting phase error values of several loops due to each type of offset. An infinite value implies the loop error

TABLE B.1 Tabulation of loop-filter function $[F(\omega)]$, loop gain function $[H(\omega)]$, and loop noise bandwidth (B_L)

Loop Filter, $F(\omega)$	*$H(\omega)$*	*B_L*
1	$\dfrac{K}{j\omega + K}$	$\dfrac{K}{4}$
$\dfrac{j\omega\tau_2 + 1}{j\omega\tau_1 + 1}$	$\dfrac{1 + j\left(\dfrac{2\zeta}{\omega_n} - \dfrac{1}{K}\right)\omega}{-\left(\dfrac{\omega}{\omega_n}\right)^2 + \dfrac{2\zeta}{\omega_n} j\omega + 1}$ $\omega_n^2 = K/\tau_1$ $2\zeta\omega_n = \dfrac{1 + K\tau_2}{\tau_1}$	$\dfrac{\omega_n}{8\zeta}\left[1 + \left(2\zeta - \dfrac{\omega_n^2}{K}\right)^2\right]$
$\dfrac{j\omega\tau_2 + 1}{j\omega\tau_1}$	$\dfrac{\left(1 + \dfrac{2\zeta}{\omega_n} j\omega\right)}{-\left(\dfrac{\omega}{\omega_n}\right)^2 + \left(\dfrac{2\zeta}{\omega_n}\right) j\omega + 1}$	$\dfrac{\omega_n}{8\zeta}(1 + 4\zeta^2)$

TABLE B.2. Steady-state phase errors for carrier dynamics and loop filters

	Phase Error	
Phase Dynamic	*$F(\omega) = 1$*	$F(\omega) = \dfrac{j\omega + b}{j\omega}$
Phase offset, ψ_0 $\psi = \psi_0$	0	0
Frequency offset, ω_d $\psi(t) = \omega_d t$	$\dfrac{\omega_d}{K}$	0
Frequency rate $\psi(t) = \dot{\omega}_d t^2/2$	∞	$\dfrac{\dot{\omega}_d}{K}$

will increase without bound, and therefore the loop cannot track that particular offset. The ability to track, however, improves as the loop order is increased, while the residual error decreases with loop gain.

A closely related parameter is the time it takes to reduce an initial phase or frequency offset to the phase-error value stated. This is often referred to as the loop *pull-in time*, or loop *acquisition time*. For a second-order loop with a carrier frequency offset of Ω rps from the expected carrier frequency, the acquisition time to reduce the initial frequency error to the zero value in Table B.2 is given by

$$T_{\text{acq}} = \frac{\Omega^2}{2\zeta\omega_n^3} \tag{B.1.9}$$

Hence frequency pull-in time increases as the square of the frequency offsets, but decreases as the cube of the loop natural frequency ω_n. This means frequency acquisition is improved by using wider bandwidth loops.

Random phase error is due to the cumulative effect of all noises entering the loop. Hence the phase error variance (about the mean errors in Table B.2) is given by

$$\begin{aligned} \sigma_e^2 = {} & \left(\frac{N_0}{A^2}\right)\frac{1}{2\pi}\int_{-\infty}^{\infty} |H(\omega)|^2\, d\omega \\ & + \left(\frac{\omega_c}{\omega_0}\right)^2 \frac{1}{2\pi}\int_{-\infty}^{\infty} S_{0s}(\omega)|1 - H(\omega)|^2\, d\omega \\ & + \frac{1}{2\pi}\int_{-\infty}^{\infty} S_c(\omega)|1 - H(\omega)|^2\, d\omega \end{aligned} \tag{B.1.10}$$

where $S_{0s}(\omega)$, $S_c(\omega)$, and N_0 are the power spectra of the loop oscillator phase noise, the carrier phase noise, and the thermal noise, respectively. Since $1 - H(j\omega)$ is an effectively high-pass filtering function, we see that the phase noise outside the loop function bandwidth contributes to the tracking error, whereas the thermal noise inside this bandwidth is important. By defining

$$B_L = \frac{1}{2\pi}\int_0^{\infty} |H(\omega)|^2\, d\omega \tag{B.1.11}$$

as the loop noise bandwidth, the thermal noise contribution can be written more compactly as

$$\sigma_{en}^2 = \frac{1}{\rho} \tag{B.1.12}$$

Here we have defined

$$\rho = \frac{A^2/2}{N_0 B_L} \tag{B.1.13}$$

as the ratio of the input carrier power to thermal noise power in the loop noise bandwidth B_L. Formulas for B_L for each type of loop in Table B.1 are listed in the last column. If the carrier being tracked is phase modulated by $\theta(t)$, then the effect of the modulation on the tracking performance of the loop must be considered. The modulation reduces the strength of the carrier component being tracked, and effectively reduces the value of A in Figure B.2, depending on the type of modulation. This reduction of A to the lower value A_m is summarized in Table B.3 for several types of common modulation formats. Reduction of A to A_m increases tracking errors in Table B.2, and converts the loop CNR ρ in Eq. (B.1.13) to

$$\rho_m = \left(\frac{A_m}{A}\right)^2 \rho \tag{B.1.14}$$

TABLE B.3 Modulation suppression on tracking carrier amplitude

Phase Modulation on Carrier	*Tracking Carrier Amplitude, A_m*
No modulation	A
Subcarrier PM, deviation Δ	$AJ_0(\Delta)$
K subcarriers PM, deviation Δ_i	$A \prod_{i=1}^{K} J_0(\Delta_i)$
Binary waveform, deviation Δ_c	$A \cos \Delta_c$
Binary waveform (Δ_c) plus K subcarriers (Δ_i)	$A \cos \Delta_c \prod_{i=1}^{K} J_0(\Delta_i)$
Random, characteristic function $C(\omega)$	$A\lvert C(1)\rvert$

Hence, the bracket serves as a *modulation suppression factor*, which effectively reduces the carrier tracking power from the power of an unmodulated carrier.

B.2 BPSK CARRIER RECOVERY

When the received carrier is phase modulated as a BPSK carrier, there is no carrier component to be tracked (i.e., $\Delta_c = \pi/2$ and $A_m = 0$ in Table B.3), and carrier recovery cannot be obtained via a standard phase lock loop. Instead a modified system must be used, which first uses a non-linearity to eliminate (wipe-off) the modulation while creating a carrier component having a phase variation proportional to that of the received carrier. Subsequent tracking of this residual carrier component then generates the desired carrier reference. The two common methods for achieving this BPSK carrier recovery are the *squaring loop* and the *Costas loop*.

The squaring loop system is shown in Figure B.3a. The received carrier waveform is filtered and squared prior to phase referencing. The squaring wipes off the modulation and generates a harmonic at twice the carrier frequency. This harmonic can then be phase tracked, and the VCO output can then be frequency divided by 2 to serve as a phase locked reference to the received carrier.

Let the received waveform be written as

$$y(t) = Am(t)\cos[\omega_c t + \psi(t)] + n(t) \tag{B.2.1}$$

where $m(t) = \pm 1$ is the BPSK modulation and $n(t)$ is again white bandpass noise of level N_0, one-sided. The squaring generates the waveform

$$\begin{aligned} y^2(t) &= A^2 \sin^2[\omega_c t + \tfrac{1}{2}\pi m(t) + \psi(t)] \\ &\quad + 2n(t)A\sin[\omega_c t + \tfrac{1}{2}\pi m(t) + \psi(t)] + n^2(t) \\ &= \frac{A^2}{2} + \frac{A^2}{2}\sin[2\omega_c t + 2\psi(t)] + n^2(t) \\ &\quad + 2n(t)A\sin[\omega_c t + \tfrac{1}{2}\pi m(t) + \psi(t)] \end{aligned} \tag{B.2.2}$$

The output contains a constant term, a carrier component at $2\omega_c$ with no modulation and a phase variation multiplied by 2, and two interference terms involving the noise. A PLL with a VCO at $2\omega_c$ can now track the carrier component in Eq. (B.2.2). If it successfully performs the tracking, its VCO output is given by $\cos[2\omega_c t + \psi_2(t)]$, where $\psi_2(t)$ is a close replica

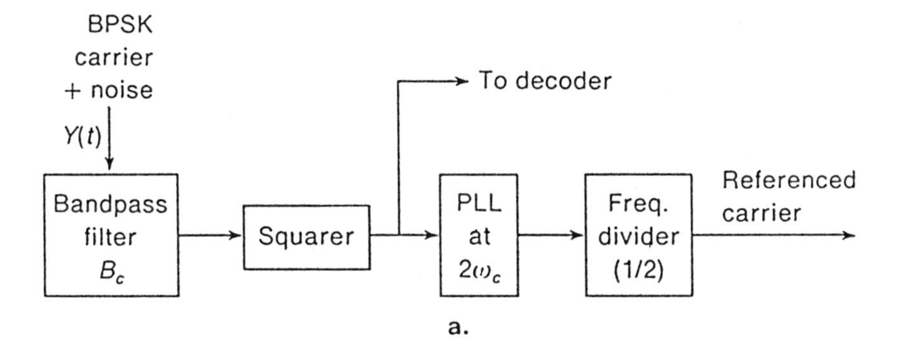

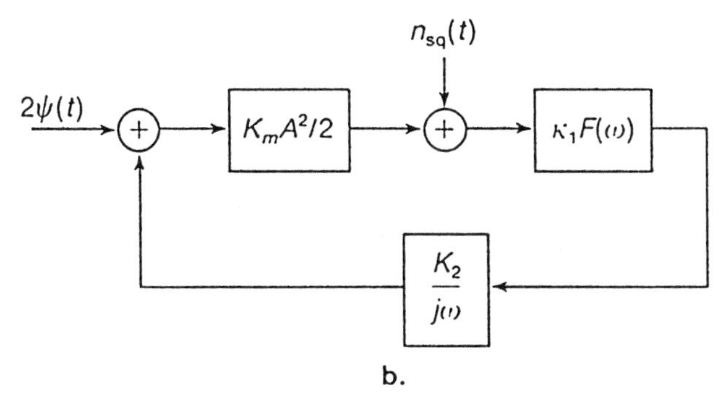

FIGURE B.3 Squaring BPSK carrier phase referencing. (a) Block diagram. (b) Equivalent loop model.

of $2\psi(t)$. By now frequency dividing the VCO output by 2, we generate the reference carrier at ω_c

$$\cos\left[\omega_c t + \frac{\psi_2(t)}{2}\right] \tag{B.2.3}$$

where the phase $\psi_2(t)/2$ is now approximately phase locked to $\psi(t)$ in Eq. (B.2.1). Note that the role of the squarer is effectively to remove the digital modulation while frequency translating the carrier to $2\omega_c$ for phase locking.

The ability to achieve accurate phase referencing with squaring systems can be determined from the equivalent loop tracking diagram in Figure B.3b. The loop is effectively driven by $2\psi(t)$ and therefore has twice the extraneous phase variation of the carrier itself. The equivalent noise $n_{sq}(t)$

entering the loop is simply the loop mixer output noise produced by the squarer noise terms [i.e., the last two terms in Eq. (B.2.2)]. These correspond to the squared carrier filtered noise and the signal–noise cross-product term. To evaluate the spectrum of $n_{sq}(t)$, we need only determine the spectrum of the squarer noise in the vicinity of $2\omega_c$. This can be determined by recalling that the squared Gaussian noise term has a spectrum given by twice the convolution of $S_n(\omega)$ with itself, and the spectrum of the cross-product term is given by the convolution of $S_n(\omega)$ with $2S_s(\omega)$, where $S_n(\omega)$, and $S_s(\omega)$ are the spectrum of the filtered noise and carrier signals in Eq. (B.2.1). In the vicinity of zero, $n_{sq}(t)$ has approximately a flat spectrum with level

$$S_{sq}(0) \approx \frac{N_0^2 B_c}{2} + \frac{N_0 A_s^2}{2} \tag{B.2.4}$$

Since the noise bandwidth B_L of the tracking loop is generally much smaller than that of the carrier filter bandwidth B_c, the mean squared noise value entering the loop bandwidth is therefore approximately $2S_{sq}(0)B_L$. The resulting carrier reference phase variance, after phase dividing by 2, is then

$$\begin{aligned} \sigma_e^2 &= \frac{1}{4}\left[\frac{2S_{sq}(0)B_L}{A^4/8}\right] \\ &= \frac{2N_0 B_L}{A^2} + \frac{2N_0^2 B_c B_L}{A^4} \\ &= \frac{N_0 B_L}{P_c}\left[1 + \frac{N_0 B_c}{2P_c}\right] \end{aligned} \tag{B.2.5}$$

where $P_c = A^2/2$ is again the carrier power. The bracket is called the *loop squaring factor* and accounts for the increases in error caused by the squaring operation. Since the bracket is greater than 1, squaring loop errors are larger than those for standard tracking loops, the amount of the increase dependent on the carrier CNR. In effect, the squaring factor exhibits the fact that the loop noise changes from a squared noise term to a signal-noise cross-product as the carrier CNR increases. It is convenient to again denote

$$\frac{E_b}{N_0} = \frac{P_c T_b}{N_0} \tag{B.2.6}$$

and write Eq. (B.2.5) as

$$\sigma_e^2 = \frac{B_L T_b}{(E_b/N_0)}\left[1 + \frac{\frac{1}{2}B_c T_b}{E_b/N_0}\right] \tag{B.2.7}$$

This relates the carrier referencing accuracy to the carrier E_b/N_0 used for the decoder. Since the reciprocal of the bracket in Eq. (B.2.7) effectively multiplies the E_b/N_0 factors, the former is often called a *squaring loss*, or a *squaring suppression factor*. Note that this factor depends on the presquaring bandwidth B_c. If this filter is too wide, excess noise will enter the loop, while if it is too narrow, the BPSK carrier may be distorted. Simon and Lindsey (1977) have shown that for a particular filter type and bit waveform, an optimal prefilter B_c bandwidth relative to the bit rate exists for each E_b/N_0. For a typical operation ($E_b/N_0 \approx 8 = 9$ dB), the required bandwidth is about $3/T_b$ Hz. For this prefilter bandwidth rms tracking accuracies to within 5° can be maintained at $E_b/N_0 = 10$ with $B_L T_b \leq 0.01$; that is, a loop bandwidth less than 1% of the bit rate. We can also see that the dc term in Eq. (B.2.2) is given by the carrier power $A^2/2$. Hence a low pass filter that extracts this term at the squarer output provides a convenient measure of carrier power for signal monitoring or automatic gain control.

A problem that occurs in phase referencing with a squaring loop is the phase ambiguity that must be resolved before the phase reference can be used for BPSK decoding. This ambiguity is caused by the one-half frequency division of the VCO output. The resulting divided down reference may be 180° out of phase with the desired reference (i.e., both the desired reference and its negative correspond to the same double frequency waveform at the VCO). This phase ambiguity must be resolved prior to BPSK decoding, since the negative reference will decode all bits exactly opposite to their true polarity. Resolving this ambiguity is usually accomplished by first decoding a known binary word.

A second way to achieve BPSK carrier referencing is by the use of the Costas, or quadrature, loop shown in Figure B.4. The system involves two parallel tracking loops operating simultaneously from the same VCO. One loop, called the in-phase loop, uses the VCO as in a PLL, and the second (quadrature) loop uses a 90° shifted VCO. The mixer outputs are each filtered and multiplied to form the nonlinearity for modulation wipe-off. The multiplied arm voltages are then filtered and used to control the VCO. The low pass filters in each arm must be wide enough to pass the data modulation without distortion. If the input to the Costas loop

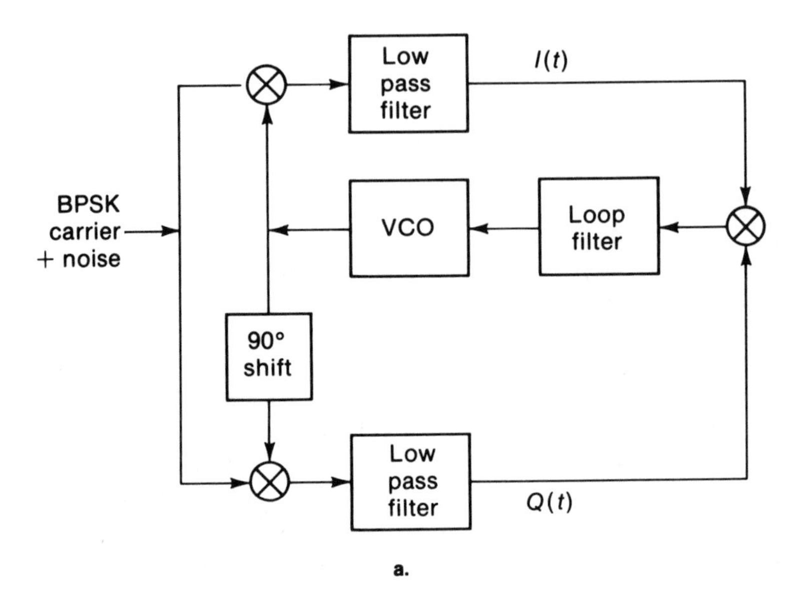

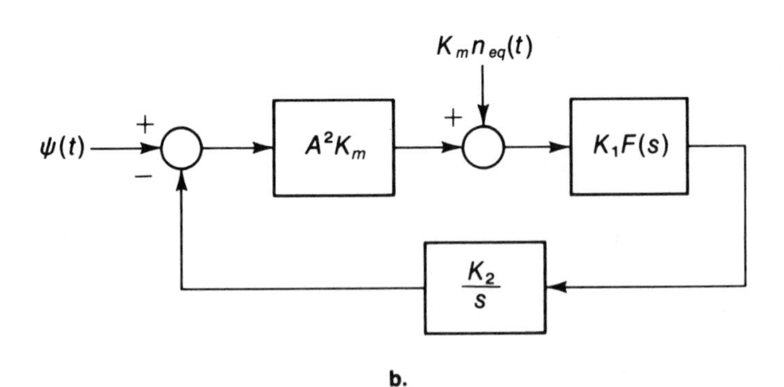

FIGURE B.4 BPSK Costas loop. (a) Block diagram. (b) Equivalent loop model.

is the waveform in Eq. (B.2.1), the in-phase mixer generates.

$$\begin{aligned} I(t) &= A_s \cos[\tfrac{1}{2}\pi m(t) + \psi(t) - \psi_o(t)] + n_{mc}(t) \\ &= A_s m(t) \cos[\psi_e(t)] + n_{mc}(t) \end{aligned} \tag{B.2.8a}$$

whereas the quadrature mixer simultaneously generates

$$Q(t) = A_s m(t) \sin[\psi_e(t)] + n_{ms}(t) \tag{B.2.8b}$$

where $\psi_e(t) = \psi(t) - \psi_o(t)$, and $\psi_o(t)$ is the phase of the VCO in Figure B.4a. The output of the multiplier is then

$$
\begin{aligned}
I(t)Q(t) &= A_s^2 m^2(t) \sin[\psi_e(t)] \cos[\psi_e(t)] + n_{sq}(t) \\
&= \tfrac{1}{2} A_s^2 \sin[2\psi_e(t)] + n_{sq}(t)
\end{aligned}
\tag{B.2.9}
$$

where

$$
\begin{aligned}
n_{sq}(t) = n_{mc}(t)n_{ms}(t) &+ n_{mc}(t) \sin[\tfrac{1}{2}\pi m(t) + \psi_e(t)] \\
&+ n_{ms}(t) \cos[\tfrac{1}{2}\pi m(t) + \psi_e(t)]
\end{aligned}
\tag{B.2.10}
$$

The second equality in Eq. (B.2.9) follows, since $m(t) = \pm 1$. When the phase error $[\psi(t) - \psi_o(t)]$ is small, the loop has the equivalent linear model shown in Figure B.4b. The Costas loop therefore tracks the phase variation $\psi(t)$ with the VCO without interference from the carrier modulation $m(t)$. The quadrature tracking loops have effectively removed the modulation, allowing the loop filter and VCO to accomplish the carrier synchronization. The mixer noises $n_{mc}(t)$ and $n_{ms}(t)$ in Eq. (B.2.8) are baseband versions of the carrier noise in Eq. (B.2.1). It can be shown (Lindsey and Simon, 1973) that $n_{sq}(t)$ in Eq. (B.2.10) has the identical spectrum as $n_{sq}(t)$ in Eq. (B.2.2) for the squaring loop. The frequency division by the factor of 2 can be accomplished directly in the loop gain, since the error term $\psi_e(t)$ must be derived from the argument of the sine term in Eq. (B.2.9). Thus the equivalent Costas quadrature loop is identical to the equivalent squaring loop system shown in Figure B.3a and generates the same mean squared carrier phase error given in Eq. (B.2.7).

An advantage of the Costas loop is that the PSK decoding is partially accomplished right within the loop. If the loop is tracking well so that $\psi_0(t) \approx \psi(t)$, we see the quadrature mixer output in Eq. (B.2.8) is

$$
I(t) \approx A_s m(t) + n_{mc}(t) \tag{B.2.11}
$$

This result is identical to that produced at the multiplier output in a coherent BPSK decoder. Hence both data demodulation and phase referencing can be accomplished directly from the Costas loop. For this reason the quadrature arm is often called the *decisioning* arm and the in-phase arm, the *tracking* arm. One can think of the multiplier of the system as allowing the bit polarity of the decisioning loop to correct the phase error orientation of the tracking loop, thereby removing the modulation. Often Costas loops are designed with the decisioning arm filter followed by a limiter (Figure B.5). At high SNR, the limiter output

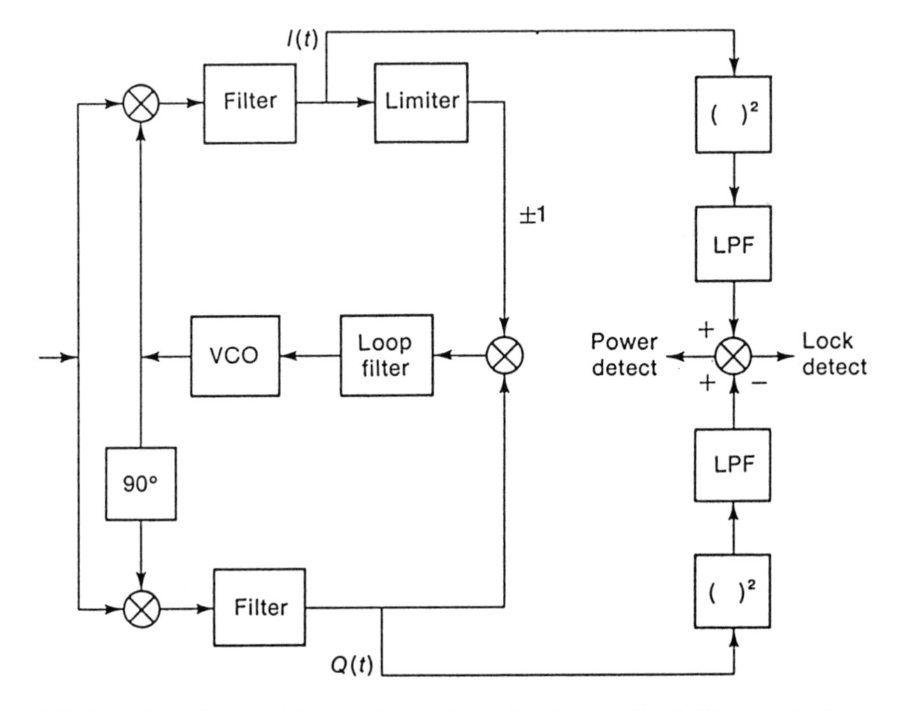

FIGURE B.5 Data-aided carrier referencing loop with AGC and lock detection.

will have a sign during each bit interval that is identical to the present data bit polarity. This limiter output then multiplies the loop error and removes the loop modulation. In effect, the data bit sign is used to aid the tracking loop, and such modified systems are often called *data aided carrier extraction* loops.

An interesting property of Costas loops is that the system generates signals that can be used for other auxiliary purposes as well. This can be seen by reexamining Eq. (B.2.8) and noting the following (the overbar denotes low pass filtering):

$$\overline{I^2(t) - Q^2(t)} = A_s^2\overline{m^2(t)}\cos(2\psi_e) \tag{B.2.12}$$

Squaring, low pass filtering, and subtracting the arm voltages (Figure B.5) produce an output that indicates phase lock [$\cos(2\psi_e) \to 1$ as $\psi_e \to 0$] and can therefore serve as a lock detector. When $\psi_e = 0$, this generates an output proportional to the average signal power, which can also be used for automatic gain control (AGC). We also note

$$\overline{I^2(t)} + \overline{Q^2(t)} = A_s^2\overline{m^2(t)} + \overline{n_{ms}^2(t)} + \overline{n_{mc}^2(t)}$$
$$= \text{total power of input} \tag{B.2.13}$$

This produces a measurement of the total RF input power and therefore can be used for RF power control. Finally,

$$\overline{I(t)\frac{dQ}{dt}} - \overline{Q(t)\frac{dI}{dt}} \cong [A_s^2\overline{m^2(t)}\cos(2\psi_e)]\frac{d\psi_e}{dt} \tag{B.2.14}$$

By differentiating in each arm and cross-multiplying, a signal proportional to the frequency error, $d\psi_e/dt$, is generated. This can be used for frequency offset measurements, or for aiding in frequency acquisition.

B.3 QPSK CARRIER REFERENCING

When digital data are modulated on the carrier with QPSK (quadraphase) encoding, carrier extraction must be accomplished by a slightly different means than with BPSK. This is due to the fact that the QPSK carrier, although still containing a suppressed carrier component, contains one of four possible phases, rather than two, during each bit interval. To extract the QPSK carrier, a fourth power, instead of a squaring, system must be used, as shown in Figure B.6a. The QPSK carrier is passed into the fourth power device, and a phase lock loop can be locked to the fourth harmonic. To see this, we write the QPSK signal as

$$y(t) = A\cos[\omega_c t + \theta(t) + \psi(t)] \tag{B.3.1}$$

where $\theta(t)$ is one of the phase angles $n\pi/2$, $n = 1, 2, 3, 4$, during each bit time. When raised to the fourth power, we obtain the following terms in the vicinity of frequency $4\omega_c$:

$$y^4(t) \cong \left(\frac{A^4}{4}\right)\cos[4\omega_c t + 4\theta(t) + 4\psi(t)] + n_4(t)$$
$$= \left(\frac{A^4}{4}\right)\cos[4\omega_c t + 4\psi(t)] + n_4(t) \tag{B.3.2}$$

The first term is the desired harmonic to be tracked, with the modulation removed. The term $n_4(t)$ represents all the carrier–noise cross-products in

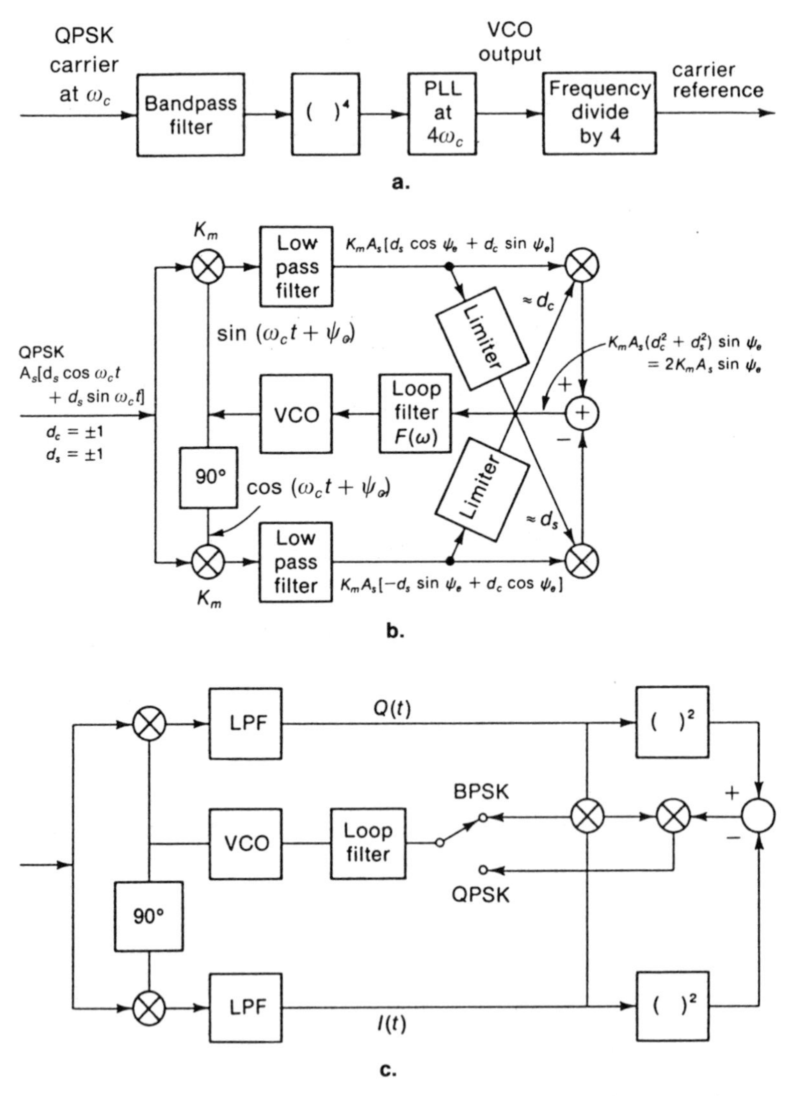

FIGURE B.6 QPSK carrier referencing. (a) Fourth-power loops. (b) Costas crossover loop. (c) Modified BPSK loop.

the vicinity of $4\omega_c$. The subsequent tracking loop at $4\omega_c$ in Figure B.6a tracks the phase variation $4\psi(t)$. Frequency division of the loop VCO reference output by a factor of 4 produces a phase locked carrier reference that is phase coherent with the suppressed QPSK carrier at ω_c. A phase ambiguity will again occur from this frequency division. In this case we have a fourth order ambiguity since frequency division by a factor of 4 can produce any of the four phase angles $i\pi/4$, $i = 1, 2, 3, 4$ on the divided reference frequency.

The contribution of the fourth order noise terms to the loop error can be computed as in the squaring loop analysis for BPSK. However, the procedure is now more complicated than it was for the squaring loop since the multiplication terms involving the QPSK carrier–noise cross-products are present in all orders up to the fourth power. It has been shown (Lindsey and Simon, 1973) that if the carrier bandpass filter in Figure B.6a has the value $B_c = 1/T_b$, where T_b is the QPSK bit time, then the fourth-power loop generates the mean squared phase error at the carrier frequency of approximately

$$\sigma_e^2 \cong \frac{N_0 B_L}{P_c}\left[1 + \frac{3.8}{E_b/N_0} + \frac{4.2}{(E_b/N_0)^2} + \frac{1}{(E_b/N_0)^3}\right] \tag{B.3.3}$$

The bracketed term now plays the same role for the fourth-power loop as the squaring factor in Eq. (B.2.7) for the squaring loop. We emphasize that Eq. (B.3.3) refers to the phase error between the received QPSK carrier and the divided-down reference carrier at ω_c. The phase error in the loop at $4\omega_c$ has a mean squared value that is $4^2 = 16$ times as large. This must be taken into account when analyzing the PLL tracking stability in Figure B.6a.

Carrier extraction for QPSK signals can also be derived from a modified form of the Costas loop involving loop crossover arms, as shown in Figure B.6b. The signal waveforms are shown throughout the figure. Note that each filter output contains data bits from both quadrature carriers that compose the QPSK subcarrier waveform. However, the sign of the output of the arm filters produced by the limiters is used to cross over and mix with the opposite arm signal. If the loop phase error is small, the output of the limiters corresponds to the cosine term of the arm filter outputs, and therefore has the sign of the data bits of the quadrature components as shown. Hence, the limiters effectively demodulate the QPSK quadrature bits, and the crossover produces a common phase error term that is cancelled after subtraction, leaving a remainder error term proportional to $\sin \psi_e$. The latter signal can then be used to generate an error signal to phase control the loop VCO, thereby closing the QPSK Costas loop. The cross-over connections of the QPSK loop can be redrawn. For example, the $\times 4$ tracking loop in Figure B.6a will generate a tracking error voltage proportional to $\sin(4\psi_e)$. This can be expanded trigonometrically as

$$\sin(4\psi_e) = 4\cos(\psi_e)\sin(\psi_e)[\sin^2(\psi_e) - \cos^2(\psi_e)] \tag{B.3.4}$$

Neglecting noise and assuming that $m(t)$ contains ideal pulses, the right side can be related to the Costas signals in Eq. (B.2.8) as

$$\sin(4\psi_e) = 4Q(t)I(t)[I^2(t) - Q^2(t)] \tag{B.3.5}$$

This suggests a modified version of the BPSK Costas loop with the connections shown in Figure B.6c for QPSK carrier recovery. Note that the system is basically an extension of the BPSK system by adding on the nonlinear section. We can easily connect from one to the other by simply reconnecting the VCO.

A more recent system for QPSK phase referencing is the demodulation–remodulation (Demod–Remod) (Braun and Lindsey, 1978; Simon and Alem, 1978; Weber and Alem, 1980) loop in Figure B.7. The system obtains its phase referencing by attempting to reconstruct a local version of the received QPSK carrier. It does this by using the reference VCO to demodulate the input via the quadrature mixer to produce the baseband. It then remodulates the demodulated baseband back as to the same VCO reference carrier, thereby creating the local QPSK carrier. If the bits are

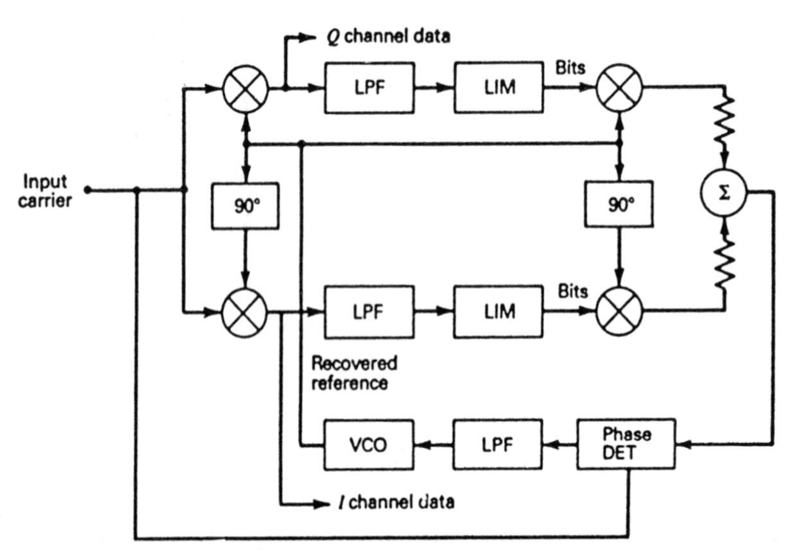

FIGURE B.7 Demand–Remod loop for QPSK carrier referencing.

demodulated correctly, the reconstructed carrier has the same modulation phase as the received carrier but differs by the phase reference error. Hence a phase comparison between the received waveform and the local reconstructed QPSK carrier generates a voltage error that corrects the local reference phase, bringing it into synchronism. Demod–Remod systems have the advantage that tracking is accomplished directly at the carrier frequency, so no divide-down ambiguities occur. However, the system requires more hardware (two more mixers than the Costas loop) and depends strongly on the accurate reconstruction of the local QPSK carrier. This means reliable bit decoding, and exacting balancing of the quadrature arms (as indicated by the gain adjustment in the output arm in Figure B.7). For this reason the Demod–Remod system requires high input SNR and, since it primarily phase adjusts, is predominantly a phase tracking subsystem. It has been shown that if the bits are decoded correctly, the tracking error voltage in the Demod–Remod system evolves identically to that of a Costas loop error voltage, and the performance of the two systems become identical at high SNR values.

Unbalanced QPSK (UQPSK) carriers in Eq. (2.3.14) cannot be phase referenced by the standard fourth-power loops, since the modulation phase states are not equally spaced. Hence UQPSK carriers require modified Costas crossover loops or modified Demod–Remod loops for phase referencing. The modification occurs in the attenuation settings of each arm to account for the quadrature power imbalance. Since the crossover loop depends on interference subtraction from the crossover arms, the UQPSK loops are extremely sensitive to gain adjustments for exact cancellation. Likewise, the Demod–Remod loop must reconstruct the UQPSK carrier with the data arms having the exact power relation and is also sensitive to gain settings.

If the UQPSK imbalance is high, the carrier closely resembles a BPSK carrier, and phase referencing can be achieved by a standard squaring system in Figure B.3a. Denote the UQPSK carrier as

$$c(t) = \sqrt{2P_c}m_c(t)\cos(\omega_c t + \psi) + \sqrt{2P_s}m_s(t)\sin(\omega_c t + \psi) \quad \text{(B.3.6)}$$

where $P_c \gg P_s$. If the input to the squarer is $c(t) + n(t)$, the squarer output (neglecting dc values) is

$$\begin{aligned}[c(t) + n(t)]^2 &= (P_c - P_s)\cos(2\omega_c t + 2\psi) \\ &\quad + 2\sqrt{P_c P_s}m_c(t)m_s(t)\sin(2\omega_c t + 2\psi) + n_{sq}(t) \quad \text{(B.3.7)}\end{aligned}$$

where $n_{sq}(t)$ is again the sum of the signal–noise and noise–noise cross-product terms in Eq. (B.2.2). This noise has the spectral level in Eq. (B.2.4)

$$S_{sq}(0) = \frac{N_0^2 B_c}{2} + N_0 P_T \tag{B.3.8}$$

with $P_T = P_c + P_s$ is the total UQPSK carrier power. The tracking term is given by the first term in Eq. (B.3.8) and has amplitude $(P_c - P_s)$. The second term represents interference due to the quadrature modulation and involves the cross-product of the two data waveforms. The power of this term, in the vicinity of $2\omega_c$, in the tracking loop bandwidth B_L, is then

$$\text{Quadrature noise} = [P_c P_s S_{cs}(0)]2B_L \tag{B.3.9}$$

where $S_{cs}(\omega)$ is the convolution of the spectral densities $S_c(\omega)$ and $S_s(\omega)$ of the two data waveforms. This means

$$S_{cs}(0) = \frac{1}{2\pi}\int_{-\infty}^{\infty} S_c(\omega)S_s(\omega)d\omega \tag{B.3.10}$$

The effective tracking loop SNR is then

$$\rho = \frac{(P_c - P_s)^2/2}{[N_0 P_T + (N_0^2 B_c/2) + P_c P_s S_{sc}(0)]2B_L} \tag{B.3.11}$$

Let $P_c = qP_T$, $P_s = (1-q)P_T$, and $P_T = P_c + P_s$, and write

$$\begin{aligned}\rho &= \frac{P_T^2[q-(1-q)]^2/2}{\{N_0 P_T[1 + (N_0 B_c/2P_T)] + P_T^2 q(1-q)S_{cs}(0)\}2B_L} \\ &= \frac{1}{4}\left(\frac{P_T}{N_0 B_L}\right)\left\{\frac{(2q-1)^2}{[1 + (N_0 B_c/2P_T)] + (P_T/N_0)S_{cs}(0)q(1-q)}\right\}\end{aligned} \tag{B.3.12}$$

The resulting UQPSK phase referencing error, after dividing by 2, is then

$$\sigma_e^2 = \frac{1}{4\rho} \tag{B.3.13}$$

Note that the bracket in Eq. (B.3.12) represents a UQPSK squaring loss and is due to both the squaring of the noise and the cross-product self-noise of the data itself. This self-noise depends on the spectral densities of the baseband waveforms and the type of digital modulation, whereas

TABLE B.4 Self-noise spectral level for UQPSK modulation

High-Rate (R_2 bps) Modulation Format	*Low-Rate (R_1 bps) Modulation Format*	$S_{cs}(0)$
NRZ	NRZ	$\frac{1}{R_2}\left(1 - \frac{R_1}{R_2}\right)$
Manchester	NRZ	$\frac{1}{6}\left[\frac{R_1}{R_2^2}\right]$
NRZ	Manchester	$\frac{1}{R_2}\left[1 - \frac{R_1}{R_2}\right]$, $R_2 \geq 2R_1$
		$\frac{1}{R_2}\left[\frac{R_2}{R_1} - \frac{1}{6}\left(\frac{R_2}{R_1}\right)^2 - 1 + \frac{1}{3}\left(\frac{R_1}{R_2}\right)\right]$, $R_2 < 2R_1$
Manchester	Manchester	$\frac{1}{2}\left(\frac{R_1}{R_2^2}\right)$, $R_2 \geq 2R_1$
		$\frac{1}{R_2}\left[2 - \left(\frac{R_2}{R_1}\right) + \frac{1}{6}\left(\frac{R_2}{R_1}\right)^2 - \frac{5}{6}\left(\frac{R_1}{R_2}\right)\right]$, $R_2 < 2R_1$

the amount of spectral overlap will depend on the individual data rates. Table B.4 summarizes some basic formulas (Yuen, 1962) for evaluating the parameter $S_{cs}(0)$ in Eq. (B.3.10) for some popular combinations of modulation formats and data rates used on each arm. For situations in which the self-noise dominates in Eq. (B.3.11) these results become important for determining and controlling the loop CNR. It may be necessary to actually place one of the data streams onto a digital subcarrier so as to further separate the spectra in Eq. (B.3.10) for increased self-noise reduction.

B.4 MPSK CARRIER REFERENCING SYSTEMS

Carrier referencing systems for higher-order MPSK phase modulations can be obtained by extending to higher powered nonlinearities or introducing more complicated crossover loops. A carrier with M possible car-

TABLE B.5 Tabulation of rms phase error σ_e^2 for MPSK carrier recovery subsystems

Carrier Modulation Format	*Mean Squared Phase Reference Error for Carrier Recovery Subsystem*
BPSK	$B_L T_b \left[\frac{1}{E_b/N_0} + \frac{1/2}{(E_b/N_0)^2} \right]$
QPSK	$B_L T_b \left[\frac{1}{E_b/N_0} + \frac{3.8}{(E_b/N_0)^2} + \frac{4.2}{(E_b/N_0)^3} + \frac{1}{(E_b/N_0)^4} \right]$
8-PSK	$B_L T_b \left[\frac{1}{E_b/N_0} + \frac{24.5}{(E_b/N_0)^2} + \frac{294}{(E_b/N_0)^3} + \frac{1837}{(E_b + N_0)^4} + \frac{5.8 \times 10^3}{(E_b/N_0)^5} + \frac{8.8 \times 10^3}{(E_b/N_0)^6} + \frac{5.04 \times 10^3}{(E_b/N_0)^7} + \frac{630}{(E_b/N_0)^8} \right]$
MPSK	$B_L T_b \left[\sum_{q=1}^{M} \binom{M}{q}^2 \left(\frac{q!}{M^2} \right) \frac{1}{(E_b/N_0)^a} \right]$

rier phases can be recovered by an Mth power device, followed by a loop tracking the Mth harmonic. The loop provides a phase coherent harmonic that can be divided by M to produce the desired reference. Since both carrier and noise are raised to the higher power, more carrier–noise cross-product terms will be generated that will again appear as noise interference to the loop tracker. Table B.5. tabulates the resulting MPSK phase reference error variance obtained in this way for 8 PSK and for the general MPSK system. Included are the previous results for BPSK and QPSK. Note the increase in error variance as the order M is increased, and the reciprocal dependence on the carrier E_b/N_0, for a given $B_L T_b$ product.

B.5 BIT TIMING

After the data modulated carrier has been demodulated to baseband via the coherent carrier reference, bit timing must be established to clock the bit or word decoding. Bit timing therefore corresponds to the operation of extracting from the demodulated baseband waveform a time coherent clock at the bit rate or word rate of the data. Bit timing subsystems generally operate in conjunction with the decoder (Figure B.8) and use the same demodulated baseband waveform to extract the bit timing clock. Timing markers provided from this clock then can be used to synchronize the decoder. Subsystems for performing bit timing from the data can be classified into two basic types. One involves squaring of the data waveform to generate timing waveforms that can be used for time synchronization. The other uses some form of transition tracking of the bit edges to generate the timing. Both types of system inherently involve modulation removal

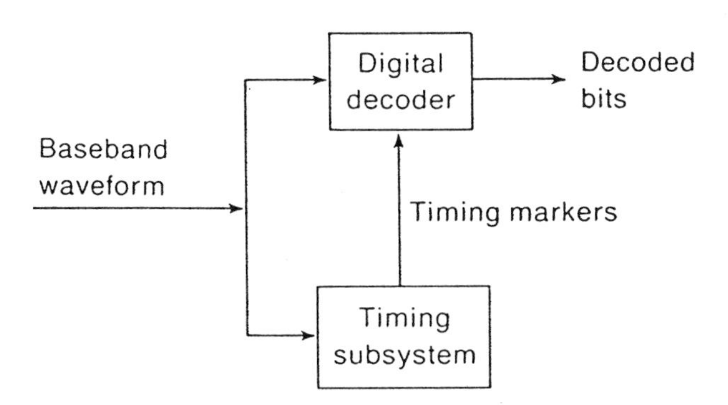

FIGURE B.8 Bit timing subsystem and digital decoder.

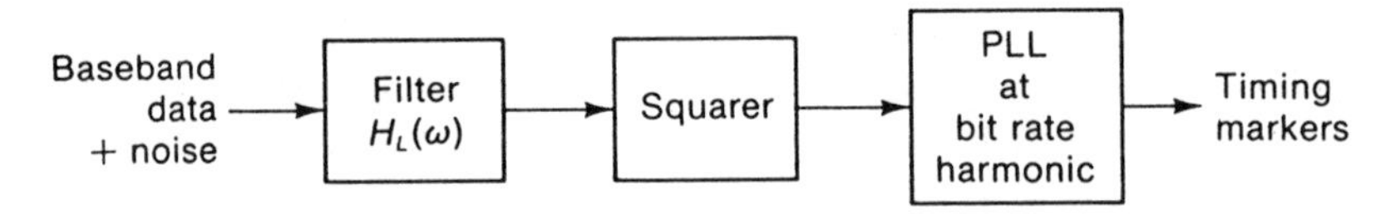

FIGURE B.9 Bit timing filter-squarer system.

and harmonic generation but the mechanics of each is accomplished in different ways.

Squaring systems (Figure B.9) contain some form of low pass waveform filtering followed by rectification or squaring to remove the modulation and generate a timing subsystem that follows. This can be seen from the diagram in Figure B.10. Figure B.10a shows a binary digital NRZ waveform composed of ideal pulses with random data bits. Figure B.10b

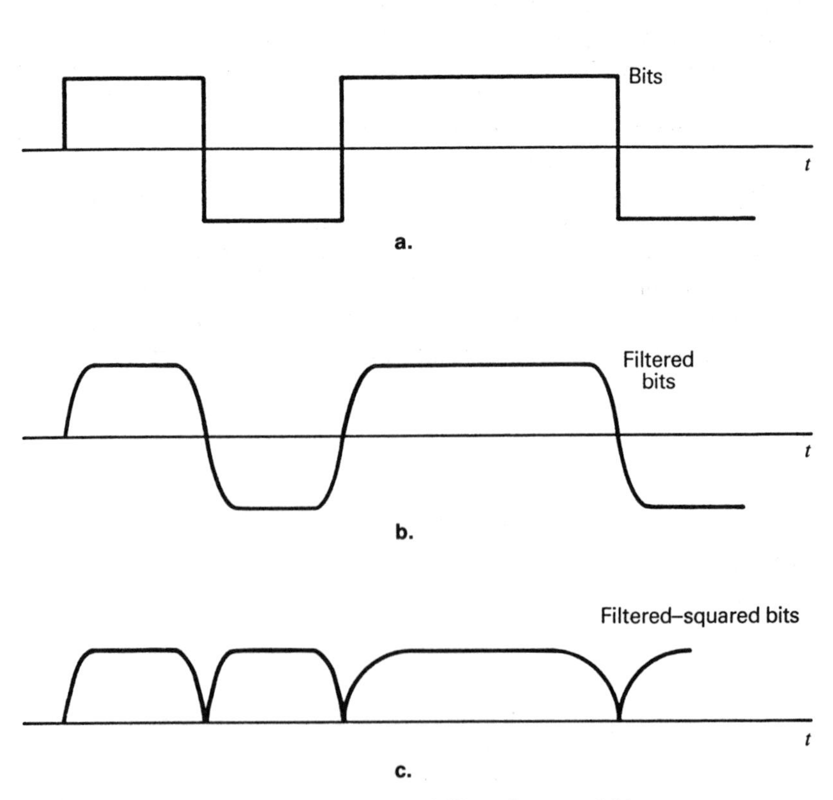

FIGURE B.10 Bits, filtered bits, and filtered-squared bits.

shows a low pass filtered version of this waveform, producing rounded pulses as shown. Squaring (or rectification) of this waveform produces the sequence in Figure B.10c in which the presence of a repetitive component is obvious. This means that the squared waveform contains harmonic power at the bit rate frequency. This harmonic is in exact time synchronism with the zero crossings of the original bit waveform. Thus tracking of the bit rate harmonic in a PLL following the squarer will produce timing markers at the zero crossings that can be used for bit timing.

The squared waveform in Figure B.10c will have a spectrum composed of a continuous portion plus discrete harmonics of the bit rate frequency $1/T_b$. If $H_L(\omega)$ is the low pass prefilter function in Figure B.9 and $P(\omega)$ the bit pulse transform, then for equal likely data bits, the first harmonic at $1/T_b$ at the input to the PLL will have power

$$P_1 = \frac{1}{T_b}\left|S_2\left(\frac{2\pi}{T_b}\right)\right|^2 \tag{B.5.1}$$

where $S_2(\omega) = [H_L(\omega)P(\omega)] \oplus [H_L(\omega)P(\omega)]$; that is, $S_2(\omega)$ is the convolution of $H_L(\omega)P(\omega)$ with itself. This component is to be tracked by the loop, while the continuous part adds spectral interference (called *self-noise*, or *data-dependent noise*) to the loop input. This self-noise can generally be neglected, relative to the receiver noise entering the loop. Note that the harmonic tracking power depends on the filtering $H_L(\omega)$.

The baseband waveform is actually the sum of the data sequence in Figure B.10a plus the baseband white noise (spectral level N_0). The filtering-squaring operation produces the harmonic in Eq. (B.5.1) plus the signal filtered noise cross-product and the squared noise component. The PLL tracks the harmonic with these cross-product and squared noise waveforms appearing as interference. The timing accuracy is therefore dependent on the phase tracking accuracy of the loop. A timing error τ in seconds is related to a loop phase error ψ_e in radians at $f = 1/T_b$ by $\tau = T_b(\psi_e/2\pi)$. Thus the timing error variance and phase error variance are related by

$$\sigma_\tau^2 = \sigma_e^2\left(\frac{T_b}{2\pi}\right)^2 \tag{B.5.2}$$

The phase error variance can be obtained by standard tracking loop analysis, with an input carrier tone having the power in Eq. (B.5.1) and considering the interference in the vicinity of $1/T_b$ entering the loop as noise. Holmes (1980) has shown that for an RC presquaring filter the squared noise term produces a timing error variance in Eq. (B.5.2) of

$$\sigma_n^2 = C_n\left(\frac{N_0^2 T_b^3 B_L}{E_b^2}\right). \tag{B.5.3}$$

where C_n is a coefficient dependent on the $H_L(\omega)$ filter bandwidth and E_b/N_0 is the baseband bit energy-to-noise level at the timing system input. For the first order RC prefilter with 3 dB frequency f_3 and NRZ pulses, C_n has the value

$$C_n = \frac{r^3\pi(1 + r^{-2})}{4(1 - e^{-2\pi r})} \tag{B.5.4}$$

where $r = f_3 T_b$. The signal–noise cross-product contributes a timing error variance of

$$\sigma_{\text{sn}}^2 = C_{\text{sn}}\left(\frac{B_L T_b^3}{E_b/N_0}\right) \tag{B.5.5}$$

where, for the RC prefilter,

$$C_{\text{sn}} = \frac{3r(4r^2 + 13r^4 - r^2 - 1)}{2\pi(1 - e^{-r})(r^2 + 1)(4r^2 + 1)(r^2 + 4)} \tag{B.5.6}$$

The total mean squared timing error for the filter–square bit synchronizer is then

$$\begin{aligned} \sigma_\tau^2 &= \sigma_n^2 + \sigma_{\text{sn}}^2 \\ &= B_L T_b^3\left[\frac{C_n}{(E_b/N_0)^2} + \frac{C_{\text{sn}}}{(E_b/N_0)}\right] \end{aligned} \tag{B.5.7}$$

Normalizing σ_τ^2 to the bit time, we can rewrite as

$$\left(\frac{\sigma_\tau}{T_b}\right)^2 = \frac{B_L T_b}{(E_b/N_0)\mathscr{S}_q} \tag{B.5.8}$$

where we have introduced the bit-timing squaring loss

$$\mathscr{S}_q = \left[C_{\text{sn}} + \frac{C_n}{(E_b/N_0)}\right]^{-1} \tag{B.5.9}$$

Note that at low E_b/N_0, Eq. (B.5.7) behaves as $(E_b/N_0)^{-2}$, and as $(E_b/N_0)^{-1}$ at high E_b/N_0. It has been shown (Holmes, 1982) that at $E_b/N_0 \approx 10$ dB, the squaring loss in Eq. (B.5.9) is minimized with a value of $r \approx \frac{3}{16}$—that is, the prefilter bandwidth should be set to about $\frac{3}{16}$ of the bit rate frequency $1/T_b$. For this case

$$C_n \simeq 0.318$$
$$C_{sn} \simeq 0.545 \tag{B.5.10}$$

When these coefficients are used in Eq. (B.5.7), the normalized rms timing error in Eq. (B.5.8) plots as shown in Figure B.11 for several values of $B_L T_b$. In general bit timing must be maintained to within about 1% of a bit time to prevent serious degradation to the PE decoding performance. We see from Figure B.11 that bit timing to within 1% of a bit time is

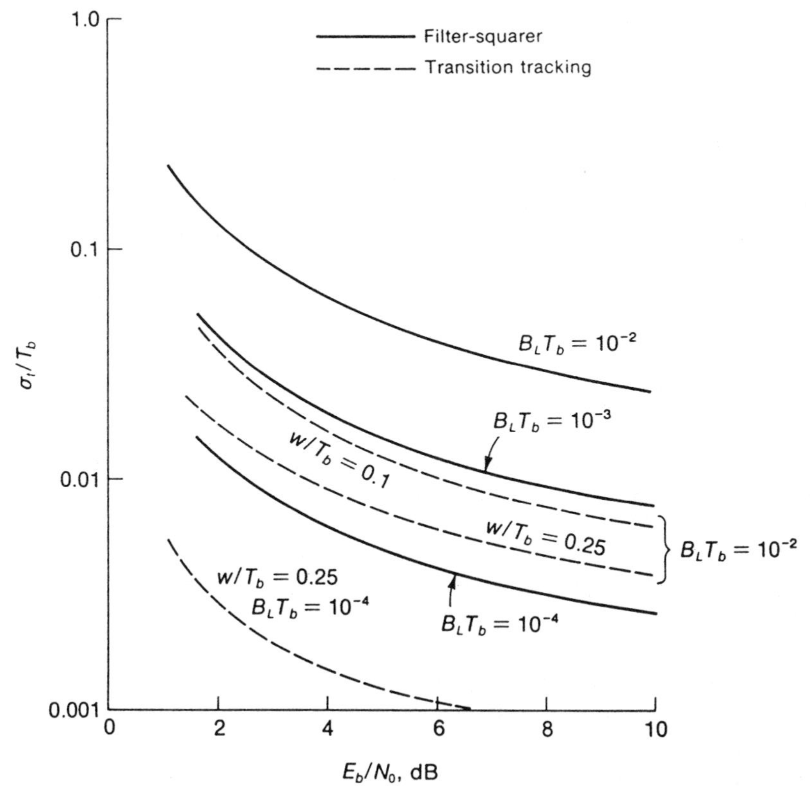

FIGURE B.11 Timing error vs. baseband E_b/N_0 for bit-timing systems.

feasible at $E_b/N_0 \approx 10$ dB, with moderate values of loop bandwidth ($B_L T_b \leq 10^{-2}$). This means, for example, that with a bit clock of 10 Mbps, a timing accuracy on the order of nanoseconds can be retained with this type of timing system.

A second way to achieve bit timing from a modulated NRZ waveform is to utilize *transition tracking* on the bit edges, in conjunction with bit decisioning. Transition tracking is obtained by integrating over the bit transitions (bit edges) to generate a timing error voltage for feedback control. To see this, consider Figure B.12a, which shows a periodic pulse sequence with a transition being integrated over a w sec integration interval. We see that if the integration is centered exactly on the pulse transition, a zero integration value is produced, because of the equal positive and negative areas. If the integration is offset by δ sec, the integrator output generates the function $\pm g(\delta)$ shown, where the $\pm$ sign depends on the particular data bits (i.e., whether the transition changes from a positive to negative value or vice versa). To rectify this bit sign, a bit decision is made in a separate decoding arm and used to multiply the integrator output. By inserting a one-bit delay in the loop input and using the present bit decision, the bit polarity will always rectify the sign of the loop error voltage, producing a timing correction voltage of the proper sign independent of the data bit. This correction voltage is filtered and fed back to control the timing location of the integration for the next transition. Thus the previous bit transitions are effectively used to remove the modulation and the loop operates to hold the integrator centered on the transitions. Bit timing markers can then be generated in synchronism with the integration for timing control. The latter system is referred to as *data-aided transition tracking* and is the bit-timing equivalent of the suppressed carrier phase referencing system of the same type. The bit delay can also be eliminated in modified versions of the system by simply hard-limiting the baseband waveform in the upper arm and using the limiter output to invert the error. As long as the noise does not cause the bit waveform polarity to change sign during a bit, the tracking error will be properly rectified.

In transition tracking a timing error voltage is generated each pulse time when a transition occurs. (In NRZ data, this is once per bit if the bit changes sign but is twice per bit for Manchester data, with a transition guaranteed every midbit time.) The loop filter acts to smooth these error samples into a continuous loop control. In this sense the transition tracker appears as a sampling (discrete) control loop with error samples generated at regular intervals. The loop therefore has the equivalent timing model shown in Figure B.12b. The input is the timing offset of the received data bits, and the feedback variable is the relative timing position of the

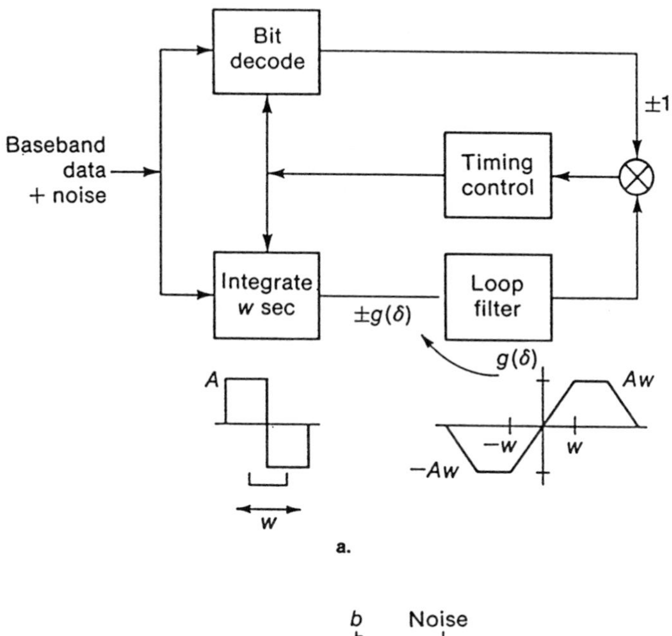

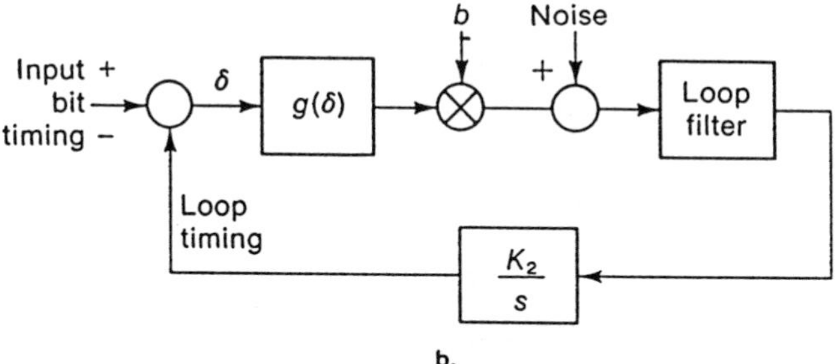

FIGURE B.12 Transition tracking bit-timing loop. (a) Block diagram. (b) Equivalent model.

integrator. The function $bg(\delta)$ converts timing errors to voltages, and the loop integrates to adjust integrator timing. The system is identical to the nonlinear sine wave phase tracking loop in Figure B.1b, except the nonlinear sine function is replaced by the nonlinear $g(\delta)$ function. Thus Figure B.12b is the bit timing equivalent to the phase locking operation discussed earlier.

The multiplication by b accounts for the bit decision to rectify the error, with $b = +1$ if the bit is correctly detected and $b = -1$ if the bit is incorrectly decoded. The noise level entering the loop is the integrated noise variance per sample ($N_0 w$) divided by the sampling bandwidth $1/T_b$, producing the effective noise level $N_0 w T_b$. Note that the loop is linear for $|\delta| \leq w/2$, and therefore the transition tracking loop has an equivalent linear model with gain given by the mean slope of $bg(\delta)$ at $\delta = 0$. Hence

$$\text{Loop gain} = \frac{A\omega}{w/2}\mathscr{E}(b)$$
$$= 2A\mathscr{E}(b) \tag{B.5.11}$$

where A is the data bit amplitude. The mean of the random variable b is given by $\mathscr{E}(b) = (1)(1 - \text{PE}) + (-1)\text{PE} = 1 - 2\text{PE}$, where PE is the bit decoding probability. Using the binary BPSK PE, Eq. (B.5.11) becomes

$$\text{Loop gain} = 2A\left[1 - 2Q\left[\left(\frac{2E_b}{N_0}\right)^{1/2}\right]\right] \tag{B.5.12}$$

With this gain inserted into the linear model, the loop can be analyzed as a standard phase lock loop with noise bandwidth B_L. The timing error variance is then

$$\sigma_\tau^2 = \frac{(N_0 w T_b)2B_L}{4A^2[\text{Erf}(E_b/N_0)^{1/2}]^2} \tag{B.5.13}$$

The variance normalized to the bit time T_b is then

$$\left(\frac{\sigma_\tau}{T_b}\right)^2 = \frac{N_0 2B_L w/T_b}{4A^2[\text{Erf}(E_b/N_0)^{1/2}]^2} \tag{B.5.14}$$

Note the direct dependence of the timing error variance on the integration width w. When a reduced integration time is used, the timing accuracy improves. However, it must be remembered that w is made smaller, the linear tracking range in Figure B.12a is also decreased, making linear operation more difficult to retain. Again defining

$$\rho = \frac{N_0 2B_L}{A^2} = \frac{B_L T_b}{E_b/N_0} \tag{B.5.15}$$

we can write Eq. (B.5.14) as

$$\left(\frac{\sigma_\tau}{T_b}\right)^2 = \frac{1}{\rho\mathscr{S}_q} \tag{B.5.16}$$

where $\mathscr{S}_q$ is now the timing squaring loss

$$\mathscr{S}_q = \left[\frac{4\,\mathrm{Erf}^2(E_b/N_0)^{1/2}}{w/T_b}\right]^{-1} \tag{B.5.17}$$

Equation (B.5.16) is also plotted in Figure B.11 for several values of $B_L T_b$ and w/T_b. Timing performance is comparable to the filter-squarer system and can be improved by decreasing w. The transition loop with NRZ data also suffers from the fact that long sequences of identical bits produce no edge transitions for timing update. Insertion of intentional transitions may be necessary to guarantee the presence of a transition within a prescribed time interval. By using Manchester data, a transition is guaranteed at the center of every bit, thereby aiding the bit timing operation.

Transition tracking performance can be improved by accumulating sequences of m error samples prior to timing correction. (This requires replacing the filter in Figure B.12 by a digital accumulator.) Such an accumulation multiplies up the noise voltage variance by m but also increases the loop gain in Eq. (B.5.12) by m. The normalized timing error variance, assuming perfect bit decisioning, is then

$$\begin{aligned}\left(\frac{\sigma_\tau}{T_b}\right)^2 &= \frac{mN_0 w}{m^2 4A^2 T_b^2} \\ &= \frac{(w/T_b)N_0/4}{mE_b/N_0}\end{aligned} \tag{B.5.18}$$

where w/T_b is the integration window expressed as a fraction of the bit time. Note the direct improvement as we accumulate over more samples. However, this improvement occurs only if the signal–error samples remain the same throughout the accumulation time. Hence m is restricted to the number of bits over which the received timing variation does not change appreciably.

REFERENCES

Braun, W., and Lindsey, W. (1978), "Carrier Synchronization for Unbalanced QPSK," *IEEE Trans. Comm.*, **COM-26** (September).

Holmes, J. (1980), "Tracking Performance of the Filter and Square Bit Synchronizer," *IEEE Trans. Comm.*, **COM-28** (August).

Holmes, J. (1982), *Coherent Spread Spectrum Systems*, Wiley, New York, 121–129.

Lindsey, S., and Simon, M. (1973), *Telecommunication Systems Engineering*, Prentice-Hall, Englewood Cliffs, NJ.
Simon, M., and Alem, W. (1978), "Tracking Performance of Unbalanced QPSK Demodulators—Biphase Costas Loop," *IEEE Trans. Comm.*, **COM-26** (August).
Simon, M., and Lindsey, W. (1977), "Optimum Performance of Suppressed Carrier Receivers with Costas Loop," *IEEE Trans. Comm.* (February).
Weber, C., and Alem, W. (1980), "Performance Analysis of Demod-Remod Coherent Receiver for QPSK," *IEEE Trans. Comm.*, **COM-28** (December).
Yuen, J. (1982), "Deep Space Telecommunications System Engineering," *JPL Publ.*, 82–76, Chap. 5.

APPENDIX C
Satellite Ranging and Position Location Systems

Ranging systems provide a measurement of range (distance) between two points. This range measurement is achieved by processing a transmitted waveform between the two points. When several such range measurements are made, the location of a target can be uniquely determined. Although ranging is not truly a part of the satellite communication subsystem, its operation is intimately related to the carrier referencing and its waveform transmission, and processing is generally integrated directly into the communication link. In many cases, ranging performance and accuracy may ultimately limit the design of the communication subsystem.

C.1 RANGING SYSTEMS

A functional block diagram of a satellite ranging system is shown in Figure C.1. Measurement of range between two points is accomplished by transmitting a signal marker from one point to the other and back, and measuring the round-trip transit time. Since the transmitted marker travels at the speed of light,

$$\begin{aligned}[\text{Range in meters}] &= \tfrac{1}{2}(3 \times 10^{8})(\text{round-trip time in s}) \\ &= (150)(\text{round-trip time in } \mu\text{s}) \qquad (\text{C.1.1})\end{aligned}$$

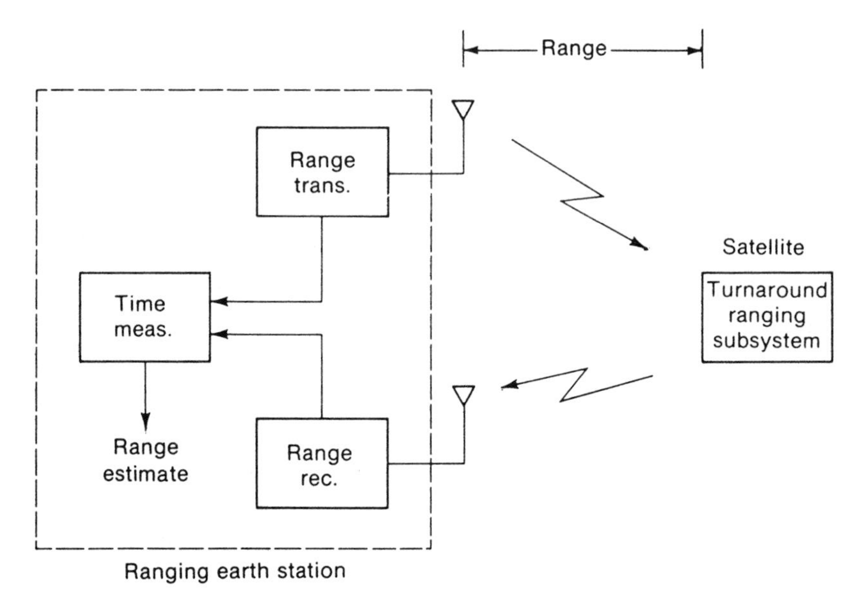

FIGURE C.1 Turnaround ranging system block diagram.

Thus, range measurements can be made from estimates of the time between the instant of marker transmission and the instant of its return. The waveform containing the marker, called the *ranging signal*, is transmitted as an RF waveform from the earth station. At the satellite, the RF waveform is reflected, or instantaneously returned, back to ground. The returned signal is processed to recover the ranging signal and time marker, and its arrival time is compared, for the two-way range estimate, to the original transmission time. This range measurement is generally continually updated by retransmitting the ranging signal periodically, and repeating the measurement. This allows a continual measurement of the satellite range as it changes in time.

Since the returned ranging signal is invariably marred by noise, a direct time comparison between transmitted and received signals will yield range accuracies much worse than desired. For this reason, the ranging receiver attempts to align a clean version of the ranging signal with the noisy returned signal so that the time difference measurement can be made with uncontaminated waveforms. Thus, the ranging system is confronted with the task of aligning a local receiver marker sequence with a received marker sequence contained within the returned ranging signal. This alignment can be made by locking together a local version of the ranging signal with the received ranging signal. An error in the clocking of the receiver marker sequence will lead directly to an error in the measurement of the time difference, and therefore an erroneous range estimate. If the rms clocking error is $\sqrt{\epsilon^2}$ seconds, then, from Eq. (C.1.1), an rms range error of

$$\text{rms range error} = \pm 1.51\sqrt{\epsilon^2}(10^8)\ \text{m} \tag{C.1.2}$$

will occur. Hence, clock timing accuracy determines the ultimate range accuracy.

The return of the RF ranging waveform from the satellite can be accomplished by simple electromagnetic reflection or by actively re-transmitting. In the former case, the satellite plays the role of a target, and the ranging transmitter "bounces" the signal off the target and monitors its return. The ranging system is then much like a conventional radar system, and received power levels depend primarily on target cross-sectional characteristics. In most satellite systems, retransmission is often necessary to produce the receiver power levels required for accurate ranging. In this case, the satellite must itself contain an electronic package that has the capability of receiving and instantaneously retransmitting the RF ranging waveform back to ground (see Figure 1.6). Active systems that make use of this instantaneous retransmission method are called *turnaround ranging systems.* The turnaround operation requires the satellite to receive, amplify, and retransmit the ranging signal. This must be done with negligible uncalibrated delay, since any retransmission delay will be interpreted as transit time and cause false range readings at the ground. The retransmission can involve simply a frequency translation of the uplink ranging waveform, or can use a local ranging signal generator that locks to the received waveform and retransmits the referenced range signal to ground.

RF ranging waveforms are formed by modulating the desired ranging signal onto an RF carrier. Short distance (ground-to-ground) ranging systems may often use merely pulse trains as the ranging signal. Satellite systems prefer continuous ranging signals for improved power efficiency. Since a locking operation may have to be performed at both ends in a turnaround ranging system, continuous ranging signals are usually constructed as periodic digital waveforms, the latter called *range codes.* This means the ranging waveforms are formed much like the address codes used in CDMA, and the code-locking operation can be accomplished by the code-acquisition subsystems described in Chapter 7. The ranging waveforms are generated from the ranging codes by PSK modulating the code chips onto a ranging subcarrier. In this case, if the chips are w sec wide, a ranging subcarrier bandwidth of

$$B_r \cong 2/w \text{ Hz} \tag{C.1.3}$$

is required to transmit the ranging waveform.

In addition to the simplicity in code alignment procedures, digital ranging codes have the advantage that the measurement of transmitted and received time differences is computationally simplified. We need only monitor the correlation between the transmitted and local received ranging codes (Figure C.2), and count the number of code chip shifts needed to align the two. Since the chips have known time width, the number of required chip shifts is proportional to the desired range. The narrower the ranging code pulse width, the higher the resolution in the eventual range estimate. If code alignment can be measured accurately to within a fraction η of one chip time w, then from Eq. (C.1.2) range can be measured to within

$$\text{Range accuracy} = \eta w(1.51)(10^8) \text{ m} \tag{C.1.4}$$

Thus, the narrower the chip width the more accurate the range measurement. For example, with $\eta = 0.1$, a microsecond chip will allow range to be measured to within 16 m. This, in turn, will require a ranging subcarrier bandwidth in Eq. (C.1.3) of about 2 MHz. Hence, ranging accuracy is achieved at the expense of ranging channel bandwidth.

Range codes differ from address codes in the manner in which the code period is selected. While address codes in CDMA usually have their code period related to bit times, the range code period is set by the range being measured. If the range is such that the round-trip travel time will exceed the range code period, the returned marker will occur after the next

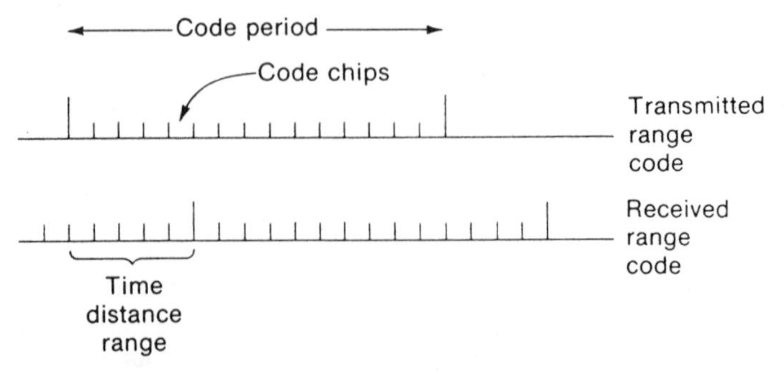

FIGURE C.2 Timing diagram for coded ranging signals.

marker is sent. This leads to ambiguities in the subsequent range measurement. Hence range code period determines the maximum range that can be measured without ambiguities. If the range code period is T_c s, then the maximum range is

$$\text{Maximum range} = 1.5T_c(10^8)\ \text{m} \tag{C.1.5}$$

Therefore, range codes are desired with as long a period as possible to increase the maximum unambiguous ranging distance. Since the code chips should also be narrow for good measurement resolution, the long period implies long-length codes (i.e., many chips per period). Typical codes used for satellite ranging usually contain about 10^5 chips, each about 1 μs in width, providing for an unambiguous range measurement out to about 10^4 km.

C.2 COMPONENT RANGE CODES

In addition to the proper period and chip width, ranging codes must have a convenient autocorrelation property that makes code alignment easy to recognize. Hence range codes should have the desirable correlation of shift register pseudo-random-noise (PRN) sequences used for address codes (see Figure 7.3). However, even though PRN codes have the desired correlation property, the use of such long-length codes for ranging will often require prohibitively long acquisition times, especially in systems that specifically require relatively fast range acquisition. For this reason, there is interest in using range codes that sacrifice some of the advantages of PRN correlation in order to reduce the initial range code acquisition time. One method for accomplishing this is to use ranging codes constructed from combinations of smaller length PRN codes, called *component codes*, Figure C.3a. Several methods exist for logically combining short periodic binary codes into a long period sequence whose length is the product of the lengths of the components. This code combining is achieved by using linear or nonlinear modulo-two logic with repeated versions of the component code set, and generally requires that the component PRN codes have lengths that are relatively prime. If the combined code has been properly constructed, alignment of an individual PRN component with the entire code, and correlation over a sufficient number of chips will produce a small average correlation value. This partial correlation value is a fraction of the full code correlation. As the next component code is aligned with the code and correlated, it also produces a partial correlation that will be added to that of the first. As

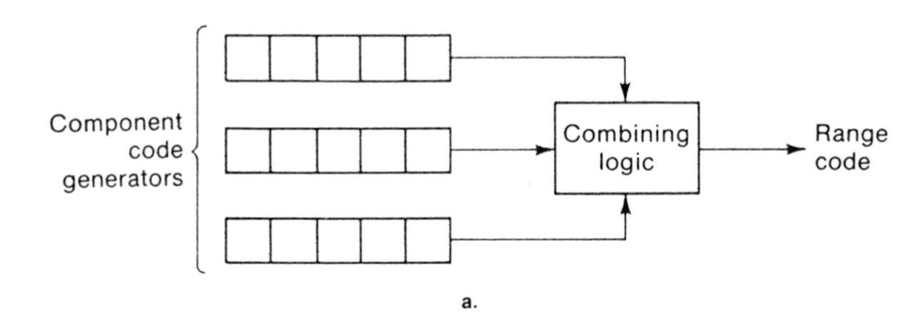

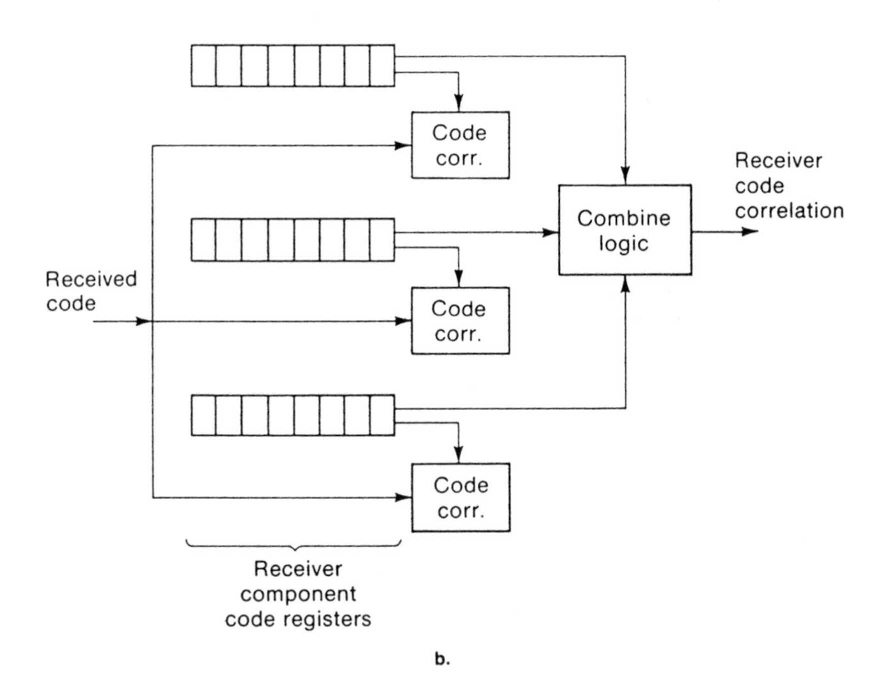

FIGURE C.3 Component code-ranging systems. (a) Range code generator block diagram. (b) Receiver component code correlator.

each component is so aligned, the correlation increases by fractional steps, until full correlation is achieved, and all components are aligned.

These partial correlation values are used by the receiver to align identical receiver components with the received code. The receiver, using the same combination logic to generate the long-length code, will therefore produce the code in exact time synchronism with the received version using the aligned components, as shown in Figure C.3b. Each partial correlation step indicates the alignment of a new component, and each component can be aligned by testing each position of its own length. Thus, a component code of length n_i will require at most n_i chip positions to be searched. Since this is true for each component, only $\sum n_i$ total positions need be searched. Hence, construction of long-length codes from shorter component sequences have the basic advantage that while the total length is determined by the product of the code length, the acquisition time is determined by their sum. The number of chips to be searched by this method is considerably less than the length. For example, component codes of length 31, 11, 15, and 7 bits can be combined into range code of length $31 \times 11 \times 15 \times 7 = 35{,}805$ chips, but only 64 chip positions theoretically need be examined for acquisition.

There are several disadvantages to using ranging codes generated from component codes. One is that the resulting combined code is not a PRN code, and will not necessarily have the desirable sharply peaked correlation in which marker location is obvious. Such non-PRN codes may have out-of-phase correlations that are large enough to cause ambiguities with the true in-phase correlation value. In this regard, some combined ranging codes may be more desirable than others, a fact that influences selection of components and combining logic. Another disadvantage is that the ranging circuitry becomes more complicated. Each range code generator must be replaced by a parallel combination of several PRN code registers, one for each component, and by logic circuitry needed to perform the combining prior to subcarrier modulation. In addition, a mechanism must be provided for separately shifting and correlating each component during acquisition.

A key point to consider with component code ranging is that only a partial correlation value is used to indicate a component code alignment. This may be somewhat difficult to recognize in a noisy environment. For example, if the partial correlation is 0.25 of full value, it is equivalent to a $\frac{1}{16}$ reduction in the effective ranging signal power value used for determining the component acquisition probability. This means the integration time for that component must be multiplied by 16 to achieve an equivalent acquisition probability. This type of adjustment must be made for each component, remembering that the probability of successful acquisition is given by the product of the acquisition probabilities for each component. Furthermore, in a practical system the integration time for each code component cannot be varied from code to code, and the longest integration time among all codes must be used for each. As an example, suppose the four component codes in column 1 of Table C.1 are combined to form a range code. We assume the system operates with a total received ranging signal power of P_r watts in additive white noise of

TABLE C.1 Tabulation of component code parameters

Component Code Length	*$\log_2$ (length)*	*Partial Correlation Value*	*Required (E_c/N_0)*
7	2.8	$\frac{1}{4}$	4.5
11	3.46	$\frac{1}{4}$	3.8
15	3.91	$\frac{1}{6}$	3.5
31	4.95	$\frac{1}{6}$	3.0

level N_0 W/Hz, one-sided. The partial correlation of the codes is listed in column 3, and we desire an acquisition probability of 0.999 for each component. The required E_c/N_0 for each code is listed in column 4, obtained from Figure 7.12 for each code length. The required code acquisition time is then determined by the maximum E_c/N_0 and the smallest partial correlation value. Hence,

$$\text{Time to acquire} = \left(\frac{N_0}{P_r}\right)\left(\frac{1}{\frac{1}{4}}\right)^2 (4.5) \tag{C.2.1}$$

where P_r is range code power. For a given value of P_r/N_0, the required acquisition time can be computed from Eq. (C.2.1). While Eq. (C.2.1) indicates an increase in the acquisition time as P_r/N_0 is reduced, it implies that acquisition will be achieved at any value of P_r/N_0 if enough time is allowed. In a practical system, however, one must be concerned with the ability to phase reference (properly demodulate the PSK-coded subcarrier) as P_r/N_0 is reduced. The discussion of the carrier referencing analysis in Appendix B is applicable here. The ability to maintain phase referencing with the partially correlated signal values and the loop noise bandwidth will determine the minimal value of P_r/N_0 for successful code acquisition.

C.3 TONE-RANGING SYSTEMS

Another ranging technique used to reduce acquisition time makes use of sinusoid or square wave harmonics. The operation is referred to as *side tone acquisition*, and the system is sometimes called *tone ranging*. Tone ranging systems take advantage of the phase relation between harmonics to quickly resolve ambiguities and avoid the multiposition search associated with coded acquisition. Consider a sequence of N harmonic signals $\{S_i(t)\}$, as in Figure C.4. (The figure shows square waves, but the discussion pertains to sine waves of the same period as well.) We show three such harmonics, each at one-half the frequency of the previous. That is, if f_i is the frequency of the harmonic $S_i(t)$

$$f_i = \frac{f_1}{2^{i-1}} \tag{C.3.1}$$

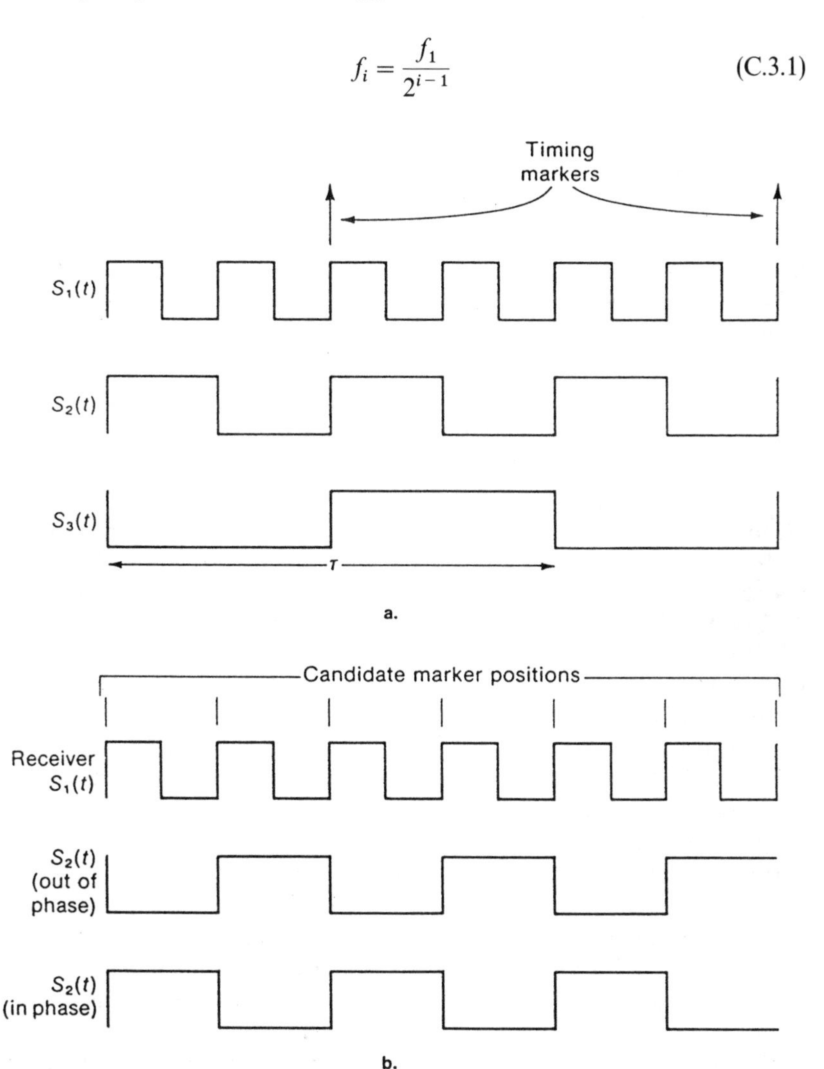

FIGURE C.4 Tone ranging waveform diagram (square waves). (a) All tones in phase. (b) Out-of-phase tones.

where f_1 is the clock (highest) frequency. At the ranging transmitter these harmonics are aligned so that the transmitter marker can be interpreted as the point in time where each harmonic simultaneously has a positive-going zero crossing. Only one such point occurs during each period of the lowest harmonic, $S_N(t)$. Hence, the lowest harmonic determines the periodicity of the markers. The set of phase-aligned harmonics are then summed to form the range signal

$$r(t) = \sum_{i=1}^{N} a_i S_i(t) \tag{C.3.2}$$

where $\{a_i\}$ are the square wave amplitudes. The range signal in Eq. (C.3.2) is periodic with period $\tau = l/f_N$ s, that is, the period of the lowest harmonic. At the receiver ranging subsystem, Figure C.5, the same harmonics are generated with the same phase relationship, but with an arbitrary time reference point. The first receiver harmonic $S_1(t)$ is first phase locked to the received ranging signal, using either a sine wave or square wave clock loop. If the loop bandwidth is less than the harmonic frequency separation, the loop will track the clock harmonic only at f_1 in Eq. (C.3.2), and the receiver and received versions of $S_1(t)$ will be brought (theoretically) into phase alignment.

Within the total acquisition period of τ sec, each positive zero crossing of $S_1(t)$ is a candidate time reference point of the received ranging signal, and, therefore, $\tau f_1 = f_1/f_N = 2^{N-1}$ ambiguity points must be resolved. This ambiguity resolution is provided by the remaining harmonics. Since the next harmonic, $S_2(t)$, is time-locked to the first harmonic [its positive zero crossings are aligned with those of $S_1(t)$], we see from Figure C.4b that only two possibilities can occur: the receiver $S_2(t)$ is exactly in phase with the received $S_2(t)$, or it is exactly 180° out of phase. If the local harmonic $S_2(t)$ is correlated over a period with the received ranging signal in Eq. (C.3.2), it will correlate with only the second harmonic, since all the harmonics are orthogonal. Thus, because of the harmonic phasing, a correlation of $S_2(t)$ with $r(t)$ will produce either a large positive in-phase correlation value or a large negative out-of-phase correlation. The in-phase or out-of-phase possibility of $S_2(t)$ can therefore be determined by a single polarity test on the correlation of $S_2(t)$ and the received ranging signal. If $S_2(t)$ is decided as being in phase, it is left alone. If it is decided as out of phase, it is shifted by 180° (shifted forward by one-half its period) to be brought into second harmonic alignment.

We now have $S_1(t)$ and $S_2(t)$ properly aligned with the corresponding components of the received ranging signal. Since the negative-going zero crossings of $S_2(t)$ cannot be marker locations, only the positive zero

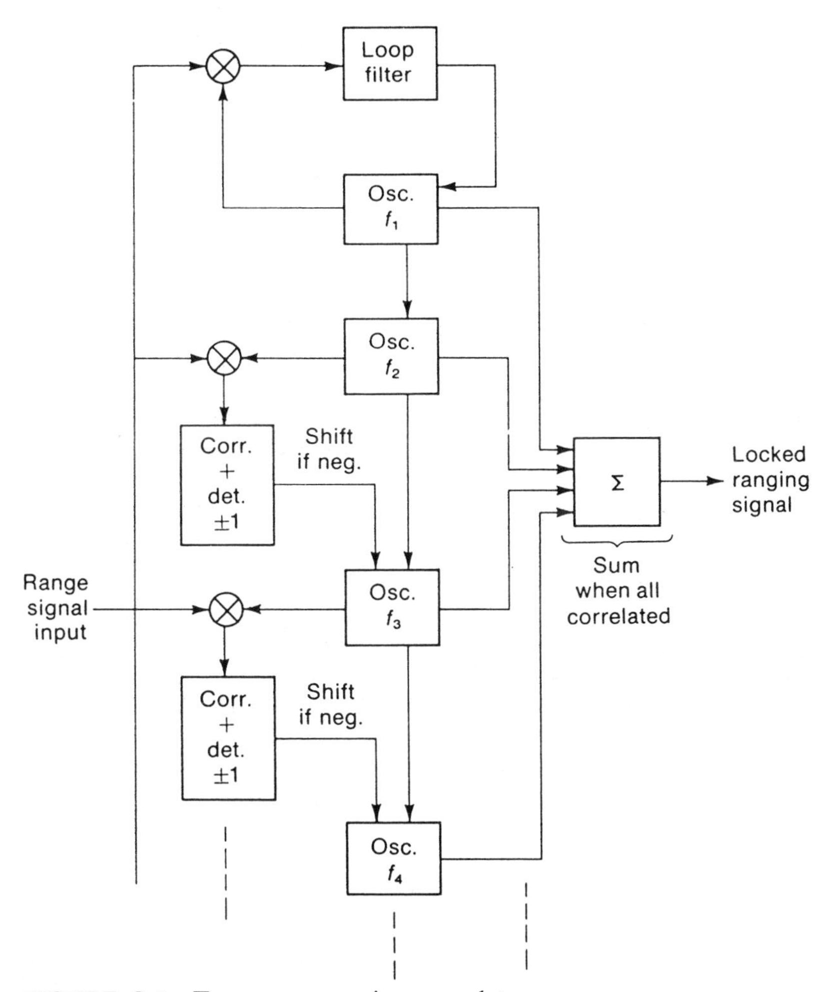

FIGURE C.5 Tone range receiver correlator.

crossings common to both $S_1(t)$ and $S_2(t)$ signify possible marker positions. This conclusion eliminates one-half of the original ambiguity points of $S_1(t)$. Thus, by a single polarity decision test on the correlation of $S_2(t)$, we now need only consider $2^{N-1}/2$ possible ambiguity points. The procedure is now repeated with $S_3(t)$. This latter is generated in phase synchronism with $S_2(t)$, and correlated with the received waveform. A polarity decision on the correlation output will determine if $S_3(t)$ is in or out of phase. This allows us to align $S_3(t)$, which again eliminates one-half the remaining ambiguity points, leaving only $2^{N-1}/2^2$ possibilities. By continuing in this manner—each time correlating, deciding, and aligning the subsequent harmonic—we continually reduce the remaining candidate points by one-half. It is important to note that each harmonic alignment must be made in sequence, since a harmonic cannot be correlated until the previous one has been aligned. After $N-1$ such decisions, only

$$\frac{f_1\tau}{2^{N-1}} = \frac{f_1}{f_N 2^{N-1}} = \frac{2^{N-1}}{2^{N-1}} = 1 \tag{C.3.3}$$

ambiguity point will remain. If all polarity decisions were correct, this last point correctly identifies the marker position. The correct marker position now allows continual generation of the receiver timing markers, and the clocking of the highest harmonic maintains the timing, so that receiver and received markers are synchronized. These timing markers can then be used for repeated range measurements by comparing to the time markers of the transmitted signal.

Note that in tone ranging the original 2^{N-1} ambiguity points are resolved by $N-1$ binary polarity decisions made in sequence. Conversely, we can say $f_1\tau$ ambiguities were resolved by $\log_2(f_1\tau)$ binary decisions. Contrast this with the PRN ML acquisition schemes, which required a correlation at each of the ambiguity points. Hence, a substantial reduction in ambiguity decisioning is achieved by tone ranging. Acquisition systems in which the number of ambiguity decisions to be made is proportional to only the log of the number of ambiguity points are referred to as *rapid acquisition systems.*

We have interpreted the operation of the tone ranging receiver by the diagram of Figure C.5. That is, the binary polarity decisions are used to align the harmonics, which are then used to reconstruct the time synchronized ranging signal in Eq. (C.3.2). However, we can also interpret the individual harmonic polarity decisions as digits of a binary word. The sequence of decisions then formulate a complete word, and the word uniquely identifies an ambiguity point. In this sense, the decisions are not used to align a harmonic, but rather to enumerate a location. Since k ambiguity points require $\log_2 k$ binary digits for unique representation, we see that rapid acquisition systems, such as tone ranging, resolve ambiguities with the minimal number of receiver decisions.

The key parameters characterizing any acquisition system are the acquisition probability, the time to acquire, and the signal power levels. Since acquisition is successful in tone ranging only if each harmonic polarity decision is correct, the probability of acquisition is then

$$\text{PAC} = \prod_{i=2}^{N} \text{PC}_i \tag{C.3.4}$$

where PC_i is the probability of correctly deciding the phase of the ith harmonic. Since this is simply an antipodal binary decision,

$$PC_i = Q\left[\left(\frac{2a_i^2 T_i}{N_0}\right)^{1/2}\right] \tag{C.3.5}$$

where T_i is the correlation time of the ith harmonic. The total acquisition time is then the sum of the $\{T_i\}$. However, practical limitations will require that each harmonic use the same integration time. Hence,

$$\text{Acquisition time} = (N-1)\left[\max_i T_i\right]$$

$$= \log_2(f_{1\tau})\left[\max_i T_i\right] \tag{C.3.6}$$

We see that the total time to acquire is directly related to the number of harmonics, which in turn is logarithmically proportional to the clock frequency and ranging period. The former defines the highest harmonic and the latter the lowest harmonic, and the number of decisions depends on the required number of harmonics that must be inserted between. The latter also determines the complexity of the acquisition receiver in Figure C.5. A basic disadvantage of tone ranging is that the total ranging power must be divided among all harmonics. Hence, the power levels a_i^2 in Eq. (C.3.2) is only a fraction of the available ranging power, and this power per harmonic varies inversely as the number of harmonics. In particular, even after successful acquisition, only a fraction of the ranging power is available for the clock, and range accuracy is degraded over that of a coded system. In some cases, the clock is given a significant fraction of the total power, at the expense of longer integration times in Eq. (C.3.5).

Transmission of tone ranging signals requires a bandwidth necessary to send all harmonics simultaneously. For sinusoidal harmonics, this requires a bandwidth extending from the lowest harmonic, f_N, to the highest f_1. The lowest tone may in fact be quite low if the maximum range is long. For example, if the range is on the order of 20,000 km, then $f_N = 1/\tau = 3 \times 10^8/2(2 \times 10^7) = 7.5$ Hz. For square waves, the upper bandedge must be at least $2f_1$ to avoid pulse rounding of the higher harmonics. The ranging signal must be modulated on the RF, and demodulated prior to ranging processing. More serious is the dynamic voltage range of the tone ranging signal, since the signals are arithmetically summed in Eq. (C.3.2). Even square waves will produce a multilevel-voltage waveform, which makes it somewhat disadvantageous compared to the convenient bipolar nature of coded signals. Hard-limiting of the tone ranging signal to produce binary signals causes harmonic power suppression that must be accounted for in Eq. (C.3.5).

The harmonics in Eq. (C.3.2) were taken as half frequencies. By using higher submultiples, the number of required harmonics, and the number of corresponding decisions, between the highest and the lowest can be reduced. If each harmonic was $1/k$ of the previous, then only $\log_k(f_1\tau)$ harmonics are required. However, each harmonic will have one of k possible phase shifts relative to its corresponding received harmonic when correlating. Hence, a decision on one of k phase values, instead of an antipodal decision, must be made. This is identical to word detecting with polyphase block-coded signalling. The harmonic detection probabilities in Eq. (C.3.4) must now be replaced by the general polyphase results of Section A.5. Recall there that the effective detection energy was reduced by the number of phase states to be distinguished. This means integration time must be increased by the same factor to compensate. Hence, use of higher order submultiple harmonics reduces the number of harmonics, but increase the length of integration time per harmonic for decisioning.

C.4 POSITION LOCATION AND NAVIGATION

The satellite ranging concept can also be used for position location by a near-earth receiver. If the range from a satellite of known location to a receiver is determined, then the receiver location is known to within a spatial sphere centered at the satellite (Figure C.6). If the range from a second satellite is also determined, the uncertainty is reduced to the intersection line of two spheres. If a third satellite range is determined, a unique receiver position in three dimensions is generated. Hence a receiver can pinpoint its own location by measuring the range from three satellites of known location. The ranges can be measures simultaneously from three separates satellites (real time position location) or can perform three separate rangings from one satellite at three positions in its orbit (non-real time).

To measure the range from a satellite, a receiver must either have turn-around ranging capability to the satellite (as in Figure C.1) or must have a perfectly synchronized clock as the satellite transmits ranging waveforms. By knowing the exact time of transmission of each satellite's ranging waveform, and observing its arrival time, the one-way range (delay time) can be determined from each satellite. The exact position of the receiver can then be (theoretically) computed. (Corrections may have to

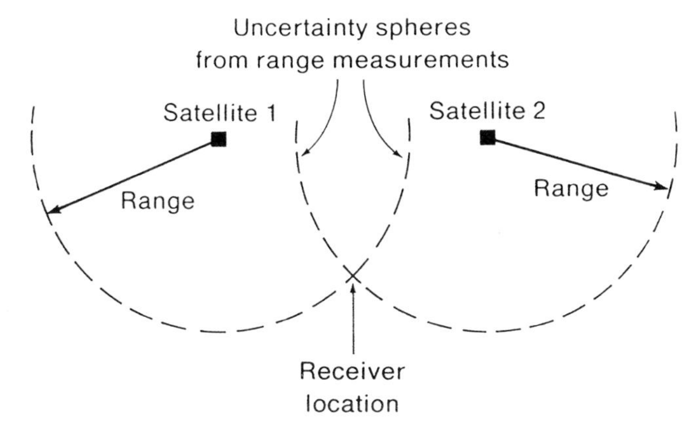

FIGURE C.6 Receiver position location using range measurements from satellites (two-dimensional case).

be inserted to account for transmission delay variations due to uncertainties in Doppler, atmospheric dispersion, receiver hardware, and even relativistic effects.) If the Doppler shifts of the RF frequencies carrying the ranging waveforms can also be measured, the instantaneous velocity of the receiver can be estimated as well. Hence the satellite transmissions can be used for both position location and navigation (referred to as *PLAN* systems).

If the receiver does not have turn-around ranging hardware, or does not have an inherent clock, position can still be determined by using range difference from four separate satellite transmissions. By knowing the location of each satellite, and having each satellite transmit at the same time (or a fixed known time offset) the received ranging waveforms can provide a measure of the range difference (difference in the arrival times) of the satellites. A measured range differential from two satellites again defines a sphere of uncertainty of the receiver, and measurements of two other corresponding range differences (relative to either of the first two) will again provide a unique receiver position. Thus a receiver position location system can be designed by having each of a multiple set of satellites repeatedly (and in synchronism) each transmit its own ranging waveform. A receiver acquires the ranging waveforms of four of the satellites by separately locking up its own range waveform correlators. (A receiver must have a waveform correlator for each satellite that can occur in its field of view.) By reading the difference in the time of arrival of the synchronized timing markets (Figure C.7), three range differentials can be determined and the receiver position can be computed. The accuracy of

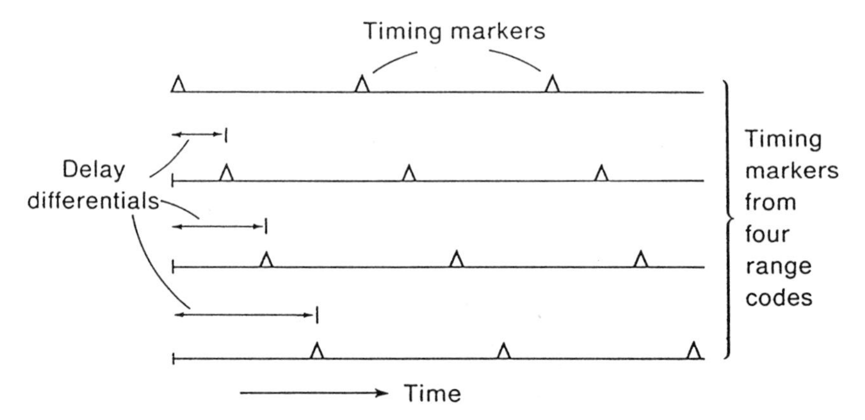

FIGURE C.7 Determination of three range differentials from four ranging codes.

the position location is therefore limited only by the accuracy to which the time of arrival can be determined (and the delay uncertainties corrected). Doppler differentials on the RF carriers from the satellites can also be used to estimate directional velocities as well.

The Global Positioning Satellite (GPS) system in Figure 1.3a is designed in this way. Since receivers are required to operate without any inherent clock, and position location must be available in real time, four transmitting satellites must be observed. The GPS network provides that at least six Navstar satellites will be observed from any near-earth location world-wide. The satellites ranging waveform are basic PRN ranging codes (called P-codes) phase modulated on L-band carriers (Table 1.5). (In fact, two separate L-band carriers are used to permit receiver correction for Doppler and dispersion.) The GPS P-codes are product codes constructed from two periodic subcodes of almost equal length of about 15×10^6 chips, each occurring at a rate of 10.23 Mcps. This produces an overall PRN P-code from each satellite having a period of about one week. There are 37 different P-codes of this length that can be assigned to any satellite. The P-codes are locked to a common clock among all the satellites to guarantee synchronized transmissions. Each P-code chip has an arrival time that can be determined to an accuracy of about 10 ns, and a receiver can compute its location to within about 30 m.

To provide rapid acquisition of the long PRN P-codes, and to efficiently separate the individual satellite signals, each P-code is synchronized to a shorter acquisition code (called CA-code). The CA-codes are a unique Gold code set of length 1023 chips, having a period of about 1 ms. The

CA-Gold code provides the low cross-correlation among the other codes of the set, and permit rapid acquisition due to its shorter length. Initial lock-up of the CA Gold code from a satellite, along with the binary transmission of a superimposed low rate "hand-over word" sequence, provides almost instantaneous acquisition of the accompanying P-code. The P-code and CA-code from a satellite are transmitted simultaneously, but are separated by placing them on the quadrature arms of a QPSK modulation at the L-band carriers. A GPS receiver therefore first acquires the CA code of a satellite, then locks up the corresponding P-code. The four P-codes from the four satellites are acquired simultaneously using a parallel correlator bank, and range differences are read from the P-code timing markers, from which receiver location is determined.

APPENDIX D
Nonlinear Amplification of Carrier Waveforms

Often a satellite transponder involves some type of nonlinearity during the high-power amplification of its carrier signals. It is therefore important in system design to understand the effects of such nonlinearities on the waveforms in the receiver and to account for them properly in system analysis. To consider this, we present some general results concerning nonlinear amplification of carrier waveforms. We concentrate only on the memoryless type of nonlinear device, in which the present value of the device output depends only on the present value of the input and not on any of its past.

Assume that the nonlinearity of the system is described by the function $g[x]$. Thus is $x(t)$ is the time process at the nonlinearity input, the output is

$$y(t) = g[x(t)] \tag{D.1}$$

as shown in Figure D.1. A common analytical procedure is to expand $y(t)$ in terms of the Fourier transform of the function $g(x)$. That is, if we denote $G(\omega)$ as the transform of $g(x)$, then $y(t)$ can be written as

$$y(t) = \frac{1}{2\pi} \int_{-\infty}^{\infty} G(\omega) e^{j\omega x(t)} \, d\omega \tag{D.2}$$

We point out that since typical amplifiers have a $g(x)$ that is real and odd

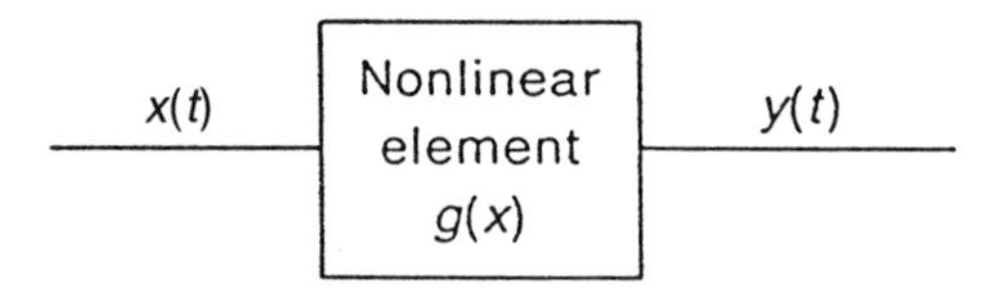

FIGURE D.1 Functional model of nonlinearity.

$[g(-x) = -g(x)]$, it is straightforward to show that its resulting $G(\omega)$ must be imaginary and odd in ω. This means Eq. (D.2) can be simplified to

$$y(t) = \frac{2}{2\pi}\int_0^\infty (jG(\omega))\ \mathrm{Im}[e^{j\omega x(t)}]d\omega \tag{D.3}$$

where Im[·] represents imaginary part. The use of Eq. (D.3) for analyzing nonlinearities in terms of their transform is particularly convenient for the type of carrier waveforms we have been considering in satellite links. Let us represent the general angle modulated carrier in the receiver (at either the front end or IF) as

$$x_c(t) = A\cos[\omega_c t + \theta(t)] \tag{D.4}$$

where $\theta(t)$ represents the phase modulation. We can then use the Jacobi–Anger identity

$$e^{j\omega\alpha\cos\beta} = \sum_{m=-\infty}^{\infty} \epsilon_m J_m(\omega\alpha)e^{jm\beta} \tag{D.5}$$

where $J_m(x)$ is the Bessel Function of order m, and $\epsilon_0 = 1$, $\epsilon_m = 2$, $m \neq 0$. Equation (D.3) becomes

$$y(t) = \sum_m^\infty h(m)\sin[m\omega_c t + m\theta(t)] \tag{D.6}$$

where

$$h(m) = \frac{\epsilon_m}{2\pi}\int_{-\infty}^{\infty} (jG(\omega))J_m(\omega A)d\omega \tag{D.7}$$

Equation (D.6) is a general expression for the output of any odd nonlinearity in Figure D.1 when the input is given by Eq. (D.4). Note that the output always appears as the sum of harmonically related, modulated carriers with amplitude variations depending on the type of nonlinear transform $G(\omega)$. We therefore see that harmonic generation is inherent in all nonlinear amplifiers. The harmonics appear at all multiples of the carrier frequency, with each harmonic containing a phase modulation having a corresponding multiplication factor. The coefficients $h(m)$ denote the amplitudes of each such harmonic, and $h^2(m)/2$ is its power. The term corresponding to $m = 1$ is the output carrier having the same frequency and phase modulation as the input carrier. The power in this output carrier is then

$$P_c = \frac{h^2(1)}{2} = \left[\frac{1}{2\pi}\int_{-\infty}^{\infty} (jG(\omega))J_m(\omega A)d\omega\right]^2 \tag{D.8}$$

A saturating (soft limiting) amplifier, as in Figure 4.26, can be conveniently modeled by the gain function

$$\begin{aligned} y(t) &= \mathrm{Erf}\left[\frac{x(t)}{b}\right], \qquad x(t) > 0 \\ &= -\mathrm{Erf}\left[\frac{|x(t)|}{b}\right], \qquad x(t) < 0 \end{aligned} \tag{D.9}$$

where $\mathrm{Erf}(x)$ is the Gaussian error function and b is a normalizing parameter determining the saturation level. The Fourier transform of Eq. (D.9) is given by

$$G(\omega) = \frac{2}{j\omega} e^{-b^2\omega^2/2} \tag{D.10}$$

The hard limiting amplifier is a special case of this with $b = 0$. When Eq. (D.10) is used in Eq. (D.8), we obtain

$$h(m) = \frac{\epsilon_m}{\pi}\int_{-\infty}^{\infty}\left(\frac{1}{\omega}\right)J_m(\omega A)e^{-b^2\omega^2/2}\,d\omega \tag{D.11}$$

By letting $x = \omega A$, so that $d\omega/\omega = dx/x$, this integral can be rewritten as

$$h(m) = \frac{\epsilon_m}{\pi}\int_{0}^{\infty}\left(\frac{1}{x}\right)J_m(Ax)e^{-b^2x^2/2A^2}\,dx \tag{D.12}$$

The power in the mth harmonic is then

$$P_m = \frac{h^2(m)}{2} = \frac{1}{2}\left[\frac{\epsilon_m}{\pi}\int_0^\infty J_m(x)e^{-(b^2/A^2)(x^2/2)}\frac{dx}{x}\right]^2 \tag{D.13}$$

and the power in the output carrier is

$$P_c = \left[\frac{1}{\pi}\int_0^\infty J_1(x)e^{-(b/A)^2x^2/2}\frac{dx}{x}\right]^2 \tag{D.14}$$

Consider now the extension of Eq. (D.4) to the case of K-modulated carriers, so that

$$x(t) = \sum_{i=1}^{K} A_i \cos[\omega_i t + \theta_i(t)] \tag{D.15}$$

where A_i, ω_i, and θ_i are the amplitudes, frequencies, and phase modulations of the individual carriers. Applying the expansion in Eq. (D.5), we now have

$$\begin{aligned} e^{j\omega x(t)} &= \prod_{i=1}^{K}\sum_{m_i}^{\infty} \epsilon_{m_i} J_{m_i}(\omega A_i) e^{j[m_i\omega_i t + jm_i\theta_i(t)]} \\ &= \sum_{m_K}\cdots\sum_{m_1}\prod_{i=1}^{K} \epsilon_{m_i} J_{m_i}(\omega A_i) e^{j[m_i\omega_i t + m_i\theta_i(t)]} \end{aligned} \tag{D.16}$$

Substituting Eq. (D.16) into Eq. (D.3) now yields

$$y(t) = \sum_{m_K}^{\infty}\cdots\sum_{m_1}^{\infty} h(m_1, \ldots, m_K)\sin\left[\sum_{i=1}^{K}(m_i\omega_i t + m_i\theta_i(t)\right] \tag{D.17}$$

where

$$h(m_1, \ldots, m_K) = \frac{1}{\pi}\int_0^\infty (jG(\omega))\left[\prod_{i=1}^{K}\epsilon_{m_i}J_{m_i}(\omega A_i)\right]d\omega \tag{D.18}$$

Equation (D.17) shows that the output will now be composed of a sum of terms, one for each integer vector $(m_1, m_2, \ldots, m_k)$, with each such term corresponding to a modulated sine wave at the combined harmonic frequencies of the input carriers, and having amplitudes $h(m_1, m_2, \ldots, m_k)$.

To model the saturating power amplifier of Eq. (D.9), we use the soft limiting characteristic of Eq. (D.9), whose transform is given in Eq. (D.10). For this model Eq. (D.18) becomes

$$h(m_1, \ldots, m_K) = \frac{1}{\pi} \int_0^\infty \prod_{i=1}^{K} J_{m_i}(xA_i) e^{-b^2x^2/2} \frac{dx}{x} \tag{D.19}$$

Equation (D.19) is useful in our discussion of satellite amplifiers in Chapters 4 and 5.

Index